# Deep Beauty

*Understanding the Quantum World Through Mathematical Innovation*

No scientific theory has caused more puzzlement and confusion than quantum theory. Physics is supposed to help us to understand the world, but quantum theory makes it seem a very strange place.

This book is about how mathematical innovation can help us gain deeper insight into the structure of the physical world. Chapters by top researchers in the mathematical foundations of physics explore new ideas, especially novel mathematical concepts, at the cutting edge of future physics. These creative developments in mathematics may catalyze the advances that enable us to understand our current physical theories, especially quantum theory. The authors bring diverse perspectives, unified only by the attempt to introduce fresh concepts that will open up new vistas in our understanding of future physics.

Hans Halvorson is Professor of Philosophy at Princeton University. He has written extensively on the foundations of quantum physics, with articles appearing in the *Journal of Mathematical Physics, Physical Review, Philosophy of Science*, and *British Journal of Philosophy of Science*, among others. He is currently working on applying the tools of category theory to questions in the foundations of mathematics. Halvorson has been the recipient of the Mellon New Directions Fellowhip (2007), the Cushing Memorial Prize in the History and Philosophy of Physics (2004), Best Article of the Year by a Recent Ph.D. (Philosophy of Science Association, 2001), and Ten Best Philosophy Articles of the Year (*The Philosopher's Annual*, 2001 and 2002).

# Deep Beauty

## *Understanding the Quantum World Through Mathematical Innovation*

Edited by
**Hans Halvorson**

*Princeton University*

CAMBRIDGE UNIVERSITY PRESS
Cambridge, New York, Melbourne, Madrid, Cape Town,
Singapore, São Paulo, Delhi, Tokyo, Mexico City

Cambridge University Press
32 Avenue of the Americas, New York, NY 10013-2473, USA

www.cambridge.org
Information on this title: www.cambridge.org/9781107005709

© Cambridge University Press 2011

This publication is in copyright. Subject to statutory exception
and to the provisions of relevant collective licensing agreements,
no reproduction of any part may take place without the written
permission of Cambridge University Press.

First published 2011

Printed in the United States of America

*A catalog record for this publication is available from the British Library.*

*Library of Congress Cataloging in Publication data*

Deep beauty: understanding the quantum world through mathematical innovation / edited by
Hans Halvorson.
   p.   cm.
Includes bibliographical references and index.
ISBN 978-1-107-00570-9 (hardback)
1. Quantum theory – Mathematics.   2. Mathematical physics.   I. Halvorson, Hans.
QC174.17.M35D44   2011
530.1201'51–dc22        2010046924

ISBN 978-1-107-00570-9 Hardback

Cambridge University Press has no responsibility for the persistence or accuracy of URLs for
external or third-party Internet Web sites referred to in this publication and does not guarantee that
any content on such Web sites is, or will remain, accurate or appropriate.

Just as the essence of perception is not sensing objects but apprehending them, even if we can only apprehend them through the mediation of sense, so the paradigm of a real world is not its sensible imaginability but its intelligible apprehensibility. I do not mean by this that anything which can be conceived by the intellect is thereby shown to exist, but I do mean that anything that concretely exists can be grasped by the intellect in its concrete existence. If therefore the universe of modern physics is one in which all attempts to make it intelligible by models of sensory type fail and which requires for its systematisation the kind of concepts that are used by quantum physics, this does not in the least imply that it is unreal or subjective. It simply means that the formulae of quantum physics express the kind of intelligibility that it has.

*Eric Mascall (1905–1993)*

# Contents

| | |
|---|---|
| *Contributors* | *page* ix |
| *Preface* | xi |
| *Acknowledgments* | xiii |
| **Introduction**<br>*Hans Halvorson* | 1 |

### I Beyond the Hilbert Space Formalism: Category Theory

| | |
|---|---|
| 1 **A Prehistory of *n*-Categorical Physics**<br>*John C. Baez and Aaron D. Lauda* | 13 |
| 2 **A Universe of Processes and Some of Its Guises**<br>*Bob Coecke* | 129 |
| 3 **Topos Methods in the Foundations of Physics**<br>*Chris J. Isham* | 187 |
| 4 **The Physical Interpretation of Daseinisation**<br>*Andreas Döring* | 207 |
| 5 **Classical and Quantum Observables**<br>*Hans F. de Groote* | 239 |
| 6 **Bohrification**<br>*Chris Heunen, Nicolaas P. Landsman, and Bas Spitters* | 271 |

### II Beyond the Hilbert Space Formalism: Operator Algebras

| | |
|---|---|
| 7 **Yet More Ado about Nothing: The Remarkable Relativistic Vacuum State**<br>*Stephen J. Summers* | 317 |
| 8 **Einstein Meets von Neumann: Locality and Operational Independence in Algebraic Quantum Field Theory**<br>*Miklós Rédei* | 343 |

## III  Behind the Hilbert Space Formalism

9 **Quantum Theory and Beyond: Is Entanglement Special?**  365
 *Borivoje Dakić and Časlav Brukner*

10 **Is Von Neumann's "No Hidden Variables" Proof Silly?**  393
 *Jeffrey Bub*

11 **Foliable Operational Structures for General Probabilistic Theories**  409
 *Lucien Hardy*

12 **The Strong Free Will Theorem**  443
 *John H. Conway and Simon Kochen*

*Index*  455

# Contributors

**John C. Baez**
Department of Mathematics, University of California, Riverside, California, United States

**Časlav Brukner**
Faculty of Physics, University of Vienna, Austria; Institute of Quantum Optics and Quantum Information, Austrian Academy of Sciences, Vienna, Austria

**Jeffrey Bub**
Philosophy Department and Institute for Physical Science and Technology, University of Maryland, College Park, Maryland, United States

**Bob Coecke**
Computing Laboratory, University of Oxford, United Kingdom

**John H. Conway**
Department of Mathematics, Princeton University, Princeton, New Jersey, United States

**Borivoje Dakić**
Faculty of Physics, University of Vienna, Austria

**Hans F. de Groote**
Formerly of the Fachbereich Informatik und Mathematik, Goethe-Universität Frankfurt a.M., Germany

**Andreas Döring**
Computing Laboratory, University of Oxford, United Kingdom

**Lucien Hardy**
Perimeter Institute, Waterloo, Ontario, Canada

**Chris Heunen**
Computing Laboratory, University of Oxford, United Kingdom

**Chris J. Isham**
The Blackett Laboratory, Imperial College, London, United Kingdom

**Simon Kochen**
Department of Mathematics, Princeton University, Princeton, New Jersey, United States

**Nicolaas P. Landsman**
Institute for Mathematics, Astrophysics, and Particle Physics, Radboud University Nijmegen, Nijmegen, The Netherlands

**Aaron D. Lauda**
Department of Mathematics, Columbia University, New York, New York, United States

**Miklós Rédei**
Department of Philosophy, Logic and Scientific Method, London School of Economics and Political Science, London, United Kingdom

**Bas Spitters**
Department of Computer Science, Radboud University Nijmegen, Nijmegen, The Netherlands

**Stephen J. Summers**
Department of Mathematics, University of Florida, Gainesville, Florida, United States

# Preface

In the fall of 2006, I received a phone call from Charles (Chuck) Harper, then the Senior Vice President and Chief Strategist of the John Templeton Foundation (JTF).[1] Chuck asked me, What are the topics that most need attention in your field (philosophy of physics) but that are not receiving funding from other sources? I told Chuck that current academic culture does not provide much encouragement for scientists to pursue "foundational" or "philosophical" issues. Whereas in previous centuries many of the great scientific minds also displayed a sharp philosophical acumen—witness Descartes, Leibniz, Bohr, Einstein, von Neumann, and Weyl—our current generation of great scientists seems to lack the time, ability, or interest to expand their scientific research into the more speculative or conceptual realms. So I told Chuck we should provide some encouragement to those scientists who wish to continue the tradition of philosophical reflection on their subjects.

This book is the direct result of Chuck's, and JTF's, taking this idea seriously— that is, the idea that "philosophical" is not an antonym of "scientific." Concretely, the support provided by JTF enabled us to bring a group of distinguished philosophers and scientists from around the globe to Princeton, New Jersey, for a two-day conference in October 2007.[2] The invited speakers were given free rein to speak on whatever topic they chose. However, it immediately became apparent that there was a common methodological theme: Although the speakers had diverse research goals, they shared in common the vision of achieving a deeper insight into foundational issues by stretching the resources provided by traditional mathematics. (For elaboration on this point, see the introduction.)

After the conference, it was agreed that the speakers would expand their ideas into chapters for a book. The book in your hands (or on your screen, or being read to you)

---

[1] Currently Chancellor for International Distance Learning and Senior Vice President, Global Programs, of the American University System, as well as President of Vision-Five.com Consulting.
[2] See http://symposia.templeton.org/deep_beauty/.

is the result. If our original intention for this book was realized—and we think that it was *in excelsis*—then its twelve essays carry the tradition of natural philosophy into the twenty-first century.

*Hans Halvorson*
Princeton University

# Acknowledgments

Many thanks to the John Templeton Foundation (JTF) for its generous financial backing for this project. I hope that this book offers high yields in the growth of knowledge.

Thanks also to the contributors to this volume for their dedication of time and expertise to this project and for their willingness to stretch themselves beyond traditional disciplinary boundaries.

Thanks to Chuck Harper, who played such an instrumental role in the early stages of the project: from that first phone call, to helping develop a funding proposal, to attending the conference, and to many other forms of encouragement both tangible and intangible.

Thanks also to Hyung Choi, Director of Mathematical and Physical Sciences at JTF, who worked with Chuck and me on the early stages of the project and who kept faith in the project to the very end.

Thanks to the team at Ellipsis Enterprises—Pamela Contractor, president and director; Matthew Bond, assistant editor; and especially Rob Schluth, senior editor and program director—for making my job as editor as painless and worry-free as possible. It is an unusual luxury when a scholar can devote his time to content alone, leaving all difficult administrative, developmental editing, and publishing preparation matters to be dealt with by a qualified staff.

Thanks to Lauren Cowles, Senior Editor, Mathematics and Computer Science at Cambridge University Press, for supporting and overseeing this book project.

This book is dedicated to two Johns: to the sage from Budapest, who lived long before this project was conceived; and to the yellow-haired boy from Princeton, whose life began in tandem with the project.

# Introduction

## Hans Halvorson

No scientific theory has caused more puzzlement and confusion than quantum theory. Beginning in 1900, the theory developed in fits and starts and found a consistent mathematical framing only when John von Neumann published his *Mathematische Grundlagen der Quantenmechanik* in 1932 [12]. But even today, we struggle to understand the world as pictured by quantum theory. Physics is supposed to help us to *understand* the world, and yet quantum theory makes it seem a very strange place.

One might be tempted to push aside our puzzlement as the result of our clinging to a primitive worldview. But our puzzlement is not merely a psychological obstacle; it is also an obstacle to the development of physics itself. This obstacle is encountered primarily in our attempts to unify quantum theory and the general theory of relativity. As argued persuasively by Chris Isham (who is represented in this volume), Lee Smolin [10], and others, the primary obstacle between us and future physics is our own failure to understand the conceptual foundations of current physical theories.

How, then, are we to make conceptual progress? What is the process by which we find a new perspective, a perspective in which previously puzzling phenomena find a place in an intelligible—and perhaps beautiful—structure?

I do not wish to make prescriptions or to claim that conceptual progress can be achieved in only one way. But this book begins with the *Ansatz* that conceptual progress might be achieved through free creations of the human intellect. And where are we to find this free creative activity? According to a distinguished tradition, beginning with the philosopher Immanuel Kant and running through the philosopher-mathematicians Gottlob Frege and L. E. J. Brouwer, the mathematical sciences are in the business of constructing new and "fruitful" concepts. Thus, this book begins from the assumption that creative developments in mathematics might catalyze the conceptual advances that enable us to understand our current physical theories (in particular, quantum theory) and thereby to promote future advances in physics.

Because the guiding theme of this book is methodological rather than thematic, its chapters are naturally written from diverse perspectives, unified only by the attempt to introduce new concepts that will aid our understanding of current physics, as well as the growth of future physics. Some of the authors are mathematicians (Conway, de Groote, Kochen, Spitters); some mathematical physicists (Baez, Coecke, Döring, Heunen, Isham, Landsman, Lauda, Summers); some theoretical physicists (Brukner, Dakić, Hardy); and some philosophers (Bub, Redéi). But regardless of their professional affiliations, each author takes an interdisciplinary approach that combines methods and ideas from physics, mathematics, and philosophy. In the remainder of this Introduction, we briefly overview the various chapters and their contribution to the ongoing task of making sense of the physical world.

## I.1 Beyond Hilbert Space

Quantum theory was born from a failure—namely, the failure of classical mechanics to provide accurate statistical predictions (e.g., in the case of blackbody radiation). Indeed, it was Einstein who saw clearly in the years between 1900 and 1905 that the framework of classical physics required a major overhaul. But unlike the theory of relativity, quantum theory did not result from a single stroke of genius. Rather, the following three decades witnessed a prolonged struggle by some of the century's greatest minds, including Niels Bohr, Arnold Sommerfeld, Max Born, Werner Heisenberg, Erwin Schrödinger, and Paul Dirac. Throughout this period, the developing "quantum" theory was not much more than a cobbled-together set of statistical rules of thumb that provided more accurate predictions than classical statistical mechanics.

In the second half of the 1920s, these struggles yielded two major mathematical advances: first, Schrödinger's introduction of the wave mechanical formalism; and second, Heisenberg's introduction of matrix mechanics. But it was only in 1932 that these two advances were unified, and these new statistical recipes were provided with a systematic theoretical underpinning. In a stroke of mathematical genius, John von Neumann axiomatized the theory of mathematical spaces equipped with linear structure and an inner product, a type of space that was finding extensive use by David Hilbert's school in Göttingen. Von Neumann labeled any such space that is topologically complete (i.e., containing limit points for all Cauchy sequences) a "Hilbert space." He then went on to show how vectors in a Hilbert space can represent the states of quantum systems and how linear operators on a Hilbert space can represent the quantities, or "observables," of the system. With von Neumann's formalism in hand, quantum theorists had a precise mathematical justification for their statistical recipes. Quantum theory had entered the domain of *mathematical physics*.

However, von Neumann's formalization of quantum theory has yielded a false sense of conceptual clarity, for von Neumann's formalization pushes back but does not solve the basic interpretive problems of quantum theory. In particular, his formalism provides accurate statistical predictions, but only if it is severely limited in its application. Indeed, we still do not know how to apply quantum mechanics to individual systems, to macroscopic systems, or, a fortiori, to "observers" like ourselves.

Furthermore, although the Hilbert space formalism of quantum theory served as the framework for some of the twentieth century's greatest scientific achievements (e.g., the standard model of particle physics), it is not clear that it will prove serviceable in the attempt to unify quantum theory and the general theory of relativity. In fact, according to some notable physicists—such as Penrose [4] and Isham (see Chapter 3 in this volume)—the Hilbert space formalism might itself be implicated in our seeming inability to find a conceptual unification of our best two physical theories.

With these facts in mind, the authors of this book engage critically with the very mathematical foundations of quantum theory. In fact, not a single contributor to this book accepts, uncritically, the "standard formalism"—the Hilbert space formalism—as a background framework with which to pursue conceptual and empirical questions. Rather, a consistent theme of this volume is that we need to think creatively, not just *within the current framework*, but *beyond* it; that is, we need to think creatively about how to transcend, or at least reenvision, the current framework.

As mentioned, the authors of this volume approach this task from a broad range of perspectives. Several of them (e.g., Baez and Lauda; Coecke; Döring; Heunen, Landsman, and Spitters; Isham) attack the problem using the tools of category theory, the theory of mathematical structures attributable primarily to Samuel Eilenberg and Saunders Mac Lane (see, for example, [5]). Others (e.g., Redéi, Summers) make extensive use of the theory of operator algebras, a theory originally developed by von Neumann himself that has found application in formalizing quantum field theory and (deformation) quantization theory. Yet others (e.g., Dakić and Brukner, Hardy) prefer to reduce mathematical assumptions to a bare minimum in the interest of displaying more vividly the physical content of quantum theory and more general probabilistic theories. Thus, although the underlying motivations are analogous, the tools employed are quite diverse.

## I.2 Categorical Approaches to Quantum Theory

In recent years, category theory has found many uses in physics and, indeed, in many of the exact sciences. This volume contains a representative sample of cutting-edge uses of category theory in the *foundations of physics*.

In this book, three sorts of category-theoretic approaches to the foundations of physics are represented: an $n$-categorical approach (Baez and Lauda), a monoidal categorical approach (Coecke), and a topos theoretical approach (Isham; Döring; Heunen, Landsman, and Spitters). Anyone who is acquainted with category theory will recognize immediately that these approaches need not be seen as opposed or even as disjoint. Indeed, they are in many ways mutually reinforcing and might even someday be unified (e.g., by some notion of a weak monoidal $n$-topos).

### I.2.1 $n$-Categorical Physics

In their magisterial "A Prehistory of $n$-Categorical Physics," Chapter 1 in this volume, John C. Baez and Aaron D. Lauda recount in this volume the ways in which $n$-category theory has entered into physics and discuss many of the ways in which $n$-categories might play a role in the physical theories of the future. But why, you might ask, should we think that $n$-categories are a good place to look for some new insight into the very basic structures of the physical world? As Baez and Lauda point out, the theory of $n$-categories is itself based on a perspective-changing idea: the idea that what might be seen as an object from one point of view might be seen as a process from another point of view. For the simplest example of this "Copernican revolution" of mathematical framework, consider the example of a group, that is, a set $G$ equipped with a binary product and an identity element $e \in G$ satisfying certain equations. Because we frequently think of categories on the model of concrete categories (i.e., categories of sets equipped with structure), it comes as a bit of a surprise to realize that a group is *itself* an example of a category. In particular, a group $G$ is a category with one object (call it whatever you wish, say $*$) and whose arrows are elements of $G$.

Such a change of perspective might seem rather minor, but we should not minimize the amount of insight that can be gained by seeing a familiar object in a new guise. For example, once we see a group as a category, we can also see a group representation as a certain sort of functor, that is to say, a functor into the category HILB of Hilbert spaces. But now these group representations themselves naturally form a category, and we can consider the arrows in this category, what are usually called "intertwiners." With this new perspective on groups, Baez and Lauda point out that Feynman diagrams and Penrose spin networks are both examples of categories of group representations with intertwiners as arrows.

Baez and Lauda go on to discuss some of the most interesting recent developments in which category theory, and $n$-category theory in particular, promises to open new vistas. Among these developments, they discuss topological quantum field theories and quantum groups. They also briefly discuss Baez's own "periodic table" of $n$-categories, which neatly characterizes the zoology of higher categories.

### I.2.2 Quantum Theory in Monoidal Categories

As briefly mentioned, the category HILB of Hilbert spaces plays a central role in quantum physics. We now expect, however, that quantum theory will play a central role in the computation theory of the future. After all, physical computers are made of objects that obey the laws of quantum mechanics.

It is well known that a quantum computer behaves differently than a classical computer; it is the differences in behavior that account, for example, for the fact that a quantum computer should be able to solve some problems more efficiently than any classical computer. But theoretical computer science is wont to abstract away from the nitty-gritty details of physical systems. In most cases, the computer scientist needs only to know the *structural* properties of the systems at his disposal; it is these structural properties that determine how such systems might be used to implement computations or other information-theoretic protocols.

It is no surprise, then, that theoretical computer scientists have led the way in describing the structural features of quantum systems. It is also no surprise that theoretical computer scientists have found it useful to use notions from category theory in describing these structures.

In "A Universe of Processes and Some of Its Guises," Chapter 2 in this volume, Bob Coecke provides a blueprint of a universe governed by quantum mechanics. Intriguingly, however, we see this universe through the eyes of a computer scientist: we do not see waves, particles, or any other concrete manifestation of physical processes. Rather, by means of a diagrammatic calculus, Coecke displays the very structures of the processes that are permitted (and forbidden) by the laws of quantum theory.

What is perhaps most striking about Coecke's approach is the sheer ratio of results to assumptions. From an extremely Spartan set of assumptions about how processes can combine (both vertically and horizontally), Coecke is able to reproduce all of the central results of quantum information science (in a broadly construed sense that includes "von Neumann measurement").

Another noteworthy aspect of Coecke's chapter is his discussion of the relation of categorical quantum mechanics (in its monoidal category guise) to other traditional approaches to the mathematical foundations of quantum mechanics (e.g., quantum logic, convex sets, $C^*$-algebras). Here, we get a "compare and contrast" from a researcher who has worked on both sides of the fence, first as a member of the Brussels school (directly descended from the Geneva school of Jauch and Piron) and more recently as a cofounder (with Samson Abramsky) and leader of the categorical approach to quantum computation. Thus, this chapter is absolutely mandatory reading for anyone interested in the fate of our attempts to *understand* the formalism of quantum theory and its utility in describing the processes that occur in our world.

### I.2.3 Quantum Theory in Toposes

What is so radical about quantum theory? Perhaps the first thing to spring to mind is *indeterminism*: quantum theory describes a world in which the future is not determined by the past. With a bit more sophistication, one might claim that the most radical feature of quantum theory is *nonlocality*: quantum theory describes a world in which subtle dependency relations exist between events that occur in distant regions of space.

Another suggestion, originally put forward by Birkhoff and von Neumann [2], and later taken up by the philosopher Hilary Putnam [8], is that quantum theory overturns the laws of classical logic. According to this proposal, the rules of classical (formal) logic—in particular, the distribution postulate (of conjunction over disjunction)—lead to conclusions in conflict with the predictions of quantum theory. Thus, the new physics requires a revolution in logic. Indeed, Putnam went on to claim that quantum theory's relation to logic is directly analogous to general relativity's relation to geometry: just as general relativity forces us to abandon Euclidean geometry, so quantum theory forces us to abandon classical logic.

But this proposal has not found many advocates—even Birkhoff, von Neumann, and Putnam eventually abandoned the idea. Neither has quantum logic catalyzed progress within physics or suggested routes toward the unification of quantum theory and general relativity. Even if quantum logic has not been shown to be wrong, it has proved to be *mathematically sterile*: it fails to link up in interesting ways with mainstream developments in mathematical physics.

The central motivating idea behind quantum logic is that the quantum revolution is a thoroughgoing *conceptual* revolution; that is, that it requires us to revise some of the *constitutive* concepts of our worldview. The idea itself is intriguing and perhaps even plausible. Thus, we turn with great interest to a recent proposal by Jeremy Butterfield and Chris Isham [3]. According to the Butterfield-Isham proposal, quantum mechanics requires us to replace not only classical logic but also the entire classical mathematical universe—as articulated in twentieth-century mathematical logic and set theory—with a more general universe of sets, namely a *topos*. It is true that such a replacement would also necessitate a replacement of classical logic but not, á la von Neumann, with a nondistributive logic. Rather, the internal logic of a topos is intuitionistic logic, where the law of excluded middle fails.

Three of the chapters in this book (Chapters 3, 4, and 6)—by Isham; Döring; and Heunen, Landsman, and Spitters—push the Butterfield-Isham idea even further. As we will see, the underlying idea of these approaches is strikingly similar to Putnam's, although it is executed within an infinitely richer and more fruitful mathematical context.

The chapters in this book represent two distinct approaches to using topos theory in the foundations of physics: the approach of Döring and Isham and the approach of Heunen, Landsman, and Spitters. (Both approaches have been developed extensively in the literature, and I refer the reader to the references within the chapters in this book.) Although there are several divergences in implementation between the Döring-Isham approach and the Heunen-Landsman-Spitters approach, the underlying idea is similar and in both cases would amount to nothing less than a Copernican revolution.

The idea of adopting a new mathematical universe is so radical and profound that one cannot appreciate it without immersing oneself in these works. (Of course, it would also help to spend some time learning background rudiments of topos theory; for this I recommend the book by Mac Lane and Moerdijk [6].) Rather than attempt to summarize the content of these chapters, I recommend that the reader begin by reading Isham's chapter, which provides a lucid motivation and discussion of the framework. The reader may then wish to proceed to the more technically demanding chapters by Heunen et al. and by Döring. Finally, in reading Döring's chapter, the reader can gain further insight by referring to de Groote's chapter,[1] Chapter 5, which carefully articulates some of the background mathematics needed to generalize familiar notions from the classical universe of sets to the quantum topos.

## I.3 Operator Algebras

Since the 1960s, it has been appreciated that the theory of operator algebras (especially $C^*$- and von Neumann algebras) provides a natural generalization of the Hilbert space formalism and is especially suitable for formalizing quantum field theories, or quantum theories with superselection rules. More recently, operator algebras have been applied to the task of clarifying conceptual issues. In this vein, see especially the work on nonlocality carried out by Summers [11] and the work on quantum logic carried out by Rédei [9]. Summers and Rédei continue this sort of foundational work in their chapters in this book (Chapters 7 and 8, respectively). Summers addresses the vacuum state in relativistic quantum field theory (QFT) in his chapter, "Yet More Ado about Nothing: The Remarkable Relativistic Vacuum State," whereas Rédei examines Einstein's notion of "separability" of physical systems in his chapter, "Einstein Meets von Neumann: Locality and Operational Independence in Algebraic Quantum Field Theory."

In his chapter, Summers aims to characterize properties of the vacuum in relativistic QFT in a mathematically precise way. He begins with the standard

---

[1] Published posthumously. See note in Chapter 5.

characterization, which involves both symmetries (the vacuum as invariant state) and energy conditions (the vacuum as lowest energy state). He then points out that these characterizations do not straightforwardly generalize to QFT on curved spacetimes. Thus, we stand in need of a more mathematically nuanced characterization of the vacuum.

According to Summers, the primary tool needed for this characterization is the Tomita-Takesaki modular theory, in particular, the geometrical interpretation of modular theory provided by Bisognano and Wichmann. However, Summers proceeds to recount a more ambitious program that he and his collaborators have undertaken, a program that would use modular symmetries as a basis from which the very structure of spacetime could be recovered. As Summers points out, such a reconstruction would have profound conceptual implications. Indeed, one is tempted to say that the success of such a program would be a partial vindication of Leibniz-Machian relationalism about spacetime. But whether or not the reconstruction supports certain philosophical views about the nature of spacetime, a clearer understanding of the vacuum is crucial for the development of future physics, especially because future physical theories will most certainly not posit a fixed-background Minkowski spacetime structure.

Summers also discusses the fact—without mentioning explicitly that it was first proved by himself and Reinhard Werner—that the vacuum state is nonlocal and indeed violates Bell's inequality maximally relative to measurements that can be performed in tangent spacetime wedges. In doing so, Summers notes the importance of making fine-grained distinctions between different types of nonlocality. This theme is treated at length in the chapter by Rédei.

Rédei begins in a historical vein by discussing Einstein's worries about quantum theory, in particular his notion of "separability" of physical systems. Although Einstein's objections to indeterminism are better known (witness: "God does not roll dice"), Einstein seems to have lost even more sleep over the issue of nonlocality. Indeed, it seems he thought that quantum nonlocality would make physics impossible!

Rédei distills from Einstein's writings a set of criteria that any theory must satisfy to be consistent with the principle of locality. He then proceeds to argue that relativistic QFT does in fact satisfy these criteria! Moreover, Rédei's arguments are far from speculative—or, as some might dismissively say, "philosophical." Rather, Rédei proceeds in a highly mathematical spirit: he translates the criteria into precise mathematical claims, and then he employs the tools of operator algebras in an attempt to demonstrate that the criteria are satisfied. The net result is a paradigm example of mathematical innovation in the service of conceptual clarification.

## I.4 Behind the Hilbert Space Formalism

We have seen that several of the chapters in this book take well-developed (or independently developing) mathematical theories and apply them in innovative ways to the foundations of physics. Such an approach is characteristic of mathematical physics. This book, however, also represents a second approach, an approach more characteristic of theoretical physics. In particular, theoretical physicists begin from explicitly physical principles, rather than from mathematical assumptions, and then attempt to formulate these physical principles in as transparent a fashion as possible, using mathematical formalism when it might help achieve that goal. The three chapters by Dakić and Brukner (Chapter 9), by Bub (Chapter 10), and by Hardy (Chapter 11) exemplify this second methodology.

In their chapter, "Quantum Theory and Beyond: Is Entanglement Special?," Boriroje Dakić and Časlav Brukner aim to clarify the fundamental physical principles underlying quantum theory; in doing so, they keep in firm view the relationship between quantum theory and potential future theories in physics. The authors begin by recounting several recent attempts to derive the formalism of quantum theory from physical principles that were motivated by Einstein's derivation of special relativity. As they note, such derivations ought to be subjected to severe critical scrutiny because thinking that quantum theory "must be true" could easily impede the development of successor theories and could easily blind us to ways in which quantum theory could be modified or superseded.

Nonetheless, Dakić and Brukner prove that quantum theory is the *unique* theory that describes entangled states and that satisfies their other physical principles. This striking result displays a sort of robustness of the central features of quantum theory: to the extent that the basic physical principles are justified, we can expect *any* future theory to incorporate, rather than supersede, quantum theory.

This same sensitivity to quantum mechanics as a potentially replaceable theory is displayed throughout the chapter by Lucien Hardy. In "Foliable Operational Structures for General Probabilistic Theories," Hardy in essence provides a parameterization of theories in terms of a crucial equation involving two variables, $K$ and $N$. In this parameterization, classical mechanics is characterized by the equation $K = N$, whereas quantum mechanics is characterized by the equation $K = N^2$. This leaves open the possibility of alternative theories, or even possible successor theories, of greater conceptual intricacy. Our past and current theories are only at the very low end of an infinite hierarchy of increasingly complex theories.

Hardy's chapter also pays special attention to the generalizability, or projectability into the future, of our theories. In particular, Hardy constructs his generalized probabilitistic framework without reliance on a notion of fixed background time. As a result, the framework stands ready for application to *relativistic* contexts. But more is true: Hardy develops his framework with an eye on synthesis of general relativity and quantum mechanics, a context in which causal structure is flexible enough that it might be adapted to contexts where even it is subject to quantum indeterminacy.

In his chapter, "Is von Neumann's 'No Hidden Variables' Proof Silly?", Jeffrey Bub, takes up the question of whether the Hilbert space formalism of quantum mechanics is complete. That is, do all states correspond to vectors (or density operators), or could there be "hidden variables"? This question was supposedly answered in the negative in 1932 by von Neumann's no hidden variables proof. If this argument were valid, there would be a strong sense in which the interpretive problems of quantum mechanics could *not* be solved by means of technical innovation—for example, by providing a more complete formalism.

But von Neumann's argument has not convinced everyone. In particular, John Bell [1] and, subsequently, David Mermin [7], argued that von Neumann's result is based on an illicit assumption—in particular, that von Neumann imposes unrealistic constraints on the mathematical representation of hidden variables. These critiques of von

Neumann's result were motivated by and, in turn, provide support for hidden variable programs, such as Bohmian mechanics.

Bub argues, however, that Bell and Mermin's criticism is off the mark. Rather, claims Bub, von Neumann states quite clearly that an operator $A + B$ has no direct physical significance in cases where $A$ and $B$ are incompatible (i.e., not simultaneously measurable). Read from this perspective, von Neumann intends to show not that hidden variables are impossible *tout court* but rather that hidden variables are inconsistent with the way that quantum mechanics uses mathematical objects to represent physical objects. But then the possibility opens that intuitive desiderata for a physical theory of micro-objects (e.g., determinism) could be satisfied only by overhauling the Hilbert space formalism.

Bub closes his chapter on this suggestive note, leaving it for the reader to judge whether it would be preferable to maintain the Hilbert space formalism, along with its puzzling interpretive consequences, or to attempt to replace it with some other formalism.

The book concludes with Chapter 12, an already famous article, "The Strong Free Will Theorem," by John H. Conway and Simon Kochen (reprinted in this volume with permission). But what has such an argument to do with the theme of the book—that is, with the theme of conceptual insight developing in tandem with mathematical insight? The careful reader will see that Conway and Kochen's argument proceeds independently of the standard formalism (i.e., Hilbert spaces) for quantum theory. That is, the authors do not take the Hilbert space formalism for granted and then draw out conceptual consequences regarding free will. Rather, they argue from simple, physically verifiable assumptions to the conclusion that if an experimenter has the freedom to choose what to measure, then particles have the freedom to choose what result to yield. The only input here from quantum mechanics is indirect: quantum mechanics predicts that Conway and Kochen's empirical assumptions are satisfied. Thus, if quantum mechanics is true, then Conway and Kochen's argument is sound.

We see, then, that Conway and Kochen's argument exemplifies the method of applying mathematical argument to the task of gaining new conceptual insight—insight, in this case, about the logical connection between certain statistical predictions (which are in fact made by quantum mechanics) and traditional metaphysical hypotheses (freedom of the will). If their argument is successful, then Conway and Kochen have provided us with insight that transcends the bounds of our current mathematical framework—hence, insight that will endure through the vicissitudes of scientific progress or revolutions.

In conclusion, the authors of this book were given carte blanche to employ as little or as much technical apparatus as they deemed necessary to advance conceptual understanding of the foundations of physics. For some of the authors, this meant employing highly sophisticated mathematical theories such as $n$-categories (Baez and Lauda), monoidal categories (Coecke), topos theory (Döring, Isham, Heunen et al.), or operator algebras (Rédei, Summers). For other authors, the emphasis lies more on examining the physical and conceptual motivation for the Hilbert space formalism (Dakić and Brukner) or on what might lie beyond the Hilbert space formalism (Bub, Hardy).

The liberty given to the authors means that for the reader, some of these chapters are technically demanding; even for those with previous technical training, these chapters should be approached with equal doses of patience and persistence. However, the technicalities seem to be demanded by the nature of the subject matter: quantum theory shows that conceptual insights and understanding do not come cheap, and the physical world does not come ready-made to be understood by the untrained human mind. Already it required the combined mathematical genius of Dirac and von Neumann, among others, to unify the various statistical recipes of the old quantum theory. The Hilbert space formalism has proved fruitful for many years and is partially responsible for some of the great advances of twentieth-century physics. But taking the Hilbert space formalism as a fixed, non-negotiable framework may also be partially responsible for our current predicament—both our troubles in interpreting quantum mechanics and the challenges of unifying quantum theory with the general theory of relativity. If this is the case, then it is imperative that we marshal the same sorts of resources that Dirac and von Neumann did; we must, indeed, employ our utmost mathematical creativity in our attempt to find an underlying intelligibility behind the physical phenomena.

It is with this aim in mind that the contributors present this collection to you, hoping to play some small role in the next quantum leap in our understanding of nature.

## References

[1] J. S. Bell. *Speakable and Unspeakable in Quantum Mechanics*. Cambridge: Cambridge University Press, 1987.

[2] G. Birkhoff and J. von Neumann. The logic of quantum mechanics. *Ann of Mathematics*, **37** (1936), 823–43.

[3] C. J. Isham and J. Butterfield. Some possible roles for topos theory in quantum theory and quantum gravity. *Foundations of Physics*, **30** (2000), 1707–35.

[4] S. Kruglinski. Roger Penrose says physics is wrong, from string theory to quantum mechanics. *Discover*, 2009 September.

[5] S. Mac Lane. *Categories for the Working Mathematician*, 2nd ed. New York: Springer-Verlag, 1998.

[6] S. Mac Lane and I. Moerdijk. *Sheaves in Geometry and Logic: A First Introduction to Topos Theory*. New York: Springer, 1992.

[7] N. D. Mermin. Hidden variables and the two theorems of John Bell. *Reviews of Modern Physics*, **65** (1993), 803–15.

[8] H. Putnam. Is logic empirical? In *Boston Studies in the Philosophy of Science,* vol. 5, eds. R. Cohen and M. P. Wartofski. Dordrecht: D. Reidel, 1968, pp. 216–41.

[9] M. Rédei. *Quantum Logic in Algebraic Approach*. Boston: Kluwer Academic Publishers, 1998.

[10] L. Smolin. *The Trouble with Physics*. New York: Houghton-Mifflin, 2006.

[11] S. Summers. On the independence of local algebras in quantum field theory. *Reviews in Mathematical Physics*, **2** (1990), 201–47.

[12] J. von Neumann. *Mathematische Grundlagen der Quantenmechanik*. Berlin: Springer Verlag, 1932.

# PART ONE
# Beyond the Hilbert Space Formalism: Category Theory

CHAPTER 1

# A Prehistory of *n*-Categorical Physics

John C. Baez and Aaron D. Lauda

## 1.1 Introduction

This chapter is a highly subjective chronology describing how physicists have begun to use ideas from *n*-category theory in their work, often without making this explicit. Somewhat arbitrarily, we start around the discovery of relativity and quantum mechanics and lead up to conformal field theory and topological field theory. In parallel, we trace a bit of the history of *n*-categories, from Eilenberg and Mac Lane's introduction of categories, to later work on monoidal and braided monoidal categories, to Grothendieck's dreams involving ∞-categories, and subsequent attempts to realize this dream. Our chronology ends at the dawn of the twenty-first century; since then, developments have been coming so thick and fast that we have not had time to put them in proper perspective.

We call this chapter a "prehistory" because *n*-categories and their applications to physics are still in their infancy. We call it "a" prehistory because it represents just one view of a multifaceted subject; many other such stories can and should be told. Ross Street's *An Australian Conspectus of Higher Categories* [228] is a good example in that it overlaps with ours, but only slightly. There are many aspects of *n*-categorical physics that our chronology fails to mention, or touches on very briefly, and other stories could redress these deficiencies. It would also be good to have a story of *n*-categories that focus on algebraic topology, one that focuses on algebraic geometry, and one that focuses on logic. For *n*-categories in computer science, we have John Power's "Why Tricategories?" [194], which although not focused on history at least explains some of the issues at stake.

What is the goal of *this* prehistory? We are scientists rather than historians of science, so we are trying to make a specific scientific point rather than accurately describe every twist and turn in a complex sequence of events. We want to show how categories and even *n*-categories have slowly come to be seen as a good way to formalize physical theories in which "processes" can be drawn as diagrams—for example, Feynman diagrams—but interpreted algebraically—for example, as linear operators. To minimize the prerequisites, we include a gentle introduction to *n*-categories (in fact, mainly just categories and bicategories). We also include a review of some relevant aspects of twentieth-century physics.

The most obvious roads to $n$-category theory start from issues internal to pure mathematics. Applications to physics became visible only much later, starting around the 1980s. So far, these applications mainly arise around theories of quantum gravity, especially string theory and "spin foam models" of loop quantum gravity. These theories are speculative and still under development, not ready for experimental tests. They may or may not succeed. So it is too early to write a real history of $n$-categorical physics or even to know whether this subject will become important. We believe it will—but so far, all we have is a "prehistory."

## 1.2 Road Map

Before we begin our chronology, to help the reader keep from getting lost in a cloud of details, it will be helpful to sketch the road ahead. Why did categories turn out to be useful in physics? The reason is ultimately very simple. A category consists of objects $x, y, z, \ldots$ and morphisms that go between objects; for example,

$$f: x \to y.$$

A good example is the category of Hilbert spaces, where the objects are Hilbert spaces and the morphisms are bounded operators. In physics, we can think of an *object* as a state space for some physical system, and a *morphism* as a process taking states of one system to states of another (perhaps the same one). In short, we use objects to describe *kinematics* and morphisms to describe *dynamics*.

Why $n$-categories? For this, we need to understand a bit about categories and their limitations. In a category, the only thing we can do with morphisms is compose them; given a morphism $f: x \to y$ and a morphism $g: y \to z$, we can compose them and obtain a morphism $gf: x \to z$. This corresponds to our basic intuition about processes—namely, that one can occur after another. Although this intuition is temporal in nature, it lends itself to a nice spatial metaphor. We can draw a morphism $f: x \to y$ as a "black box" with an input of type $x$ and an output of type $y$:

Composing morphisms then corresponds to feeding the output of one black box into another:

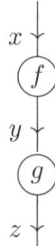

This sort of diagram might be sufficient to represent physical processes if the universe were one-dimensional, with no dimensions of space, and just one dimension of time. But, in reality, processes can occur not just in *series* but also in *parallel*—side by side, as it were:

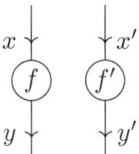

To formalize this algebraically, we need something more than a category: at the very least, a monoidal category, which is a special sort of bicategory. The term *bicategory* hints at the two ways of combining processes, in series and in parallel.

Similarly, the mathematics of bicategories might be sufficient for physics if the universe were only two-dimensional, with one dimension of space and one dimension of time. But, in our universe, it is also possible for physical systems to undergo a special sort of process in which they switch places:

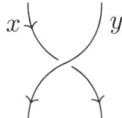

To depict this geometrically requires a third dimension, hinted at here by the crossing lines. To formalize it algebraically, we need something more than a monoidal category, at the very least a braided monoidal category, which is a special sort of tricategory.

This escalation of dimensions can continue. In the diagrams Feynman used to describe interacting particles, we can continuously interpolate between this way of switching two particles:

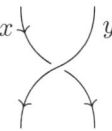

and this:

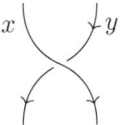

This requires four dimensions, one of time and three of space. To formalize this algebraically, we need a symmetric monoidal category, which is a special sort of tetracategory.

More general $n$-categories, including those for higher values of $n$, may also be useful in physics. This is especially true in string theory and spin foam models of quantum gravity. These theories describe strings, graphs, and their higher-dimensional generalizations propagating in spacetimes that may themselves have more than four dimensions.

So in abstract the idea is simple. We can use $n$-categories to formalize *algebraically* physical theories in which processes can be depicted *geometrically* using $n$-dimensional diagrams. But the development of this idea has been long and convoluted. It is also far from finished. In our chronology, we describe its development up to the year 2000. To keep the tale from becoming unwieldy, we have been ruthlessly selective in our choice of topics.

In particular, we can roughly distinguish two lines of thought leading toward $n$-categorical physics, one beginning with quantum mechanics, the other with general relativity. Given that a major challenge in physics is reconciling quantum mechanics and general relativity, it is natural to hope that these lines of thought will eventually merge. We are not sure yet how this will happen, but the two lines have already been interacting throughout the twentieth century. Our chronology will focus on the first. But before we start, let us give a quick sketch of both.

The first line of thought starts with quantum mechanics and the realization that in this subject, *symmetries* are crucial. Taken abstractly, the symmetries of any system form a group $G$. But to describe how these symmetries act on states of a quantum system, we need a unitary representation $\rho$ of this group on some Hilbert space $H$. This sends any group element $g \in G$ to a unitary operator $\rho(g)\colon H \to H$.

The theory of $n$-categories allows for drastic generalizations of this idea. We can see any group $G$ as a category with one object in which all the morphisms are invertible; the morphisms of this category are just the elements of the group, whereas composition is multiplication. There is also a category Hilb, in which objects are Hilbert spaces and morphisms are linear operators. A representation of $G$ can be seen as a map from the first category to the second:

$$\rho\colon G \to \text{Hilb}.$$

Such a map between categories is called a *functor*. The functor $\rho$ sends the one object of $G$ to the Hilbert space $H$, and it sends each morphism $g$ of $G$ to a unitary operator $\rho(g)\colon H \to H$. In short, it realizes elements of the abstract group $G$ as actual transformations of a specific physical system.

The advantage of this viewpoint is that now the group $G$ can be replaced by a more general category. Topological quantum field theory (TQFT) provides the most famous example of such a generalization but, in retrospect, the theory of Feynman diagrams provides another, and so does Penrose's theory of spin networks.

More dramatically, both $G$ and Hilb may be replaced by a more general sort of $n$-category. This allows for a rigorous treatment of physical theories in which physical processes are described by $n$-dimensional diagrams. The basic idea, however, is always the same: *a physical theory is a map sending abstract processes to actual transformations of a specific physical system.*

The second line of thought starts with Einstein's theory of general relativity, which explains gravity as the curvature of spacetime. Abstractly, the presence of curvature means that as a particle moves through spacetime from one point to another, its internal state transforms in a manner that depends nontrivially on the path it takes. Einstein's great insight was that this notion of curvature completely subsumes the older idea of gravity as a force. This insight was later generalized to electromagnetism and the other forces of nature; we now treat them all as various kinds of curvature.

In the language of physics, theories where forces are explained in terms of curvature are called *gauge theories*. Mathematically, the key concept in a gauge theory is that of a connection on a bundle. The idea here is to start with a manifold $M$ describing spacetime. For each point $x$ of spacetime, a bundle gives a set $E_x$ of allowed internal states for a particle at this point. A connection then assigns to each path $\gamma$ from $x \in M$ to $y \in M$ a map $\rho(\gamma) \colon E_x \to E_y$. This map, called *parallel transport*, says how a particle starting at $x$ changes state if it moves to $y$ along the path $\gamma$.

Category theory lets us see that a connection is also a kind of functor. There is a category called the path groupoid of $M$, denoted $\mathcal{P}_1(M)$, whose objects are points of $M$, in which the morphisms are paths and composition amounts to concatenating paths. Similarly, any bundle $E$ gives a transport category, denoted $\mathrm{Trans}(E)$, where the objects are the sets $E_x$ and the morphisms are maps between these. A connection gives a functor:

$$\rho \colon \mathcal{P}_1(M) \to \mathrm{Trans}(P).$$

This functor sends each object $x$ of $P_1(M)$ to the set $E_x$ and sends each path $\gamma$ to the map $\rho(\gamma)$.

So, the second line of thought, starting from general relativity, leads to a picture strikingly similar to the first! Just as a unitary group representation is a functor sending abstract symmetries to transformations of a specific physical system, a connection is a functor sending paths in spacetime to transformations of a specific physical system, a particle. And just as unitary group representations are a special case of physical theories described as maps between $n$-categories, when we go from point particles to higher-dimensional objects, we meet higher-gauge theories, which use maps between $n$-categories to describe how such objects change state as they move through spacetime [16]. In short, the first and second lines of thought are evolving in parallel, and intimately linked, in ways that still need to be understood.

Sadly, we do not have much room for general relativity, gauge theories, or higher-gauge theories in our chronology. Our focus is on group representations as applied to quantum mechanics, Feynman diagrams as applied to quantum field theory, how these diagrams became better understood with the rise of $n$-categories, and how higher-dimensional generalizations of Feynman diagrams arise in string theory, loop quantum gravity, TQFT, and the like.

## 1.3 Chronology

### 1.3.1 Maxwell (1876)

In his book *Matter and Motion*, Maxwell [166] wrote:

> Our whole progress up to this point may be described as a gradual development of the doctrine of relativity of all physical phenomena. Position we must evidently acknowledge to be relative, for we cannot describe the position of a body in any terms which do not express relation. The ordinary language about motion and rest does not so completely exclude the notion of their being measured absolutely, but the reason of this is, that in our ordinary language we tacitly assume that the earth is at rest.... There are no landmarks

in space; one portion of space is exactly like every other portion, so that we cannot tell where we are. We are, as it were, on an unruffled sea, without stars, compass, sounding, wind or tide, and we cannot tell in what direction we are going. We have no log which we can case out to take a dead reckoning by; we may compute our rate of motion with respect to the neighboring bodies, but we do not know how these bodies may be moving in space.

Readers less familiar with the history of physics may be surprised to see these words, written three years before Einstein was born. In fact, the relative nature of velocity was already known to Galileo, who also used a boat analogy to illustrate this. However, Maxwell's equations describing light made relativity into a hot topic. First, it was thought that light waves needed a medium to propagate in, the luminiferous aether, which would then define a rest frame. Second, Maxwell's equations predicted that waves of light move at a fixed speed in a vacuum regardless of the velocity of the source! This seemed to contradict the relativity principle. It took the genius of Lorentz, Poincaré, Einstein, and Minkowski to realize that this behavior of light is compatible with relativity of motion if we assume space and time are united in a geometrical structure that we now call *Minkowski spacetime*. But when this realization came, the importance of the relativity principle was highlighted, and with it the importance of *symmetry groups* in physics.

### 1.3.2 Poincaré (1894)

In 1894, Poincaré invented the *fundamental group*: for any space $X$ with a basepoint $*$, homotopy classes of loops based at $*$ form a group $\pi_1(X)$. This hints at the unification of *space* and *symmetry*, which was later to become one of the main themes of $n$-category theory. In 1945, Eilenberg and Mac Lane described a kind of inverse to the process taking a space to its fundamental group. Since the work of Grothendieck in the 1960s, many have come to believe that homotopy theory is secretly just the study of certain vast generalizations of groups, called *n-groupoids*. From this point of view, the fundamental group is just the tip of an iceberg.

### 1.3.3 Lorentz (1904)

Already in 1895, Lorentz had invented the notion of local time to explain the results of the Michelson–Morley experiment, but in 1904 he extended this work and gave formulas for what are now called *Lorentz transformations* [151].

### 1.3.4 Poincaré (1905)

In his opening address to the Paris Congress in 1900, Poincaré asked, "Does the aether really exist?" In 1904, he gave a talk at the International Congress of Arts and Science in St. Louis, in which he noted that "...as demanded by the relativity principle the observer cannot know whether he is at rest or in absolute motion."

On June 5, 1905, he wrote a paper entitled "Sur la dynamique de l'electron" [191] in which he stated: "It seems that this impossibility of demonstrating absolute motion is a general law of nature." He named the Lorentz transformations after Lorentz and showed that these transformations, together with the rotations, form a group. This is now called the *Lorentz group*.

### 1.3.5 Einstein (1905)

Einstein's first paper on relativity, "On the electrodynamics of moving bodies" [76], was received on June 30, 1905. In the first paragraph, he points out problems that arise from applying the concept of absolute rest to electrodynamics. In the second, he continues:

> Examples of this sort, together with the unsuccessful attempts to discover any motion of the earth relative to the 'light medium,' suggest that the phenomena of electrodynamics as well as of mechanics possess no properties corresponding to the idea of absolute rest. They suggest rather that, as already been shown to the first order of small quantities, the same laws of electrodynamics and optics hold for all frames of reference for which the equations of mechanics hold good. We will raise this conjecture (the purport of which will hereafter be called the 'Principle of Relativity') to the status of a postulate, and also introduce another postulate, which is only apparently irreconcilable with the former, namely, that light is always propagated in empty space with a definite velocity $c$ which is independent of the state of motion of the emitting body.

From these postulates, he derives formulas for the transformation of coordinates from one frame of reference to another in uniform motion relative to the first and shows that these transformations form a group.

### 1.3.6 Minkowski (1908)

In a famous address delivered at the 80th Assembly of German Natural Scientists and Physicians on September 21, 1908, Hermann Minkowski declared:

> The views of space and time which I wish to lay before you have sprung from the soil of experimental physics, and therein lies their strength. They are radical. Henceforth space by itself, and time by itself, are doomed to fade away into mere shadows, and only a kind of union of the two will preserve an independent reality.

He formalized special relativity by treating space and time as two aspects of a single entity, *spacetime*. In simple terms, we may think of this as $\mathbb{R}^4$, where a point $\mathbf{x} = (t, x, y, z)$ describes the time and position of an event. Crucially, this $\mathbb{R}^4$ is equipped with a bilinear form, the *Minkowski metric*:

$$\mathbf{x} \cdot \mathbf{x}' = tt' - xx' - yy' - zz',$$

which we use as a replacement for the usual dot product when calculating times and distances. With this extra structure, $\mathbb{R}^4$ is now called *Minkowski spacetime*. The group of all linear transformations

$$T \colon \mathbb{R}^4 \to \mathbb{R}^4$$

preserving the Minkowski metric is called the *Lorentz group* and denoted O(3, 1).

### 1.3.7 Heisenberg (1925)

In 1925, Werner Heisenberg came up with a radical new approach to physics in which processes were described using matrices [170]. What makes this especially remarkable is that Heisenberg, like most physicists of his day, had not heard of matrices! His idea was that given a system with some set of states, say $\{1, \ldots, n\}$, a process $U$ would be described by a bunch of complex numbers $U_j^i$ specifying the amplitude for any state $i$ to turn into any state $j$. He composed processes by summing over all possible intermediate states:

$$(VU)_k^i = \sum_j V_k^j U_j^i.$$

Later, he discussed his theory with his thesis advisor, Max Born, who informed him that he had reinvented matrix multiplication.

Heisenberg never liked the term *matrix mechanics* for his work because he thought it sounded too abstract. However, it is an apt indication of the *algebraic* flavor of quantum physics.

### 1.3.8 Born (1928)

In 1928, Max Born figured out what Heisenberg's mysterious amplitudes actually meant: the absolute value squared $|U_j^i|^2$ gives the *probability* for the initial state $i$ to become the final state $j$ via the process $U$. This spelled the end of the deterministic worldview built into Newtonian mechanics [98]. More shockingly still, given that amplitudes are complex, a sum of amplitudes can have a smaller absolute value than those of its terms. Thus, quantum mechanics exhibits destructive interference, allowing more ways for something to happen may reduce the chance that it does!

### 1.3.9 Von Neumann (1932)

In 1932, John von Neumann published a book on the foundations of quantum mechanics [177], which helped crystallize the now-standard approach to this theory. We hope that the experts will forgive us for omitting many important subtleties and caveats in the following sketch.

Every quantum system has a Hilbert space of states, $H$. A *state* of the system is described by a unit vector $\psi \in H$. Quantum theory is inherently probabilistic in that if we put the system in some state $\psi$ and immediately check to see whether it is in the state $\phi$, we get the answer "yes" with probability equal to $|\langle \phi, \psi \rangle|^2$.

A reversible process that our system can undergo is called a *symmetry*. Mathematically, any symmetry is described by a unitary operator $U \colon H \to H$. If we put the system in some state $\psi$ and apply the symmetry $U$, it will then be in the state $U\psi$. If we then check to see whether it is in some state $\phi$, we get the answer "yes" with probability $|\langle \phi, U\psi \rangle|^2$. The underlying complex number $\langle \phi, U\psi \rangle$ is called a *transition amplitude*. In particular, if we have an orthonormal basis $e^i$ of $H$, the numbers

$$U_j^i = \langle e^j, Ue^i \rangle$$

are Heisenberg's matrices!

Thus, Heisenberg's matrix mechanics is revealed to be part of a framework in which unitary operators describe physical processes. But operators also play another role in quantum theory. A real-valued quantity that we can measure by doing experiments on our system is called an *observable*. Examples include energy, momentum, angular momentum, and the like. Mathematically, any observable is described by a self-adjoint operator $A$ on the Hilbert space $H$ for the system in question. Thanks to the probabilistic nature of quantum mechanics, we can obtain various different values when we measure the observable $A$ in the state $\psi$, but the average or expected value will be $\langle \psi, A\psi \rangle$.

If a group $G$ acts as symmetries of some quantum system, we obtain a *unitary representation* of $G$, meaning a Hilbert space $H$ equipped with unitary operators

$$\rho(g) \colon H \to H,$$

one for each $g \in G$, such that

$$\rho(1) = 1_H$$

and

$$\rho(gh) = \rho(g)\rho(h).$$

Often, the group $G$ will be equipped with a topology. Then, we want symmetry transformations close to the identity to affect the system only slightly, so we demand that if $g_i \to 1$ in $G$, then $\rho(g_i)\psi \to \psi$ for all $\psi \in H$. Professionals use the term *strongly continuous* for representations with this property, but we shall simply call them *continuous*, given that we never discuss any other sort of continuity.

Continuity turns out to have powerful consequences, such as the Stone–von Neumann theorem, which states that if $\rho$ is a continuous representation of $\mathbb{R}$ on $H$, then

$$\rho(s) = \exp(-isA)$$

for a unique self-adjoint operator $A$ on $H$. Conversely, any self-adjoint operator gives a continuous representation of $\mathbb{R}$ this way. In short, there is a correspondence between observables and one-parameter groups of symmetries. This links the two roles of operators in quantum mechanics: self-adjoint operators for observables, and unitary operators for symmetries.

### 1.3.10 Wigner (1939)

We have already discussed how the Lorentz group $O(3, 1)$ acts as symmetries of spacetime in special relativity, concluding that it is the group of all linear transformations

$$T \colon \mathbb{R}^4 \to \mathbb{R}^4$$

preserving the Minkowski metric. However, the full symmetry group of Minkowski spacetime is larger, inlcuding translations as well. So, the really important group in special relativity is the so-called Poincaré group:

$$\mathbf{P} = O(3, 1) \ltimes \mathbb{R}^4$$

generated by Lorentz transformations and translations.

Some subtleties appear when we take some findings from particle physics into account. Although time reversal

$$(t, x, y, z) \mapsto (-t, x, y, z)$$

and parity

$$(t, x, y, z) \mapsto (t, -x, -y, -z)$$

are elements of **P**, not every physical system has them as symmetries. So, it is better to exclude such elements of the Poincaré group by working with the connected component of the identity, $\mathbf{P}_0$. Furthermore, when we rotate an electron a full turn, its state vector does not come back to where it started. Rather, it gets multiplied by $-1$. If we rotate it two full turns, it gets back to where it started. To deal with this, we should replace $\mathbf{P}_0$ by its universal cover, $\tilde{\mathbf{P}}_0$. For lack of a snappy name, in what follows, we call *this* group the *Poincaré group*.

We have seen that in quantum mechanics, physical systems are described by continuous unitary representations of the relevant symmetry group. In relativistic quantum mechanics, this symmetry group is $\tilde{\mathbf{P}}_0$. The Stone–von Neumann theorem then associates observables to one-parameter subgroups of this group. The most important observables in physics—energy, momentum, and angular momentum—all arise this way!

For example, time translation

$$g_s \colon (t, x, y, z) \mapsto (t + s, x, y, z)$$

gives rise to an observable $A$ with

$$\rho(g_s) = \exp(-isA)$$

and this observable is the *energy* of the system, also known as the *Hamiltonian*. If the system is in a state described by the unit vector $\psi \in H$, the expected value of its energy is $\langle \psi, A\psi \rangle$. In the context of special relativity, the energy of a system is always greater than or equal to that of the vacuum (the empty system, as it were). The energy of the vacuum is zero, so it makes sense to focus attention on continuous unitary representations of the Poincaré group with

$$\langle \psi, A\psi \rangle \geq 0.$$

These are usually called *positive-energy representations*.

In a famous 1939 paper, Eugene Wigner [241] classified the positive-energy representations of the Poincaré group. All these representations can be built as direct sums of irreducible ones, which serve as candidates for describing elementary particles, the building blocks of matter. To specify one of these representations, we need to give a number $m \geq 0$ called the *mass* of the particle, a number $j = 0, \frac{1}{2}, 1, \ldots$ called its *spin*, and sometimes a little extra data.

For example, the photon has spin 1 and mass 0, and the electron has spin $\frac{1}{2}$ and mass equal to about $9 \cdot 10^{-31}$ kilograms. Nobody knows why particles have the masses they do—this is one of the main unsolved problems in physics—but they all fit nicely into Wigner's classification scheme.

### 1.3.11 Eilenberg–Mac Lane (1945)

Eilenberg and Mac Lane [75] invented the notion of a "category" while working on algebraic topology. The idea is that whenever we study mathematical gadgets of any sort—sets, or groups, or topological spaces, or positive-energy representations of the Poincaré group, or whatever—we should also study the structure-preserving maps between these gadgets. We call the gadgets *objects* and the maps *morphisms*. The identity map is always a morphism, and we can compose morphisms in an associative way.

Eilenberg and Mac Lane thus defined a *category* $C$ to consist of:

- a collection of *objects*
- for any pair of objects $x, y$, a set of $\hom(x, y)$ of *morphisms* from $x$ to $y$, written $f : x \to y$

equipped with:

- for any object $x$, an *identity morphism* $1_x : x \to x$
- for any pair of morphisms $f : x \to y$ and $g : y \to z$, a morphism $gf : x \to z$ called the *composite* of $f$ and $g$

such that:

- for any morphism $f : x \to y$, the *left and right unit laws* hold: $1_y f = f = f 1_x$
- for any triple of morphisms $f : w \to x, g : x \to y, h : y \to z$ the *associative law* holds: $(hg)f = h(gf)$

Given a morphism $f : x \to y$, we call $x$ the *source* of $f$ and $y$ the *target* of $y$

Eilenberg and Mac Lane did much more than just define the concept of category. They also defined maps between categories, which they called *functors*. These send objects to objects, morphisms to morphisms, and preserve all the structure in sight. More precisely, given categories $C$ and $D$, a *functor* $F : C \to D$ consists of:

- a function $F$ sending objects in $C$ to objects in $D$
- for any pair of objects $x, y \in \mathrm{Ob}(C)$, a function also called $F$ sending morphisms in $\hom(x, y)$ to morphisms in $\hom(F(x), F(y))$

such that:

- $F$ preserves identities: for any object $x \in C$, $F(1_x) = 1_{F(x)}$
- $F$ preserves composition: for any pair of morphisms $f : x \to y$, $g : y \to z$ in $C$, $F(gf) = F(g)F(f)$

Many of the famous invariants in algebraic topology are actually functors, and this is part of how we convert topology problems into algebra problems and solve them. For example, the fundamental group is a functor

$$\pi_1 : \mathrm{Top}_* \to \mathrm{Grp}.$$

from the category of topological spaces equipped with a basepoint to the category of groups. In other words, not only does any topological space with basepoint $X$ have a fundamental group $\pi_1(X)$, but also any continuous map $f : X \to Y$ preserving the basepoint gives a homomorphism $\pi_1(f)\colon \pi_1(X) \to \pi_1(Y)$, in a way that gets along with composition. So, to show that the inclusion of the circle in the disc

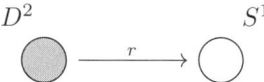

does not admit a retraction—that is, a map

$$D^2 \xrightarrow{r} S^1$$

such that this diagram commutes

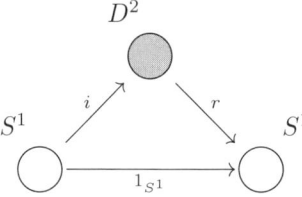

we simply hit this question with the functor $\pi_1$ and note that the homomorphism

$$\pi_1(i)\colon \pi_1(S^1) \to \pi_1(D^2)$$

cannot have a homomorphism

$$\pi_1(r)\colon \pi_1(D^2) \to \pi_1(S^1)$$

for which $\pi_1(r)\pi_1(i)$ is the identity because $\pi_1(S^1) = \mathbb{Z}$ and $\pi_1(D^2) = 0$.

However, Mac Lane later wrote that the real point of this paper was not to define categories, nor to define functors between categories, but rather to define natural transformations between functors! These can be drawn as follows:

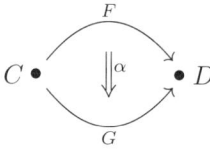

Given functors $F, G\colon C \to D$, a *natural transformation* $\alpha\colon F \Rightarrow G$ consists of:

- a function $\alpha$ mapping each object $x \in C$ to a morphism $\alpha_x\colon F(x) \to G(x)$

such that:

- for any morphism $f\colon x \to y$ in $C$, this diagram commutes:

$$\begin{array}{ccc} F(x) & \xrightarrow{F(f)} & F(y) \\ \alpha_x \downarrow & & \downarrow \alpha_y \\ G(x) & \xrightarrow{G(f)} & G(y) \end{array}$$

The commuting square here conveys the ideas that $\alpha$ not only gives a morphism $\alpha_x \colon F(x) \to G(x)$ for each object $x \in C$, but does so naturally—that is, in a way that is compatible with all the morphisms in $C$.

The most immediately interesting natural transformations are the natural isomorphisms. When Eilenberg and Mac Lane were writing their paper, there were many different recipes for computing the homology groups of a space, and they wanted to formalize the notion that these different recipes give groups that are not only isomorphic but also naturally so. In general, we say a morphism $g \colon y \to x$ is an *isomorphism* if it has an inverse; that is, a morphism $f \colon x \to y$ for which $fg$ and $gf$ are identity morphisms. A *natural isomorphism* between functors $F, G \colon C \to D$ is, then, a natural transformation $\alpha \colon F \Rightarrow G$ such that $\alpha_x$ is an isomorphism for all $x \in C$. Alternatively, we can define how to compose natural transformations and say a natural isomorphism is a natural transformation with an inverse.

Invertible functors are also important; but, here, an important theme known as *weakening* intervenes for the first time. Suppose we have functors $F \colon C \to D$ and $G \colon D \to C$. It is unreasonable to demand that if we apply first $F$ and then $G$, we get back exactly the object with which we started. In practice, all we really need, and all we typically get, is a naturally isomorphic object. So we say a functor $F \colon C \to D$ is an *equivalence* if it has a *weak inverse*; that is, a functor $G \colon D \to C$ such that there exist natural isomorphisms $\alpha \colon GF \Rightarrow 1_C$, $\beta \colon FG \Rightarrow 1_D$.

In the first applications to topology, the categories involved were mainly quite large as, for example, the category of all topological spaces or all groups. In fact, these categories are even large in the technical sense, meaning that their collection of objects is not a set but rather a proper class. But later applications of category theory to physics often involved small categories.

For example, any group $G$ can be thought of as a category with one object and only invertible morphisms because the morphisms are the elements of $G$, and composition is multiplication in the group. A representation of $G$ on a Hilbert space is then the same as a functor

$$\rho \colon G \to \mathrm{Hilb},$$

where Hilb is the category with Hilbert spaces as objects and bounded linear operators as morphisms. Although this viewpoint may seem like overkill, it is a prototype for the idea of describing theories of physics as functors, in which abstract physical processes (e.g., symmetries) are represented in a concrete way (e.g., as operators). However, this idea came long after the work of Eilenberg and Mac Lane: it was born sometime around Lawvere's 1963 thesis and came to maturity in Atiyah's 1988 definition of TQFT.

### 1.3.12 Feynman (1947)

After World War II, many physicists who had been working in the Manhattan Project to develop the atomic bomb returned to work on particle physics. In 1947, a small conference on this subject was held at Shelter Island, attended by luminaries such as Bohr, Oppenheimer, von Neumann, Weisskopf, and Wheeler. Feynman presented his work on quantum field theory, but it seems nobody understood it except Schwinger, who was later to share the Nobel Prize with him and Tomonaga. Apparently, it was a bit too far out for most of the audience.

Feynman described a formalism in which time evolution for quantum systems was described using an integral over the space of all classical histories, known as a *Feynman path integral*. These are notoriously hard to make rigorous. But he also described a way to compute these perturbatively as a sum over diagrams, now known as *Feynman diagrams*. For example, in quantum electrodynamics, the amplitude for an electron to absorb a photon is given by:

All of these diagrams describe ways for an electron and photon to come in and an electron to go out. Lines with arrows pointing downward stand for electrons. Lines with arrows pointing upward stand for positrons. The positron is the antiparticle of an electron, and Feynman realized that this could be thought of as an electron going backward in time. The wiggly lines stand for photons. The photon is its own antiparticle, so we do not need arrows on these wiggly lines.

Mathematically, each of the diagrams shown previously is shorthand for a linear operator

$$f: H_e \otimes H_\gamma \to H_e,$$

where $H_e$ is the Hilbert space for an electron and $H_\gamma$ is a Hilbert space for a photon. We take the tensor product of group representations when combining two systems, so $H_e \otimes H_\gamma$ is the Hilbert space for a photon together with an electron.

As already mentioned, elementary particles are described by certain special representations of the Poincaré group—the irreducible positive-energy ones. So $H_e$ and $H_\gamma$ are representations of this sort. We can tensor these to obtain positive-energy representations describing collections of elementary particles. Moreover, each Feynman diagram describes an *intertwining operator*, an operator that commutes with the action of the Poincaré group. This expresses the fact that if we, say, rotate our laboratory before doing an experiment, we just get a rotated version of the result we otherwise would get.

So Feynman diagrams are *a notation for intertwining operators between positive-energy representations of the Poincaré group*. However, they are so powerfully evocative that they are much more than a mere trick! As Feynman recalled later [169]:

> The diagrams were intended to represent physical processes and the mathematical expressions used to describe them. Each diagram signified a mathematical expression. In these diagrams I was seeing things that happened in space and time. Mathematical quantities were being associated with points in space and time. I would see electrons going along, being scattered at one point, then going over to another point and getting scattered there, emitting a photon and the photon goes there. I would make little pictures of all that was going on; these were physical pictures involving the mathematical terms.

Feynman first published papers containing such diagrams in 1949 [80,81]. However, his work reached many physicists through expository articles published even earlier by one of the few people who understood what he was up to, Freeman Dyson [72,73]. (For more on the history of Feynman diagrams, see the book by Kaiser [112].)

The general context for such diagrammatic reasoning came much later, from category theory. The idea is that we can draw a morphism $f \colon x \to y$ as an arrow going down:

but then we can switch to a style of drawing in which the objects are depicted not as dots but rather as wires, and the morphisms are drawn not as arrows but rather as black boxes with one input wire and one output wire:

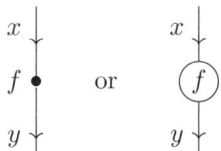

This is starting to look a bit like a Feynman diagram! However, to get really interesting Feynman diagrams, we need black boxes with many wires going in and many wires going out. The mathematics necessary for this was formalized later, in Mac Lane's 1963 paper on monoidal categories (discussed herein) and Joyal and Street's 1980s work on string diagrams [110].

### 1.3.13 Yang–Mills (1953)

In modern physics, the electromagnetic force is described by a U(1) gauge field. Most mathematicians prefer to call this a connection on a principal U(1) bundle. Jargon aside, this means that if we carry a charged particle around a loop in spacetime, its state will be multiplied by some element of U(1)—that is, a phase—thanks to the presence of the electromagnetic field. Moreover, everything about electromagnetism can be understood in these terms!

In 1953, Chen Ning Yang and Robert Mills [245] formulated a generalization of Maxwell's equations in which forces other than electromagnetism can be described by connections on $G$-bundles for groups other than U(1). With a vast amount of work by many great physicists, this ultimately led to the Standard Model, a theory in which *all forces other than gravity* are described using a connection on a principal $G$-bundle where

$$G = \mathrm{U}(1) \times \mathrm{SU}(2) \times \mathrm{SU}(3).$$

Although everyone would like to understand more deeply this curious choice of $G$, at present, it is purely a matter of fitting the experimental data.

In the Standard Model, elementary particles are described as irreducible positive-energy representations of $\tilde{P}_0 \times G$. Perturbative calculations in this theory can be done using souped-up Feynman diagrams, which are a notation for intertwining operators between positive-energy representations of $\tilde{P}_0 \times G$.

Although efficient, the mathematical jargon in the previous paragraphs does little justice to how physicists actually think about these things. For example, Yang and Mills *did not know about bundles and connections* when formulating their theory. Yang later wrote [247]:

> What Mills and I were doing in 1954 was generalizing Maxwell's theory. We knew of no geometrical meaning of Maxwell's theory, and we were not looking in that direction. To a physicist, gauge potential is a concept rooted in our description of the electromagnetic field. Connection is a geometrical concept which I only learned around 1970.

### 1.3.14 Mac Lane (1963)

In 1963, Mac Lane published a paper describing the notion of a monoidal category, [137]. The idea was that in many categories, there is a way to take the tensor product of two objects or of two morphisms. A famous example is the category Vect, where the objects are vector spaces and the morphisms are linear operators. This becomes a monoidal category with the usual tensor product of vector spaces and linear maps. Other examples include the category Set with the cartesian product of sets, and the category Hilb with the usual tensor product of Hilbert spaces. We will also be interested in FinVect and FinHilb, where the objects are *finite-dimensional* vector spaces (resp. Hilbert spaces) and the morphisms are linear maps. We will also get many examples from categories of representations of groups. The theory of Feynman diagrams, for example, turns out to be based on the symmetric monoidal category of positive-energy representations of the Poincaré group!

In a monoidal category, given morphisms $f: x \to y$ and $g: x' \to y'$, there is a morphism,

$$f \otimes g: x \otimes x' \to y \otimes y'.$$

We can also draw this as follows:

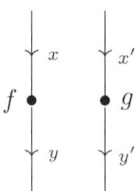

This sort of diagram is sometimes called a *string diagram*; the mathematics of these was formalized later [110], but we cannot resist using them now, given that they are so intuitive. Notice that the diagrams we could draw in a mere category were intrinsically one-dimensional because the only thing we could do is compose morphisms, which we draw by sticking one on top of another. In a monoidal category, the string diagrams become two-dimensional because now we can also tensor morphisms, which we draw by placing them side by side.

This idea continues to work in higher dimensions as well. The kind of category suitable for three-dimensional diagrams is called a *braided monoidal category*. In such a category, every pair of objects $x, y$ is equipped with an isomorphism, called the *braiding*, that switches the order of factors in their tensor product:

$$B_{x,y} \colon x \otimes y \to y \otimes x.$$

We can draw this process of switching as a diagram in three dimensions:

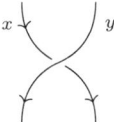

and the braiding $B_{x,y}$ satisfies axioms that are related to the topology of three-dimensional space.

All the examples of monoidal categories given above are also braided monoidal categories. Indeed, many mathematicians would shamelessly say that given vector spaces $V$ and $W$, the tensor product $V \otimes W$ is equal to the tensor product $W \otimes V$. But this is not really true; if you examine the fine print, you will see that they are just isomorphic, via this braiding:

$$B_{V,W} \colon v \otimes w \mapsto w \otimes v.$$

Actually, all the preceding examples are not just braided but also symmetric monoidal categories. This means that if you switch two things and then switch them again, you get back where you started:

$$B_{x,y} B_{y,x} = 1_{x \otimes y}.$$

Because all the braided monoidal categories Mac Lane knew satisfied this extra axiom, he considered only symmetric monoidal categories. In diagrams, this extra axiom says that:

In four or more dimensions, any knot can be untied by just this sort of process. Thus, the string diagrams for symmetric monoidal categories should really be drawn in four or more dimensions! But we can cheat and draw them in the plane, as we have in the preceding example.

It is worth taking a look at Mac Lane's precise definitions because they are a bit subtler than our summary suggests, and these subtleties are actually very interesting.

First, he demanded that a monoidal category have a unit for the tensor product, which he called the *unit object*, or 1. For example, the unit for tensor product in Vect is the ground field, and the unit for the Cartesian product in Set is the one-element set. (*Which* one-element set? Choose your favorite one!)

Second, Mac Lane did not demand that the tensor product be associative on the nose:

$$(x \otimes y) \otimes z = x \otimes (y \otimes z)$$

but only up a specified isomorphism called the *associator*:

$$a_{x,y,z} \colon (x \otimes y) \otimes z \to x \otimes (y \otimes z).$$

Similarly, he did not demand that 1 act as the unit for the tensor product on the nose but rather only up to specified isomorphisms called the *left* and *right unitors*:

$$\ell_x \colon 1 \otimes x \to x, \qquad r_x \colon x \otimes 1 \to x.$$

The reason is that in real life, it is usually too much to expect equations between objects in a category; usually, we just have isomorphisms, and this is good enough! Indeed, this is a basic moral of category theory, that equations between objects are bad; we should instead specify isomorphisms.

Third, and most subtle of all, Mac Lane demanded that the associator and left and right unitors satisfy certain coherence laws, which let us work with them as smoothly as if they *were* equations. These laws are called the *pentagon* and *triangle identities*.

Here is the actual definition. A *monoidal category* consists of:

- a category $M$
- a functor called the *tensor product* $\otimes \colon M \times M \to M$, where we write $\otimes(x, y) = x \otimes y$ and $\otimes(f, g) = f \otimes g$ for objects $x, y \in M$ and morphisms $f, g$ in $M$
- an object called the *identity object* $1 \in M$
- natural isomorphisms called the *associator*:

$$a_{x,y,z} \colon (x \otimes y) \otimes z \to x \otimes (y \otimes z),$$

the *left unit law:*

$$\ell_x \colon 1 \otimes x \to x,$$

and the *right unit law:*

$$r_x \colon x \otimes 1 \to x$$

such that the following diagrams commute for all objects $w, x, y, z \in M$:

- the *pentagon identity:*

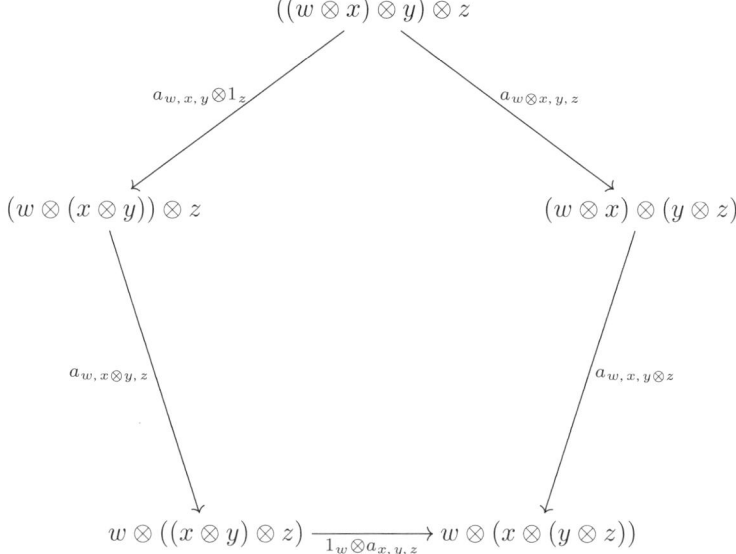

governing the associator; and

- the *triangle identity:*

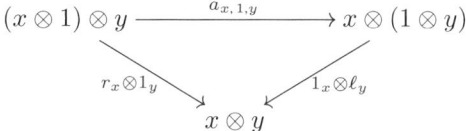

governing the left and right unitors.

The pentagon and triangle identities are the least obvious part of this definition—but also the truly brilliant part. The point of the pentagon identity is that when we have a tensor product of four objects, there are five ways to parenthesize it and, at first glance, the associator gives two different isomorphisms from $w \otimes (x \otimes (y \otimes z))$ to $((w \otimes x) \otimes y) \otimes z$. The pentagon identity says these are in fact the same! Of course, when we have tensor products of even more objects, there are even more ways to parenthesize them and even more isomorphisms between them built from the associator. However, Mac Lane showed that the pentagon identity implies these isomorphisms are all the same. If we also assume the triangle identity, all isomorphisms with the same source and target built from the associator and left and right unit laws are equal.

In fact, the pentagon was also introduced in 1963 by James Stasheff [225], as part of an infinite sequence of polytopes called *associahedra*. Stasheff defined a concept of $A_\infty$-space which is roughly a topological space having a product that is associative up to homotopy, where this homotopy satisfies the pentagon identity up homotopy, that homotopy satisfies yet another identity up to homotopy, and so on, ad infinitum. The $n$th of these identities is described by the *n*-dimensional associahedron. The first identity is just the associative law, which plays a crucial role in the definition of *monoid*, a set with associative product and identity element. Mac Lane realized that the second, the pentagon identity, should play a similar role in the definition of monoidal category. The higher ones show up in the theory of monoidal bicategories, monoidal tricategories, and so on.

With the concept of monoidal category in hand, one can define a *braided monoidal category* to consist of:

- a monoidal category $M$
- a natural isomorphism called the *braiding*:

$$B_{x,y} \colon x \otimes y \to y \otimes x$$

such that these two diagrams commute, called the *hexagon identities*:

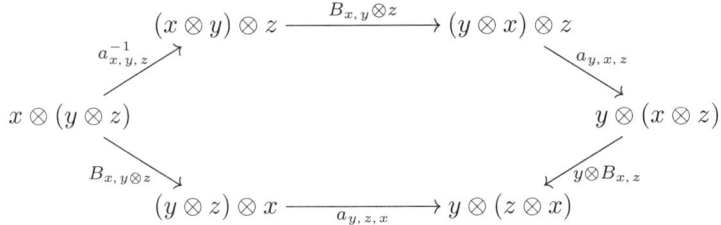

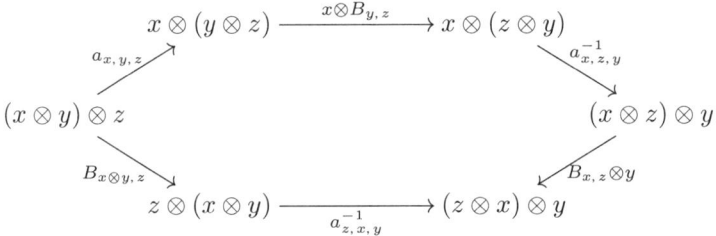

The first hexagon equation says that switching the object $x$ past $y \otimes z$ all at once is the same as switching it past $y$ and then past $z$ (with some associators thrown in to move the parentheses). The second one is similar: it says switching $x \otimes y$ past $z$ all at once is the same as doing it in two steps.

We define a *symmetric monoidal category* to be a braided monoidal category $M$ for which the braiding satisfies $B_{x,y} = B_{y,x}^{-1}$ for all objects $x$ and $y$. A monoidal, braided monoidal, or symmetric monoidal category is called *strict* if $a_{x,y,z}$, $\ell_x$, and $r_x$ are always identity morphisms. In this case, we have

$$(x \otimes y) \otimes z = x \otimes (y \otimes z),$$

$$1 \otimes x = x, \qquad x \otimes 1 = x.$$

Mac Lane showed that every monoidal or symmetric monoidal category is equivalent to a strict one in a certain precise sense. The same is true for braided monoidal categories. However, the examples that turn up in nature, like Vect, are rarely strict.

### 1.3.15 Lawvere (1963)

The famous category theorist F. William Lawvere began his graduate work under Clifford Truesdell, an expert on continuum mechanics, that very practical branch of classical field theory that deals with fluids, elastic bodies, and the like. In the process, Lawvere got very interested in the foundations of physics, particularly the notion of physical theory, and his research took an abstract turn. Because Truesdell had worked with Eilenberg and Mac Lane during World War II, he sent Lawvere to visit Eilenberg at Columbia University, and that is where Lawvere wrote his thesis.

In 1963, Lawvere finished a thesis on functorial semantics [145]. This is a general framework for theories of mathematical or physical objects in which a theory is described by a category $C$ and a model of this theory is described by a functor $Z \colon C \to D$. Typically, $C$ and $D$ are equipped with extra structure, and $Z$ is required to preserve this structure. The category $D$ plays the role of an environment in which the models live; often, we take $D = \mathrm{Set}$.

Variants of this idea soon became important in topology, especially PROPs and operads. In the late 1960s and early 1970s, Mac Lane [138], Boardmann and Vogt [36], May [167], and others used these variants to study homotopy-coherent algebraic structures; that is, structures with operations satisfying laws only up to homotopy, with the homotopies themselves obeying certain laws, but only up to homotopy, ad infinitum. The easiest examples are Stasheff's $A_\infty$-spaces, which we mentioned in the previous section. The laws governing $A_\infty$-spaces are encoded in associahedra such as the pentagon. In later work, it was seen that the associahedra form an operad. By the 1990s, operads had become important both in mathematical physics [150, 165] and the theory of $n$-categories [148]. Unfortunately, explaining this line of work would take us far afield.

Other outgrowths of Lawvere's vision of functorial semantics include the definitions of conformal field theory and TQFT, propounded by Segal and Atiyah in the late 1980s. We will have much more to say about these. In keeping with physics terminology, these later authors use the word *theory* for what Lawvere called a model—namely, a structure-preserving functor $Z \colon C \to D$. There is, however, a much more important difference. Lawvere focused on classical physics and took $C$ and $D$ to be categories with cartesian products. Segal and Atiyah focused on quantum physics and took $C$ and $D$ to be symmetric monoidal categories, not necessarily cartesian.

### 1.3.16 Bénabou (1967)

In 1967, Bénabou [31] introduced the notion of a bicategory or, as it is sometimes now called, a weak 2-category. The idea is that besides objects and morphisms, a bicategory has 2-morphisms going between morphisms, like this:

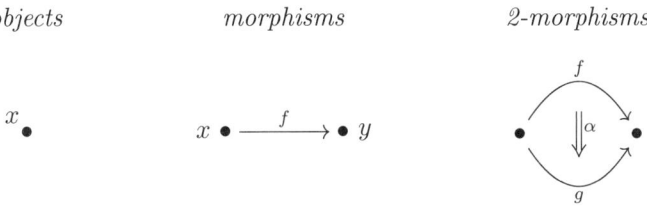

In a bicategory, we can compose morphisms as in an ordinary category, but also we can compose 2-morphisms in two ways, vertically and horizontally:

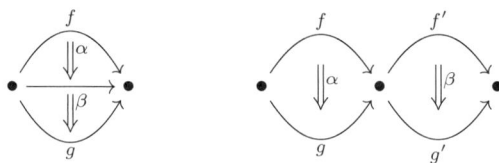

There are also identity morphisms and identity 2-morphisms and various axioms governing their behavior. Most important, the usual laws for composition of morphisms—the left and right unit laws and associativity—hold only *up to specified 2-isomorphisms*. (A *2-isomorphism* is a 2-morphism that is invertible with respect to vertical composition.) For example, given morphisms $h: w \to x$, $g: x \to y$, and $f: y \to z$, we have a 2-isomorphism called the *associator*:

$$a_{f,g,h}: (fg)h \to f(gh).$$

As in a monoidal category, this should satisfy the pentagon identity.

Bicategories are everywhere once you know how to look. For example, there is a bicategory Cat in which:

- the objects are categories
- the morphisms are functors
- the 2-morphisms are natural transformations

This example is unusual because composition of morphisms happens to satisfy the left and right unit laws and associativity on the nose, as equations. A more typical example is Bimod, in which:

- the objects are rings
- the morphisms from $R$ to $S$ are $R - S$-bimodules
- the 2-morphisms are bimodule homomorphisms

Here, composition of morphisms is defined by tensoring. Given an $R - S$-bimodule $M$ and an $S - T$-bimodule, we can tensor them over $S$ to get an $R - T$-bimodule. In this example, the laws for composition hold only up to specified 2-isomorphisms.

Another class of examples comes from the fact that a monoidal category is secretly a bicategory with one object! The correspondence involves a kind of "reindexing," as shown in the following table:

| **Monoidal Category** | **Bicategory** |
|---|---|
| — | Objects |
| Objects | Morphisms |
| Morphisms | 2-Morphisms |
| Tensor product of objects | Composite of morphisms |
| Composite of morphisms | Vertical composite of 2-morphisms |
| Tensor product of morphisms | Horizontal composite of 2-morphisms |

In other words, to see a monoidal category as a bicategory with only one object, we should call the objects of the monoidal category *morphisms* and call its morphisms *2-morphisms*.

A good example of this trick involves the monoidal category Vect. Start with Bimod and choose your favorite object—say, the ring of complex numbers. Then, take all those bimodules of this ring that are complex vector spaces and all the bimodule homomorphisms between these. You now have a sub-bicategory with just one object—or, in other words, a monoidal category! This is Vect.

The fact that a monoidal category is secretly just a degenerate bicategory eventually stimulated a lot of interest in higher categories, and people began to wonder what kinds of degenerate higher categories give rise to braided and symmetric monoidal categories. The impatient reader can jump ahead to 1995, when the pattern underlying all of these monoidal structures and their higher-dimensional analogs became more clear.

### 1.3.17 Penrose (1971)

In general, relativity people had been using index-ridden expressions for a long time. For example, suppose we have a binary product on a vector space $V$:

$$m \colon V \otimes V \to V.$$

A normal person would abbreviate $m(v \otimes w)$ as $v \cdot w$ and write the associative law as

$$(u \cdot v) \cdot w = u \cdot (v \cdot w).$$

A mathematician might show off by writing

$$m(m \otimes 1) = m(1 \otimes m)$$

instead. But physicists would choose a basis $e^i$ of $V$ and set

$$m(e^i \otimes e^j) = \sum_k m_k^{ij} e^k$$

or

$$m(e^i \otimes e^j) = m_k^{ij} e^k$$

for short, using the Einstein summation convention to sum over any repeated index that appears once as a superscript and once as a subscript. Then, they would write the associative law as follows:

$$m_p^{ij} m_l^{pk} = m_l^{iq} m_q^{jk}.$$

Mathematicians would mock them for this but, until Penrose came along, there was really no better completely general way to manipulate tensors. Indeed, before Einstein introduced his summation convention in 1916, things were even worse. He later joked to a friend [182]:

> I have made a great discovery in mathematics; I have suppressed the summation sign every time that the summation must be made over an index which occurs twice....

In 1971, Penrose [186] introduced a new notation where tensors are drawn as black boxes, with superscripts corresponding to wires coming in from above and subscripts corresponding to wires going out from below. For example, he might draw $m \colon V \otimes V \to V$ as:

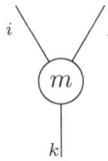

and the associative law as:

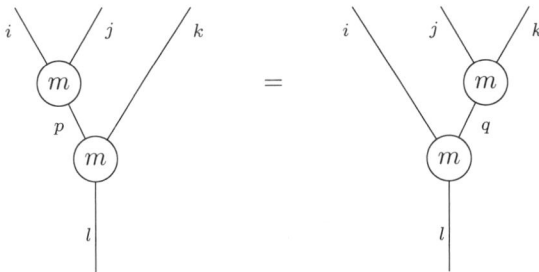

In this notation, we sum over the indices labeling internal wires—by which we mean wires that are the output of one box and an input of another. This is just the Einstein summation convention in disguise, so the previous picture is merely an artistic way of drawing this:

$$m^{ij}_p m^{pk}_l = m^{iq}_l m^{jk}_q.$$

But, it has an enormous advantage: *no ambiguity is introduced if we leave out the indices* because the wires tell us how the tensors are hooked together:

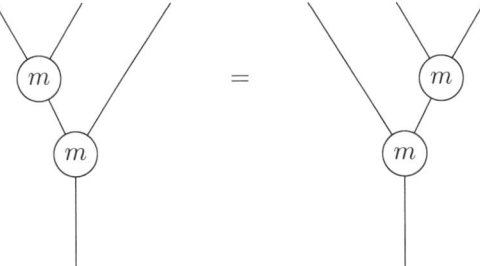

This is a more vivid way of writing the mathematician's equation

$$m(m \otimes 1_V) = m(1_V \otimes m)$$

because tensor products are written horizontally and composition vertically instead of trying to compress them into a single line of text.

In modern language, what Penrose had noticed here was that FinVect, the category of finite-dimensional vector spaces and linear maps, is a symmetric monoidal category, so we can draw morphisms in it using string diagrams. But, he probably was not thinking about categories; he was probably more influenced by the analogy to Feynman diagrams.

Indeed, Penrose's pictures are very much like Feynman diagrams but simpler. Feynman diagrams are pictures of morphisms in the symmetric monoidal category of positive-energy representations of the Poincaré group! It is amusing that this complicated example was considered long before Vect, but that is how it often works. Simple ideas rise to consciousness only when difficult problems make them necessary.

Penrose also considered some examples more complicated than FinVect but simpler than full-fledged Feynman diagrams. For any compact Lie group $K$, there is a symmetric monoidal category Rep($K$). Here, the objects are finite-dimensional continuous unitary representations of $K$—that is a bit of a mouthful, so we will just call them *representations*. The morphisms are *intertwining operators* between representations; that is, operators $f : H \to H'$ with

$$f(\rho(g)\psi) = \rho'(g)f(\psi)$$

for all $g \in K$ and $\psi \in H$, where $\rho(g)$ is the unitary operator by which $g$ acts on $H$ and $\rho'(g)$ is the one by which $g$ acts on $H'$. The category Rep($K$) becomes symmetric monoidal with the usual tensor product of group representations,

$$(\rho \otimes \rho')(g) = \rho(g) \otimes \rho(g')$$

and the obvious braiding.

As a category, Rep($K$) is easy to describe. Every object is a direct sum of finitely many *irreducible* representations—that is, representations that are not themselves a direct sum in a nontrivial way. So, if we pick a collection $E_i$ of irreducible representations, one from each isomorphism class, we can write any object $H$ as

$$H \cong \bigoplus_i H^i \otimes E_i,$$

where the $H^i$ is the finite-dimensional Hilbert space describing the multiplicity with which the irreducible $E_i$ appears in $H$,

$$H^i = \hom(E_i, H).$$

Then, we use Schur's Lemma, which describes the morphisms between irreducible representations:

- When $i = j$, the space $\hom(E_i, E_j)$ is one-dimensional: all morphisms from $E_i$ to $E_j$ are multiples of the identity.
- When $i \neq j$, the space $\hom(E_i, E_j)$ is zero-dimensional: all morphisms from $E$ to $E'$ are zero.

So every representation is a direct sum of irreducibles, and every morphism between irreducibles is a multiple of the identity (possibly zero). Given that composition is linear in each argument, this means there is only one way that composition of morphisms can possibly work. So, the category is completely pinned down as soon as we know the set of irreducible representations.

One nice thing about Rep($K$) is that every object has a dual. If $H$ is some representation, the dual vector space $H^*$ also becomes a representation, with

$$(\rho^*(g)f)(\psi) = f(\rho(g)\psi)$$

for all $f \in H^*$, $\psi \in H$. In our string diagrams, we can use little arrows to distinguish between $H$ and $H^*$. A downward-pointing arrow labeled by $H$ stands for the object $H$, and an upward-pointing arrow stands for $H^*$. For example,

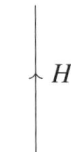

is the string diagram for the identity morphism $1_{H^*}$. This notation is meant to remind us of Feynman's idea of antiparticles as particles going backward in time.

The dual pairing

$$e_H \colon H^* \otimes H \to \mathbb{C}$$
$$f \otimes v \mapsto f(v)$$

is an intertwining operator, as is the operator

$$i_H \colon \mathbb{C} \to H \otimes H^*$$
$$c \mapsto c\, 1_H$$

where we think of $1_H \in \mathrm{hom}(H, H)$ as an element of $H \otimes H^*$. We can draw these operators as a cup:

and a cap:

Note that if no edges reach the bottom (or top) of a diagram, it describes a morphism to (or from) the trivial representation of $G$ on $\mathbb{C}$—because this is the tensor product of *no* representations.

The cup and cap satisfy the *zig-zag identities*:

$$\cap\cup = |$$

$$\cup\cap = |$$

These identities are easy to check. For example, the first zig-zag gives a morphism from $H$ to $H$ that we can compute by feeding in a vector $\psi \in H$:

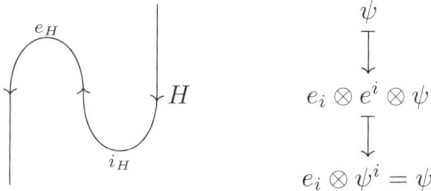

$$\psi \mapsto e_i \otimes e^i \otimes \psi \mapsto e_i \otimes \psi^i = \psi$$

So, indeed, this is the identity morphism. But the beauty of these identities is that they let us straighten out a portion of a string diagram as if it were actually a piece of string! Algebra is becoming topology.

Furthermore, we have:

$$\bigcirc\!\!\bigcirc = \dim(H)$$

This requires a little explanation. A closed diagram—one with no edges coming in and no edges coming out—denotes an intertwining operator from the trivial representation to itself. Such a thing is just multiplication by some number. The preceding equation says the operator on the left is multiplication by $\dim(H)$. We can check this as follows:

$$1 \mapsto e^i \otimes e_i \mapsto e_i \otimes e^i \mapsto \delta^i_i = \dim(H)$$

So, *a loop gives a dimension*. This explains a big problem that plagues Feynman diagrams in quantum field theory—namely, the divergences or infinities that show up in diagrams containing loops, like this:

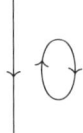

or, more subtly, like this:

These infinities come from the fact that most positive-energy representations of the Poincaré group are infinite dimensional. The reason is that this group is noncompact. For a compact Lie group, all the irreducible continuous representations are finite dimensional.

So far, we have been discussing representations of compact Lie groups quite generally. In his theory of spin networks [187, 188], Penrose worked out all the details for SU(2), the group of $2 \times 2$ unitary complex matrices with determinant 1. This group is important because it is the universal cover of the three-dimensional rotation group. This lets us handle particles like the electron, which does not come back to its original state after one full turn—but does after two!

The group SU(2) is the subgroup of the Poincaré group whose corresponding observables are the components of angular momentum. Unlike the Poincaré group, it is compact. As already mentioned, we can specify an irreducible positive-energy representation of the Poincaré group by choosing a mass $m \geq 0$, a spin $j = 0, \frac{1}{2}, 1, \frac{3}{2}, \ldots$ and sometimes a little extra data. Irreducible unitary representations of SU(2) are simpler. For these, we just need to choose a spin. The group SU(2) has one irreducible unitary representation of each dimension. Physicists call the representation of dimension $2j + 1$ the spin-$j$ representation, or simply $j$ for short.

Every representation of SU(2) is isomorphic to its dual, so we can pick an isomorphism

$$\sharp \colon j \to j^*$$

for each $j$. Using this, we can stop writing little arrows on our string diagrams. For example, we get a new cup

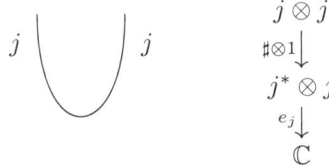

and similarly a new cap. These satisfy an interesting relation:

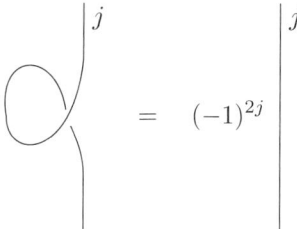

Physically, this means that when we give a spin-$j$ particle a full turn, its state transforms trivially when $j$ is an integer:

$$\psi \mapsto \psi,$$

but it picks up a sign when $j$ is an integer plus $\frac{1}{2}$,

$$\psi \mapsto -\psi.$$

Particles of the former sort are called *bosons*; those of the latter sort are called *fermions*.

The funny minus sign for fermions also shows up when we build a loop with our new cup and cap:

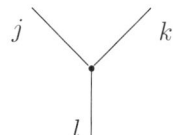 $= (-1)^{2j} (2j+1)$

Rather than the usual dimension of the spin-$j$ representation, we get the dimension times a sign, depending on whether this representation is bosonic or fermionic! This is sometimes called the *superdimension* because its full explanation involves what physicists call supersymmetry. Alas, we have no space to discuss this here; we must hasten on to Penrose's theory of spin networks!

Spin networks are a nice notation for morphisms between tensor products of irreducible representations of SU(2). The key underlying fact is that:

$$j \otimes k \cong |j-k| \oplus |j-k|+1 \oplus \cdots \oplus j+k.$$

Thus, the space of intertwining operators hom($j \otimes k, l$) has dimension 1 or 0, depending on whether or not $l$ appears in this direct sum. We say the triple $(j, k, l)$ is *admissible* when this space has dimension 1. This happens when the triangle inequalities are satisfied,

$$|j-k| \leq l \leq j+k$$

and also $j + k + l \in \mathbb{Z}$.

For any admissible triple $(j, k, l)$, we can choose a nonzero intertwining operator from $j \otimes k$ to $l$, which we draw as follows:

Using the fact that a closed diagram gives a number, we can normalize these intertwining operators so that the theta network takes a convenient value; say:

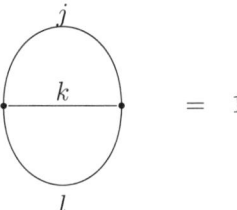

When the triple $(j, k, l)$ is not admissible, we define

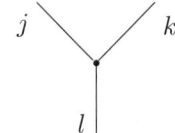

to be the zero operator, so that

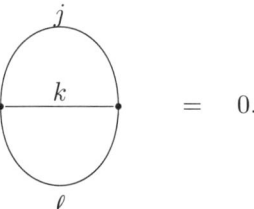

We can then build more complicated intertwining operators by composing and tensoring the ones we have described so far. For example, this diagram shows an intertwining operator from the representation $2 \otimes \frac{3}{2} \otimes 1$ to the representation $\frac{5}{2} \otimes 2$:

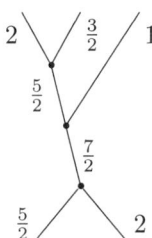

A diagram of this sort is called a *spin network*. The resemblance to a Feynman diagram is evident. There is a category where the morphisms are spin networks and a functor from this category to Rep(SU(2)). A spin network with no edges coming in from the top and no edges coming out at the bottom is called *closed*. A closed spin network determines an intertwining operator from the trivial representation of SU(2) to itself, and thus a complex number. (For more details, see the article by Major [159].)

Penrose noted that spin networks satisfy several interesting rules. For example, we can deform a spin network in various ways without changing the operator it describes. We have already seen the zig-zag identity, which is an example of this. Other rules involve changing the topology of the spin network. The most important of these is the *binor identity* for the spin-$\frac{1}{2}$ representation:

$$\underset{\frac{1}{2}\ \frac{1}{2}}{\overset{\frac{1}{2}\ \frac{1}{2}}{\times}} = \underset{\frac{1}{2}\ \frac{1}{2}}{\overset{\frac{1}{2}\ \frac{1}{2}}{\asymp}} + \underset{}{\overset{\frac{1}{2}\ \ \frac{1}{2}}{||}}$$

We can use this to prove something we have already seen:

$$\overset{\frac{1}{2}}{\varphi} = \overset{\frac{1}{2}}{\cap\!\cup} + \overset{\frac{1}{2}}{\bigcirc}\bigg| = -\bigg|^{\frac{1}{2}}$$

Physically, this says that turning a spin-$\frac{1}{2}$ particle around 360 degrees multiplies its state by $-1$.

There are also interesting rules involving the spin-1 representation, which imply some highly nonobvious results. For example, every planar graph with three edges meeting at each vertex, no edge-loops, and every edge labeled by the spin-1 representation

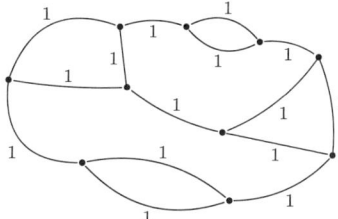

evaluates to a nonzero number [189]. But Penrose showed that this fact is equivalent to the four-color theorem!

By now, Penrose's diagrammatic approach to the finite-dimensional representations of SU(2) has been generalized to many compact simple Lie groups. A good treatment of this material is the free online book by Cvitanović [62]. His book includes a brief history of diagrammatic methods that makes a nice companion to the present work. Much of the work in his book was done in the 1970s. However, the huge burst of work on diagrammatic methods for algebra came later, in the 1980s, with the advent of quantum groups.

### 1.3.18 Ponzano–Regge (1968)

Sometimes history turns around and goes back in time, like an antiparticle. This seems like the only sensible explanation of the revolutionary work of Ponzano and Regge [193], who applied Penrose's theory of spin networks *before it was invented* to relate tetrahedron-shaped spin networks to gravity in three-dimensional spacetime. Their work eventually led to a theory called the *Ponzano–Regge model*, which allows for an exact solution of many problems in three-dimensional quantum gravity [38].

In fact, Ponzano and Regge's article on this topic appeared in the proceedings of a conference on spectroscopy, because the $6j$ symbol is important in chemistry. But for our purposes, the $6j$ symbol is just the number obtained by evaluating this spin network:

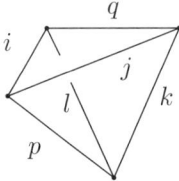

depending on six spins $i, j, k, l, p, q$.

In the Ponzano–Regge model of three-dimensional quantum gravity, spacetime is made of tetrahedra, and we label the edges of tetrahedra with spins to specify their *lengths*. To compute the amplitude for spacetime to have a particular shape, we multiply a bunch of amplitudes (i.e., complex numbers), one for each tetrahedron, one for each triangle, and one for each edge. The most interesting ingredient in this recipe is the amplitude for a tetrahedron. This is given by the $6j$ symbol.

But we have to be a bit careful! Starting from a tetrahedron whose edge lengths are given by spins,

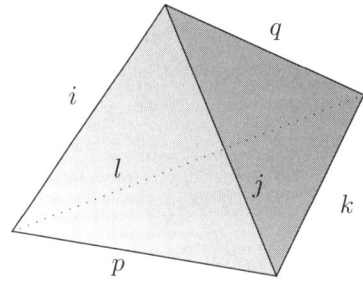

we compute its amplitude using the *Poincaré dual* spin network, which has:

- one vertex at the center of each face of the original tetrahedron
- one edge crossing each edge of the original tetrahedron

It looks like this:

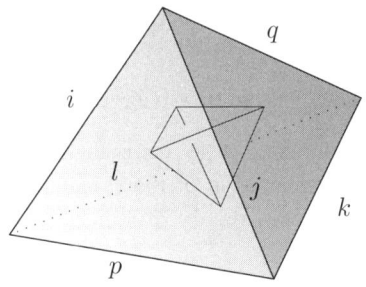

Its edges inherit spin labels from the edges of the original tetrahedron:

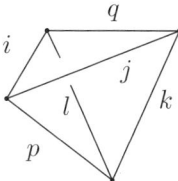

*Voilà!* The $6j$ symbol!

It is easy to get confused because the Poincaré dual of a tetrahedron just happens to be another tetrahedron. But there are good reasons for this dualization process. For example, the $6j$ symbol vanishes if the spins labeling three edges meeting at a vertex violate the triangle inequalities because then these spins will be inadmissible. For example, we need

$$|i - j| \le p \le i + j$$

or the intertwining operator

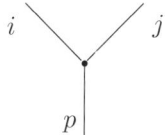

will vanish, forcing the $6j$ symbols to vanish as well. But in the original tetrahedron, these spins label the three sides of a triangle:

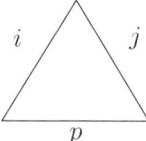

So *the amplitude for a tetrahedron vanishes if it contains a triangle that violates the triangle inequalities!*

This is exciting because it suggests that the representations of SU(2) somehow know about the geometry of tetrahedra. Indeed, there are other ways for a tetrahedron to be impossible besides having edge lengths that violate the triangle inequalities. The $6j$ symbol does not vanish for all these tetrahedra, but it is exponentially damped—very much as a particle in quantum mechanics can tunnel through barriers that would be impenetrable classically, but with an amplitude that decays exponentially with the width of the barrier.

In fact, the relation between Rep(SU(2)) and three-dimensional geometry goes much deeper. Regge and Ponzano found an excellent asymptotic formula for the $6j$ symbol that depends entirely on geometrically interesting aspects of the corresponding tetrahedron—for example, its volume, the dihedral angles of its edges, and so on. But what is truly amazing is that this asymptotic formula also matches what one would want from a theory of quantum gravity in three-dimensional spacetime!

More precisely, the Ponzano–Regge model is a theory of Riemannian quantum gravity in three dimensions. Gravity in our universe is described with a Lorentzian metric on four-dimensional spacetime, where each tangent space has the Lorentz group acting on it. But we can imagine gravity in a universe where spacetime is three-dimensional and the metric is Riemannian, so each tangent space has the rotation group SO(3) acting on it. The quantum description of gravity in this universe should involve the double cover of this group, SU(2)—essentially because it should describe not just how particles of integer spin transform as they move along paths but also particles of half-integer spin. And it seems the Ponzano–Regge model is the right theory to do this.

A rigorous proof of Ponzano and Regge's asymptotic formula was not given until 1999, by Justin Roberts [203]. Physicists are still finding wonderful surprises in the Ponzano–Regge model. For example, if we study it on a 3-manifold with a Feynman diagram removed, with edges labeled by suitable representations, it describes not only pure quantum gravity but also *matter!* The series of papers by Freidel and Louapre explain this in detail [86].

Besides its meaning for geometry and physics, the $6j$ symbol also has a purely category-theoretic significance: it is a concrete description of the associator in Rep(SU(2)). The associator gives a linear operator

$$a_{i,j,k}\colon (i \otimes j) \otimes k \to i \otimes (j \otimes k).$$

The $6j$ symbol is a way of expressing this operator as several numbers. The idea is to use our basic intertwining operators to construct operators:

$$S\colon (i \otimes j) \otimes k \to l, \qquad T\colon l \to i \otimes (j \otimes k),$$

namely:

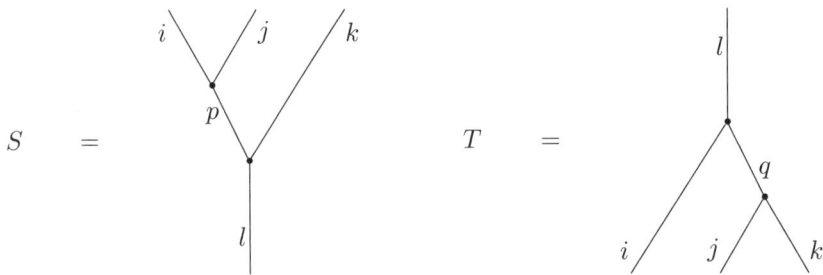

Using the associator to bridge the gap between $(i \otimes j) \otimes k$ and $i \otimes (j \otimes k)$, we can compose $S$ and $T$ and take the trace of the resulting operator, obtaining a number. These numbers encode everything there is to know about the associator in the monoidal category Rep(SU(2)). Moreover, these numbers are just the $6j$ symbols:

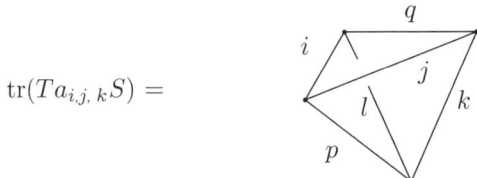

This can be proved by gluing the pictures for $S$ and $T$ together and warping the resulting spin network until it looks like a tetrahedron! We leave this as an exercise for the reader.

The upshot is a remarkable and mysterious fact: the associator in the monoidal category of representations of SU(2) encodes information about three-dimensional quantum gravity! This fact will become less mysterious when we see that three-dimensional quantum gravity is almost a TQFT. In our discussion of Barrett and Westbury's 1992 article on TQFTs, we will see that a large class of three-dimensional TQFTs can be built from monoidal categories.

### 1.3.19 Grothendieck (1983)

In his 600–page letter entitled *Pursuing Stacks*, Grothendieck fantasized about $n$-categories for higher $n$—even $n = \infty$—and their relation to homotopy theory [100]. The rough idea of an $\infty$-category is that it should be a generalization of a category that has objects, morphisms, 2-morphisms, and so on forever. In the fully general, weak $\infty$-categories, all the laws governing composition of $j$-morphisms should hold only up to specified $(j + 1)$-morphisms, which in turn satisfy laws of their own but only up to specified $(j + 2)$-morphisms, and so on. Furthermore, all of these higher morphisms that play the role of laws should be equivalences—where a $k$-morphism is an equivalence if it is invertible *up to equivalence*. The circularity here is not necessarily vicious, but it hints at how tricky $\infty$-categories, can be.

Grothendieck believed that among the weak $\infty$-categories, there should be a special class, the weak $\infty$-groupoids, in which all $j$-morphisms ($j \geq 1$) are equivalences. He also believed that every space $X$ should have a weak $\infty$-groupoid $\Pi_\infty(X)$ called its *fundamental $\infty$-groupoid*, in which:

- the objects are points of $X$
- the morphisms are paths in $X$
- the 2-morphisms are paths of paths in $X$
- the 3-morphisms are paths of paths of paths in $X$
- and so forth

Moreover, $\Pi_\infty(X)$ should be a complete invariant of the homotopy type of $X$, at least for nice spaces like CW complexes. In other words, two nice spaces should have equivalent fundamental $\infty$-groupoids if and only if they are homotopy equivalent.

The preceding sketch of Grothendieck's dream is phrased in terms of a globular approach to $n$-categories, where the $n$-morphisms are modeled after $n$-dimensional discs:

| Objects | Morphisms | 2-Morphisms | 3-Morphisms | ... |
|---|---|---|---|---|
| • | •⟶• | ⇓ between two arrows | ⇛ between two 2-morphisms | Globes |

However, he also imagined other approaches based on $j$-morphisms with different shapes, such as simplices:

| Objects | Morphisms | 2-Morphisms | 3-Morphisms | ... | |
|---|---|---|---|---|---|
| • | •⟶• | (triangle) | (tetrahedron) | | Simplices |

In fact, simplicial weak $\infty$-groupoids had already been developed in a 1957 paper by Kan [113]; these are now called *Kan complexes*. In this framework, $\Pi_\infty(X)$ is indeed a complete invariant of the homotopy type of any nice space $X$. So the real challenge is to define *weak $\infty$-categories* in the simplicial and other approaches, then define weak $\infty$-groupoids as special cases of these, and prove their relation to homotopy theory.

Great progress toward fulfilling Grothendieck's dream has been made in recent years. We cannot possibly do justice to the enormous body of work involved, so we simply offer a quick thumbnail sketch. Starting around 1977, Street began developing a simplicial approach to $\infty$-categories [226, 227] based on ideas from the physicist Roberts [202]. Thanks in large part to the recently published work of Verity, this approach has really begun to take off [237–239].

In 1995, Baez and Dolan initiated another approach to weak $n$-categories, the opetopic approach [12]:

| Objects | Morphisms | 2-Morphisms | 3-Morphisms | ... | |
|---|---|---|---|---|---|
| • | •⟶• | (opetope 2) | (opetope 3) | | Opetopes |

The idea here is that an $(n + 1)$-dimensional opetope describes a way of gluing together $n$-dimensional opetopes. The opetopic approach was corrected and clarified by various authors [43–46, 146, 162], and by now it has been developed by Makkai [160] into a full-fledged foundation for mathematics. We have already mentioned how in category theory, it is considered a mistake to assert equations between objects. Instead, one should specify an isomorphism between them. Similarly, in $n$-category theory, it is a mistake to assert an equation between $j$-morphisms for any $j < n$; one should instead specify an equivalence. In Makkai's approach to the foundations of mathematics based on weak $\infty$-categories, *equality plays no role, so this mistake is impossible to make.* Instead of stating equations, one must always specify equivalences.

Also starting around 1995, Tamsamani [230] and Simpson [220–223] developed a multisimplicial approach to weak $n$-categories. In a 1998 paper, Batanin [25, 229] initiated a globular approach to weak $\infty$-categories. Penon [185] gave a related, very compact definition of $\infty$-category, which was later improved by Batanin, Cheng, and Makkai [28, 51]. There is also a topologically motivated approach using operads attributable to Trimble [175], which was studied and generalized by Cheng and Gurski [47, 49]. Yet another theory is due to Joyal, with contributions by Berger [32, 106].

This great diversity of approaches raises the question of when two definitions of $n$-category count as equivalent. In *Pursuing Stacks*, Grothendieck proposed the following answer. Suppose that for all $n$, we have two different definitions of weak $n$-category—say, $n$-category$_1$ and $n$-category$_2$. Then, we should try to construct the $(n+1)$-category$_1$ of all $n$-categories$_1$ and the $(n+1)$-category$_1$ of all $n$-categories$_2$ and see whether these are equivalent as objects of the $(n+2)$-category$_1$ of all $(n+1)$-categories$_1$. If so, we may say the two definitions are equivalent as seen from the viewpoint of the first definition.

Of course, this strategy for comparing definitions of weak $n$-category requires a lot of work. Nobody has carried it out for any pair of significantly different definitions. There is also some freedom of choice involved in constructing the two $(n+1)$-categories$_1$ in question. One should do it in a "reasonable" way, but what does that mean? And what if we get a different answer when we reverse the roles of the two definitions?

A somewhat less strenuous strategy for comparing definitions is suggested by homotopy theory. Many different approaches to homotopy theory are in use, and although they are superficially very different, there is by now a well-understood sense in which they are fundamentally the same. Different approaches use objects from different categories to represent topological spaces or, more precisely, the homotopy-invariant information in topological spaces, called their *homotopy types*. These categories are not equivalent, but each one is equipped with a class of morphisms called *weak equivalences*, which play the role of homotopy equivalences. Given a category $C$ equipped with a specified class of weak equivalences, under mild assumptions, one can throw in inverses for these morphisms and obtain a category called the *homotopy category* Ho($C$). Two categories with specified equivalences may be considered the same for the purposes of homotopy theory if their homotopy categories are equivalent in the usual sense of category theory. The same strategy —or more sophisticated variants—can be applied to comparing definitions of $n$-category, so long as one can construct a *category of $n$-categories*.

Starting around 2000, work began on comparing different approaches to $n$-category theory [32, 45, 47, 161, 224]. There has also been significant progress toward achieving Grothendieck's dream of relating weak $n$-groupoids to homotopy theory [26, 27, 33, 54, 183, 184, 231]. But $n$-category theory is still far from mature. This is *one* reason the present chapter is just a prehistory.

Luckily, Leinster has written a survey of definitions of $n$-category [147] and also a textbook on the role of operads and their generalizations in higher category theory [149]. Cheng and Lauda have prepared an illustrated guidebook of higher categories for those who like to visualize things [50]. (The book by Baez and May [14] provides more background for readers who want to learn the subject. For applications to algebra, geometry, and physics, try the conference proceedings edited by Getzler and Kapranov [93] and by Davydov et al. [63].)

### 1.3.20 String Theory (1980s)

In the 1980s, there was a huge outburst of work on string theory. There is no way to summarize it all here, so we shall content ourselves with a few remarks about its relation to $n$-categorical physics. For a general overview, the reader can start with the introductory text by Zweibach [249] and then turn to the book by Green, Schwarz, and Witten [97], which was written in the 1980s, or the book by Polchinski [192], which covers more recent developments.

String theory goes beyond ordinary quantum field theory by replacing zero-dimensional point particles by one-dimensional objects, either circles, called *closed strings*, or intervals, called *open strings*. So, in string theory, the essentially one-dimensional Feynman diagrams depicting worldlines of particles are replaced by two-dimensional diagrams depicting string worldsheets:

This is a hint that as we pass from ordinary quantum field theory to string theory, the mathematics of *categories* is replaced by the mathematics of *bicategories*. However, this hint took a while to be recognized.

To compute an operator from a Feynman diagram, only the topology of the diagram matters, including the specification of which edges are inputs and which are outputs. In string theory, we need to equip the string worldsheet with a conformal structure, which is a recipe for measuring angles. More precisely, a *conformal structure* on a surface is an orientation together with an equivalence class of Riemannian metrics, where two metrics counts as equivalent if they give the same answers whenever we use them to compute angles between tangent vectors.

A conformal structure is also precisely what we need to do *complex analysis* on a surface. The power of complex analysis is what makes string theory so much more tractable than theories of higher-dimensional membranes.

### 1.3.21 Joyal–Street (1985)

Around 1985, Joyal and Street introduced braided monoidal categories [108]. The story is nicely told in Street's *Conspectus* [228], so here we focus on the mathematics.

As we have seen, braided monoidal categories are just like Mac Lane's symmetric monoidal categories, but without the law

$$B_{x,y} = B_{y,x}^{-1}.$$

The point of dropping this law becomes clear if we draw the isomorphism $B_{x,y} \colon x \otimes y \to y \otimes x$ as a little braid:

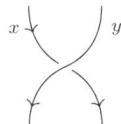

Then, its inverse is naturally drawn as

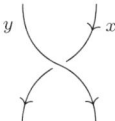

because then the equation $B_{x,y} B_{x,y}^{-1} = 1$ makes topological sense,

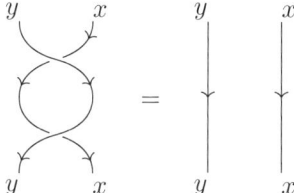

and similarly for $B_{x,y}^{-1} B_{x,y} = 1$:

In fact, these equations are familiar in knot theory, where they describe ways of changing a two-dimensional picture of a knot (or braid or tangle) without changing it as a three-dimensional topological entity. Both of these equations are called the *second Reidemeister move*.

Conversely, the law $B_{x,y} = B_{y,x}^{-1}$ would be drawn as

and this is *not* a valid move in knot theory. In fact, using this move, all knots become trivial. So it makes some sense to drop it, and this is just what the definition of braided monoidal category does.

Joyal and Street constructed a very important braided monoidal category called *Braid*. Every object in this category is a tensor product of copies of a special object $x$, which we draw as a point. So we draw the object $x^{\otimes n}$ as a row of $n$ points. The unit for the tensor product, $I = x^{\otimes 0}$, is drawn as a blank space. All the morphisms in Braid are *endomorphisms*: they go from an object to itself. In particular, a morphism

$f\colon x^{\otimes n} \to x^{\otimes n}$ is an $n$-strand braid:

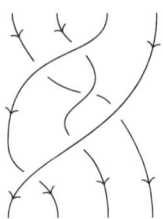

and composition is defined by stacking one braid on top of another. We tensor morphisms in Braid by setting braids side by side. The braiding is defined in an obvious way, for example, the braiding

$$B_{2,3}\colon x^{\otimes 2} \otimes x^{\otimes 3} \to x^{\otimes 3} \otimes x^{\otimes 2}$$

looks like this:

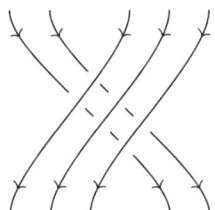

Joyal and Street showed that Braid is the free braided monoidal category on one object. This and other results of theirs justify the use of string diagrams as a technique for doing calculations in braided monoidal categories. They published an article on this in 1991, aptly titled "The Geometry of Tensor Calculus" [110].

Let us explain more precisely what it means that Braid is the free braided monoidal category on one object. For starters, Braid is a braided monoidal category containing a special object $x$, the point. But when we say Braid is the *free* braided monoidal category on this object, we are saying much more. Intuitively, this means two things. First, every object and morphism in Braid can be built from $x$ using operations that are part of the definition of braided monoidal category. Second, every equation that holds in Braid follows from the definition of braided monoidal category.

To make this precise, consider a simpler but related example. The group of integers $\mathbb{Z}$ is the free group on one element—namely, the number 1. Intuitively speaking, this means that every integer can be built from the integer 1 using operations built into the definition of group, and every equation that holds in $\mathbb{Z}$ follows from the definition of group. For example, $(1 + 1) + 1 = 1 + (1 + 1)$ follows from the associative law.

To make these intuitions precise, it is good to use the idea of a universal property. Namely, for any group $G$ containing an element $g$, there exists a unique homomorphism

$$\rho\colon \mathbb{Z} \to G$$

such that

$$\rho(1) = g.$$

The uniqueness clause here says that every integer is built from 1 using the group operations, which is why knowing what $\rho$ does to 1 determines $\rho$ uniquely. The existence clause says that every equation between integers follows from the definition of a group.

If there were extra equations, these would block the existence of homomorphisms to groups where these equations failed to hold.

So, when we say that Braid is the "free" braided monoidal category on the object $x$, we mean something *roughly* like this: for any braided monoidal category $C$, and any object $c \in C$, there is a unique map of braided monoidal categories

$$Z \colon \mathrm{Braid} \to C$$

such that

$$Z(x) = c.$$

This will not be precise until we define a map of braided monoidal categories. The correct concept is that of a braided monoidal functor. But we also need to weaken the universal property. To say that $Z$ is unique means that any two candidates sharing the desired property are *equal*. But this is too strong; it is bad to demand equality between functors. Instead, we should say that any two candidates are *isomorphic*. For this, we need the concept of braided monoidal natural isomorphism.

Once we have these concepts in hand, the correct theorem is as follows. For any braided monoidal category $C$, and any object $x \in C$, there exists a braided monoidal functor

$$Z \colon \mathrm{Braid} \to C$$

such that

$$Z(x) = c.$$

Moreover, given two such braided monoidal functors, there is a braided monoidal natural isomorphism between them.

Now we just need to define the necessary concepts. The definitions are a bit scary at first sight, but they illustrate the idea of *weakening*—that is, replacing equations by isomorphisms that satisfy equations of their own. They will also be needed for the definition of TQFTs, which we will present in our discussion of Atiyah's 1988 paper [5].

To begin with, a functor $F \colon C \to D$ between monoidal categories is *monoidal* if it is equipped with

- a natural isomorphism $\Phi_{x,y} \colon F(x) \otimes F(y) \to F(x \otimes y)$
- an isomorphism $\phi \colon 1_D \to F(1_C)$

such that:

- the following diagram commutes for any objects $x, y, z \in C$:

$$\begin{array}{ccccc}
(F(x) \otimes F(y)) \otimes F(z) & \xrightarrow{\Phi_{x,y} \otimes 1_{F(z)}} & F(x \otimes y) \otimes F(z) & \xrightarrow{\Phi_{x \otimes y, z}} & F((x \otimes y) \otimes z) \\
{\scriptstyle a_{F(x), F(y), F(z)}} \downarrow & & & & \downarrow {\scriptstyle F(a_{x,y,z})} \\
F(x) \otimes (F(y) \otimes F(z)) & \xrightarrow[1_{F(x)} \otimes \Phi_{y,z}]{} & F(x) \otimes F(y \otimes z) & \xrightarrow[\Phi_{x, y \otimes z}]{} & F(x \otimes (y \otimes z))
\end{array}$$

- the following diagrams commute for any object $x \in C$:

$$\begin{array}{ccc}
1 \otimes F(x) & \xrightarrow{\ell_{F(x)}} & F(x) \\
\phi \otimes 1_{F(x)} \downarrow & & \uparrow F(\ell_x) \\
F(1) \otimes F(x) & \xrightarrow{\Phi_{1,x}} & F(1 \otimes x)
\end{array}$$

$$\begin{array}{ccc}
F(x) \otimes 1 & \xrightarrow{r_{F(x)}} & F(x) \\
1_{F(x)} \otimes \phi \downarrow & & \uparrow F(r_x) \\
F(x) \otimes F(1) & \xrightarrow{\Phi_{x,1}} & F(x \otimes 1)
\end{array}$$

Note that we do not require $F$ to preserve the tensor product or unit on the nose. Instead, it is enough that it preserve them *up to specified isomorphisms*, which must in turn satisfy some plausible equations called coherence laws. This is typical of weakening.

A functor $F \colon C \to D$ between braided monoidal categories is *braided monoidal* if it is monoidal and it makes the following diagram commute for all $x, y \in C$:

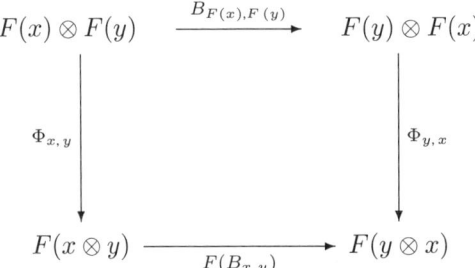

This condition says that $F$ preserves the braiding as best it can, given the fact that it only preserves tensor products up to a specified isomorphism. A *symmetric monoidal functor* is just a braided monoidal functor that happens to go between symmetric monoidal categories. No extra condition is involved here.

Having defined monoidal, braided monoidal, and symmetric monoidal functors, let us next do the same for natural transformations. Recall that a monoidal functor $F \colon C \to D$ is really a triple $(F, \Phi, \phi)$ consisting of a functor $F$, a natural isomorphism $\Phi_{x,y} \colon F(x) \otimes F(y) \to F(x \otimes y)$, and an isomorphism $\phi \colon 1_D \to F(1_C)$. Suppose that $(F, \Phi, \phi)$ and $(G, \Gamma, \gamma)$ are monoidal functors from the monoidal category $C$ to the monoidal category $D$. Then, a natural transformation $\alpha \colon F \Rightarrow G$ is *monoidal* if the diagrams

$$\begin{array}{ccc}
F(x) \otimes F(y) & \xrightarrow{\alpha_x \otimes \alpha_y} & G(x) \otimes G(y) \\
\Phi_{x,y} \downarrow & & \downarrow \Gamma_{x,y} \\
F(x \otimes y) & \xrightarrow{\alpha_{x \otimes y}} & G(x \otimes y)
\end{array}$$

and

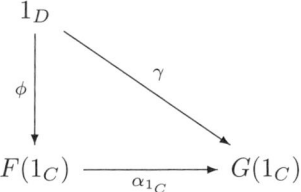

commute. There are no extra conditions required of *braided monoidal* or *symmetric monoidal* natural transformations.

The reader, having suffered through these definitions, is entitled to see an application besides Joyal and Street's algebraic description of the category of braids. At the end of our discussion of Mac Lane's 1963 paper on monoidal categories, we said that in a certain sense, every monoidal category is equivalent to a strict one. Now we can make this precise! Suppose $C$ is a monoidal category. Then, there is a strict monoidal category $D$ that is *monoidally equivalent* to $C$. That is, there are monoidal functors $F: C \to D$, $G: D \to C$ and monoidal natural isomorphisms $\alpha: FG \Rightarrow 1_D$, $\beta: GF \Rightarrow 1_C$.

This result allows us to work with strict monoidal categories, even though most monoidal categories found in nature are not strict. We can take the monoidal category we are studying and replace it by a monoidally equivalent strict one. The same sort of result is true for braided monoidal and symmetric monoidal categories.

A similar result holds for bicategories. They are all equivalent to *strict 2-categories*; that is, bicategories in which all the associators and unitors are identity morphisms. However, the pattern breaks down when we get to tricategories because not every tricategory is equivalent to a strict 3-category! At this point, the necessity for weakening becomes clear.

### 1.3.22 Jones (1985)

A *knot* is a circle smoothly embedded in $\mathbb{R}^3$:

More generally, a *link* is a collection of disjoint knots. In topology, we consider two links to be the same, or isotopic, if we can deform one smoothly without its strands crossing until it looks like the other. Classifying links up to isotopy is a challenging task that has spawned many interesting theorems and conjectures. To prove these, topologists are always looking for link invariants—that is, quantities they can compute from a link, which are equal on isotopic links.

In 1985, Jones [104] discovered a new link invariant, now called the *Jones polynomial*. To everyone's surprise, he defined this using some mathematics with no previously known connection to knot theory, which was the operator algebras developed in the

1930s by Murray and von Neumann [177] as part of a general formalism for quantum theory. Shortly thereafter, the Jones polynomial was generalized by many authors obtaining a large family of so-called quantum invariants of links.

Of all these new link invariants, the easiest to explain is the Kauffman bracket [118]. The Kauffman bracket can be thought of as a simplified version of the Jones polynomial. It is also a natural development of Penrose's 1971 work on spin networks [119].

As we have seen, Penrose gave a recipe for computing a number from any spin network. The relevant case here is a spin network with only two edges meeting at each vertex and with every edge labeled by the spin $\frac{1}{2}$. For spin networks like this, we can compute the number by repeatedly using the binor identity,

$$\chi \;=\; \asymp \;+\; \mid\mid$$

and this formula for the unknot:

$$\bigcirc \;=\; -2$$

The Kauffman bracket obeys modified versions of these identities. These involve a parameter that we will call $q$:

$$\chi \;=\; q\, \asymp \;+\; q^{-1}\, \mid\mid$$

and

$$\bigcirc \;=\; -(q^2 + q^{-2})$$

Among knot theorists, identities of this sort are called *skein relations*.

Penrose's original recipe is unable to detect linking or knotting because it also satisfies this identity:

$$\chi \;=\; \chi$$

coming from the fact that Rep(SU(2)) is a *symmetric* monoidal category. The Kauffman bracket arises from a more interesting braided monoidal category, the category of representations of the quantum group associated to SU(2). This quantum group depends on a parameter $q$, which conventionally is related to a quantity we are calling $q$ by a mildly annoying formula. To keep our story simple, we identify these two parameters.

When $q = 1$, the category of representations of the quantum group associated to SU(2) reduces to Rep(SU(2)), and the Kauffman bracket reduces to Penrose's original recipe. At other values of $q$, this category is not symmetric, and the Kauffman bracket detects linking and knotting.

In fact, all of the quantum invariants of links discovered around this time turned out to come from braided monoidal categories—namely, categories of representations of quantum groups. When $q = 1$, these quantum groups reduce to ordinary groups, their categories of representations become symmetric, and the quantum invariants of links become boring.

A basic result in knot theory says that given diagrams of two isotopic links, we can get from one to the other by warping the page on which they are drawn, together with a finite sequence of steps where we change a small portion of the diagram. There are three such steps, called the *first Reidemeister move*:

the *second Reidemeister move*:

and the *third Reidemeister move*:

Kauffman gave a beautiful, purely diagrammatic argument that his bracket was invariant under the second and third Reidemeister moves. We leave it as a challenge to the reader to find this argument, which looks very simple *after one has seen it*. However the bracket is not invariant under the first Reidemeister move, but it transforms in a simple way, as this calculation shows:

$$\vcenter{\hbox{[figure]}} = q \vcenter{\hbox{[figure]}} + q^{-1} \vcenter{\hbox{[figure]}} = -q^{-3} \vcenter{\hbox{[figure]}}$$

where we used the skein relations and did a little algebra. So, while the Kauffman bracket is not an isotopy invariant of links, it comes close; we shall later see that it is an invariant of framed links, made from ribbons. And, with a bit of tweaking, it gives the Jones polynomial, which *is* an isotopy invariant.

This and other work by Kauffman helped elevate string diagram techniques from a curiosity to a mainstay of modern mathematics. His book *Knots and Physics* was especially influential in this respect [120]. Meanwhile, the work of Jones led researchers toward a wealth of fascinating connections between von Neumann algebras, higher categories, and quantum field theory in two- and three-dimensional spacetime.

### 1.3.23 Freyd–Yetter (1986)

Among the many quantum invariants of links that appeared after the Jones polynomial, one of the most interesting is the HOMFLY-PT polynomial, which, it later became clear, arises from the category of representations of the quantum group associated to $SU(n)$. This polynomial got its curious name because it was independently discovered by many mathematicians, some of whom teamed up to write a paper about it for the *Bulletin of the American Mathematical Society* in 1985: Freyd, Yetter, Hoste, Lickorish, Millet and Ocneanu [89]. The "PT" refers to Przytycki and Traczyk, who published separately [195].

Different authors of this article took different approaches. Freyd and Yetter's approach is particularly germane to our story because they used a category in which morphisms are tangles. A *tangle* is a generalization of a braid that allows strands to double back and also allows closed loops:

So, a link is just a tangle with no strands coming in on top, and none leaving at the bottom. The advantage of tangles is that we can take a complicated link and chop it into simple pieces, which are tangles.

Shortly after Freyd heard Street give a talk on braided monoidal categories and the category of braids, Freyd and Yetter found a similar purely algebraic description of the category of oriented tangles [88]. A tangle is oriented if each strand is equipped with a smooth nowhere-vanishing field of tangent vectors, which we can draw as little arrows. We have already seen what an orientation is good for: it lets us distinguish between representations and their duals—or, in physics, particles and antiparticles.

There is a precisely defined but also intuitive notion of when two oriented tangles count as the same. Roughly speaking, this occurs whenever we can go from the first to the second by smoothly moving the strands without moving their ends or letting the strands cross. In this case, we say these oriented tangles are *isotopic*.

The category of oriented tangles has isotopy classes of oriented tangles as morphisms. We compose tangles by sticking one on top of the other. Just like Joyal and Street's category of braids, Tang is a braided monoidal category, where we tensor tangles by placing them side by side, and the braiding is defined using the fact that a braid is a special sort of tangle.

In fact, Freyd and Yetter gave a purely algebraic description of the category of oriented tangles as a compact braided monoidal category. Here, a monoidal category $C$ is *compact* if every object $x \in C$ has a *dual*; that is, an object $x^*$ together with morphisms called the *unit*

$$\bigcap \;\; = \;\; \begin{array}{c} 1 \\ i_x \downarrow \\ x \otimes x^* \end{array}$$

and the *counit*

$$\bigcup \;\; = \;\; \begin{array}{c} x^* \otimes x \\ e_x \downarrow \\ 1 \end{array}$$

satisfying the *zig-zag identities*

We have already seen these identities in our discussion of Penrose's work. Indeed, some classic examples of compact *symmetric* monoidal categories include FinVect, where $x^*$ is the usual dual of the vector space $x$, and Rep($K$) for any compact Lie group $K$, where $x^*$ is the dual of the representation $x$. But the zig-zag identities clearly hold in the category of oriented tangles, too, and this example is braided but not symmetric.

There are some important subtleties that our sketch has overlooked so far. For example, for any object $x$ in a compact braided monoidal category, this string diagram describes an isomorphism $d_x \colon x \to x^{**}$:

But if we think of this diagram as an oriented tangle, it is isotopic to a straight line. This suggests that $d_x$ should be an identity morphism. To implement this idea, Freyd and Yetter used braided monoidal categories where each object has a *chosen* dual, and this equation holds: $x^{**} = x$. Then they imposed the equation $d_x = 1_x$, which says that

$$\text{[diagram]} \quad = \quad \text{[diagram]}$$

This seems sensible, but in category theory, it is always dangerous to impose equations between objects, like $x^{**} = x$. Indeed, the danger becomes clear when we remember that Penrose's spin networks *violate* this rule. Instead, they satisfy

$$\text{[diagram with } j\text{]} \quad = \quad (-1)^{2j+1} \text{[diagram with } j\text{]}$$

The Kauffman bracket violates the rule in an even more complicated way. As mentioned in our discussion of Jones's 1985 paper, the Kauffman bracket satisfies

$$\text{[diagram]} \quad = \quad -q^{-3} \text{[diagram]}$$

So, although Freyd and Yetter's theorem is correct, it needs some fine-tuning to cover all the interesting examples.

For this reason, Street's student Shum [219] considered tangles where each strand is equipped with both an orientation and a *framing*—a nowhere vanishing smooth field of unit normal vectors. We can draw a framed tangle as made of ribbons, where one edge of each ribbon is black and the other is red. The black edge is the actual tangle, and the normal vector field points from the black edge to the red edge. But in string diagrams, we usually avoid drawing the framing by using a standard choice, the *blackboard framing*, where the unit normal vector points at right angles to the page, toward the reader.

There is an evident notion of when two framed oriented tangles count as the same, or isotopic. Any such tangle is isotopic to one where we use the blackboard framing, so we lose nothing by making this choice. And, with this choice, the following framed tangles are not isotopic:

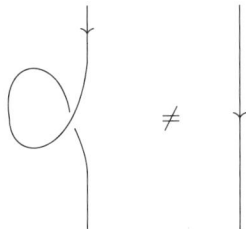

The problem is that if we think of these tangles as ribbons and pull the left one tight, it has a 360-degree twist in it.

What is the framing good for in physics? The preceding picture is the answer. We can think of each tangle as a physical process involving particles. The presence of the framing means that the left-hand process is topologically different from the right-hand process, in which a particle just sits there unchanged.

This is worth pondering in more detail. Consider the left-hand picture:

Reading this from top to bottom, it starts with a single particle. Then, a virtual particle–antiparticle pair is created on the left. Then, the new virtual particle and the original particle switch places by moving clockwise around each other. Finally, the original particle and its antiparticle annihilate each other. So this is all about *a particle that switches places with a copy of itself*.

But we can also think of this picture as a ribbon. If we pull it tight, we get a ribbon that is topologically equivalent—that is, isotopic. It has a 360-degree clockwise twist in it. This describes *a particle that rotates a full turn*:

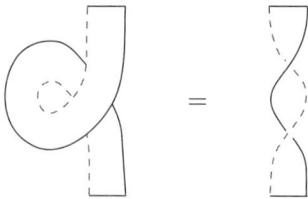

So, as far as topology is concerned, we can express the concept of rotating a single particle a full turn in terms of switching two identical particles—at least, in situations where creation and annihilation of particle–antiparticle pairs are possible. This fact is quite remarkable. As emphasized by Feynman [82], it lies at the heart of the famous "spin-statistics theorem" in quantum field theory. We have already seen that in theories of physics where spacetime is four-dimensional, the phase of a particle is multiplied by either 1 or $-1$ when we rotate it a full turn: 1 for bosons, $-1$ for fermions. The spin-statistics theorem says that switching two identical copies of this particle has the same effect on their phase: 1 for bosons, $-1$ for fermions.

The story becomes even more interesting in theories of physics where spacetime is three-dimensional. In this situation, space is two-dimensional, so we can distinguish between clockwise and counterclockwise rotations. Now, the spin-statistics theorem says that rotating a single particle a full turn clockwise gives the same phase as switching two identical particles of this type by moving them clockwise around each other. Rotating a particle a full turn clockwise need not have the same effect as rotating it counterclockwise, so this phase need not be its own inverse. In fact, it can be *any* unit complex number. This allows for exotic particles that are neither bosons nor fermions. In 1982, such particles were dubbed *anyons* by Frank Wilczek [242].

Anyons are not just mathematical curiosities. Superconducting thin films appear to be well described by theories in which the dimension of spacetime is three: two dimensions for the film, one for time. In such films, particle-like excitations arise, which act like anyons to a good approximation. The presence of these quasiparticles causes the film to respond in a surprising way to magnetic fields when current is running through it. This is called the *fractional quantum Hall effect* [78].

In 1983, Robert Laughlin [142] published an explanation of the fractional quantum Hall effect in terms of anyonic quasiparticles. He won the Nobel Prize for this work in 1998, along with Horst Störmer and Daniel Tsui, who observed this effect in the laboratory [95]. By now, we have an increasingly good understanding of anyons in terms of a quantum field theory called *Chern–Simons theory*, which also explains knot invariants such as the Kauffman bracket. (For more on this, see our discussion of Witten's 1989 paper on Chern–Simons theory.)

But we are getting ahead of ourselves! Let us return to the work of Shum. She constructed a category in which the objects are finite collections of oriented points in the unit square. By "oriented," we mean that each point is labeled either $x$ or $x^*$. We call a point labeled by $x$ *positively oriented*, and one labeled by $x^*$ *negatively oriented*. The morphisms in Shum's category are isotopy classes of framed oriented tangles. As usual, composition is defined by gluing the top of one tangle to the bottom of the other. We shall call this category $1\,Tang_2$. The reason for this curious notation is that the tangles themselves have dimension 1, but they live in a space—or spacetime, if you prefer—of dimension $1 + 2 = 3$. The number 2 is called the *codimension*. It turns out that varying these numbers leads to some very interesting patterns.

Shum's theorem gives a purely algebraic description of $1\text{Tang}_2$ in terms of ribbon categories. We have already seen that in a compact braided monoidal category $C$, every object $x \in C$ comes equipped with an isomorphism to its double dual, which we denoted $d_x: x \to x^{**}$. A ribbon category is a compact braided monoidal category in which each object $x$ is also equipped with another isomorphism, $c_x: x^{**} \to x$, which must satisfy a short list of axioms. We call this a *ribbon structure*. Composing this ribbon structure with $d_x$, we get an isomorphism

$$b_x = c_x d_x : x \to x.$$

Now, the point is that we can draw a string diagram for $b_x$, which is very much like the diagram for $d_x$, but with $x$ as the output instead of $x^{**}$:

This is the composite of $d_x$ which we know how to draw, and $c_x$, which we leave invisible, given that we do not know how to draw it.

In modern language, Shum's theorem says that $1\text{Tang}_2$ is the free-ribbon category on one object—namely, the positively oriented point, $x$. The definition of ribbon category is designed to make it obvious that $1\text{Tang}_2$ is a ribbon category. But in what sense is it free on one object? For this, we define a ribbon functor to be a braided monoidal functor between ribbon categories that preserves the ribbon structure. Then, the statement is this. First, given any ribbon category $C$ and any object $c \in C$, there is a ribbon functor

$$Z: 1\text{Tang}_2 \to C$$

such that

$$Z(x) = c.$$

Second, $Z$ is unique up to a braided monoidal natural isomorphism.

For a thorough account of Shum's theorem and related results, see Yetter's book [247]. We emphasized some technical aspects of this theorem because they are rather strange. As we shall see, the theme of $n$-categories with duals becomes increasingly important as our history winds to its conclusion, but duals remain a bit mysterious. Shum's theorem is the first hint of this. To avoid the equation between objects $x^{**} = x$, it seems we are forced to introduce an isomorphism $c_x: x^{**} \to x$ with no clear interpretation as a string diagram. We will see similar mysteries later.

Shum's theorem should remind the reader of Joyal and Street's theorem stating that Braid is the free braided monoidal category on one object. They are the first in a long line of results that describe interesting topological structures as free structures on one object, which often corresponds to a point. This idea has been dubbed "the primacy of the point."

### 1.3.24 Drinfel'd (1986)

In 1986, Vladimir Drinfel'd won the Fields Medal for his work on quantum groups [71]. This was the culmination of a long line of work on exactly solvable problems in low-dimensional physics, which we can only briefly sketch.

Back in 1926, Heisenberg [101] considered a simplified model of a ferromagnet like iron, consisting of spin-$\frac{1}{2}$ particles—electrons in the outermost shell of the iron atoms—sitting in a cubical lattice and interacting only with their nearest neighbors. In 1931, Bethe [35] proposed an ansatz that let him exactly solve for the eigenvalues of the Hamiltonian in Heisenberg's model, at least in the even simpler case of a *one-dimensional* crystal. This was subsequently generalized by Onsager [178], C. N. and C. P. Yang [246], Baxter [30], and many others.

The key turns out to be something called the Yang–Baxter equation. It is easiest to understand this in the context of two-dimensional quantum field theory. Consider a Feynman diagram in which two particles come in and two go out:

This corresponds to some operator

$$B: H \otimes H \to H \otimes H$$

where $H$ is the Hilbert space of states of the particle. It turns out that the physics simplifies immensely, leading to exactly solvable problems, if:

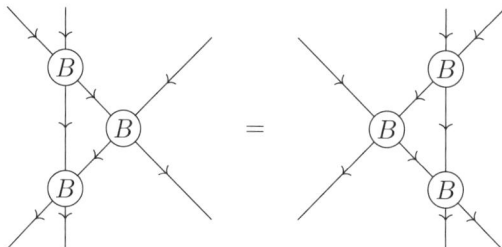

This says we can slide the lines around in a certain way without changing the operator described by the Feynman diagram. In terms of algebra:

$$(B \otimes 1)(1 \otimes B)(B \otimes 1) = (1 \otimes B)(B \otimes 1)(1 \otimes B).$$

This is the *Yang–Baxter equation*; it makes sense in any monoidal category.

In their 1985 paper, Joyal and Street noted that given any object $x$ in a braided monoidal category, the braiding

$$B_{x,x}: x \otimes x \to x \otimes x$$

is a solution of the Yang–Baxter equation. If we draw this equation using string diagrams, it looks like the third Reidemeister move in knot theory:

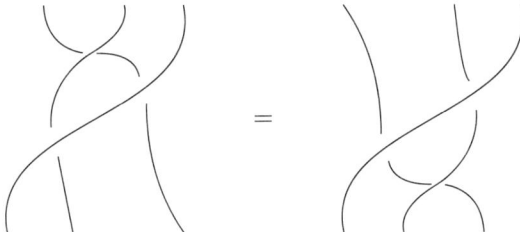

Joyal and Street also showed that given any solution of the Yang–Baxter equation in any monoidal category, we can build a braided monoidal category.

Mathematical physicists enjoy exactly solvable problems, so after the work of Yang and Baxter, a kind of industry developed, devoted to finding solutions of the Yang–Baxter equation. The Russian school, led by Faddeev, Sklyanin, Takhtajan, and others, was especially successful [79]. Eventually, Drinfel'd discovered how to get solutions of the Yang–Baxter equation from any simple Lie algebra. The Japanese mathematician Jimbo did this as well, at about the same time [103].

What they discovered was that the universal enveloping algebra $U\mathfrak{g}$ of any simple Lie algebra $\mathfrak{g}$ can be deformed in a manner depending on a parameter $q$, giving a one-parameter family of Hopf algebras $U_q\mathfrak{g}$. Given that Hopf algebras are mathematically analogous to groups, and in some physics problems, the parameter $q$ is related to Planck's constant $\hbar$ by $q = e^{\hbar}$, the Hopf algebras $U_q\mathfrak{g}$ are called *quantum groups*. There is, by now, an extensive theory of these [42, 121, 158].

Moreover, these Hopf algebras have a special property implying that any representation of $U_q\mathfrak{g}$ on a vector space $V$ comes equipped with an operator

$$B: V \otimes V \to V \otimes V$$

satisfying the Yang–Baxter equation. We shall say more about this in our discussion of a 1989 paper by Reshetikhin and Turaev [200].

This work led to a far more thorough understanding of exactly solvable problems in two-dimensional quantum field theory [77]. It was also the first big, *explicit* intrusion of category theory into physics. As we shall see, Drinfel'd's constructions can be nicely explained in the language of braided monoidal categories. This led to the widespread adoption of this language, which was then applied to other problems in physics. Everything beforehand looks category-theoretic in retrospect.

### 1.3.25 Segal (1988)

In an attempt to formalize some of the key mathematical structures underlying string theory, Graeme Segal [214] proposed axioms describing a conformal field theory. *Roughly*, these say that it is a symmetric monoidal functor

$$Z: 2\mathrm{Cob}_\mathbb{C} \to \mathrm{Hilb}$$

with some nice extra properties. Here, $2\text{Cob}_\mathbb{C}$ is the category whose morphisms are string worldsheets, like this:

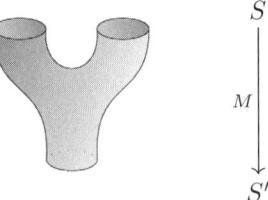

We compose these morphisms by gluing them end to end, like this:

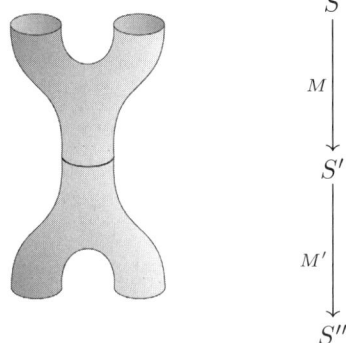

A bit more precisely, an object of $2\text{Cob}_\mathbb{C}$ is a union of parametrized circles, and a morphism $M\colon S \to S'$ is a two-dimensional cobordism equipped with some extra structure. Here, an $n$-dimensional cobordism is roughly an $n$-dimensional compact oriented manifold with boundary, $M$, whose boundary has been written as the disjoint union of two $(n-1)$-dimensional manifolds $S$ and $S'$, called the *source* and *target*.

In the case of $2\text{Cob}_\mathbb{C}$, we need these cobordisms to be equipped with a conformal structure and a parametrization of each boundary circle. The parametrization lets us give the composite of two cobordisms a conformal structure built from the conformal structures on the two parts.

In fact, we are glossing over many subtleties here; we hope this sketch gets the idea across. In any event, $2\text{Cob}_\mathbb{C}$ is a symmetric monoidal category, where we tensor objects or morphisms by setting them side by side:

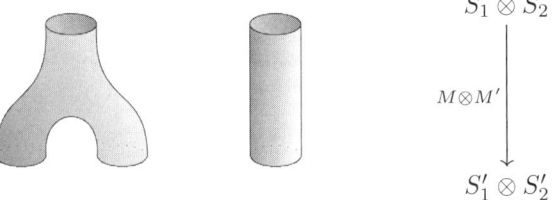

Similarly, Hilb is a symmetric monoidal category with the usual tensor product of Hilbert spaces. A basic rule of quantum physics is that the Hilbert space for a disjoint union of two physical systems should be the tensor product of their Hilbert spaces. This suggests that a conformal field theory, viewed as a functor $Z\colon 2\text{Cob}_\mathbb{C} \to \text{Hilb}$, should preserve tensor products—at least up to a specified isomorphism. So we should demand that $Z$ be a monoidal functor. A bit more reflection along these lines leads us to demand that $Z$ be a symmetric monoidal functor.

There is more to the full definition of a conformal field theory than merely a symmetric monoidal functor $Z: 2\mathrm{Cob}_\mathbb{C} \to \mathrm{Hilb}$. For example, we also need a positive-energy condition reminiscent of the condition we already met for representations of the Poincaré group. Indeed, there is a profusion of different ways to make the idea of conformal field theory precise, starting with Segal's original definition. But the different approaches are nicely related, and the subject of conformal field theory is full of deep results, interesting classification theorems, and applications to physics and mathematics. A good introduction is the book by Di Francesco, Mathieu, and Senechal [65].

### 1.3.26 Atiyah (1988)

Shortly after Segal proposed his definition of conformal field theory, Atiyah [5] modified it by dropping the conformal structure and allowing cobordisms of an arbitrary fixed dimension. He called the resulting structure a *topological quantum field theory*, or TQFT for short. One of his goals was to formalize some work by Witten [243] on invariants of four-dimensional manifolds coming from a quantum field theory, sometimes called the *Donaldson theory*, which is related to Yang–Mills theory. These invariants have led to a revolution in our understanding of four-dimensional topology—but, ironically, Donaldson theory has never been successfully dealt with using Atiyah's axiomatic approach. We will say more about this in our discussion of Crane and Frenkel's 1994 paper. For now, let us simply explain Atiyah's definition of a TQFT.

In modern language, an *n-dimensional TQFT* is a symmetric monoidal functor

$$Z: n\mathrm{Cob} \to \mathrm{FinVect}.$$

Here, FinVect stands for the category of finite-dimensional complex vector spaces and linear operators between them; $n$Cob is the category with:

- compact oriented $(n-1)$-dimensional manifolds as objects
- oriented $n$-dimensional cobordisms as morphisms

Taking the disjoint union of manifolds makes $n$Cob into a monoidal category. The braiding in $n$Cob can be drawn like this:

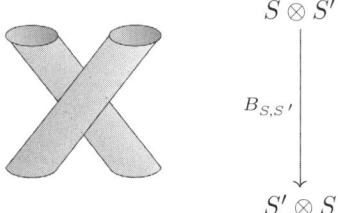

but, because we are interested in abstract cobordisms, not embedded in any ambient space, this braiding will be symmetric.

Physically, a TQFT describes a featureless universe that looks *locally* the same in every state. In such an imaginary universe, the only way to distinguish different states is by doing global observations—for example, by carrying a particle around a noncontractible loop in space. Thus, TQFTs appear to be very simple toy models of physics that ignore most of the interesting features of what we see around us. It is precisely for this reason that TQFTs are more tractable than full-fledged quantum field theories. In what follows, we shall spend quite a bit of time explaining how TQFTs are related to $n$-categories. If $n$-categorical physics is ever to blossom, we must someday go further. There are some signs that this may be starting [212], but attempting to discuss this would lead us out of our prehistory.

Mathematically, the study of TQFTs quickly leads to questions involving duals. In our explanation of the work of Freyd and Yetter, we mentioned compact monoidal categories, where every object has a dual. One can show that $n$Cob is compact, with the dual $x^*$ of an object $x$ being the same manifold equipped with the opposite orientation. Similarly, FinVect is compact with the usual notion of dual for vector spaces. The categories Vect and Hilb are not compact because we can always define the dimension of an object in a compact braided monoidal category by

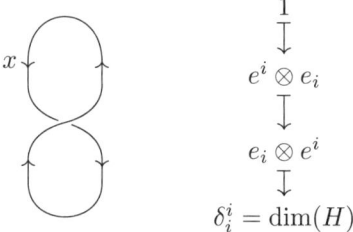

but this diverges for an infinite-dimensional vector space, or Hilbert space. As we have seen, the infinities that plague ordinary quantum field theory arise from his fact.

As a category, FinVect is equivalent to FinHilb, the category of finite-dimensional complex Hilbert spaces and linear operators. However, FinHilb and also Hilb have something in common with $n$Cob that Vect lacks: they have duals for morphisms. In $n$Cob, given a morphism

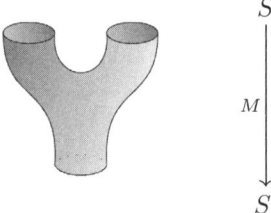

we can reverse its orientation and switch its source and target to obtain a morphism going backwards in time:

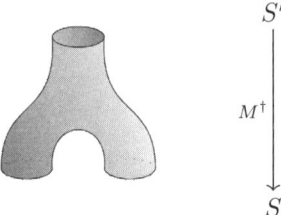

Similarly, given a linear operator $T: H \to H'$ between Hilbert spaces, we can define an operator $T^\dagger: H' \to H$ by demanding that

$$\langle T^\dagger \phi, \psi \rangle = \langle \phi, T\psi \rangle$$

for all vectors $\psi \in H, \phi \in H'$.

Isolating the common properties of these constructions, we say a category *has duals for morphisms* if, for any morphism $f: x \to y$, there is a morphism $f^\dagger: y \to x$ such that

$$(f^\dagger)^\dagger = f, \quad (fg)^\dagger = g^\dagger f^\dagger, \quad 1_x^\dagger = 1_x.$$

We then say morphism $f$ is *unitary* if $f^\dagger$ is the inverse of $f$. In the case of Hilb, this is just a unitary operator in the usual sense.

As we have seen, symmetries in quantum physics are described not just by group representations on Hilbert spaces but also by *unitary* representations. This is a hint of the importance of duals for morphisms in physics. We can always think of a group $G$ as a category with one object and with all morphisms invertible. This becomes a category with duals for morphisms by setting $g^\dagger = g^{-1}$ for all $g \in G$. A representation of $G$ on a Hilbert space is the same as a functor $\rho: G \to$ Hilb, and this representation is unitary precisely when

$$\rho(g^\dagger) = \rho(g)^\dagger.$$

The same sort of condition shows up in many other contexts in physics. So, quite generally, given any functor $F: C \to D$ between categories with duals for morphisms, we say $F$ is *unitary* if $F(f^\dagger) = F(f)^\dagger$ for every morphism in $C$. It turns out that the physically most interesting TQFTs are the *unitary* ones, which are *unitary* symmetric monoidal functors

$$Z: n\text{Cob} \to \text{FinHilb}.$$

Although categories with duals for morphisms play a crucial role in this definition, the 1989 article by Doplicher and Roberts [70], and also the 1995 paper by Baez and Dolan [11], they seem to have been a bit neglected by category theorists until 2005, when Selinger [218] introduced them under the name of dagger categories as part of his work on the foundations of quantum computation. Perhaps one reason for this neglect is that their definition implicitly involves an equation between objects—something normally shunned in category theory.

To see this equation between objects explicitly, note that a category with duals for morphisms, or *dagger category*, may be defined as a category $C$ equipped with a contravariant functor, $\dagger: C \to C$ such that

$$\dagger^2 = 1_C$$

and $x^\dagger = x$ for every object $x \in C$. Here, by a *contravariant* functor, we mean one that reverses the order of composition; this is just a way of saying that $(fg)^\dagger = g^\dagger f^\dagger$.

Contravariant functors are well accepted in category theory, but it raises eyebrows to impose equations between objects, like $x^\dagger = x$. This is not just a matter of fashion. Such equations cause real trouble. If $C$ is a dagger category, and $F: C \to D$ is an equivalence of categories, we cannot use $F$ to give $D$ the structure of a dagger category, precisely because of this equation. Nonetheless, the concept of dagger category seems crucial in quantum physics. So there is a tension that remains to be resolved here.

The reader may note that this is not the first time an equation between objects has obtruded in the study of duals. We have already seen one in our discussion of Freyd and Yetter's 1986 paper. In that case, the problem involved duals for objects rather than morphisms. And in that case, Shum found a way around the problem. When it comes to duals for morphisms, no comparable fix is known. However, in our discussion of the Doplicher and Roberts 1989 paper, we see that the two problems are closely connected.

### 1.3.27 Dijkgraaf (1989)

Shortly after Atiyah defined TQFTs, Dijkgraaf gave a purely algebraic characterization of two-dimensional TQFTs in terms of commutative Frobenius algebras [67].

Recall that a two-dimensional TQFT is a symmetric monoidal functor $Z\colon 2\mathrm{Cob} \to \mathrm{Vect}$. An object of 2Cob is a compact oriented one-dimensional manifold—a disjoint union of copies of the circle $S^1$. A morphism of 2Cob is a two-dimensional cobordism between such manifolds. Using Morse theory, we can chop any two-dimensional cobordism $M$ into elementary building blocks that contain only a single critical point. These are called the *birth of a circle*, the *upside-down pair of pants*, the *death of a circle* and the *pair of pants*:

Every two-dimensional cobordism is built from these by composition, tensoring, and the other operations present in any symmetric monoidal category. So we say that 2Cob is generated as a symmetric monoidal category by the object $S^1$ and these morphisms. Moreover, we can list a complete set of relations that these generators satisfy:

$$\tag{1.1}$$

$$\tag{1.2}$$

$$\tag{1.3}$$

$$\tag{1.4}$$

2Cob is completely described as a symmetric monoidal category by means of these generators and relations.

Applying the functor $Z$ to the circle gives a vector space $F = Z(S^1)$, and applying it to the following cobordisms gives certain linear maps:

$$i \colon \mathbb{C} \to F \qquad m \colon F \otimes F \to F \qquad \varepsilon \colon F \to \mathbb{C} \qquad \Delta \colon F \to F \otimes F$$

This means that our two-dimensional TQFT is completely determined by choosing a vector space $F$ equipped with linear maps $i, m, \varepsilon, \Delta$ satisfying the relations drawn as the preceding pictures.

Surprisingly, all this stuff amounts to a well-known algebraic structure, a commutative Frobenius algebra. For starters, Equation 1.1

$$\begin{array}{c} F \otimes F \otimes F \\ {\scriptstyle 1_F \otimes m} \downarrow \\ F \otimes F \\ {\scriptstyle \mu} \downarrow \\ F \end{array} \qquad = \qquad \begin{array}{c} F \otimes F \otimes F \\ {\scriptstyle m \otimes 1_F} \downarrow \\ F \otimes F \\ {\scriptstyle m} \downarrow \\ F \end{array}$$

says that the map $m$ defines an associative multiplication on $F$. The second relation says that the map $i$ gives a unit for the multiplication on $F$. This makes $F$ into an *algebra*. The upside-down versions of these relations appearing in Equation 1.2 say that $F$ is also a *coalgebra*. An algebra that is also a coalgebra, where the multiplication and comultiplication are related by Equation 1.3, is called a *Frobenius algebra*. Finally, Equation 1.4 is the commutative law for multiplication.

In 1996, Abrams [1] was able to construct a category of two-dimensional TQFTs and prove that it is equivalent to the category of commutative Frobenius algebras. This makes precise the sense in which a two-dimensional topological quantum field theory is a commutative Frobenius algebra. It implies that when one has a commutative Frobenius algebra in the category FinVect, one immediately gets a symmetric monoidal functor $Z \colon$ 2Cob $\to$ Vect; hence, a two-dimensional TQFT. This perspective is explained in great detail in the book by Kock [124].

In modern language, the essence of Abrams's result is contained in the following theorem: 2Cob *is the free symmetric monoidal category on a commutative Frobenius algebra*. To make this precise, we first define a commutative Frobenius algebra in *any* symmetric monoidal category, using the same diagrams shown herein. Next, suppose $C$ is any symmetric monoidal category and $c \in C$ is a commutative Frobenius algebra in $C$. Then, first there exists a symmetric monoidal functor

$$Z \colon 2\text{Cob} \to C$$

with

$$Z(S^1) = c$$

and such that $Z$ sends the multiplication, unit, cocomultiplication, and counit for $S^1$ to those for $c$. Second, $Z$ is unique up to a symmetric monoidal natural isomorphism.

This result should remind the reader of Joyal and Street's algebraic characterization of the category of braids and Shum's characterization of the category of framed oriented tangles. It is a bit more complicated because the circle is a bit more complicated than the point. The idea of an extended TQFT, which we shall describe later, strengthens the concept of a TQFT so as to restore the "primacy of the point."

### 1.3.28 Doplicher–Roberts (1989)

In 1989, Sergio Doplicher and John Roberts published a paper [69] showing how to reconstruct a compact topological group $K$—for example, a compact Lie group—from its category of finite-dimensional continuous unitary representations, Rep($K$). They then used this to show one could start with a fairly general quantum field theory and *compute* its gauge group instead of putting the group in by hand [70].

To do this, they actually needed some extra structure on Rep($K$). For our purposes, the most interesting thing they needed was its structure as a symmetric monoidal category with duals. Let us define this concept.

In our discussion of Atiyah's 1988 paper on TQFTs [5], we explained what it means for a category to be a dagger category or to have duals for morphisms. When such a category is equipped with extra structure, it makes sense to demand that this extra structure be compatible with this duality. For example, we can demand that an isomorphism $f : x \to y$ be *unitary*, meaning

$$f^\dagger f = 1_x, \qquad f f^\dagger = 1_y.$$

So, we say a monoidal category $C$ *has duals for morphisms* if its underlying category has duals for morphisms, the duality preserves the tensor product,

$$(f \otimes g)^\dagger = f^\dagger \otimes g^\dagger$$

and, moreover, all the relevant isomorphisms are unitary (the associators $a_{x,y,z}$, and the left and right unitors $\ell_x$ and $r_x$). We say a braided or symmetric monoidal category *has duals for morphisms* if all of these conditions hold and in addition the braiding $B_{x,y}$ is unitary. There is an easy way to make 1Tang$_2$ into a braided monoidal category with duals for morphisms. Both $n$Cob and FinHilb are symmetric monoidal categories with duals for morphisms.

Besides duals for morphisms, we may consider duals for objects. In our discussion of Freyd and Yetter's 1986 work, we said a monoidal category has duals for objects, or is compact, if for each object $x$, there is an object $x^*$ together with a unit $i_x : 1 \to x \otimes x^*$ and counit $e_x : x^* \otimes x \to 1$ satisfying the zig-zag identities.

Now, suppose that a braided monoidal category has both duals for morphisms and duals for objects. Then, there is yet another compatibility condition we can—and should—demand. Any object has a counit, shaped like a cup:

and taking the dual of this morphism, we get a kind of cap:

Combining these with the braiding, we get a morphism like this:

This looks just like the morphism $b_x \colon x \to x$ that we introduced in our discussion of Freyd and Yetter's 1989 article [88]—only now it is the result of combining duals for objects and duals for morphisms! Some string diagram calculations suggest that $b_x$ should be unitary. So, we say a braided monoidal category *has duals* if it has duals for objects, duals for morphisms, and the twist isomorphism $b_x \colon x \to x$, constructed as shown previously, is unitary for every object $x$.

In a symmetric monoidal category with duals, one can show that $b_x^2 = 1_x$. In physics, this leads to the boson/fermion distinction mentioned earlier because a boson is any particle that remains unchanged when rotated a full turn, whereas a fermion is any particle whose phase gets multiplied by $-1$ when rotated a full turn. Both $n$Cob and Hilb are symmetric monoidal categories with duals, and both are bosonic in the sense that $b_x = 1_x$ for every object. The same is true for Rep$(K)$ for any compact group $K$. This features prominently in the paper by Doplicher and Roberts.

In recent years, interest has grown in understanding the foundations of quantum physics with the help of category theory. One reason is that in theoretical work on quantum computation, there is a fruitful overlap between the category theory used in quantum physics and that used in computer science. In a 2004 paper on this subject, Abramsky and Coecke [2] introduced symmetric monoidal categories with duals under the name of *strongly compact closed categories*. These entities were later dubbed *dagger compact categories* by Selinger [218], and this name seems to have caught on. What we are calling symmetric monoidal categories with duals for morphisms, he calls *dagger symmetric monoidal categories*.

### 1.3.29 Reshetikhin–Turaev (1989)

We have mentioned how Jones's discovery (in 1985) of a new invariant of knots led to a burst of work on related invariants. Eventually, it was found that all of these so-called quantum invariants of knots can be derived in a systematic way from quantum groups. A particularly clean treatment using braided monoidal categories can be found in a paper by Nikolai Reshetikhin and Vladimir Turaev [199]. This is a good point to summarize a bit of the theory of quantum groups in its modern form.

The first thing to realize is that a quantum group is not a group; it is a special sort of algebra. What quantum groups and groups have in common is that their categories of representations have similar properties. The category of finite-dimensional representations of a group is a symmetric monoidal category with duals for objects. The category of finite-dimensional representations of a quantum group is a *braided* monoidal category with duals for objects.

As we saw in our discussion of Freyd and Yetter's 1986 work, the category $1\text{Tang}_2$ of tangles in three dimensions is the *free* braided monoidal category with duals on one object $x$. So, if $\text{Rep}(A)$ is the category of finite-dimensional representations of a quantum group $A$, any object $V \in \text{Rep}(A)$ determines a braided monoidal functor

$$Z \colon 1\text{Tang}_2 \to \text{Rep}(A)$$

with

$$Z(x) = V.$$

This functor gives an invariant of tangles—namely, a linear operator for every tangle and, in particular, a number for every knot or link.

So what sort of algebra has representations that form a braided monoidal category with duals for objects? This turns out to be one of a family of related questions with related answers. The more extra structure we put on an algebra, the nicer its category of representations becomes:

| Algebra | Category |
|---|---|
| Bialgebra | Monoidal category |
| Quasitriangular bialgebra | Braided monoidal category |
| Triangular bialgebra | Symmetric monoidal category |
| Hopf algebra | Monoidal category with duals for objects |
| Quasitriangular Hopf algebra | Braided monoidal category with duals for objects |
| Triangular Hopf algebra | Symmetric monoidal category with duals for objects |

Algebras and their categories of representations.

For each sort of algebra $A$ in the left-hand column, its category of representations $\text{Rep}(A)$ becomes a category of the sort listed in the right-hand column. In particular, a quantum group is a kind of quasitriangular Hopf algebra.

In fact, the correspondence between algebras and their categories of representations works both ways. Under some mild technical assumptions, we can recover $A$ from $\text{Rep}(A)$ together with the forgetful functor $F \colon \text{Rep}(A) \to \text{Vect}$ sending each representation to its underlying vector space. The theorems guaranteeing this are called *Tannaka–Krein reconstruction theorems* [109]. They are reminiscent of the Doplicher–Roberts reconstruction theorem, which allows us to recover a compact topological group $G$ from its category of representations. However, they are easier to prove, and they came earlier.

So someone who strongly wishes to avoid learning about quasitriangular Hopf algebras can get away with it, at least for a while, if they know enough about braided monoidal categories with duals for objects. The latter subject is ultimately more fundamental. Nonetheless, it is very interesting to see how the correspondence between algebras and their categories of representations works. So let us sketch how any bialgebra has a monoidal category of representations and then give some examples coming from groups and quantum groups.

First, recall that an *algebra* is a vector space $A$ equipped with an associative multiplication

$$m \colon A \otimes A \to A$$
$$a \otimes b \mapsto ab$$

together with an element $1 \in A$ satisfying the left and right unit laws, $1a = a = a1$ for all $a \in A$. We can draw the multiplication using a string diagram:

We can also describe the element $1 \in A$ using the unique operator $i \colon \mathbb{C} \to A$ that sends the complex number 1 to $1 \in A$. Then, we can draw this operator using a string diagram:

In this notation, the associative law looks like this:

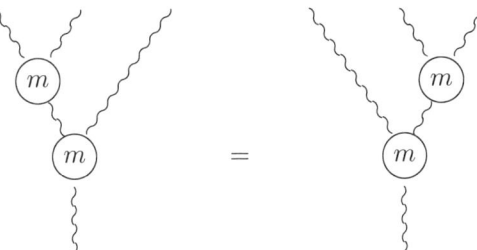

whereas the left and right unit laws look like this:

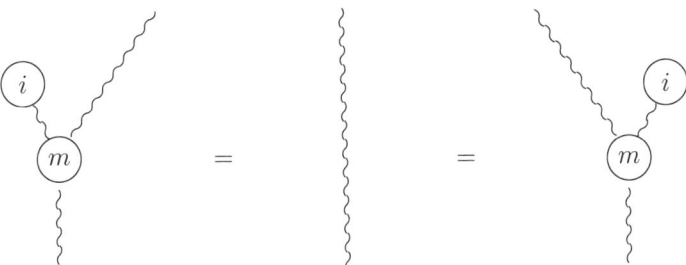

A representation of an algebra is a lot like a representation of a group, except that instead of writing $\rho(g)v$ for the action of a group element $g$ on a vector $v$, we write $\rho(a \otimes v)$ for the action of an algebra element $a$ on a vector $v$. More precisely, a *representation* of an algebra $A$ is a vector space $V$ equipped with an operator

$$\rho \colon A \otimes V \to V$$

satisfying these two laws:

$$\rho(1 \otimes v) = v, \qquad \rho(ab \otimes v) = \rho(a \otimes \rho(b \otimes v)).$$

Using string diagrams, we can draw $\rho$ as follows:

Note that wiggly lines refer to the object $A$ and straight lines refer to $V$. Then, the two laws obeyed by $\rho$ look very much like associativity and the left unit law:

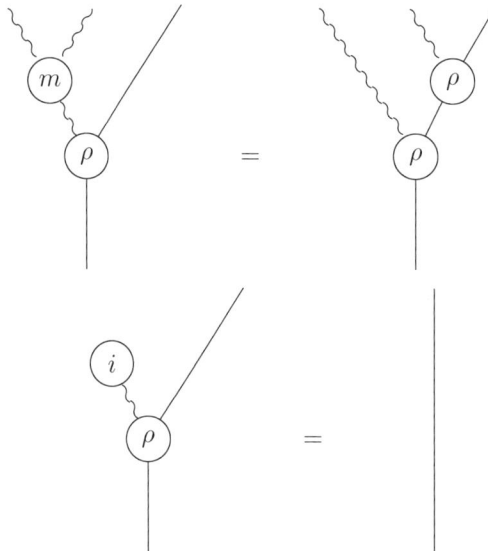

To make the representations of an algebra into the objects of a category, we must define morphisms between them. Given two algebra representations—say, $\rho \colon A \otimes V \to V$ and $\rho' \colon A \otimes V' \to V'$— we define an *intertwining operator* $f \colon V \to V'$ to be a linear operator such that

$$f(\rho(a \otimes v)) = \rho'(a \otimes f(v)).$$

This closely resembles the definition of an intertwining operator between group representations. It says that acting by $a \in A$ and then applying the intertwining operator is the same as applying the intertwining operator and then acting by $a$.

With these definitions, we obtain a category Rep($A$) with finite-dimensional representations of $A$ as objects and intertwining operators as morphisms. However, unlike group representations, there is no way, in general, to define the tensor product of algebra representations! For this, we need $A$ to be a bialgebra. To understand what this means, first recall from our discussion of Dijkgraaf's 1989 thesis that a *coalgebra* is just like an algebra, only upside down. More precisely, it is a vector space equipped with a *comultiplication*:

and the *counit:*

satisfying the *coassociative law*:

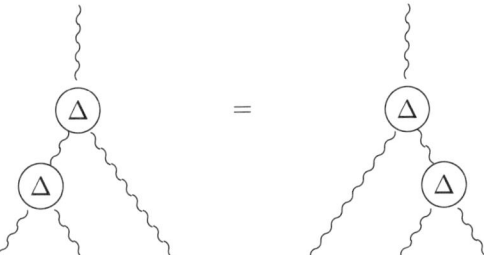

and left/right *counit laws*:

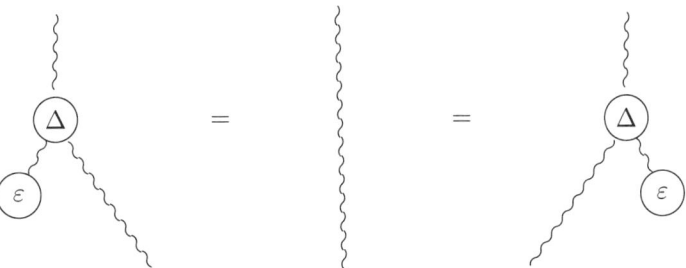

A *bialgebra* is a vector space equipped with an algebra and coalgebra structure that are compatible in a certain way. We have already seen that a Frobenius algebra is both an algebra and a coalgebra, with the multiplication and comultiplication obeying the compatibility conditions in Equation 1.3. A bialgebra obeys *different* compatibility conditions. These can be drawn using string diagrams, but it is more enlightening to note that they are precisely the conditions we need to make the category of representations of an algebra $A$ into a *monoidal* category. The idea is that the comultiplication

$\Delta \colon A \to A \otimes A$ lets us duplicate an element $A$ so that it can act on both factors in a tensor product of representations, say $\rho$ and $\rho'$:

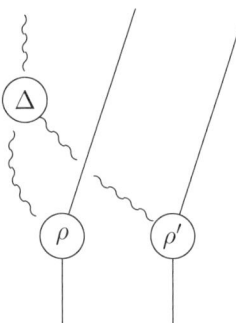

This gives Rep($A$) a tensor product. Similarly, we use the counit to let $A$ act on $\mathbb{C}$ as follows:

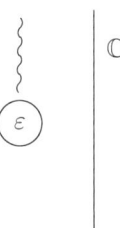

We can then write down equations saying that Rep($A$) is a monoidal category with the same associator and unitors as in Vect, and with $\mathbb{C}$ as its unit object. These equations are then the definition of bialgebra.

As we have seen, the category of representations of a compact Lie group $K$ is also a monoidal category. In this sense, bialgebras are a generalization of such groups. Indeed, there is a way to turn any group of this sort into a bialgebra $A$, and when the group is simply connected, this bialgebra has an equivalent category of representations,

$$\text{Rep}(K) \simeq \text{Rep}(A).$$

So, as far as its representations are concerned, there is really no difference. But a big advantage of bialgebras is that we can often deform them to obtain new bialgebras that *do not* come from groups.

The most important case is when $K$ is not only simply connected and compact, but also *simple*, which for Lie groups means that all of its normal subgroups are finite. We have already been discussing an example, SU(2). Groups of this sort were classified by Élie Cartan in 1894 and, by the mid-1900s, their theory had grown to one of the most enormous and beautiful edifices in mathematics. The fact that one can deform them to get interesting bialgebras, called *quantum groups*, opened a brand new wing in this edifice, and the experts rushed in.

A basic fact about groups of this sort is that they have complex forms. For example, SU(2) has the complex form SL(2), consisting of $2 \times 2$ complex matrices with determinant 1. This group contains SU(2) as a subgroup. The advantage of SU(2) is that it is compact, which implies that its finite-dimensional continuous representations can always be made unitary. The advantage of SL(2) is that it is a complex manifold, with all of the group operations being analytic functions; this allows us to define analytic representations of this group. For our purposes, another advantage of SL(2) is that its Lie algebra is a complex vector space. Luckily, we do not have to choose one group over the other because the finite-dimensional continuous unitary representations of SU(2) correspond precisely to the finite-dimensional analytic representations of SL(2). And, as emphasized by Hermann Weyl, *every* simply connected, compact, simple Lie group $K$ has a complex Lie group $G$ for which this relation holds!

These facts let us say a bit more about how to get a bialgebra with the same representations as our group $K$. First, we take the complex form $G$ of the group $K$ and consider its Lie algebra, $\mathfrak{g}$. Then we let $\mathfrak{g}$ freely generate an algebra in which these relations hold:

$$xy - yx = [x, y]$$

for all $x, y \in \mathfrak{g}$. This algebra is called the *universal enveloping algebra* of $\mathfrak{g}$, denoted $U\mathfrak{g}$. It is, in fact, a bialgebra, and we have an equivalence of monoidal categories:

$$\text{Rep}(K) \simeq \text{Rep}(U\mathfrak{g}).$$

What Drinfel'd discovered is that we can deform $U\mathfrak{g}$ and get a *quantum group* $U_q\mathfrak{g}$. This is a family of bialgebras depending on a complex parameter $q$, with the property that $U_q\mathfrak{g} \cong U\mathfrak{g}$ when $q = 1$. Moreover, these bialgebras are unique, up to changes of the parameter $q$ and other inessential variations.

In fact, quantum groups are much better than mere bialgebras; they are quasitriangular Hopf algebras. This is just an intimidating way of saying that $\text{Rep}(U_q\mathfrak{g})$ is not merely a monoidal category but, in fact, it is a braided monoidal category with duals for objects. And this, in turn, is just an intimidating way of saying that any representation of $U_q\mathfrak{g}$ gives an invariant of framed oriented tangles! Reshetikhin and Turaev's paper explained exactly how this works.

If all of this seems too abstract, take $K = \text{SU}(2)$. From what we have already said, these categories are equivalent:

$$\text{Rep}(\text{SU}(2)) \simeq \text{Rep}(U\mathfrak{sl}(2)),$$

where $\mathfrak{sl}(2)$ is the Lie algebra of SL(2). So, we get a braided monoidal category with duals for objects, $\text{Rep}(U_q\mathfrak{sl}(2))$, which reduces to $\text{Rep}(\text{SU}(2))$ when we set $q = 1$. This is why $U_q\mathfrak{sl}(2)$ is often called *quantum SU(2)*, especially in the physics literature.

Even better, the quantum group $U_q\mathfrak{sl}(2)$ has a two-dimensional representation that reduces to the usual spin-$\frac{1}{2}$ representation of SU(2) at $q = 1$. Using this representation to get a tangle invariant, we obtain the Kauffman bracket—at least, up to some minor normalization issues that we shall ignore here. So Reshetikhin and Turaev's paper massively generalized the Kauffman bracket and set it into its proper context, the representation theory of quantum groups!

In our discussion of Kontsevich's 1993 paper [131], we will sketch how to actually get our hands on quantum groups.

### 1.3.30 Witten (1989)

In the 1980s, there was a lot of work on the Jones polynomial [125], leading up to the result we just sketched, a beautiful description of this invariant in terms of representations of quantum SU(2). Most of this early work on the Jones polynomial used two-dimensional pictures of knots and tangles—the string diagrams that we have been discussing here. This was unsatisfying in one respect: researchers wanted an intrinsically three-dimensional description of the Jones polynomial.

In his paper, "Quantum field theory and the Jones polynomial" [244], Witten gave such a description using a gauge field theory in three-dimensional spacetime, called *Chern–Simons theory*. He also described how the category of representations of SU(2) could be deformed into the category of representations of quantum SU(2) using a conformal field theory called the *Wess–Zumino–Witten model*, which is closely related to Chern–Simons theory. We say a little about this in our discussion of Kontsevich's 1993 paper.

### 1.3.31 Rovelli–Smolin (1990)

Around 1986, Abhay Ashtekar discovered a new formulation of general relativity, which made it more closely resemble gauge theories such as Yang–Mills theory [3]. In 1990, Rovelli and Smolin [209] published a paper that used this to develop a new approach to the old and difficult problem of quantizing gravity—that is, treating it as a quantum rather than a classical field theory. This approach is usually called *loop quantum gravity*, but in its later development, it came to rely heavily on Penrose's spin networks [8, 210]. It reduces to the Ponzano–Regge model in the case of three-dimensional quantum gravity; the difficult and so far unsolved challenge is finding a correct treatment of four-dimensional quantum gravity in this approach, if one exists.

As we have seen, spin networks are mathematically like Feynman diagrams with the Poincaré group replaced by SU(2). However, Feynman diagrams describe *processes* in ordinary quantum field theory, whereas spin networks describe *states* in loop quantum gravity. For this reason, it seemed natural to explore the possibility that some sort of two-dimensional diagrams going between spin networks are needed to describe processes in loop quantum gravity. These were introduced by Reisenberger and Rovelli in 1996 [198], and further formalized and dubbed spin foams in 1997 [8,9]. As we shall see, just as Feynman diagrams can be used to do computations in categories like the category of Hilbert spaces, spin foams can be used to do computations in bicategories like the bicategory of 2-Hilbert spaces.

For a review of loop quantum gravity and spin foams with plenty of references for further study, start with the article by Rovelli [207]. Then try his book [208] and the book by Ashtekar [4].

### 1.3.32 Kashiwara and Lusztig (1990)

Every matrix can be written as a sum of a lower triangular matrix, a diagonal matrix, and an upper triangular matrix. Similarly, for every simple Lie algebra $\mathfrak{g}$, the quantum group $U_q\mathfrak{g}$ has a triangular decomposition:

$$U_q\mathfrak{g} \cong U_q^-\mathfrak{g} \otimes U_q^0\mathfrak{g} \otimes U_q^+\mathfrak{g}.$$

If one is interested in the braided monoidal category of finite dimensional representations of $U_q\mathfrak{g}$, then it turns out that one needs only to understand the lower triangular part $U_q^-\mathfrak{g}$ of the quantum group. Using a sophisticated geometric approach, Lusztig [154, 155] defined a basis for $U_q^-\mathfrak{g}$ called the *canonical basis*, which has remarkable properties. Using algebraic methods, Kashiwara [115–117] defined a global crystal basis for $U_q^-(\mathfrak{g})$, which was later shown by Grojnowski and Lusztig [99] to coincide with the canonical basis.

What makes the canonical basis so interesting is that given two basis elements $e^i$ and $e^j$, their product $e^i e^j$ can be expanded in terms of basis elements

$$e^i e^j = \sum_k m_k^{ij} e^k$$

where the constants $m_k^{ij}$ are *polynomials* in $q$ and $q^{-1}$, and these polynomials have *natural numbers* as coefficients. If we had chosen a basis at random, we would only expect these constants to be rational functions of $q$, with rational numbers as coefficients.

The appearance of natural numbers here hints that quantum groups are just shadows of more interesting structures where the canonical basis elements become objects of a category, multiplication becomes the tensor product in this category, and addition becomes a direct sum in this category. Such a structure could be called a *categorified* quantum group. Its existence was explicitly conjectured in a paper by Crane and Frenkel [58], which we discuss later herein. Indeed, this was already visible in Lusztig's geometric approach to studying quantum groups using so-called perverse sheaves [156].

For a simpler example of this phenomenon, recall our discussion of Penrose's 1971 paper. We saw that if $K$ is a compact Lie group, the category Rep($K$) has a tensor product and direct sums. If we pick one irreducible representation $E^i$ from each isomorphism class, then every object in Rep($K$) is a direct sum of these objects $E^i$, which thus act as a kind of basis for Rep($K$). As a result, we have

$$E^i \otimes E^j \cong \bigoplus_k M_k^{ij} \otimes E^k$$

for certain finite-dimensional vector spaces $M_k^{ij}$. The dimensions of these vector spaces, say

$$m_k^{ij} = \dim(M_k^{ij}),$$

are *natural numbers*. We can define an algebra with one basis vector $e^i$ for each $E^i$, and with a multiplication defined by

$$e^i e^j = \sum_k m_k^{ij} e^k.$$

This algebra is called the *representation ring* of $K$, and denoted $R(K)$. It is associative because the tensor product in Rep($K$) is associative up to isomorphism.

In fact, representation rings were discovered before categories of representations. Suppose that someone had handed us such a ring and asked us to explain it. Then, the fact that it had a basis where the constants $m_k^{ij}$ are natural numbers would be a clue that it came from a monoidal category with direct sums!

The special properties of the canonical basis are a similar clue, but here there is an extra complication. Instead of natural numbers, we are getting polynomials in $q$ and $q^{-1}$ with natural number coefficients. We shall give an explanation of this later, in our discussion of Khovanov's 1999 article.

### 1.3.33 Kapranov–Voevodsky (1991)

Around 1991, Kapranov and Voevodsky made available a preprint in which they initiated work on 2-vector spaces and what we now call braided monoidal bicategories [114]. They also studied a higher-dimensional analogue of the Yang–Baxter equation called the *Zamolodchikov tetrahedron equation*. Recall from our discussion of Joyal and Street's 1985 paper that any solution of the Yang–Baxter equation gives a braided monoidal category. Kapranov and Voevodsky argued that, similarly, any solution of the Zamolodchikov tetrahedron equation gives a braided monoidal bicategory.

The basic idea of a braided monoidal bicategory is straightforward: it is like a braided monoidal category but with a bicategory replacing the underlying category. This lets us weaken equational laws involving 1-morphisms, replacing them by specified 2-isomorphisms. To obtain a useful structure, we also need to impose equational laws on these 2-isomorphisms—so-called *coherence laws*. This is the tricky part, which is why Kapranov and Voevodsky's original definition of semistrict braided monoidal 2-category required a number of fixes [15, 61, 64], leading ultimately to the fully general concept of braided monoidal bicategory introduced by McCrudden [168].

However, their key insight was striking and robust. As we have seen, any object in a braided monoidal category gives an isomorphism

$$B = B_{x,x} \colon x \otimes x \to x \otimes x$$

satisfying the Yang–Baxter equation

$$(B \otimes 1)(1 \otimes B)(B \otimes 1) = (1 \otimes B)(B \otimes 1)(1 \otimes B),$$

which in pictures corresponds to the third Reidemeister move. In a braided monoidal bicategory, the Yang–Baxter equation holds only up to a 2-isomorphism

$$Y \colon (B \otimes 1)(1 \otimes B)(B \otimes 1) \Rightarrow (1 \otimes B)(B \otimes 1)(1 \otimes B),$$

which, in turn, satisfies the Zamolodchikov tetrahedron equation.

This equation is best understood using diagrams. If we think of $Y$ as the surface in 4-space traced out by the process of performing the third Reidemeister move,

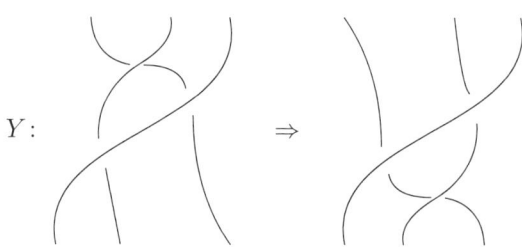

then the *Zamolodchikov tetrahedron equation* says the surface traced out by first performing the third Reidemeister move on a threefold crossing, and then sliding the result under a fourth strand is isotopic to that traced out by first sliding the threefold crossing under the fourth strand, and then performing the third Reidemeister move. So, this octagon commutes:

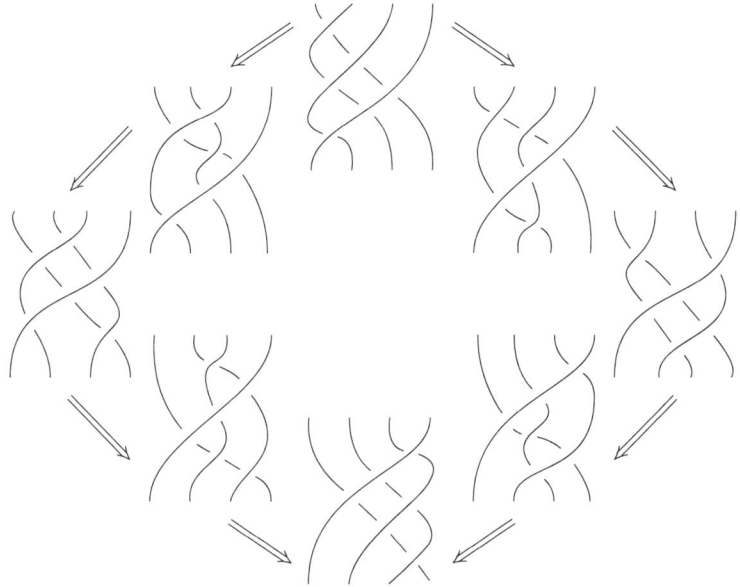

Just as the Yang–Baxter equation relates two different planar projections of three lines in $\mathbb{R}^3$, the Zamolodchikov tetrahedron relates two different projections onto $\mathbb{R}^3$ of four lines in $\mathbb{R}^4$. This suggests that solutions of the Zamolodchikov equation can give invariants of two-dimensional tangles in four-dimensional space (roughly, surfaces embedded in 4-space) just as solutions of the Yang–Baxter equation can give invariants of tangles (roughly, curves embedded in 3-space). Indeed, this was later confirmed [13, 40, 41].

Drinfel'd's work on quantum groups naturally gives solutions of the Yang–Baxter equation in the category of vector spaces. This suggested to Kapranov and Voevodsky the idea of looking for solutions of the Zamolodchikov tetrahedron equation in some bicategory of 2-vector spaces. They defined 2-vector spaces using the following analogy:

| $\mathbb{C}$ | Vect |
|---|---|
| $+$ | $\oplus$ |
| $\times$ | $\otimes$ |
| $0$ | $\{0\}$ |
| $1$ | $\mathbb{C}$ |

Analogy between ordinary linear algebra and higher linear algebra.

So, just as a finite-dimensional vector space may be defined as a set of the form $\mathbb{C}^n$, they defined a *2-vector space* to be a category of the form Vect$^n$. And, just as a linear operator $T: \mathbb{C}^n \to \mathbb{C}^m$ may be described using an $m \times n$ matrix of complex numbers, they defined a *linear functor* between 2-vector spaces to be an $m \times n$ matrix of vector spaces! Such matrices indeed act to give functors from $\text{Vect}^n$ to $\text{Vect}^m$. We can also add and multiply such matrices in the usual way, but with $\oplus$ and $\otimes$ taking the place of $+$ and $\times$.

Finally, there is a new layer of structure. Given two linear functors $S, T \colon \text{Vect}^n \to \text{Vect}^m$, Kapranov and Voevodsky defined a *linear natural transformation* $\alpha \colon S \Rightarrow T$ to be an $m \times n$ matrix of linear operators

$$\alpha_{ij} \colon S_{ij} \to T_{ij}$$

going between the vector spaces that are the matrix entries for $S$ and $T$. This new layer of structure winds up making 2-vector spaces into the objects of a *bicategory*.

Kapranov and Voevodsky called this bicategory 2Vect. They also defined a tensor product for 2-vector spaces, which turns out to make 2Vect into a monoidal bicategory. The Zamolodchikov tetrahedron equation makes sense in any monoidal bicategory, and any solution gives a *braided* monoidal bicategory. Conversely, any object in a braided monoidal bicategory gives a solution of the Zamolodchikov tetrahedron equation. These results hint that the relation among quantum groups, solutions of the Yang–Baxter equation, braided monoidal categories, and three-dimensional topology is not a freak accident because all of these concepts may have higher-dimensional analogues! To reach these higher-dimensional analogues, it seems we need to take concepts and systematically boost their dimension by making the following replacements:

| Elements | Objects |
|---|---|
| Equations between elements | Isomorphisms between objects |
| Sets | Categories |
| Functions | Functors |
| Equations between functions | Natural isomorphisms between functors |

Analogy between set theory and category theory.

In their 1994 paper, Crane and Frenkel called this process of dimension boosting *categorification* [58]. We have already seen, for example, that the representation category $\text{Rep}(K)$ of a compact Lie group is a categorification of its representation ring $R(K)$. The representation ring is a vector space; the representation category is a 2-vector space. In fact, the representation ring is an algebra, and as we see in our discussion of Barrett and Westbury's 1993 papers [22, 23], the representation category is a 2-algebra.

### 1.3.34 Reshetikhin–Turaev (1991)

In 1991, Reshetikhin and Turaev [200] published a paper in which they constructed invariants of 3-manifolds from quantum groups. These invariants were later seen to be part of a full-fledged three-dimensional TQFT. Their construction made rigorous ideas from Witten's 1989 paper [244] on Chern–Simons theory and the Jones polynomial, so this TQFT is now usually called the *Witten–Reshetikhin–Turaev theory*.

Their construction uses representations of a quantum group $U_q\mathfrak{g}$ but not the whole category $\text{Rep}(U_q\mathfrak{g})$. Instead, they use a special subcategory, which can be constructed when $q$ is a suitable root of unity. This subcategory has many nice properties. For example, it is a braided monoidal category with duals and also a 2-vector space with a *finite* basis of irreducible objects. These, and some extra properties, are summarized by saying that this subcategory is a modular tensor category. Such categories were later intensively studied by Turaev [233] and many others [18]. In this work, the Witten–Reshetikhin–Turaev construction was generalized to obtain a three-dimensional TQFT from any modular tensor category. Moreover, it was shown that any quantum group $U_q\mathfrak{g}$ gives rise to a modular tensor category when $q$ is a suitable root of unity.

However, it was later seen that in most cases, there is a four-dimensional TQFT of which the Witten–Reshetikhin–Turaev TQFT in three dimensions is merely a kind of side effect. So, for the purposes of understanding the relation between $n$-categories and TQFTs in various dimensions, it is better to postpone further treatment of the Witten–Reshetikhin–Turaev theory until our discussion of Turaev's 1992 article on the four-dimensional aspect of this theory [232].

### 1.3.35 Turaev–Viro (1992)

In 1992, the topologists Turaev and Viro [235] constructed another invariant of 3-manifolds—which we now know is part of a full-fledged three-dimensional TQFT—from the modular category arising from quantum SU(2). Their construction was later generalized to all modular tensor categories and, indeed, beyond. By now, any three-dimensional TQFT arising via this construction is called a *Turaev–Viro model*.

The relation between the Turaev–Viro model and the Witten–Reshetikhin–Turaev theory is subtle and interesting, but for our limited purposes, a few words will suffice. Briefly, it later became clear that a sufficiently nice *braided monoidal* category lets us construct a *four-dimensional* TQFT, which has a Witten–Reshetikhin–Turaev TQFT in three dimensions as a kind of shadow. Conversely, Barrett and Westbury discovered that we only need a sufficiently nice *monoidal* category to construct a *three-dimensional* TQFT—and the Turaev–Viro models are among these. This outlook makes certain patterns clearer; we shall explain these patterns further in sections to come.

When writing their original paper, Turaev and Viro did not know about the Ponzano–Regge model of quantum gravity. However, their construction amounts to taking the Ponzano–Regge model and curing it of its divergent sums by replacing SU(2) by the corresponding quantum group. Despite the many technicalities involved, the basic idea is simple. The Ponzano–Regge model is not a three-dimensional TQFT because it assigns divergent values to the operator $Z(M)$ for many cobordisms $M$. The reason is that computing this operator involves triangulating $M$, labeling the edges by spins $j = 0, \frac{1}{2}, 1, \ldots$, and summing over spins. Given that there are infinitely many choices of the spins, the sum may diverge. And because the spin labeling an edge describes its length, this divergence arises physically from the fact that we are summing over geometries that can be *arbitrarily large*.

Mathematically, spins correspond to irreducible representations of SU(2). There are, of course, infinitely many of these. The same is true for the quantum group $U_q\mathfrak{sl}(2)$. But, in the modular tensor category, we keep only *finitely many* of the irreducible representations of $U_q\mathfrak{sl}(2)$ as objects, corresponding to the spins $j = 0, \frac{1}{2}, 1, \ldots, \frac{k}{2}$, where $k$ depends on the root of unity $q$. This cures the Ponzano–Regge model of its infinities. Physically, introducing the parameter $q$ corresponds to introducing a nonzero cosmological constant in three-dimensional quantum gravity. The cosmological constant endows the vacuum with a constant energy density and forces spacetime to curl up instead of remaining flat. This puts an upper limit on the size of spacetime, avoiding the divergent sum over arbitrarily large geometries.

We postpone a detailed description of the Turaev–Viro model until our discussion of Barrett and Westbury's 1993 paper. As mentioned, this paper strips Turaev and Viro's construction down to its bare essentials, building a three-dimensional TQFT from any sufficiently nice monoidal category; the braiding is inessential. But the work of Barrett and Westbury is a categorified version of Fukuma, Hosono, and Kawai's work [92] on two-dimensional TQFTs, so we should first discuss that.

### 1.3.36 Fukuma–Hosono–Kawai (1992)

Fukuma, Hosono, and Kawai found a way to construct two-dimensional TQFTs from semisimple algebras [92]. Although they did not put it this way, they essentially created a "recipe" to turn any two-dimensional cobordism

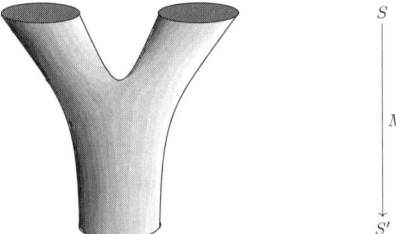

into a string diagram and use that diagram to define an operator between vector spaces:

$$\tilde{Z}(M) \colon \tilde{Z}(S) \to \tilde{Z}(S')$$

This gadget $\tilde{Z}$ is not quite a TQFT, but with a little extra work, it gives a TQFT, which we will call $Z$.

The recipe begins as follows. Triangulate the cobordism $M$:

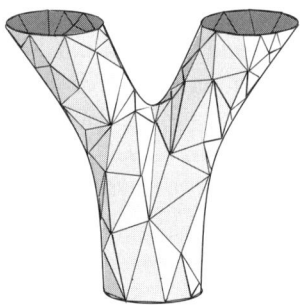

This picture already looks a bit like a string diagram, but never mind that. Instead, take the Poincaré dual of the triangulation, drawing a string diagram with

- one vertex in the center of each triangle of the original triangulation
- one edge crossing each edge of the original triangulation.

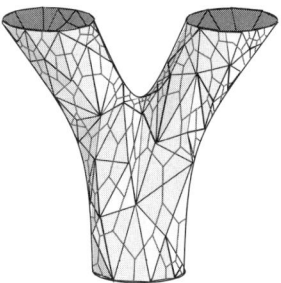

We then need a way to evaluate this string diagram and get an operator.

For this, fix an associative algebra $A$. Then, using Poincaré duality, each triangle in the triangulation can be reinterpreted as a string diagram for multiplication in $A$:

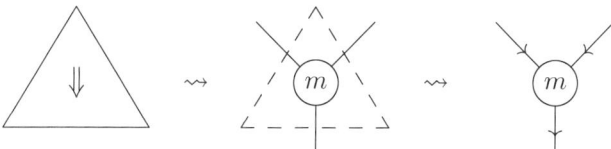

Actually, there is a slight subtlety here. The string diagram comes with some extra information, specifically little arrows on the edges, which tell us which edges are coming in and which are going out. To avoid the need for this extra information, let us equip $A$ with an isomorphism to its dual vector space $A^*$. Then, we can take any triangulation of $M$ and read it as a string diagram for an operator $\tilde{Z}(M)$. If our triangulation gives the manifold $S$ some number of edges, say $n$, and gives $S'$ some other number of edges, say $n'$, then we have

$$\tilde{Z}(M)\colon \tilde{Z}(S) \to \tilde{Z}(S')$$

where

$$\tilde{Z}(S) = A^{\otimes n}, \qquad \tilde{Z}(S') = A^{\otimes n'}.$$

We would like this operator $\tilde{Z}(M)$ to be well defined and independent of our choice of triangulation for $M$. And now a miracle occurs. In terms of triangulations, the associative law

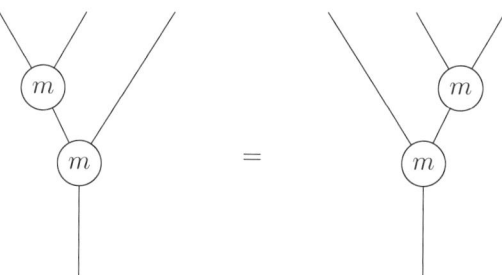

can be redrawn as follows:

This equation is already famous in topology! It is the *2–2 move*, one of two so-called Pachner moves for changing the triangulation of a surface without changing the surface's topology. The other is the *1–3 move*:

By repeatedly using these moves, we can go between any two triangulations of $M$ that restrict to the same triangulation of its boundary.

The associativity of the algebra $A$ guarantees that the operator $\tilde{Z}(M)$ does not change when we apply the 2–2 move. To ensure that $\tilde{Z}(M)$ is also unchanged by the 1–3 move, we require $A$ to be semisimple. There are many equivalent ways of defining this concept. For example, given that we are working over the complex numbers, we can define an algebra $A$ to be *semisimple* if it is isomorphic to a finite direct sum of matrix algebras. A more conceptual definition uses the fact that any algebra $A$ comes equipped with a bilinear form

$$g(a, b) = \mathrm{tr}(L_a L_b),$$

where $L_a$ stands for left multiplication by $a$

$$\begin{aligned} L_a \colon A &\to A \\ x &\mapsto ax \end{aligned}$$

and tr stands for the trace. We can reinterpret $g$ as a linear operator $g \colon A \otimes A \to \mathbb{C}$, which we can draw as a cup:

We say $g$ is *nondegenerate* if we can find a corresponding cap that satisfies the zig-zag equations. Then, we say the algebra $A$ is *semisimple* if $g$ is nondegenerate. In this case, the map $a \mapsto g(a, \cdot)$ gives an isomorphism $A \cong A^*$, which lets us avoid writing little arrows on our string diagram. Even better, with the chosen cap and cup, we get the equation

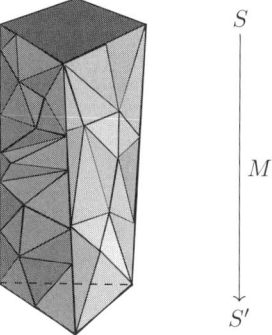

where each circle denotes the multiplication $m: A \otimes A \to A$. This equation then turns out to imply the 1–3 move! Proving this is a good workout in string diagrams and Poincaré duality.

So, starting from a semisimple algebra $A$, we obtain an operator $\tilde{Z}(M)$ from any triangulated two-dimensional cobordism $M$. Moreover, this operator is invariant under both Pachner moves. But how does this construction give us a two-dimensional TQFT? It is easy to check that

$$\tilde{Z}(MM') = \tilde{Z}(M)\,\tilde{Z}(M'),$$

which is a step in the right direction. We have seen that $\tilde{Z}(M)$ is the same regardless of which triangulation we choose for $M$, as long as we fix the triangulation of its boundary. Unfortunately, it depends on the triangulation of the boundary. After all, if $S$ is the circle triangulated with $n$ edges, then $\tilde{Z}(S) = A^{\otimes n}$. So we need to deal with this problem.

Given two different triangulations of the same 1-manifold, say $S$ and $S'$, we can always find a triangulated cobordism $M: S \to S'$, which is a *cylinder*, meaning it is homeomorphic to $S \times [0, 1]$, with $S$ and $S'$ as its two ends. For example:

This cobordism gives an operator $\tilde{Z}(M): \tilde{Z}(S) \to \tilde{Z}(S')$, and because this operator is independent of the triangulation of the interior of $M$, we obtain a canonical operator from $\tilde{Z}(S)$ to $\tilde{Z}(S')$. In particular, when $S$ and $S'$ are equal as triangulated manifolds, we get an operator

$$p_S: \tilde{Z}(S) \to \tilde{Z}(S).$$

This operator is not the identity, but a simple calculation shows that it is a *projection*, meaning

$$p_S^2 = p_S.$$

In physics jargon, this operator acts as a projection onto the space of physical states. And if we define $Z(S)$ to be the range of $p_S$, and $Z(M)$ to be the restriction of $\tilde{Z}(M)$ to $Z(S)$, we can check that $Z$ is a TQFT!

How does this construction relate to the construction of two-dimensional TQFTs from commutative Frobenius algebras explained in our discussion of Dijkgraaf's 1989 thesis? To answer this, we need to see how the commutative Frobenius algebra $Z(S^1)$ is related to the semisimple algebra $A$. In fact, $Z(S^1)$ turns out to be the *center* of $A$, the set of elements that commute with all other elements of $A$.

The proof is a nice illustration of the power of string diagrams. Consider the simplest triangulated cylinder from $S^1$ to itself. We get this by taking a square, dividing it into two triangles by drawing a diagonal line, and then curling it up to form a cylinder:

This gives a projection

$$p = p_{S^1} \colon \tilde{Z}(S^1) \to \tilde{Z}(S^1)$$

whose range is $Z(S^1)$. Because we have triangulated $S^1$ with a single edge in this picture, we have $\tilde{Z}(S^1) = A$. So, the commutative Frobenius algebra $Z(S^1)$ sits inside $A$ as the range of the projection $p \colon A \to A$.

Let us show that the range of $p$ is precisely the center of $A$. First, take the triangulated cylinder-and draw the Poincaré dual-string diagram: Erasing everything except this

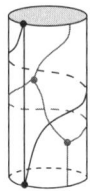

string diagram, we obtain a kind of formula for $p$:

$$p = \vcenter{\hbox{⧖}}$$

where the little circles stand for multiplication in $A$. To see that $p$ maps $A$ *onto* into its center, it suffices to check that if $a$ lies in the center of $A$, then $pa = a$. This is a nice string-diagram calculation:

In the second step, we use the fact that $a$ is in the center of $A$; in the last step, we use semisimplicity. Similarly, to see that $p$ maps $A$ *into* its center, it suffices to check that for any $a \in A$, the element $pa$ commutes with every other element of $A$. In string-diagram notation, this says that:

The proof is as follows:

### 1.3.37 Barrett–Westbury (1993)

In 1993, Barrett and Westbury completed a paper that greatly extended the work in Turaev and Viro's paper from the previous year [235]. Unfortunately, it reached publication much later, so everyone speaks of the Turaev–Viro model. Barrett and Westbury showed that in order to construct three-dimensional TQFTs, we need only a nice monoidal category, not a braided monoidal category. More technically, we do not need a modular tensor category; a spherical category will suffice [22]. Their construction can be seen as a categorified version of the Fukuma–Hosono–Kawai construction, and we shall present it from that viewpoint.

The key to the Fukuma–Hosono–Kawai construction was getting an operator from a triangulated two-dimensional cobordism and checking its invariance under the 2–2 and 1–3 Pachner moves. In both of these moves, the before and after pictures can be seen as the front and back of a tetrahedron:

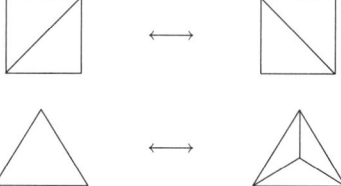

All of this has an analogue one dimension up. For starters, there are also Pachner moves in three dimensions. The *2–3 move* takes us from two tetrahedra attached along a triangle to three sharing an edge, or vice versa:

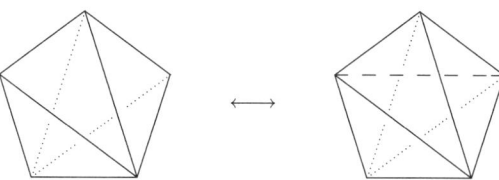

On the left side we see two tetrahedra sharing a triangle, the tall isosceles triangle in the middle. On the right, we see three tetrahedra sharing an edge, the dashed horizontal line. The *1–4 move* lets us split one tetrahedron into four or merge four back into one:

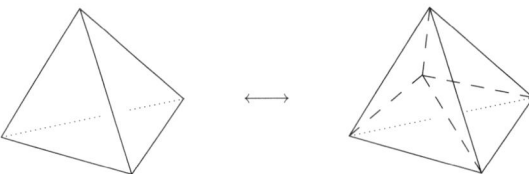

Given a three-dimensional cobordism $M: S \to S'$, repeatedly applying these moves lets us go between any two triangulations of $M$ that restrict to the same triangulation of its boundary. Moreover, for both of these moves, the before and after pictures can be seen as the front and back of a *4-simplex*, the four-dimensional analogue of a tetrahedron.

Fukuma, Hosono, and Kawai [92] constructed two-dimensional TQFTs from certain monoids— namely, semisimple algebras. As we have seen, the key ideas were the following:

- A triangulated two-dimensional cobordism gives an operator by letting each triangle correspond to multiplication in a semisimple algebra.
- Because the multiplication is associative, the resulting operator is invariant under the 2–2 Pachner move.
- Given that the algebra is semisimple, the operator is also invariant under the 1–3 move.

In a very similar way, Barrett and Westbury constructed three-dimensional TQFTs from certain monoidal categories called *spherical categories*. We can think of a spherical category as a categorified version of a semisimple algebra. The key ideas are these:

- A triangulated compact two-dimensional manifold gives a vector space by letting each triangle correspond to tensor product in a spherical category.
- A triangulated three-dimensional cobordism gives an operator by letting each tetrahedron correspond to the associator in the spherical category.
- Because the associator satisfies the pentagon identity, the resulting operator is invariant under the 2–3 Pachner move.
- Given that the spherical category is semisimple, the operator is also invariant under the 1–4 move.

The details are a bit elaborate, so let us just sketch some of the simplest, most beautiful aspects. Recall from our discussion of Kapranov and Voevodsky's 1991 paper that categorifying the concept of vector space gives the concept of 2-vector space. Just as there is a category Vect of vector spaces, there is a bicategory 2Vect of 2-vector spaces, with

- 2-vector spaces as objects
- linear functors as morphisms
- linear natural transformations as 2-morphisms

In fact, 2Vect is a monoidal bicategory, with a tensor product satisfying

$$\text{Vect}^m \otimes \text{Vect}^n \simeq \text{Vect}^{mn}.$$

This lets us define a *2-algebra* to be a 2-vector space $A$ that is also a monoidal category for which the tensor product extends to a linear functor

$$m\colon A \otimes A \to A,$$

and for which the associator and unitors extend to linear natural transformations. We have already seen a nice example of a 2-algebra— namely, Rep($K$)—for a compact Lie group $K$. Here, the tensor product is the usual tensor product of group representations.

Now let us fix a 2-algebra $A$. Given a triangulated compact two-dimensional manifold $S$, we can use Poincaré duality to reinterpret each triangle as a picture of the multiplication $m\colon A \otimes A \to A$:

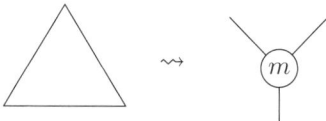

As in the Fukuma–Hosono–Kawai model, this lets us turn the triangulated manifold into a string diagram. And, as before, if $A$ is semisimple—or, more precisely, if $A$ is a spherical category—we do not need to write little arrows on the edges of this string diagram for it to make sense. But because everything is categorified, this string diagram now describes a linear *functor*. Given that $S$ has no boundary, this string diagram starts and ends with no edges, so it describes a linear functor from $A^{\otimes 0}$ to itself. Just as the tensor product of *zero* copies of a vector space is defined to be $\mathbb{C}$, the tensor product of no copies of 2-vector space is defined to be Vect. But a linear functor from Vect to itself is given by a $1 \times 1$ matrix of vector spaces—that is, a vector space! This recipe gives us a vector space $\tilde{Z}(S)$ for any compact 2d manifold $S$.

Next, from a triangulated three-dimensional cobordism $M\colon S \to S'$, we wish to obtain a linear operator $\tilde{Z}(M)\colon \tilde{Z}(S) \to \tilde{Z}(S')$. For this, we can use Poincaré duality to reinterpret each tetrahedron as a picture of the associator. The front and back of the tetrahedron correspond to the two functors that the associator goes between:

A more three-dimensional view is helpful here. Starting from a triangulated three-dimensional cobordism $M\colon S \to S'$, we can use Poincaré duality to build a piecewise-linear cell complex, or 2-complex, for short. This is a two-dimensional generalization of a graph; just as a graph has vertices and edges, a 2-complex has vertices, edges, and polygonal faces. The 2-complex dual to the triangulation of a three-dimensional cobordism has

- one vertex in the center of each tetrahedron of the original triangulation
- one edge crossing each triangle of the original triangulation
- and one face crossing each edge of the original triangulation

We can interpret this 2-complex as a higher-dimensional analogue of a string diagram and use this to compute an operator $\tilde{Z}(M)\colon \tilde{Z}(S) \to \tilde{Z}(S')$. This outlook is stressed in spin foam models [8, 9], of which the Turaev–Viro–Barrett–Westbury model is the simplest and most successful.

Each tetrahedron in $M$ gives a little piece of the 2-complex, which looks like this:

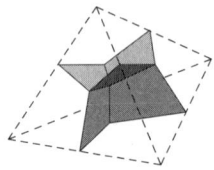

If we look at the string diagrams on the front and back of this picture, we see that they describe the two linear functors that the associator goes between:

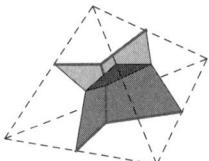
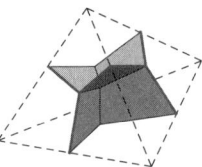

This is just a deeper look at something we already saw in our discussion of Ponzano and Regge's 1968 paper. There, we saw a connection among the tetrahedron, the $6j$ symbols, and the associator in Rep(SU(2)). Now we are seeing that for any spherical category, a triangulated three-dimensional cobordism gives a two-dimensional cell complex built out of pieces that we can interpret as associators. So, just as triangulated 2-manifolds give us linear functors, triangulated three-dimensional cobordisms give us linear natural transformations.

More precisely, recall that every compact triangulated 2-manifold $S$ gave a linear functor from Vect to Vect, or $1 \times 1$ matrix of vector spaces, which we reinterpreted as a vector space $\tilde{Z}(S)$. Similarly, every triangulated three-dimensional cobordism $M: S \to S'$ gives a linear natural transformation between such linear functors. This amounts to a $1 \times 1$ matrix of linear operators, which we can reinterpret as a linear operator $\tilde{Z}(M): \tilde{Z}(S) \to \tilde{Z}(S')$.

The next step is to show that $\tilde{Z}(M)$ is invariant under the 2–3 and 1–4 Pachner moves. If we can do this, the rest is easy; we can follow the strategy we have already seen in the Fukuma–Hosono–Kawai construction and obtain a three-dimensional TQFT.

At this point, another miracle comes to our rescue: the pentagon identity gives invariance under the 2–3 move! The 2–3 move goes from two tetrahedra to three, but each tetrahedron corresponds to an associator, so we can interpret this move as an equation between a natural transformation built from two associators and one built from three. And this equation is just the pentagon identity.

To see why, ponder the pentagon of pentagons in Figure 1.1. This depicts five ways to parenthesize a tensor product of objects $w, x, y, z$ in a monoidal category. Each corresponds to a triangulation of a pentagon. (The repeated appearance of the number five here is just a coincidence.) We can go between these parenthesized tensor products using the associator. In terms of triangulations, each use of the associator corresponds to a 2–2 move. We can go from the top of the picture to the lower right in

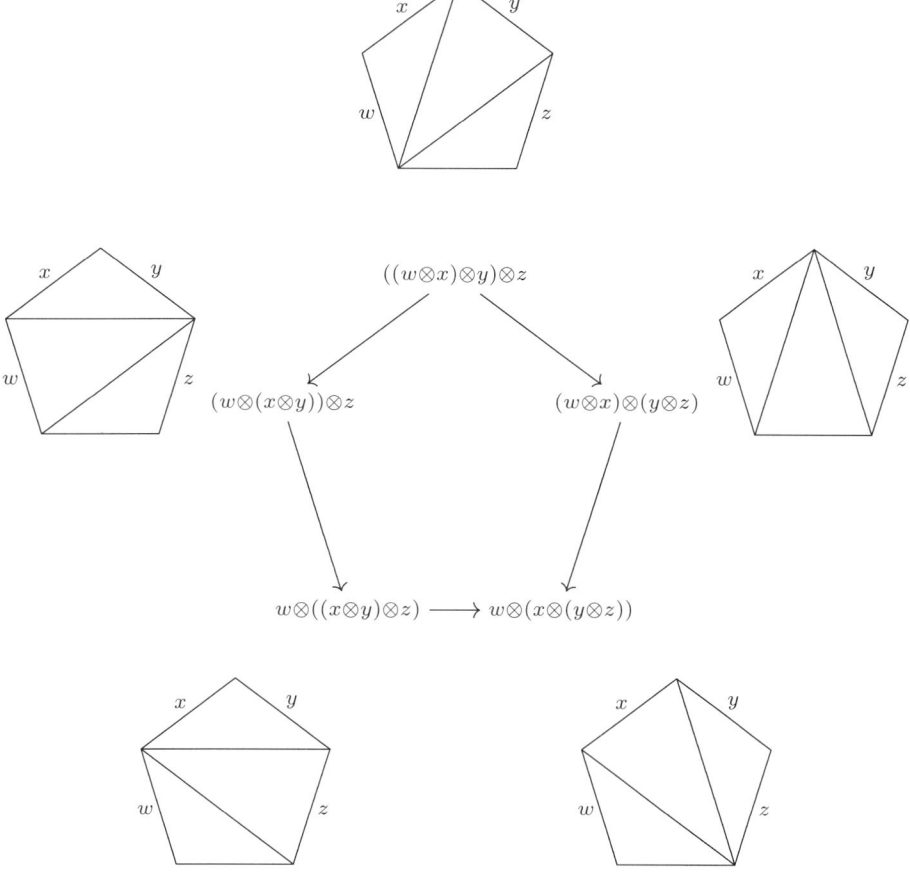

**Figure 1.1.** Deriving the 2–3 Pachner move from the pentagon identity.

two ways, one using two steps and one using three. The two-step method builds up this picture:

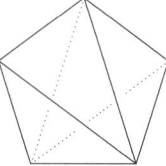

which shows two tetrahedra attached along a triangle. The three-step method builds up this picture:

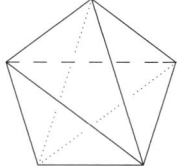

which shows three tetrahedra sharing a common edge. The pentagon identity thus yields the 2–3 move:

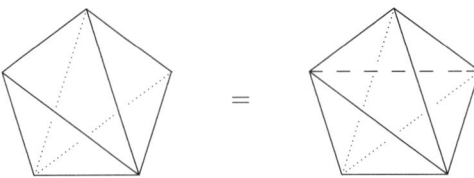

The other axioms in the definition of spherical category then yield the 1-4 move, and so we get a TQFT.

At this point, it is worth admitting that the link between the associative law and the 2–2 move, and that between the pentagon identity and the 2–3 move, are not really miracles in the sense of unexplained surprises. This is just the beginning of a pattern that relates the $n$-dimensional simplex and the $(n-1)$-dimensional Stasheff associahedron. An elegant explanation of this can be found in Street's 1987 article, "The algebra of oriented simplexes" [226]—the same one in which he proposed a simplicial approach to weak $\infty$-categories. Because there are also Pachner moves in every dimension [181], the Fukuma–Hosono–Kawai model and the Turaev–Viro–Barrett–Westbury model should be just the first of an infinite series of constructions building $(n+1)$-dimensional TQFTs from semisimple $n$-algebras. But this is largely open territory, apart from some important work in four dimensions, which we turn to next.

### 1.3.38 Turaev (1992)

As we already mentioned, the Witten–Reshetikhin–Turaev construction of three-dimensional TQFTs from modular tensor categories is really just a spinoff of a way to get *four-dimensional* TQFTs from modular tensor categories. This became apparent in 1991, when Turaev released a preprint [234] on building four-dimensional TQFTs from modular tensor categories. In 1992, he published an article with more details [232], and his book explains the ideas even more thoroughly [233]. His construction amounts to a four-dimensional analogue of the Turaev–Viro–Barrett–Westbury construction. Namely, from a four-dimensional cobordism $M: S \to S'$, one can compute a linear operator $\tilde{Z}(M): \tilde{Z}(S) \to \tilde{Z}(S')$ with the help of a two-dimensional CW complex sitting inside $M$. As already mentioned, we think of this complex as a higher-dimensional analogue of a string diagram.

In 1993, following work by the physicist Ooguri [179], Crane and Yetter [60] gave a different construction of four-dimensional TQFTs from the modular tensor category associated to quantum $SU(2)$. This construction used a triangulation of $M$. It was later generalized to a large class of modular tensor categories [59]; and, thanks to the work of Justin Roberts [204], it is clear that Turaev's construction is related to the Crane–Yetter construction by Poincaré duality, following a pattern we have seen already.

At this point, the reader, seeking simplicity amid these complex historical developments, should feel a bit puzzled. We have seen that

- The Fukuma–Hosono–Kawai construction gives two-dimensional TQFTs from sufficiently nice monoids (semisimple algebras).
- The Turaev–Viro–Bartlett–Westbury construction gives three-dimensional TQFTs from sufficiently nice monoidal categories (spherical categories).

Given this, it would be natural to expect some similar construction gives four-dimensional TQFTs from sufficiently nice monoidal bicategories.

Indeed, this is true. Mackaay [157] proved it in 1999. But how does this square with the following fact? The Turaev–Crane–Yetter construction gives four-dimensional TQFTs from sufficiently nice braided monoidal categories (modular tensor categories). The answer is very nice. It turns out that braided monoidal categories are a *special case* of monoidal bicategories.

We should explain this because it is part of a fundamental pattern called the *periodic table* of $n$-categories. As a warmup, let us see why a commutative monoid is the same as a monoidal category with only one object. This argument goes back to work of Eckmann and Hilton [74], published in 1962. A categorified version of their argument shows that a braided monoidal category is the same as a monoidal bicategory with only one object. This seems to have first been noticed by Joyal and Tierney [111], around 1984.

Suppose, first, that $C$ is a category with one object $x$. Then, composition of morphisms makes the set of morphisms from $x$ to itself, denoted $\hom(x, x)$, into a *monoid*, a set with an associative multiplication and an identity element. Conversely, any monoid gives a category with one object in this way.

But now suppose that $C$ is a monoidal category with one object $x$. Then, this object must be the unit for the tensor product. As before, $\hom(x, x)$ becomes a monoid using composition of morphisms. But now we can also tensor morphisms. By Mac Lane's coherence theorem, we may assume, without loss of generality, that $C$ is a strict monoidal category. Then, the tensor product is associative, and we have $1_x \otimes f = f = f \otimes 1_x$ for every $f \in \hom(x, x)$. So $\hom(x, x)$ becomes a monoid in a second way, with the same identity element.

However, the fact that the tensor product is a functor implies the *interchange law*:

$$(ff') \otimes (gg') = (f \otimes g)(f' \otimes g').$$

This lets us carry out the following remarkable argument, called the *Eckmann–Hilton argument*:

$$\begin{aligned}
f \otimes g &= (1f) \otimes (g1) \\
&= (1 \otimes g)(f \otimes 1) \\
&= gf \\
&= (g \otimes 1)(1 \otimes f) \\
&= (g1) \otimes (1f) \\
&= g \otimes f.
\end{aligned}$$

In short, composition and tensor product are equal, and they are both commutative! So $\hom(x, x)$ is a commutative monoid. Conversely, one can show that any commutative monoid can be thought of as the morphisms in a monoidal category with just one object.

In fact, Eckmann and Hilton came up with their argument in work on topology, and its essence is best revealed by a picture. Let us draw the composite of morphisms by putting one on top of the other and draw their tensor product by putting them side by side. We have often done this using string diagrams, but just for a change, let us draw morphisms as squares. Then, the Eckmann–Hilton argument goes as follows:

$$\underset{f \otimes g}{\boxed{\begin{array}{c|c} f & g \end{array}}} = \underset{(1\otimes g)(f\otimes 1)}{\boxed{\begin{array}{c|c} f & 1 \\ \hline 1 & g \end{array}}} = \underset{gf}{\boxed{\begin{array}{c} f \\ \hline g \end{array}}} = \underset{(g\otimes 1)(1\otimes f)}{\boxed{\begin{array}{c|c} 1 & f \\ \hline g & 1 \end{array}}} = \underset{g \otimes f}{\boxed{\begin{array}{c|c} g & f \end{array}}}$$

We can categorify this whole discussion. For starters, we noted in our discussion of Bénabou's 1967 paper that if $C$ is a bicategory with one object $x$, then $\hom(x, x)$ is a monoidal category—and, conversely, any monoidal category arises in this way. Then, the Eckmann–Hilton argument can be used to show that a monoidal bicategory with one object is a braided monoidal category. Given that categorification amounts to replacing equations with isomorphisms, each step in the argument now gives an isomorphism:

$$\begin{aligned} f \otimes g &\cong (1f) \otimes (g1) \\ &\cong (1 \otimes g)(f \otimes 1) \\ &\cong gf \\ &\cong (g \otimes 1)(1 \otimes f) \\ &\cong (g1) \otimes (1f) \\ &\cong g \otimes f. \end{aligned}$$

Composing these, we obtain an isomorphism from $f \otimes g$ to $g \otimes f$, which we can think of as a braiding:

$$B_{f,g} \colon f \otimes g \to g \otimes f.$$

We can even go further and check that this makes $\hom(x, x)$ into a braided monoidal category.

A picture makes this plausible. We can use the third dimension to record the process of the Eckmann–Hilton argument. If we compress $f$ and $g$ to small discs for clarity, it looks like this:

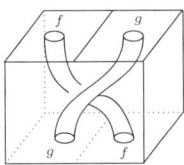

This clearly looks like a braiding!

In these pictures, we are moving $f$ around $g$ clockwise. There is an alternate version of the categorified Eckmann–Hilton argument that amounts to moving $f$ around $g$

counterclockwise:

$$\begin{aligned} f \otimes g &\cong (f1) \otimes (1g) \\ &\cong (f \otimes 1)(1 \otimes g) \\ &\cong fg \\ &\cong (1 \otimes f)(g \otimes 1) \\ &\cong (1g) \otimes (f1) \\ &\cong g \otimes f. \end{aligned}$$

This gives the following picture:

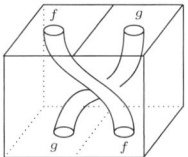

This picture corresponds to a *different* isomorphism from $f \otimes g$ to $g \otimes f$—namely, the reverse braiding:

$$B_{g,f}^{-1} \colon f \otimes g \to g \otimes f.$$

This is a great example of how different proofs of the same equation may give different isomorphisms when we categorify them.

The four-dimensional TQFTs constructed from modular tensor categories were a bit disappointing in that they gave invariants of four-dimensional manifolds that were already known and were unable to shed light on the deep questions of four-dimensional topology. The reason could be that braided monoidal categories are rather degenerate examples of monoidal bicategories. In their 1994 work, Crane and Frenkel [58] began the search for more interesting monoidal bicategories coming from the representation theory of *categorified* quantum groups. As of now, it is still unknown whether these give more interesting four-dimensional TQFTs.

### 1.3.39 Kontsevich (1993)

In his famous paper of 1993, Kontsevich [131] arrived at a deeper understanding of quantum groups, based on ideas of Witten, but making less explicit use of the path utilizing a more integral approach to quantum field theory.

In a nutshell, the idea is this. Fix a compact, simply connected simple Lie group $K$ and finite-dimensional representations $\rho_1, \ldots, \rho_n$. Then, there is a way to attach a vector space $Z(z_1, \ldots z_n)$ to any choice of distinct points $z_1, \ldots, z_n$ in the plane, and a way to attach a linear operator

$$Z(f) \colon Z(z_1, \ldots, z_n) \to Z(z'_1, \ldots, z'_n)$$

to any $n$-strand braid going from the points $(z_1, \ldots, z_n)$ to the points $z'_1, \ldots, z'_n$. The trick is to imagine each strand of the braid as the worldline of a particle in three-dimensional spacetime. As the particles move, they interact with each other via a gauge field, satisfying the equations of Chern–Simons theory. So we use parallel transport to describe how their internal states change. As usual, in quantum theory, this process is described by a linear operator, and this operator is $Z(f)$. Given that Chern–Simons theory describes a gauge field with zero curvature, this operator depends only on the topology of the braid. So, with some work, we get a braided monoidal category from these data. With more work, we can get operators not just for braids but also tangles—and, thus, a braided monoidal category with duals for objects. Finally, using a Tannaka–Krein reconstruction theorem, we can show that this category is the category of finite-dimensional representations of a quasitriangular Hopf algebra, the quantum group associated to $G$.

### 1.3.40 Lawrence (1993)

In 1993, Lawrence wrote an influential paper on extended TQFTs [143], which she further developed in later work [144]. As we have seen, many TQFTs can be constructed by first triangulating a cobordism, attaching a piece of algebraic data to each simplex, and then using these to construct an operator. For the procedure to give a TQFT, the resulting operator must remain the same when we change the triangulation by a Pachner move. Lawrence tackled the question of precisely what is going on here. Her approach was to axiomatize a structure with operations corresponding to ways of gluing together $n$-dimensional simplexes, satisfying relations that guarantee invariance under the Pachner moves.

The use of simplexes is not ultimately the essential point here. The essential point is that we can build any $n$-dimensional spacetime out of a few standard building blocks, which can be glued together locally in a few standard ways. This lets us describe the topology of spacetime purely combinatorially, by saying how the building blocks have been assembled. This reduces the problem of building TQFTs to an essentially *algebraic* problem, although it is one of a novel sort.

(Here we are glossing over the distinction among topological, piecewise-linear, and smooth manifolds. Despite the term *TQFT*, our description is really suited to the case of *piecewise-linear* manifolds, which can be chopped into simplexes or other polyhedra. Luckily, there is no serious difference between piecewise-linear and smooth manifolds in dimensions below 7, and both of these agree with topological manifolds below dimension 4.)

Not every TQFT need arise from this sort of recipe; we loosely use the term *extended TQFT* for those that do. The idea is that whereas an ordinary TQFT gives operators only for $n$-dimensional manifolds with boundary (or, more precisely, cobordisms), an extended one assigns some sort of data to $n$-dimensional manifolds *with corners*—for example, simplexes and other polyhedra. This is a physically natural requirement, so it is believed that the most interesting TQFTs are extended ones.

In ordinary algebra, we depict multiplication by setting symbols side by side on a line; multiplying $a$ and $b$ gives $ab$. In category theory, we visualize morphisms as arrows, which we glue together end to end in a one-dimensional way. In studying TQFTs, we need higher-dimensional algebra to describe how to glue pieces of spacetime together.

The idea of higher-dimensional algebra had been around for several decades, but by this time, it began to really catch on. For example, in 1992, Brown wrote a popular exposition of higher-dimensional algebra, aptly titled "Out of line" [37]. It became clear that $n$-categories should provide a very general approach to higher-dimensional algebra because they have ways of composing $n$-morphisms that mimic ways of gluing together $n$-dimensional simplexes, globes, or other shapes. Unfortunately, the theory of $n$-categories was still in its early stages of development, limiting its potential as a tool for studying extended TQFTs.

For this reason, a partial implementation of the idea of extended TQFT became of interest—see, for example, Crane's 1995 paper [57]. Instead of working with the symmetric monoidal category $n$Cob, he began to grapple with the symmetric monoidal bicategory $n$Cob$_2$, where, roughly speaking

- objects are compact, oriented $(n-2)$-dimensional manifolds
- morphisms are $(n-1)$-dimensional cobordisms
- 2-morphisms are $n$-dimensional cobordisms between cobordisms

His idea was that a once-extended TQFT should be a symmetric monoidal functor

$$Z \colon n\mathrm{Cob}_2 \to 2\mathrm{Vect}.$$

In this approach, cobordisms between cobordisms are described using manifolds with corners. The details are still a bit tricky. It seems that the first precise general construction of $n$Cob$_2$ as a bicategory was given by Morton [174] in 2006, and in 2009, Schommer-Pries proved that 2Cob$_2$ was a symmetric monoidal bicategory [211]. Lurie's [152] more powerful approach goes in a somewhat different direction, as we explain in our discussion of Baez and Dolan's 1995 paper.

Because two-dimensional TQFTs are completely classified by the result in Dijkgraaf's 1989 thesis [67], the concept of once-extended TQFT may seem like "overkill" in dimension 2. But this would be a short-sighted attitude. Around 2001, motivated in part by work on $D$-branes in string theory, Moore and Segal [171, 215] introduced once-extended two-dimensional TQFTs under the name of open-closed topological string theories. However, they did not describe these using the bicategory 2Cob$_2$. Instead, they considered a symmetric monoidal category 2Cob$^{\mathrm{ext}}$ whose objects include not just compact one-dimensional manifolds like the circle (closed strings) but also one-dimensional manifolds with boundary like the interval (open strings). Here are morphisms that generate 2Cob$^{\mathrm{ext}}$ as a symmetric monoidal category:

Using these, Moore and Segal showed that a once-extended two-dimensional TQFT gives a Frobenius algebra for the interval and a commutative Frobenius algebra for the circle. The operations in these Frobenius algebras account for all but the last two morphisms shown. The last two give a projection from the first Frobenius algebra to the second and an inclusion of the second into the center of the first.

Later, Lauda and Pfeiffer [140] gave a detailed proof that 2Cob$^{\mathrm{ext}}$ is the free symmetric monoidal category on a Frobenius algebra, equipped with a projection into its center, satisfying certain relations. Using this, they showed [141] that the Fukuma–Hosono–Kawai construction can be extended to obtain symmetric monoidal functors $Z\colon 2\mathrm{Cob}^{\mathrm{ext}} \to \mathrm{Vect}$. Fjelstad, Fuchs, Runkel, and Schweigert have gone in a different direction, describing full-fledged, open-closed conformal field theories using Frobenius algebras [83, 90, 91].

Once-extended TQFTs should be even more interesting in dimension 3. At least in a rough way, we can see how the Turaev–Viro–Barrett–Westbury construction should generalize to give examples of such theories. Recall that this construction starts with a 2-algebra $A \in 2\text{Vect}$ satisfying some extra conditions.

- A triangulated compact one-dimensional manifold $S$ gives a 2-vector space $\tilde{Z}(S)$, built by tensoring one copy of $A$ for each edge in $S$.
- A triangulated two-dimensional cobordism $M: S \to S'$ gives a linear functor $\tilde{Z}(M): \tilde{Z}(S) \to \tilde{Z}(S')$, built out of one multiplication functor $m: A \otimes A \to A$ for each triangle in $M$.
- A triangulated three-dimensional cobordism between cobordisms $\alpha: M \Rightarrow M'$ gives a linear natural transformation $\tilde{Z}(\alpha): \tilde{Z}(M) \Rightarrow \tilde{Z}(M')$, built out of one associator for each tetrahedron in $\alpha$.

From $\tilde{Z}$, we should then be able to construct a once-extended three-dimensional TQFT:

$$Z: 3\text{Cob}_2 \to 2\text{Vect}.$$

However, to the best of our knowledge, this construction has not been carried out. The work of Kerler and Lyubashenko constructs the Witten–Reshetikhin–Turaev theory as a kind of extended three-dimensional TQFT using a somewhat different formalism: double categories instead of bicategories [123].

### 1.3.41 Crane–Frenkel (1994)

In 1994, Louis Crane and Igor Frenkel wrote a paper entitled "Four-dimensional topological quantum field theory, Hopf categories, and the canonical bases" [58]. In this paper, they discussed algebraic structures that provide TQFTs in various low dimensions:

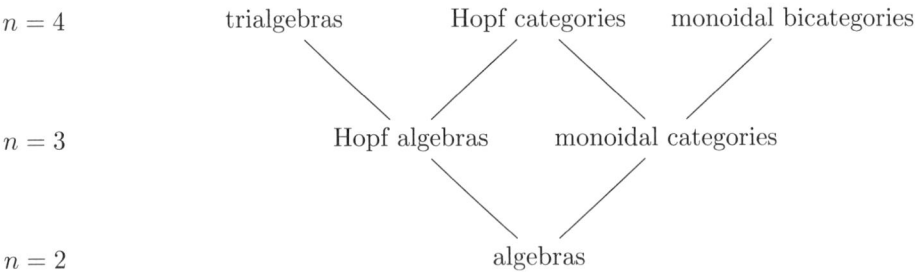

This chart is a bit schematic, so let us expand on it a bit. In our discussion of Fukuma, Hosono, and Kawai's 1992 work [92], we saw how they constructed two-dimensional TQFTs from certain algebras—namely, semisimple algebras. In our discussion of Barrett and Westbury's paper from the same year, we saw how they constructed three-dimensional TQFTs from certain monoidal categories—namely, spherical categories. But any Hopf algebra has a monoidal category of representations, and we can use Tannaka–Krein reconstruction to recover a Hopf algebra from its category of representations. This suggests that we might be able to construct three-dimensional TQFTs directly from certain Hopf algebras. Indeed, this is the case, as was shown by Kuperberg [136] and Chung, Fukuma, and Shapere [53]. Indeed, there is a beautiful direct relation between three-dimensional topology and the Hopf algebra axioms.

Crane and Frenkel speculated on how this pattern continues in higher dimensions. To anyone who understands the dimension-boosting nature of categorification, it is natural to guess that one can construct four-dimensional TQFTs from certain monoidal bicategories. Indeed, as mentioned, this was later shown by Mackaay [157], who was greatly influenced by the Crane–Frenkel paper. But this, in turn, suggests that we could obtain monoidal bicategories by considering 2-representations of categorified Hopf algebras, or Hopf categories—and that perhaps we could construct four-dimensional TQFTs *directly* from certain Hopf categories.

This may be true. In 1997, Neuchl [176] gave a definition of Hopf categories and showed that a Hopf category has a monoidal bicategory of 2-representations on 2-vector spaces. In 1999, Carter, Kauffman, and Saito [39] found beautiful relations between four-dimensional topology and the Hopf category axioms.

Crane and Frenkel also suggested that there should be some kind of algebra whose category of representations was a Hopf category. They called this a *trialgebra*. They sketched the definition; in 2004 Pfeiffer [190] gave a more precise treatment, showing that any trialgebra has a Hopf category of representations.

However, *defining* these structures is just the first step toward constructing interesting four-dimensional TQFTs. As Crane and Frenkel put it, "To proceed any further we need a miracle, namely, the existence of an interesting family of Hopf categories."

Many of the combinatorial constructions of three-dimensional TQFTs input a Hopf algebra, or the representation category of a Hopf algebra, and produce a TQFT. However, the most interesting class of three-dimensional TQFTs come from Hopf algebras that are deformed universal enveloping algebras $U_q \mathfrak{g}$. The question is, Where can one find an interesting class of Hopf categories that will give invariants that are useful in four-dimensional topology?

Topology, in four dimensions, is very different from lower dimensions. It is the first dimension where homeomorphic manifolds can fail to be diffeomorphic. In fact, there exist *exotic* $\mathbb{R}^4$'s, manifolds homeomorphic to $\mathbb{R}^4$, but not diffeomorphic to it. This is the only dimension in which exotic $\mathbb{R}^n$s exist! The discovery of exotic $\mathbb{R}^4$s relied on invariants coming from quantum field theory that can distinguish between homeomorphic four-dimensional manifolds that are not diffeomorphic. Indeed, this subject, known as *Donaldson theory* [68], is what motivated Witten to invent the term *topological quantum field theory* in the first place [243]. Later, Seiberg and Witten revolutionized this subject with a streamlined approach [216, 217], and Donaldson theory was rebaptized *Seiberg–Witten theory*. There are, by now, some good introductory texts on these matters [85, 172, 173]. The book by Scorpan [213] is especially inviting.

But this mystery remains: How—*if at all!*—can the 4-manifold invariants coming from quantum field theory be computed using Hopf categories, trialgebras, or related structures? Although such structures would give TQFTs suitable for piecewise-linear manifolds, there is no essential difference between piecewise-linear and smooth manifolds in dimension 4. Unfortunately, interesting examples of Hopf categories seem hard to construct.

Luckily, Crane and Frenkel did more than sketch the definition of a Hopf category. They also conjectured where examples might arise:

> The next important input is the existence of the canonical bases, for a special family of Hopf algebras, namely, the quantum groups. These bases are actually an indication of the existence of a family of Hopf categories, with structures closely related to the quantum groups.

Crane and Frenkel suggested that the existence of the Lusztig–Kashiwara canonical bases for upper triangular part of the enveloping algebra and the Lusztig canonical bases for the entire quantum groups give strong evidence that quantum groups are the shadows of a much richer structure, which we might call a *categorified quantum group*.

Lusztig's geometric approach produces monoidal categories associated to quantum groups, categories of perverse sheaves. Crane and Frenkel hoped that these categories could be given a combinatorial or algebraic formulation revealing a Hopf category structure. Recently, some progress has been made toward fulfilling Crane and Frenkel's hopes. In particular, these categories of perverse sheaves have been reformulated into an algebraic language related to the categorification of $U_q^+ \mathfrak{g}$ [129, 236]. The entire quantum group $U_q \mathfrak{sl}_n$ has been categorified by Khovanov and Lauda [130, 139], and they also gave a conjectural categorification of the entire quantum group $U_q \mathfrak{g}$ for every simple Lie algebra $\mathfrak{g}$. Categorified representation theory, or 2-representation theory, has taken off, thanks largely to the foundational work of Chuang and Rouquier [52, 206].

There is much more that needs to be understood. In particular, categorification of quantum groups at roots of unity has received only a little attention [128], and the Hopf category structure has not been fully developed. Furthermore, these approaches have not yet obtained braided monoidal bicategories of 2-representations of categorified quantum groups; neither have they constructed four-dimensional TQFTs.

### 1.3.42 Freed (1994)

In 1994, Freed published an important paper [84] that exhibited how higher-dimensional algebraic structures arise naturally from the Lagrangian formulation of topological quantum field theory. Among many other things, this paper clarified the connection between quasitriangular Hopf algebras and three-dimensional TQFTs. It also introduced an informal concept of 2-Hilbert space, categorifying the concept of Hilbert space. This was later made precise, at least in the finite-dimensional case [7, 23], so it is now tempting to believe that much of the formalism of quantum theory can be categorified. The subtleties of analysis involved in understanding infinite-dimensional 2-Hilbert spaces remain challenging, with close connections to the theory of von Neumann algebras [10].

### 1.3.43 Kontsevich (1994)

In a lecture at the 1994 International Congress of Mathematicians in Zürich, Kontsevich [132] proposed the "homological mirror symmetry conjecture," which led to a burst of work relating string theory to higher categorical structures. A detailed discussion of this work would drastically increase the size of this chapter, so we content ourselves with a few elementary remarks.

We have already mentioned the concept of an $A_\infty$ space, a topological space equipped with a multiplication that is associative up to a homotopy that satisfies the pentagon equation up to a homotopy ... and so on, forever, in a manner governed by the Stasheff polytopes [225]. This concept can be generalized to any context that allows for a notion of homotopy between maps. In particular, it generalizes to the world of homological algebra, which is a simplified version of the world of homotopy theory. In homological algebra, the structure that takes the place of a topological space is a *chain complex*, a sequence of abelian groups and homomorphisms

$$V_0 \xleftarrow{d_1} V_1 \xleftarrow{d_2} V_2 \xleftarrow{d_3} \cdots$$

with $d_i d_{i+1} = 0$. In applications to physics, we focus on the case where the $V_i$ are vector spaces and the $d_i$ are linear operators. Regardless of this, we can define maps between chain complexes, called *chain maps*, and homotopies between chain maps, called *chain homotopies*.

| Topological spaces | Chain complexes |
|---|---|
| Continuous maps | Chain maps |
| Homotopies | Chain homotopies |

Analogy between homotopy theory and homological algebra.

(For a very readable introduction to these matters, see the book by Rotman [205]; for a more strenuous one that goes further, try the book with the same title by Weibel [240].)

The analogy between homotopy theory and homological algebra ultimately arises from the fact that whereas homotopy types can be seen as $\infty$-groupoids, chain complexes can be seen as $\infty$-groupoids that are strict and also abelian. The process of turning a topological space into a chain complex, so important in algebraic topology, thus amounts to taking a $\infty$-groupoid and simplifying it by making it strict and abelian.

Because this fact is less widely appreciated than it should be, let us quickly sketch the basic idea. Given a chain complex $V$, each element of $V_0$ corresponds to an object in the corresponding $\infty$-groupoid. Given objects $x, y \in V_0$, a morphism $f \colon x \to y$ corresponds to an element $f \in V_1$ with

$$d_1 f + x = y.$$

Given morphisms $f, g \colon x \to y$, a 2-morphism $\alpha \colon f \Rightarrow g$ corresponds to an element $\alpha \in V_2$ with

$$d_2 \alpha + f = g,$$

and so on. The equation $d_i d_{i+1} = 0$ then says that an $(i + 1)$-morphism can only go between two $i$-morphisms that share the same source and target—just as we expect in the globular approach to $\infty$-categories.

The analogue of an $A_\infty$ space in the world of chain complexes is called an $A_\infty$ algebra [122, 134, 165]. More generally, one can define a structure, called an $A_\infty$ *category*, that has a set of objects, a chain complex hom($x, y$), for any pair of objects, and a composition map that is associative up to a chain homotopy that satisfies the pentagon identity up to a chain homotopy . . . and so on. Just as a monoid is the same as a category with one object, an $A_\infty$ algebra is the same as an $A_\infty$ category with one object.

Kontsevich used the language of $A_\infty$ categories to formulate a conjecture about mirror symmetry, a phenomenon already studied by string theorists. *Mirror symmetry* refers to the observation that various pairs of superficially different string theories seem, in fact, to be isomorphic. In Kontsevich's conjecture, each of these theories is an open-closed, topological string theory. We already introduced this concept near the end of our discussion of Lawrence's 1993 paper. Recall that such a theory is designed to describe processes involving open strings (intervals) and closed strings (circles). The basic building blocks of such processes are these:

In the simple approach we discussed, the space of states of the open string is a *Frobenius algebra*. The space of states of the closed string is a *commutative Frobenius algebra*, typically the center of the Frobenius algebra for the open string. In the richer approach developed by Kontsevich and subsequent authors, notably Costello [56], states of the open string are instead described by an $A_\infty$ category with some extra structure mimicking that of a Frobenius algebra. The space of states of the closed string is obtained from this using a subtle generalization of the concept of center. To get some sense of this, let us ignore the Frobenius aspects and simply regard the space of states of an open string as an algebra. Multiplication in this algebra describes the process of two open strings colliding and merging together:

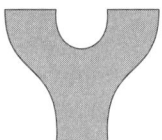

The work in question generalizes this simple idea in two ways. First, it treats an algebra as a special case of an $A_\infty$ algebra— namely, one for which only the 0th vector space in its underlying chain complex is nontrivial. Second, it treats an $A_\infty$ algebra as a special case of an $A_\infty$ category—namely, an $A_\infty$ category with just one object.

How should we understand a general $A_\infty$ category as describing the states of an open-closed topological string? First, the different objects of the $A_\infty$ category correspond to different boundary conditions for an open string. In physics, these boundary conditions are called *D-branes* because they are thought of as membranes in spacetime on which the open strings begin or end. The "$D$" stands for Dirichlet, who studied boundary conditions back in the mid-1800s. (A good introduction to $D$-branes from a physics perspective can be found in Polchinski's book [192].)

For any pair of $D$-branes $x$ and $y$, the $A_\infty$ category gives a chain complex hom($x, y$). What is the physical meaning of this? It is the space of states for an open string that starts on the $D$-brane $x$ and ends on the $D$-brane $y$. Composition describes a process in which open strings in the states $g \in \hom(x, y)$ and $f \in \hom(y, z)$ collide and stick together to form an open string in the state $fg \in \hom(x, z)$.

However, note that the space of states $\hom(x, y)$ is not a mere vector space. It is a chain complex—so it is secretly a strict $\infty$-groupoid! This lets us talk about states that are not *equal* but still *isomorphic*. In particular, composition in an $A_\infty$ category is associative only up to isomorphism; the states $(fg)h$ and $f(gh)$ are not usually equal, merely isomorphic via an associator,

$$a_{f,g,h}: (fg)h \to f(gh).$$

In the language of chain complexes, we write this as follows:

$$da_{f,g,h} + (fg)h = f(gh).$$

This is just the first of an infinite list of equations that are part of the usual definition of an $A_\infty$ category. The next one says that the associator satisfies the pentagon identity up to $d$ of something, and so on.

Kontsevich formulated his homological mirror symmetry conjecture as the statement that two $A_\infty$ categories are equivalent. The conjecture remains unproved in general, but many special cases are known. Perhaps more important, the conjecture has become part of an elaborate web of ideas relating gauge theory to the Langlands program—which itself is a vast generalization of the circle of ideas that gave birth to Wiles's proof of Fermat's last theorem. (For a good introduction to all of this, see the survey by Edward Frenkel [87].)

### 1.3.44 Gordon–Power–Street (1995)

In 1995, Gordon, Power, and Street introduced the definition and basic theory of tricategories—or, in other words, weak 3-categories [94]. Among other things, they defined a monoidal bicategory to be a tricategory with one object. They then showed that a monoidal bicategory with one object is the same as a braided monoidal category. This is a precise working out of the categorified Eckmann–Hilton argument sketched in our discussion of Turaev's 1992 paper.

So, a tricategory with just one object and one morphism is the same as a braided monoidal category. There is also, however, a notion of strict 3-category, a tricategory where all the relevant laws hold as equations, not merely up to equivalence. Not surprising, a strict 3-category with one object and one morphism is a braided monoidal category where all the braiding, associators, and unitors are *identity* morphisms. This rules out the possibility of nontrivial braiding, which occurs in categories of braids or tangles. As a consequence, not every tricategory is equivalent to a strict 3-category.

All of this stands in violent contrast to the story one dimension down, where a generalization of Mac Lane's coherence theorem can be used to show that every bicategory is equivalent to a strict 2-category. So, although it was already known in some quarters [111], Gordon, Power, and Street's book made the need for weak $n$-categories clear to all. In a world where all tricategories were equivalent to strict 3-categories, there would be no knots!

Gordon, Power, and Street did, however, show that every tricategory is equivalent to a semistrict 3-category, in which some but not all of the laws hold as equations. They called these semistrict 3-categories Gray-categories because their definition relies on John Gray's prescient early work [96]. Constructing a workable theory of semistrict $n$-categories for all $n$ remains a major challenge.

### 1.3.45 Baez–Dolan (1995)

In 1995, Baez and Dolan [11] outlined a program for understanding extended TQFTs in terms of $n$-categories. A key part of this is the periodic table of $n$-categories. Because this only involves weak $n$-categories, let us drop the qualifier "weak" for the rest of this section and take it as given. Also, just for the sake of definiteness, let us take a globular approach to $n$-categories:

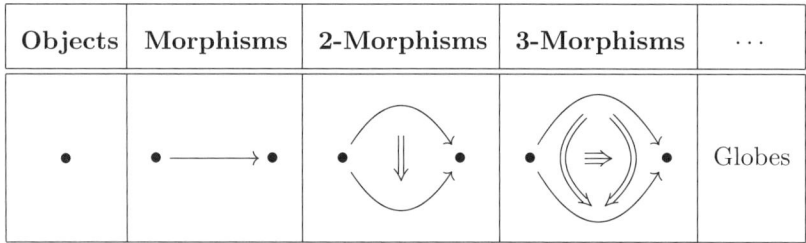

So, in this section, *2-category* will mean *bicategory* and *3-category* will mean *tricategory*. (Recently, this sort of terminology has been catching on because the use of Greek prefixes to name weak $n$-categories becomes inconvenient as the value of $n$ becomes large.)

We have already seen the beginning of a pattern involving these concepts:

- A category with one object is a monoid.
- A 2-category with one object is a monoidal category.
- A 3-category with one object is a monoidal 2-category.

The idea is that we can take an $n$-category with one object and think of it as an $(n-1)$-category by ignoring the object, renaming the morphisms *objects*, renaming the 2-morphisms *morphisms*, and so on. Our ability to compose morphisms in the original $n$-category gets reinterpreted as an ability to tensor objects in the resulting $(n-1)$-category, so we get a monoidal $(n-1)$-category.

However, we can go further because we can consider a monoidal $n$-category with one object. We have already looked at two cases of this, and we can imagine more:

- A monoidal category with one object is a commutative monoid.
- A monoidal 2-category with one object is a braided monoidal category.
- A monoidal 3-category with one object is a braided monoidal 2-category.

Here, the Eckmann–Hilton argument comes into play, as explained in our discussion of Turaev's 1992 paper. The idea is that given a monoidal $n$-category $C$ with one object, this object must be the unit for the tensor product, $1 \in C$. We can focus attention on hom(1, 1), which is an $(n-1)$-category. Given $f, g \in$ hom(1, 1), there are *two* ways to combine them: we can compose them or tensor them. As we have seen, we can visualize these operations as putting together little squares in two ways: vertically or horizontally:

These operations are related by an interchange morphism

$$(ff') \otimes (gg') \to (f \otimes g)(f' \otimes g'),$$

which is an equivalence (i.e., invertible in a suitably weakened sense). This allows us to carry out the Eckmann–Hilton argument and get a braiding on hom(1, 1):

$$B_{f,g}: f \otimes g \to g \otimes f.$$

Next, consider braided monoidal $n$-categories with one object. Here, the pattern seems to go like this:

- A braided monoidal category with one object is a commutative monoid.
- A braided monoidal 2-category with one object is a symmetric monoidal category.
- A braided monoidal 3-category with one object is a sylleptic monoidal 2-category.
- A braided monoidal 4-category with one object is a sylleptic monoidal 3-category.

The idea is that given a braided monoidal $n$-category with one object, we can think of it as an $(n-1)$-category with *three* ways to combine objects, all related by interchange equivalences. We should visualize these as the three obvious ways of putting together little cubes, side by side, one in front of the other, and one on top of the other.

In the first case listed the third operation doesn not give anything new. Like a monoidal category with one object, a braided monoidal category with one object is merely a commutative monoid. In the next case, we do get something new: a braided monoidal 2-category with one object is a *symmetric* monoidal category. The reason is that the third monoidal structure allows us to interpolate between the Eckmann–Hilton argument that gives the braiding by moving $f$ and $g$ around clockwise and the argument that gives the reverse braiding by moving them around counterclockwise. We obtain the equation

$$\begin{array}{c}f \quad g \\ \times \end{array} = \begin{array}{c}g \quad f \\ \times \end{array}$$

that characterizes a symmetric monoidal category.

In the case after this, instead of an equation, we obtain a *2-isomorphism* that describes the *process* of interpolating between the braiding and the reverse braiding:

$$S_{f,g}: \quad \begin{array}{c}x \quad y \\ \times\end{array} \Rightarrow \begin{array}{c}y \quad x \\ \times\end{array}$$

The reader should endeavor to imagine these pictures as drawn in four-dimensional space, so that there is room to push the top strand in the left-hand picture up into the fourth dimension, slide it behind the other strand, and then push it back down, getting the right-hand picture. Day and Street [64] later dubbed this 2-isomorphism $s_{f,g}$ the *syllepsis* and formalized the theory of sylleptic monoidal 2-categories. The definition of a fully weak sylleptic monoidal 2-category was introduced later by Street's student McCrudden [168].

To understand better the patterns at work here, it is useful to define a $k$-tuply monoidal $n$-category to be an $(n + k)$-category with just one $j$-morphism for $j < k$. A chart of these follows. This is called the *periodic table* because, like Mendeleyev's original periodic table, it guides us in extrapolating the behavior of $n$-categories from simple cases to more complicated ones. It is not really periodic in any obvious way.

|       | $n = 0$ | $n = 1$ | $n = 2$ |
|-------|---------|---------|---------|
| $k = 0$ | Sets | Categories | 2-Categories |
| $k = 1$ | Monoids | Monoidal categories | Monoidal 2-categories |
| $k = 2$ | Commutative monoids | Braided monoidal categories | Braided monoidal 2-categories |
| $k = 3$ | " | Symmetric monoidal categories | Sylleptic monoidal 2-categories |
| $k = 4$ | " | " | Symmetric monoidal 2-categories |
| $k = 5$ | " | " | " |
| $k = 6$ | " | " | " |

The Periodic Table:
Hypothesized table of $k$-tuply monoidal $n$-categories.

The periodic table should be taken with a grain of salt. For example, a claim stating that 2-categories with one object and one morphism are the same as commutative monoids needs to be made more precise. Its truth may depend on whether we consider commutative monoids as forming a category, or a 2-category, or a 3-category! This has been investigated by Cheng and Gurski [48]. There have also been attempts to craft an approach that avoids such subtleties [17].

But please ignore such matters for now; just stare at the table. The most notable feature is that the $n$th column of the periodic table seems to stop changing when $k$ reaches $n + 2$. Baez and Dolan called this the stabilization hypothesis. The idea is that adding extra monoidal structures ceases to matter at this point. Simpson later proved a version of this hypothesis in his approach to $n$-categories [223]. So let us assume that the stabilization hypothesis is true and call a $k$-tuply monoidal $n$-category with $k \geq n + 2$ a stable $n$-category.

In fact, stabilization is just the simplest of the many intricate patterns lurking in the periodic table. For example, the reader will note that the syllepsis

$$s_{f,g}: B_{f,g} \Rightarrow B_{g,f}^{-1}$$

is somewhat reminiscent of the braiding itself:

$$B_{f,g}: f \otimes g \to g \otimes f.$$

Indeed, this is the beginning of a pattern that continues as we zig-zag down the table starting with monoids. To go from monoids to commutative monoids, we add the equation $fg = gf$. To go from commutative monoids to braided monoidal categories, we then replace this equation by an isomorphism, the braiding $B_{f,g}: f \otimes g \to g \otimes f$. But the braiding engenders another isomorphism with the same source and target, the reverse braiding $B_{g,f}^{-1}$. To go from braided monoidal categories to symmetric monoidal categories, we add the equation $B_{f,g} = B_{g,f}^{-1}$. To go from symmetric monoidal categories to sylleptic monoidal 2-categories, we then replace this equation by a 2-isomorphism, the syllepsis $s_{f,g}: B_{f,g} \Rightarrow B_{g,f}^{-1}$. But this engenders another 2-isomorphism with the same source and target, the reverse syllepsis. Geometrically speaking, this is because we can also deform the left braid to the right one here

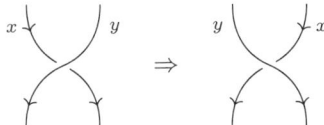

by pushing the top strand *down* into the fourth dimension and then behind the other strand. To go from sylleptic monoidal 2-categories to symmetric ones, we add an equation stating that the syllepsis equals the reverse syllepsis, and so on, forever! As we zig-zag down the diagonal, we meet ways of switching between ways of switching between... ways of switching things.

This is still just the tip of the iceberg; the patterns that arise further from the bottom edge of the periodic table are vastly more intricate. To give just a taste of their subtlety, consider the remarkable story told in Kontsevich's 1999 paper "Operads and Motives and Deformation Quantization" [133]. Kontsevich had an amazing realization: quantization of ordinary classical mechanics problems can be carried out in a systematic way using ideas from string theory. A thorough and rigorous approach to this issue required proving a conjecture by Deligne. However, early attempts to prove Deligne's conjecture had a flaw, first noted by Tamarkin, whose simplest manifestation—translated into the language of $n$-categories—involves an operation that first appears for braided monoidal 6-categories!

For this sort of reason, one would really like to see precisely which features are being added as we march down any column of the periodic table. Batanin's approach to $n$-categories offers a beautiful answer based on the combinatorics of trees [27]. Unfortunately, explaining this here would take us too far afield. The slides of a lecture Batanin delivered in 2006 give a taste of the richness of his work [29].

Baez and Dolan also emphasized the importance of $n$-categories with duals at all levels: duals for objects, duals for morphisms, ... and so on, up to $n$-morphisms. Unfortunately, they were only able to precisely define this notion in some simple cases. For example, in our discussion of Doplicher and Roberts's 1989 article, we defined monoidal, braided monoidal, and symmetric monoidal categories with duals—meaning duals for both objects and morphisms. We noted that tangles in three-dimensional space can be seen as morphisms in the free braided monoidal category on one object. This is part of a larger pattern:

- The category of framed one-dimensional tangles in two-dimensional space, $1\text{Tang}_1$, is the free monoidal category with duals on one object.
- The category of framed one-dimensional tangles in three-dimensional space, $1\text{Tang}_2$, is the free braided monoidal category with duals on one object.
- The category of framed one-dimensional tangles in four-dimensional space, $1\text{Tang}_3$, is the free symmetric monoidal category with duals on one object.

A technical point: here we are using *framed* to mean "equipped with a trivialization of the normal bundle." This is how the word is used in homotopy theory, as opposed to knot theory. In fact, a framing in this sense determines an orientation, so a framed one-dimensional tangle in three-dimensional space is what ordinary knot theorists would call a *framed oriented tangle*.

Based on these and other examples, Baez and Dolan formulated the tangle hypothesis. This concerns a conjectured $n$-category $n\text{Tang}_k$ where

- objects are collections of framed points in $[0, 1]^k$
- morphisms are framed one-dimensional tangles in $[0, 1]^{k+1}$
- 2-morphisms are framed two-dimensional tangles in $[0, 1]^{k+2}$
- and so on up to dimension $(n-1)$
- and, finally, $n$-morphisms are isotopy classes of framed $n$-dimensional tangles in $[0, 1]^{n+k}$

For short, we call the $n$-morphisms $n$-*tangles* in $(n + k)$ dimensions. Figure 1.2 may help the reader see how simple these actually are. It shows a typical $n$-tangle in $(n + k)$ dimensions for various values of $n$ and $k$. This figure is a close relative of the periodic table. The number $n$ is the *dimension* of the tangle, and $k$ is its *codimension*—that is, the number of *extra* dimensions of space.

The tangle hypothesis says that $n\text{Tang}_k$ is the free $k$-tuply monoidal $n$-category with duals on one object. As usual, the one object, $x$, is simply a point. More precisely, $x$ can be any point in $[0, 1]^k$ equipped with a framing that makes it positively oriented.

Combining the stabilization hypothesis and the tangle hypothesis, we obtain an interesting conclusion: the $n$-category $n\text{Tang}_k$ stabilizes when $k$ reaches $n + 2$. This idea is backed up by a well-known fact in topology: any two embeddings of a compact $n$-dimensional manifold in $\mathbb{R}^{n+k}$ are isotopic if $k \geq n + 2$. In simple terms, when $k$ is this large, there is enough room to untie any $n$-dimensional knot!

So, we expect that when $k$ is this large, the $n$-morphisms in $n\text{Tang}_k$ correspond to abstract $n$-tangles, not embedded in any ambient space. But this is precisely how we think of cobordisms. So, for $k \geq n + 2$, we should expect that $n\text{Tang}_k$ is a stable $n$-category where

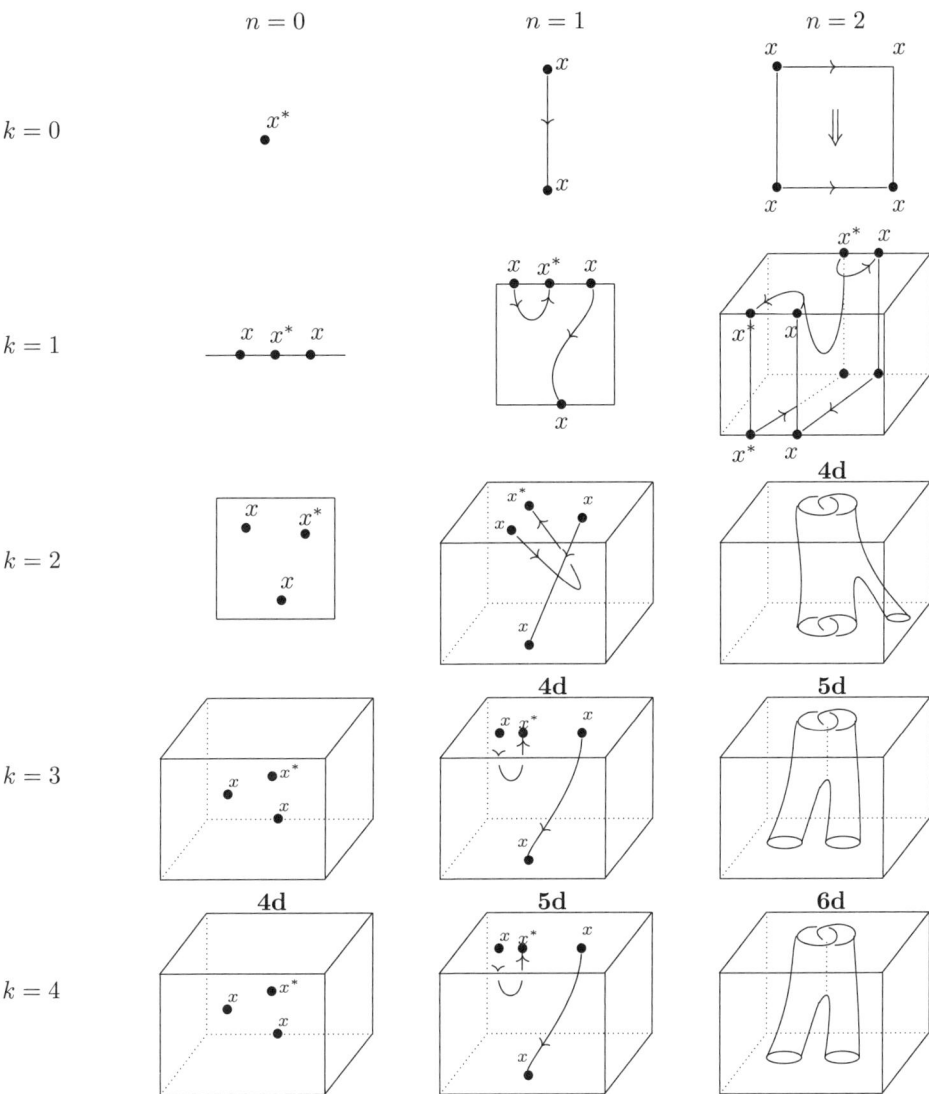

**Figure 1.2.** Examples of $n$-tangles in $(n+k)$-dimensional space.

- objects are compact framed zero-dimensional manifolds
- morphisms are framed one-dimensional cobordisms
- 2-morphisms are framed 2-dimensional cobordisms between cobordisms
- 3-morphisms are framed 3-dimensional cobordisms between cobordisms between cobordisms

and so on up to dimension $n$, where we take equivalence classes. Let us call this $n$-category $n\mathrm{Cob}_n$ because it is a further elaboration of the 2-category $n\mathrm{Cob}_2$ in our discussion of Lawrence's 1993 paper [143].

The cobordism hypothesis summarizes these ideas. It says that $n\mathrm{Cob}_n$ is the free stable $n$-category with duals on one object $x$—namely, the positively oriented point.

We have already sketched how once-extended, $n$-dimensional TQFTs can be treated as symmetric monoidal functors:

$$Z: n\mathrm{Cob}_2 \to 2\mathrm{Vect}.$$

This suggests that fully extended, $n$-dimensional TQFTs should be something similar, but with $n\mathrm{Cob}_n$ replacing $n\mathrm{Cob}_2$. Similarly, we should replace $2\mathrm{Vect}$ by some sort of $n$-category, such as something deserving the name $n\mathrm{Vect}$ or, even better, $n\mathrm{Hilb}$.

This leads to the extended TQFT hypothesis, which says that a unitary extended TQFT is a map between stable $n$-categories

$$Z: n\mathrm{Cob}_n \to n\mathrm{Hilb}$$

that preserves all levels of duality. Given that $n\mathrm{Hilb}$ should be a stable $n$-category with duals, and $n\mathrm{Cob}_n$ should be the free such thing on one object, we should be able to specify a unitary extended TQFT simply by choosing an object $H \in n\mathrm{Hilb}$ and saying that

$$Z(x) = H$$

where $x$ is the positively oriented point. This is the "primacy of the point" in a very dramatic form.

What progress has there been on making these hypotheses precise and proving them? In 1998, Baez and Langford [13] came close to proving that $2\mathrm{Tang}_2$, the 2-category of 2-tangles in four-dimensional space, was the free braided monoidal 2-category with duals on one object. (In fact, they proved a similar result for oriented but unframed 2-tangles.) In 2009, Schommer-Pries [211] came close to proving that $2\mathrm{Cob}_2$ was the free symmetric monoidal 2-category with duals on one object. (In fact, he gave a purely algebraic description of $2\mathrm{Cob}_2$ as a symmetric monoidal 2-category, but not explicitly using the language of duals.)

But, the really exciting development is the paper that Jacob Lurie [152] put on the arXiv in 2009 entitled "On the Classification of Topological Field Theories," which outlines a precise statement and proof of the cobordism hypothesis for all $n$.

Lurie's version makes use not of $n$-categories but rather of $(\infty, n)$-categories. These are $\infty$-categories such that every $j$-morphism is an equivalence for $j > n$. This helps avoid the problems with duality that we mentioned in our discussion of Atiyah's 1988 article. There are many approaches to $(\infty, 1)$-categories, including the $A_\infty$ categories mentioned in our discussion of Kontsevich's 1994 lecture [132]. Prominent alternatives include Joyal's quasicategories [107], first introduced in the early 1970s under another name by Boardmann and Vogt [36], as well as Rezk's complete Segal spaces [201]. For a comparison of some approaches, see the survey by Bergner [34]. Another good source of material on quasicategories is Lurie's enormous book on higher topos theory [153]. The study of $(\infty, n)$-categories for higher $n$ is still in its infancy. At this moment, Lurie's paper is the best place to start, although he attributes the definition he uses to Barwick, who promises a two-volume book on the subject [24].

### 1.3.46 Khovanov (1999)

In 1999, Mikhail Khovanov found a way to categorify the Jones polynomial [126]. We have already seen a way to categorify an algebra that has a basis $e^i$ for which

$$e^i e^j = \sum_k m_k^{ij} e^k,$$

where the constants $m_k^{ij}$ are *natural numbers*. Namely, we can think of these numbers as dimensions of *vector spaces* $M_k^{ij}$. Then, we can seek a 2-algebra with a basis of irreducible objects $E^i$ such that

$$E^i \otimes E^j = \sum_k M_k^{ij} \otimes E^k.$$

We say this 2-algebra categorifies our original algebra or, more technically, we say that taking the Grothendieck group of the 2-algebra gives back our original algebra. In this simple example, taking the Grothendieck group just means forming a vector space with one basis element $e^i$ for each object $E^i$ in our basis of irreducible objects.

The Jones polynomial and other structures related to quantum groups present more challenging problems. Here, instead of natural numbers, we have polynomials in $q$ and $q^{-1}$. Sometimes, as in the theory of canonical bases, these polynomials have natural number coefficients. Elsewhere, as in the Jones polynomial, they have integer coefficients. How can we generalize the concept of dimension so it can be a polynomial of this sort?

In fact, problems like this were already tackled by Emmy Noether in the late 1920s, in her work on homological algebra [66]. We have already defined the concept of a *chain complex*, but this term is used in several slightly different ways, so now let us change our definition a bit and say that a *chain complex* $V$ is a sequence of vector spaces and linear maps

$$\cdots \xleftarrow{d_{-1}} V_{-1} \xleftarrow{d_0} V_0 \xleftarrow{d_1} V_1 \xleftarrow{d_2} V_2 \xleftarrow{d_3} \cdots$$

with $d_i d_{i+1} = 0$. If the vector spaces are finite-dimensional, and only finitely many are nonzero, we can define the *Euler characteristic* of the chain complex by

$$\chi(V) = \sum_{i=-\infty}^{\infty} (-1)^i \dim(V_i).$$

The Euler characteristic is a remarkably robust invariant in that we can change the chain complex in many ways without changing its Euler characteristic. This explains why the number of vertices, minus the number of edges, plus the number of faces, is equal to 2 for every convex polyhedron!

We may think of the Euler characteristic as a generalization of dimension that can take on arbitrary *integer* values. In particular, any vector space gives a chain complex for which only $V_0$ is nontrivial and, in this case, the Euler characteristic reduces to the ordinary dimension. But, given any chain complex $V$, we can shift it to obtain a new chain complex $sV$ with

$$sV_i = V_{i+1},$$

and we have
$$\chi(sV) = -\chi(V).$$
So, shifting a chain complex is like taking its negative.

But what about polynomials in $q$ and $q^{-1}$? For these, we need to generalize vector spaces a bit further, as indicated here:

| Vector spaces | Natural numbers |
|---|---|
| Chain complexes | Integers |
| Graded vector spaces | Polynomials in $q^{\pm 1}$ with natural number coefficients |
| Graded chain complexes | Polynomials in $q^{\pm 1}$ with integer coefficients |

Algebraic structures and the values of their dimensions.

A *graded vector space* $W$ is simply a series of vector spaces $W_i$, where $i$ ranges over all integers. The *Hilbert–Poincaré series* $\dim_q(W)$ of a graded vector space is given by

$$\dim_q(W) = \sum_{i=-\infty}^{\infty} \dim(W_i) q^i.$$

If the vector spaces $W_i$ are finite-dimensional, and only finitely many are nonzero, $\dim_q(W)$ is a polynomial in $q$ and $q^{-1}$ with natural number coefficients. Similarly, a *graded chain complex* $W$ is a series of chain complexes $W_i$, and its *graded Euler characteristic* $\chi(W)$ is given by

$$\chi_q(W) = \sum_{i=-\infty}^{\infty} \chi(W_i) q^i.$$

When everything is finite enough, this is a polynomial in $q$ and $q^{-1}$ with integer coefficients.

Khovanov found a way to assign a graded chain complex to any link in such a way that its graded Euler characteristic is the Jones polynomial of that link, apart from a slight change in normalizations. This new invariant can distinguish links that have the same Jones polynomial [19]. Even better, it can be extended to an invariant of tangles in three-dimensional space, and also *2-tangles in four-dimensional space!*

To make this a bit more precise, note that we can think of a 2-tangle in four-dimensional space as a morphism $\alpha \colon S \to T$ going from one tangle in three-dimensional space, namely $S$, to another, namely $T$. For example:

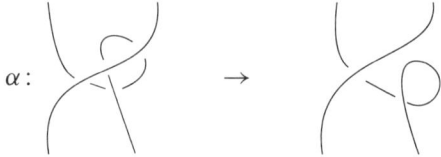

In its most recent incarnation, Khovanov homology makes use of a certain monoidal category $C$. Its precise definition takes a bit of work [20], but its objects are built using graded chain complexes, and its morphisms are built using maps between these. Khovanov homology assigns to each tangle $T$ in three-dimensional space an object $Z(T) \in C$ and assigns to each 2-tangle in four-dimensional space $\alpha \colon T \Rightarrow T'$ a morphism $Z(\alpha) \colon Z(T) \to Z(T')$.

What is especially nice is that $Z$ is a monoidal functor. This means we can compute the invariant of a 2-tangle by breaking it into pieces, computing the invariant for each piece, and then composing and tensoring the results. Actually, in the original construction due to Jacobsson [102] and Khovanov [127], $Z(\alpha)$ was only well defined up to a scalar multiple. But later, using the streamlined approach introduced by Bar-Natan [20], this problem was fixed by Clark, Morrison, and Walker [55].

So far, we have been treating 2-tangles as morphisms. But, in fact, we know they should be 2-morphisms. There should be a braided monoidal bicategory $2\text{Tang}_2$ where, roughly speaking:

- objects are collections of framed points in the square $[0, 1]^2$
- morphisms are framed oriented tangles in the cube $[0, 1]^3$
- 2-morphisms are framed oriented 2-tangles in $[0, 1]^4$

The tangle hypothesis asserts that $2\text{Tang}_2$ is the *free* braided monoidal bicategory with duals on one object $x$—namely, the positively oriented point. Indeed, a version of this claim ignoring framings is already known to be true [13].

This suggests that Khovanov homology could be defined in a way that takes advantage of this universal property of $2\text{Tang}_2$. For this, we would need to see the objects and morphisms of the category $C$ as morphisms and 2-morphisms of some braided monoidal bicategory with duals—say, $\overline{C}$—equipped with a special object $c$. Then, Khovanov homology could be seen as the essentially unique braided monoidal functor preserving duals, say

$$\overline{Z} \colon 2\text{Tang}_2 \to \overline{C},$$

with the property that

$$\overline{Z}(x) = c.$$

This would be yet another triumph of the primacy of the point.

It is worth mentioning that the authors in this field have chosen to study higher categories with duals in a manner that does not distinguish between source and target. This makes sense because duality allows one to convert input to output and vice versa. In 1999, Jones introduced planar algebras [105], which can be thought of as a formalism for handling certain categories with duals. In his work on Khovanov homology, Bar-Natan introduced a structure, called a *canopolis* [20], which is a kind of categorified planar algebra. The relation between these ideas and other approaches to $n$-category theory deserves to be clarified and generalized to higher dimensions.

One exciting aspect of Khovanov's homology theory is that it breathes new life into Crane and Frenkel's dream of understanding the special features of smooth four-dimensional topology in a purely combinatorial way, using categorification. For example, Rasmussen [196] has used Khovanov homology to give a purely combinatorial proof of the Milnor conjecture—a famous problem in topology that had been solved earlier in the 1990s using ideas from quantum field theory—namely, Donaldson theory

[135]. And as the topologist Gompf later pointed out [197], Rasmussen's work can also be used to prove the existence of an exotic $\mathbb{R}^4$.

In outline, the argument goes as follows. A knot in $\mathbb{R}^3$ is said to be *smoothly slice* if it bounds a smoothly embedded disc in $\mathbb{R}^4$. It is said to be *topologically slice* if it bounds a topologically embedded disc in $\mathbb{R}^4$ and this embedding extends to a topological embedding of some thickening of the disc. Gompf had shown that if there is a knot that is topologically but not smoothly slice, there must be an exotic $\mathbb{R}^4$. However, Rasmussen's work can be used to find such a knot!

Before this, all proofs of the existence of exotic $\mathbb{R}^4$s had involved ideas from quantum field theory: either Donaldson theory or its modern formulation, Seiberg–Witten theory. This suggests that a purely combinatorial approach to Seiberg–Witten theory is within reach. Indeed, Ozsváth and Szabó have already introduced a knot homology theory, called *Heegaard-Floer homology*, that has a conjectured relationship to Seiberg–Witten theory [180]. Now that there is a completely combinatorial description of Heegaard–Floer homology [163, 164], one cannot help but be optimistic that some version of Crane and Frenkel's dream will become a reality.

In summary, the theory of $n$-categories is beginning to shed light on some remarkably subtle connections among physics, topology, and geometry. Unfortunately, this work has not yet led to concrete successes in elementary particle physics or quantum gravity. But, given the profound yet simple ways that $n$-categories unify and clarify our thinking about mathematics and physics, we can hope that what we have seen so far is just the beginning.

## Acknowledgments

We thank the denizens of the $n$-Category Café, including Toby Bartels, Michael Batanin, David Ben-Zvi, Rafael Borowiecki, Greg Egan, Alex Hoffnung, Urs Schreiber, and Zoran Škoda, for many discussions and corrections. Greg Friedman also made many helpful corrections. J. B. thanks the Department of Pure Mathematics and Mathematical Statistics at the University of Cambridge for inviting him to give a series of lectures on this topic in July 2004 and Derek Wise for writing up notes for a course on this topic at U. C. Riverside during the 2004–2005 academic year. We both thank Hans Halvorson and everyone involved in the *Deep Beauty* symposium. A. L. was partially supported by the NSF grants, DMS-0739392 and DMS-0855713, and J. B. was partially supported by the NSF grant PHY-0653646 and the FQXi grant RFP2-08-04.

## References

[1] L. Abrams. Two-dimensional topological quantum field theories and Frobenius algebras. *Journal of Knot Theory and its Ramifications*, **5** (1996), 569–87.

[2] S. Abramsky and B. Coecke. A categorical semantics of quantum protocols. *Proceedings of the 19th IEEE Conference on Logic in Computer Science (LiCS'04)*. Washington, DC: IEEE Computer Science Press, 2004, pp. 415–25. arXiv:quant-ph/0402130.

[3] A. Ashtekar. New variables for classical and quantum gravity. *Physical Review Letters*, **57** (1986), 2244–7.

# REFERENCES

[4] A. Ashtekar. *Lectures on Nonperturbative Canonical Gravity*. Singapore: World Scientific, 1991.

[5] M. Atiyah. Topological quantum field theories. *Institut Hautes Études Sciences Publication Mathematics*, **68** (1988), 175–86.

[6] J. C. Baez. Spin network states in gauge theory. *Advances in Mathematics*, **117** (1996), 253–72. arXiv:gr-qc/9411007.

[7] J. C. Baez. Higher-dimensional algebra II: 2-Hilbert spaces. *Advances in Mathematics*, **127** (1997), 125–89. arXiv:q-alg/9609018.

[8] J. C. Baez. Spin foam models. *Classical and Quantum Gravity*, **15** (1998), 1827–58. arXiv:gr-qc/9709052.

[9] J. C. Baez. An introduction to spin foam models of $BF$ theory and quantum gravity. In *Geometry and Quantum Physics*, eds. H. Gausterer and H. Grosse. Berlin: Springer-Verlag, 2000, pp. 25–93. arXiv:gr-qc/9905087.

[10] J. C. Baez, A. Baratin, L. Freidel, and D. Wise. Infinite-dimensional representations of 2-groups. arXiv:0812.4969.

[11] J. C. Baez and J. Dolan. Higher-dimensional algebra and topological quantum field theory. *Journal of Mathematical Physics*, **36** (1995), 6073–105. arXiv:q-alg/9503002.

[12] J. C. Baez and J. Dolan. Higher-dimensional algebra III: $n$-categories and the algebra of opetopes. *Advances in Mathematics*, **135** (1998), 145–206. arXiv:q-alg/9702014.

[13] J. C. Baez and L. Langford. Higher-dimensional algebra IV: 2-tangles. *Advances in Mathematics*, **180** (2003), 705–64. arXiv:math/9811139.

[14] J. C. Baez and J. P. May. To appear in *Towards Higher Categories*, eds. $n$-Categories Foundations and Applications Berlin: Springer, 2009. Also available at http://ncatlab.org/johnbaez/show/Towards+Higher+Categories.

[15] J. C. Baez and M. Neuchl. Higher-dimensional algebra I: Braided monoidal 2-categories. *Advances in Mathematics*, **121** (1996), 196–244.

[16] J. C. Baez and U. Schreiber. Higher gauge theory. In *Categories in Algebra, Geometry and Mathematical Physics*, eds. A. Davydov et al. *Contemporary Mathematics*, vol. **431**, Providence, RI: AMS, 2007, pp. 7–30.

[17] J. C. Baez and M. Shulman. Lectures on $n$-categories and cohomology, section 5.6: Pointedness versus connectedness. To appear in *Towards Higher Categories*, eds. J. Baez and May. arXiv:math/0608420.

[18] B. Bakalov and A. Kirillov, Jr. *Lectures on Tensor Categories and Modular Functors*. Providence, RI: AMS, 2001. Preliminary version available at http://www.math.sunysb.edu/~kirillov/tensor/tensor.html.

[19] D. Bar-Natan. On Khovanov's categorification of the Jones polynomial. *Algebraic and Geometric Topology*, **2** (2002), 337–70. arXiv:math/0201043.

[20] D. Bar-Natan. Khovanov's homology for tangles and cobordisms. *Geometry and Topology*, **9** (2005), 1443–99. Also available as arXiv:math/0410495.

[21] J. W. Barrett and B. W. Westbury. Invariants of piecewise-linear 3-manifolds. *Transactions of the AMS*, **348** (1996), 3997–4022. arXiv:hep-th/9311155.

[22] J. W. Barrett and B. W. Westbury. Spherical categories. *Advances in Mathematics*, **143** (1999) 357–75. arXiv:hep-th/9310164.

[23] B. Bartlett. On unitary 2-representations of finite groups and topological quantum field theory, Ph.D. thesis, U. Sheffield, 2008. arXiv:0901.3975.

[24] C. Barwick. *Weakly Enriched M-Categories*. In preparation.

[25] M. Batanin. Monoidal globular categories as natural environment for the theory of weak $n$-categories. *Advances in Mathematics*, **136** (1998), 39–103.

[26] M. Batanin. The symmetrisation of $n$-operads and compactification of real configuration spaces. *Advances in Mathematics*, **211** (2007), 684–725. arXiv:math/0301221.

[27] M. Batanin. The Eckmann–Hilton argument and higher operads. *Advances in Mathematics*, **217** (2008), 334–85. arXiv:math/0207281.

[28] M. Batanin. On Penon method of weakening algebraic structures. *Journal of Pure and Applied Algebra*, **172** (2002), 1–23.

[29] M. Batanin. Configuration spaces from combinatorial, topological, and categorical perspectives. Lecture at Macquarie University, September 27, 2006. Available at http://www.maths.mq.edu.au/~street/BatanAustMSMq.pdf.

[30] R. J. Baxter. Solvable eight vertex model on an arbitrary planar lattice. *Philosophical Transactions of the Royal Society of London*, **289** (1978), 315–46.

[31] J. Bénabou. Introduction to bicategories. In *Reports of the Midwest Category Seminar*. Berlin: Springer-Verlag, 1967, pp. 1–77.

[32] C. Berger. A cellular nerve for higher categories. *Advances in Mathematics*, **169** (2002), 118–75. Also available at http://math.ucr.edu.fr/~cberger/.

[33] C. Berger. Iterated wreath product of the simplex category and iterated loop spaces. *Advances in Mathematics*, **213** (2007), 230–70. arXiv:math/0512575.

[34] J. Bergner. A survey of $(\infty, 1)$-categories. To appear in *Towards Higher Categories*, eds. $n$-categories foundation and applications Berlin: Springer, 2009. Baez and May. arXiv:math/0610239.

[35] H. Bethe. On the theory of metals. 1. Eigenvalues and eigenfunctions for the linear atomic chain. *Zeitschrift fur Physik*, **71** (1931), 205–26.

[36] J. M. Boardmann and R. M. Vogt. *Homotopy Invariant Structures on Topological Spaces*. Lecture Notes in Mathematics, vol. 347. Berlin: Springer-Verlag, 1973.

[37] R. Brown. Out of line. *Royal Institute Proceedings*, **64** (1992), 207–43. Also available at http://www.bangor.ac.uk/~mas010/outofline/out-home.html.

[38] S. Carlip. Quantum gravity in 2+1 dimensions: The case of a closed universe. *Living Reviews in Relativity*, **8**, No. 1 (2005). Available at http://www.livingreviews.org/lrr-2005-1.

[39] J. S. Carter, L. H. Kauffman, and M. Saito. Structures and diagrammatics of four dimensional topological lattice field theories. *Advances in Mathematics*, **146** (1999), 39–100. arXiv:math/9806023.

[40] J. S. Carter, J. H. Rieger, and M. Saito. A combinatorial description of knotted surfaces and their isotopies. *Advances in Mathematics*, **127** (1997), 1–51.

[41] J. S. Carter and M. Saito. *Knotted Surfaces and Their Diagrams*. Providence, RI: AMS, 1998.

[42] V. Chari and A. Pressley. *A Guide to Quantum Groups*. Cambridge: Cambridge U. Press, 1995.

[43] E. Cheng. The category of opetopes and the category of opetopic sets. *Theory and Applications of Categories*, **11** (2003), 353–74. Available at http://www.tac.mta.ca/tac/volumes/11/16/11-16abs.html. arXiv:math/0304284.

[44] E. Cheng. Weak $n$-categories: Opetopic and multitopic foundations. *Journal of Pure and Applied Algebra*, **186** (2004), 109–37. arXiv:math/0304277.

[45] E. Cheng. Weak $n$-categories: Comparing opetopic foundations. *Journal of Pure Applied Algebra*, **186** (2004), 219–31. arXiv:math/0304279.

[46] E. Cheng. Opetopic bicategories: Comparison with the classical theory (2003). arXiv:math/0304285.

[47] E. Cheng. Comparing operadic theories of $n$-category. arXiv:0809.2070.

[48] E. Cheng and N. Gurski. The periodic table of $n$-categories for low dimensions I: Degenerate categories and degenerate bicategories (2008). arXiv:0708.1178.

[49] E. Cheng and N. Gurski. Toward an $n$-category of cobordisms. *Theory and Application of Categories*, **18** (2007), 274–302. Available at http://www.tac.mta.ca/tac/volumes/18/10/18-10abs.html.

[50] E. Cheng and A. Lauda. *Higher-Dimensional Categories: An Illustrated Guidebook*. Available at http://www.cheng.staff.shef.ac.uk/guidebook/index.html.

## REFERENCES

[51] E. Cheng and M. Makkai. A note on the Penon definition of $n$-category. To appear in *Cahiors Top. Géom. Diff.*

[52] J. Chuang and R. Rouquier. Derived equivalences for symmetric groups and $\mathfrak{sl}_2$-categorification. *Annals of Mathematics*, **167** (2008), 245–98. arXiv:math/0407205.

[53] S. Chung, M. Fukuma, and A. Shapere. Structure of topological lattice field theories in three dimensions. *International Journal of Modern Physics*, **A9** (1994), 1305–60. arXiv:hep-th/9305080.

[54] D. -C. Cisinski. Batanin higher groupoids and homotopy types. *In Categories in algebra, geometry and mathematical physics*, eds. A. Davydov, et al. *Contemporary Mathematics*, vol. **431**. Providence, RI: A. M. S., 2007, pp. 171–86. arXiv:math/0604442.

[55] D. Clark, S. Morrison, and K. Walker. Fixing the functoriality of Khovanov homology. *Geometric Topology*, **13** (2009), 1499–582. arXiv:math/0701339.

[56] K. Costello. Topological conformal field theories and Calabi–Yau categories. *Advances in Mathematics*, **210** (2007). arXiv:math.QA/0412149.

[57] L. Crane. Clock and category: Is quantum gravity algebraic? *Journal of Mathematical Physics*, **36** (1995), 6180–93. arXiv:gr-qc/9504038.

[58] L. Crane and I. B. Frenkel. Four-dimensional topological quantum field theory, Hopf categories, and the canonical bases. *Journal of Mathematical Physics*, **35** (1994), 5136–54. arXiv:hep-th/9405183.

[59] L. Crane, L. Kauffman, and D. Yetter. State-sum invariants of 4-manifolds I. arXiv:hep-th/9409167.

[60] L. Crane and D. Yetter. A categorical construction of 4D topological quantum field theories. In *Quantum Topology*. Singapore: World Scientific, 1993, pp. 120–30. arXiv:hep-th/9301062.

[61] S. E. Crans. Generalized centers of braided and sylleptic monoidal 2-categories. *Advances in Mathematics*, **136** (1998), 183–223.

[62] P. Cvitanović. *Group Theory*. Princeton, NJ: Princeton University Press, 2003. Available at http://birdtracks.eu/.

[63] A. Davydov, M. Batanin, M. Johnson, S. Lack, and A. Neeman, eds. Categories in algebra, geometry and mathematical physics. *Contemporary Mathematics*, **431**. Providence, RI: AMS, 2007.

[64] B. Day and R. Street. Monoidal bicategories and Hopf algebroids. *Advances in Mathematics*, **129** (1997), 99–157.

[65] P. Di Francesco, P. Mathieu, and D. Senechal. *Conformal Field Theory*. Berlin: Springer-Verlag, 1997.

[66] A. Dick. *Emmy Noether: 1882–1935*, trans. H. I. Blocher. Boston: Birkhäuser, 1981.

[67] R. Dijkgraaf. *A Geometric Approach to Two-Dimensional Conformal Field Theory*. Ph.D. thesis. Utrecht: University of Utrecht, 1989.

[68] S. K. Donaldson and P. B. Kronheimer. *The Geometry of Four-Manifolds*. Oxford: Oxford U. Press, 1991.

[69] S. Doplicher and J. Roberts. A new duality theory for compact groups. *Inventiones Mathematicae*, **98** (1989), 157–218.

[70] S. Doplicher and J. Roberts. Why there is a field algebra with a compact gauge group describing the superselection in particle physics. *Communications in Mathematical Physics*, **131** (1990), 51–107.

[71] V. G. Drinfel'd. Quantum groups. In *Proceedings of the International Congress of Mathematicians*, Vol. 1 (Berkeley, CA, 1986). Providence, RI: AMS, 1987, pp. 798–820.

[72] F. J. Dyson. The radiation theories of Tomonaga, Schwinger, and Feynman. *Physics Reviews*, **75** (1949), 486–502.

[73] F. J. Dyson. The S matrix in quantum electrodynamics. *Physical Review*, **75** (1949), 1736–55.

[74] B. Eckmann and P. Hilton. Group-like structures in categories. *Mathematische Annalen*, **145** (1962), 227–55.

[75] S. Eilenberg and S. Mac Lane. General theory of natural equivalences. *Transactions of the AMS*, **58** (1945), 231–94.

[76] A. Einstein. Zur Elektrodynamik bewegter Körper. *Annalen der Physik*, **17** (1905), 891–921. Reprinted in *The Principle of Relativity*, trans. W. Perrett and G. B. Jeffrey. New York: Dover Publications, 1923.

[77] P. Etinghof and F. Latour. *The Dynamical Yang–Baxter Equation, Representation Theory, and Quantum Integrable Systems*. Oxford: Oxford U. Press, 2005.

[78] Z. F. Ezawa. *Quantum Hall Effects*, Singapore: World Scientific, 2008.

[79] L. D. Faddeev, E. K. Sklyanin, and L. A. Takhtajan. The quantum inverse problem method. *Theoretical and Mathematical Physics*, **40** (1980), 688–706.

[80] R. Feynman. The theory of positrons. *Physics Review*, **76** (1949), 749–59.

[81] R. Feynman. Space-time approach to quantum electrodynamics. *Physical Review*, **76** (1949), 769–89.

[82] R. Feynman. The reason for antiparticles. In *Elementary Particles and the Laws of Physics: The 1986 Dirac Memorial Lectures*. Cambridge: Cambridge U. Press, 1986, pp. 1–60.

[83] J. Fjelstad, J. Fuchs, I. Runkel, and C. Schweigert. Topological and conformal field theory as Frobenius algebras. In *Categories in algebra, geometry and mathematical physics*, eds. A. Davydov, et al. *Contemporary Mathematics*, **431**, Providence, RI: AMS, 2007. arXiv:math/0512076.

[84] D. S. Freed. Higher algebraic structures and quantization. *Communications in Mathematical Physics*, **159** (1994), 343–98. arXiv:hep-th/9212115.

[85] D. S. Freed and Karen Uhlenbeck. *Instantons and Four-Manifolds*. Berlin: Springer-Verlag, 1994.

[86] L. Freidel and D. Louapre. Ponzano–Regge model revisited. I: Gauge fixing, observables and interacting spinning particles. *Classical and Quantum Gravity*, **21** (2004), 5685–726. arXiv:hep-th/0401076.

[87] E. Frenkel. Gauge theory and Langlands duality. *Séminaire Bourbaki*, **61** (2009). arXiv:0906.2747.

[88] P. Freyd and D. Yetter. Braided compact closed categories with applications to low dimensional topology. *Advances in Mathematics*, **77** (1989), 156–82.

[89] P. Freyd, D. Yetter, J. Hoste, W. B. R. Lickorish, K. Millett, and A. Oceanu. A new polynomial invariant of knots and links. *Bulletin of the AMS*, **12** (1985), 239–46.

[90] J. Fuchs, I. Runkel, and C. Schweigert. Categorification and correlation functions in conformal field theory. arXiv:math/0602079.

[91] J. Fuchs and C. Schweigert. Category theory for conformal boundary conditions. *Fields Institute Communications*, **39** (2003), 25–71. arXiv:math.CT/0106050.

[92] M. Fukuma, S. Hosono, and H. Kawai. Lattice topological field theory in two dimensions. *Communications in Mathematical Physics*, **161** (1994), 157–75. arXiv:hepth/9212154.

[93] E. Getzler and M. Kapranov, eds. Higher category theory. *Contemporary Mathematics*, vol. **230**, Providence, RI: AMS, 1998.

[94] R. Gordon, A. J. Power, and R. Street. Coherence for tricategories. *Memoirs of the American Mathematical Society*, **558** (1995).

[95] A. C. Gossard, D. C. Tsui, and H. L. Störmer. Two-dimensional magnetotransport in the extreme quantum limit. *Physical Review Letters*, **48** (1982), 1559–62.

[96] J. W. Gray. Formal category theory: Adjointness for 2-Categories. Lecture Notes in Mathematics, vol. **391**. Berlin: Springer-Verlag, 1974.

[97] M. B. Green, J. H. Schwarz, and E. Witten. *Superstring Theory* (2 volumes). Cambridge: Cambridge University Press, 1987.

[98] N. Greenspan. *The End of the Certain World: The Life and Science of Max Born*. New York: Basic Books, 2005.

[99] I. Grojnowski and G. Lusztig. A comparison of bases of quantized enveloping algebras: In Linear algebraic groups and their representations. *Contemporary Mathematics*, **153**, 1993, pp. 11–9.

[100] A. Grothendieck. *Pursuing Stacks*, letter to D. Quillen, 1983. To appear in eds. G. Maltsiniotis, M. Künzer, and B. Toen, *Documents Mathématiques*, Society Mathematics Paris, France.

[101] W. Heisenberg. Multi-body problem and resonance in the quantum mechanics. *Zeitschrift für Physik*, **38** (1926), 411–26.

[102] M. Jacobsson. An invariant of link cobordisms from Khovanov homology. *Algebraic and Geometric Topology*, **4** (2004), 1211–51. arXiv:math/0206303.

[103] M. Jimbo. A $q$-difference analogue of $U(\mathfrak{g})$ and the Yang–Baxter equation. *Letter in Mathematical Physics*, **10** (1985), 63–9.

[104] V. Jones. A polynomial invariant for knots via von Neumann algebras. *Bulletin of the AMS*, **12** (1985), 103–11.

[105] V. Jones. Planar algebras, I. arXiv:math/9909027.

[106] A. Joyal. Disks, duality and $\theta$-categories. In preparation.

[107] A. Joyal. Quasi-categories and Kan complexes. *Journal of Pure Applied Algebra*, **175** (2002), 207–22.

[108] A. Joyal and R. Street. Braided monoidal categories, *Macquarie Math Reports* 860081 (1986). Available at http://rutherglen.ics.mq.edu.au/~street/.

[109] A. Joyal and R. Street. An introduction to Tannaka duality and quantum groups, in Part II of *Category Theory, Proceedings, Como 1990*, eds. A. Carboni, et al. Lecture Notes in Mathematics, vol. **1488**. Berlin: Springer-Verlag, 1991, pp. 411–92. Available at http://www.maths.mq.edu.au/~street/.

[110] A. Joyal and R. Street. The geometry of tensor calculus I. *Advances in Mathematics*, **88** (1991), 55–113.

[111] A. Joyal and M. Tierney. Algebraic homotopy types. Handwritten lecture notes, 1984.

[112] D. Kaiser. *Drawing Theories Apart: The Dispersion of Feynman Diagrams in Postwar Physics*. Chicago: U. Chicago Press, 2005.

[113] D. M. Kan. On c.s.s. complexes. *Annals of Mathematics*, **79** (1957), 449–76.

[114] M. Kapranov and V. Voevodsky. 2-Categories and Zamolodchikov tetrahedra equations. In Algebraic groups and their generalizations. *Proceedings of the Symposium Pure Mathematics*, vol. **56**, Providence, RI: AMS, 1994, pp. 177–260.

[115] M. Kashiwara. Crystalizing the $q$-analogue of universal enveloping algebras. *Communications in Mathematical Physics*, **133** (1990), 249–60.

[116] M. Kashiwara. On crystal bases of the $q$-analogue of universal enveloping algebras. *Duke Mathematical Journal*, **63** (1991), 465–516.

[117] M. Kashiwara. Global crystal bases of quantum groups. *Duke Mathematical Journal*, **69** (1993), 455–85.

[118] L. H. Kauffman. State models and the Jones polynomial. *Topology*, **26** (1987), 395–407.

[119] L. H. Kauffman. Spin networks and the bracket polynomial. *Banach Center Publications*, **42** (1998), 187–204. Also available at http://www.math.uic.edu/~kauffman/spinnet.pdf.

[120] L. H. Kauffman. *Knots and Physics*. Singapore: World Scientific, 2001.

[121] C. Kassel. *Quantum Groups*. Berlin: Springer-Verlag, 1995.

[122] B. Keller. Introduction to $A_\infty$-algebras and modules. *Homology, Homotopy and Applications*, **3** (2001), 1–35. Available at http://www.intlpress.com/HHA/v3/n1/.

[123] T. Kerler and V. V. Lyubashenko. Non-Semisimple Topological Quantum Field Theories for 3-Manifolds with Corners. *Lecture Notes in Mathematics*, vol. **1765**. Berlin: Springer-Verlag, 2001.

[124] J. Kock. *Frobenius Algebras and 2d Topological Quantum Field Theories*. London Mathematical Society Student Texts, vol. **59**. Cambridge: Cambridge U. Press, 2004.

[125] T. Kohno, ed. *New Developments in the Theory of Knots*. Singapore: World Scientific, 1990.

[126] M. Khovanov. A categorification of the Jones polynomial. *Duke Mathematical Journal*, **101** (2000), 359–426. arXiv:math/9908171.

[127] M. Khovanov. An invariant of tangle cobordisms. *Transactions of the AMS*, **358** (2006), 315–27. arXiv:math/0207264.

[128] M. Khovanov. Hopfological algebra and categorification at a root of unity: The first steps. arXiv:math/0509083.

[129] M. Khovanov and A. Lauda. A diagrammatic approach to categorification of quantum groups I. *Representation Theory*, **13** (2009), 309–47. arXiv:0803.4121.

[130] M. Khovanov and A. Lauda. A diagrammatic approach to categorification of quantum groups III. arXiv:0807.3250.

[131] M. Kontsevich. Vassiliev's knot invariants. *Adv. Soviet Math.*, **16** (1993), 137–50.

[132] M. Kontsevich. Homological algebra of mirror symmetry. *Proceedings of the International Congress of Mathematicians (Zürich, 1994)*. Boston: Birkhäuser, 1995, pp. 120–39. arXiv:alg-geom/9411018.

[133] M. Kontsevich. Operads and motives in deformation quantization. *Letters in Mathematical Physics*, **48** (1999), 35–72. arXiv:math/9904055.

[134] M. Kontsevich and Y. Soibelman. Notes on $A_\infty$-algebras, $A_\infty$ categories and non-commutative geometry I. arXiv:math/0606241.

[135] P. B. Kronheimer and T. S. Mrowka. Gauge theory for embedded surfaces I. *Topology*, **32** (1993), 773–826. Available at http://www.math.harvard.edu/~kronheim/papers.html.

[136] G. Kuperberg. Involutory Hopf algebras and 3-manifold invariants. *International Journal of Mathematics*, **2** (1991), 41–66. arXiv:math/9201301.

[137] S. Mac Lane. Natural associativity and commutativity. *Rice University Studies*, **49** (1963), 28–46.

[138] S. Mac Lane. Categorical algebra. *Bulletin of the AMS*, **71** (1965), 40–106. Available at http://projecteuclid.org/euclid.bams/1183526392.

[139] A. D. Lauda. A categorification of quantum sl(2). arXiv:0803.3652.

[140] A. D. Lauda and H. Pfeiffer. Open-closed strings: Two-dimensional extended TQFTs and Frobenius algebras. arXiv:math.AT/0510664.

[141] A. D. Lauda and H. Pfeiffer. State sum construction of two-dimensional open-closed topological quantum field theories. arXiv:math.QA/0602047.

[142] R. B. Laughlin. Anomalous quantum Hall effect: An incompressible quantum fluid with fractionally charged excitations. *Physical Review Letters*, **50** (1983), 1395.

[143] R. J. Lawrence. Triangulations, categories and extended topological field theories. In *Quantum Topology*. Singapore: World Scientific, 1993, pp. 191–208.

[144] R. J. Lawrence. Algebras and triangle relations. *Journal of Pure Applied Algebra*, **100** (1995), 43–72.

[145] F. W. Lawvere. *Functorial Semantics of Algebraic Theories*, Ph.D. thesis, Columbia University, 1963. Reprinted in *Theory and Applications of Categories*, **5** (2004), 1–121. Available at http://www.tac.mta.ca/tac/reprints/articles/5/tr5abs.html.

[146] T. Leinster. Structures in higher-dimensional category theory. arXiv:math/0109021.

[147] T. Leinster. A survey of definitions of $n$-category. *Theory and Applications of Categories*, **10** (2002), 1–70. Available at http://www.tac.mta.ca/tac/volumes/10/1/10-01abs.html. arXiv:math/0107188.

[148] T. Leinster. *Higher Operads, Higher Categories*. Cambridge: Cambridge University Press, 2003. arXiv:math.CT/0305049.

[149] T. Leinster. Higher operads, higher categories. London Mathematical Society Lecture Note Series, vol. **298**. Cambridge: Cambridge University Press, 2004. arXiv:math/0305049.

[150] J. -L. Loday, J. D. Stasheff, and A. A. Voronov, eds., *Operads: Proceedings of Renaissance Conferences*. Providence, RI: AMS, 1997.

[151] H. A. Lorentz. Electromagnetic phenomena in a system moving with any velocity less than that of light. *Proceedings of the Academy of Science Amsterdam*, **IV** (1904), 669–78.

[152] J. Lurie. On the classification of topological field theories. arXiv:0905.0465.

[153] J. Lurie. *Higher Topos Theory*. Princeton, NJ: Princeton University Press, 2009. arXiv:math/0608040.

[154] G. Lusztig. Canonical bases arising from quantized enveloping algebras. *Journal of the AMS*, **3** (1990), 447–98.

[155] G. Lusztig. Quivers, perverse sheaves, and quantized enveloping algebras. *Journal of the AMS*, **4** (1991), 365–421.

[156] G. Lusztig. *Introduction to Quantum Groups*. Boston: Birkhäuser, 1993.

[157] M. Mackaay. Spherical 2-categories and 4-manifold invariants. *Advances in Mathematics*, **143** (1999), 288–348. arXiv:math/9805030.

[158] S. Majid. *Foundations of Quantum Group Theory*. Cambridge: Cambridge University Press, 1995.

[159] S. Major. A spin network primer. *American Journal of Physics*, **67** (1999), 972–80. arXiv:gr-qc/9905020.

[160] M. Makkai. The multitopic $\omega$-category of all multitopic $\omega$-categories. Available at http://www.math.mcgill.ca/makkai.

[161] M. Makkai. On comparing definitions of "weak $n$-category." Available at http://www.math.mcgill.ca/makkai.

[162] M. Makkai, C. Hermida, and J. Power. On weak higher-dimensional categories I, II. *Journal of Pure Applied Algebra*, **157** (2001), 221–77.

[163] C. Manolescu, P. Ozsváth, and S. Sarkar. A combinatorial description of knot Floer homology. *Annals of Mathematics*, **169** (2009), 633–60. arXiv:math/0607691.

[164] C. Manolescu, P. Ozsváth, Z. Szabó, and D. Thurson. On combinatorial link Floer homology. *Geometry and Topology*, **11** (2007), 2339–412. arXiv:math/0610559.

[165] M. Markl, S. Shnider, and J. D. Stasheff. *Operads in Algebra, Topology and Physics*, Providence, RI: AMS, 2002.

[166] J. C. Maxwell. *Matter and Motion*. London: Society for Promoting Christian Knowledge, 1876. Reprinted in New York: Dover, 1952.

[167] J. P. May. *The Geometry of Iterated Loop Spaces*. Lecture Notes in Mathematics, vol. **271**. Berlin: Springer-Verlag, 1972.

[168] P. McCrudden. Balanced coalgebroids. *Theory and Applications of Categories*, **7** (2000), 71–147. Available at http://www.tac.mta.ca/tac/volumes/7/n6/7-06abs.html.

[169] J. Mehra. *The Beat of a Different Drum: The Life and Science of Richard Feynman*. Oxford: Clarendon Press, 1994.

[170] J. Mehra. *The Formulation of Matrix Mechanics and Its Modifications 1925–1926*. Berlin: Springer-Verlag, 2000.

[171] G. Moore and G. B. Segal. *Lectures on Branes, K-Theory and RR Charges*. In Lecture Notes from the Clay Institute School on Geometry and String Theory held at the Isaac Newton Institute, Cambridge, UK, 2001–2002. Available at http://www.physics.rutgers.edu/~gmoore/clay1/clay1.html and at http://online.itp.ucsb.edu/online/mp01/moore1/.

[172] J. W. Morgan. *The Seiberg–Witten Equations and Applications to the Topology of Smooth Four-Manifolds*. Princeton, NJ: Princeton University Press, 1996.

[173] J. W. Morgan and R. Friedman, eds. *Gauge Theory and the Topology of Four-Manifolds*. Providence, RI: AMS, 1997.

[174] J. Morton. Double bicategories and double cospans. To appear in *Journal of Homotopy and Related Structures*. arXiv:math/0611930.

[175] nLab, Trimble n-category. Available at http://ncatlab.org/nlab/show/Trimble+n-category.

[176] M. Neuchl. *Representation Theory of Hopf Categories*, Ph.D. dissertation, Department of Mathematics, University of Munich, 1997. Available at http://math.ucr.edu/home/baez/neuchl.ps.

[177] J. von Neumann. *Mathematische Grundlager der Quantenmechanik*. Berlin: Springer-Verlag, 1932.

[178] L. Onsager. Crystal statistics. 1. A Two-dimensional model with an order disorder transition. *Physical Review*, **65** (1944), 117–49.

[179] H. Ooguri. Topological lattice models in four dimensions. *Modern Physics Letters A*, **7** (1992), 2799–810.

[180] P. Ozsváth and Z. Szabó. Holomorphic disks and topological invariants for closed three-manifolds. *Annals of Mathematics*, **159** (2004), 1027–158. arXiv:math/0101206.

[181] U. Pachner, P. L. homeomorphic manifolds are equivalent by elementary shelling. *European Journal of Combinatorics*, **12** (1991), 129–45.

[182] A. Pais. *Subtle Is the Lord: The Science and the Life of Albert Einstein*. Oxford: Oxford University Press, 1982, section 12c.

[183] S. Paoli. Semistrict models of connected 3-types and Tamsamani's weak 3-groupoids. arXiv:0607330.

[184] S. Paoli. Semistrict Tamsamani $n$-groupoids and connected $n$-types. arXiv:0701655.

[185] J. Penon. Approche polygraphique des $\infty$-categories non strictes. *Cahiers Top. Géom. Diff.*, **40** (1999), 31–80.

[186] R. Penrose. Applications of negative dimensional tensors. In *Combinatorial Mathematics and its Applications*, ed. D. Welsh. New York: Academic Press, 1971, pp. 221–44.

[187] R. Penrose. Angular momentum: An approach to combinatorial spacetime. In *Quantum Theory and Beyond*, ed. T. Bastin. Cambridge: Cambridge University Press, 1971, pp. 151–80.

[188] R. Penrose. On the nature of quantum geometry. In *Magic Without Magic*, ed. J. Klauder. New York: Freeman, 1972, pp. 333–54.

[189] R. Penrose. Combinatorial quantum theory and quantized directions. In *Advances in Twistor Theory*, eds. L. Hughston and R. Ward. Boston: Pitman Advanced Publishing Program, 1979, pp. 301–17.

[190] H. Pfeiffer. 2-Groups, trialgebras and their Hopf categories of representations. *Advances in Mathematics*, **212** (2007), 62–108. arXiv:math.QA/0411468.

[191] H. Poincaré. Sur la dynamique de l'electron. *Comptes Rendus de l'Academie des Sciences*, **140** (1905), 1504–8.

[192] J. Polchinski. *String Theory* (2 vol.). Cambridge: Cambridge University Press, 1998.

[193] G. Ponzano and T. Regge. Semiclassical limits of Racah coefficients. In *Spectroscopic and Group Theoretical Methods in Physics: Racah Memorial Volume*, ed. F. Bloch. Amsterdam: North-Holland, 1968, pp. 75–103.

[194] A. J. Power. Why tricategories? *Information and Computation*, **120** (1995), 251–62. Available at http://www.lfcs.inf.ed.ac.uk/reports/94/ECS-LFCS-94-289/.

[195] J. Przytycki and P. Traczyk. Conway algebras and skein equivalence of links. *Proceedings of the AMS*, **100** (1987), 744–8.

[196] J. Rasmussen. Khovanov homology and the slice genus. To appear in *Inventiones Mathematicae*. arXiv:math/0402131.

[197] J. Rasmussen. Knot polynomials and knot homologies. arXiv:math/0504045.

[198] M. Reisenberger and C. Rovelli. "Sum over surfaces" form of loop quantum gravity. *Physical Review*, **D56** (1997), 3490–508. arXiv:gr-qc/9612035.

[199] N. Reshetikhin and V. Turaev. Ribbon graphs and their invariants derived from quantum groups. *Communications in Mathematical Physics*, **127** (1990), 1–26.

[200] N. Reshetikhin and V. Turaev. Invariants of 3-manifolds via link-polynomials and quantum groups. *Inventiones Mathematicae*, **103** (1991), 547–97.

[201] C. Rezk. A model for the homotopy theory of homotopy theory. *Transaction American Mathematical Society*, **353** (2001), 973–1007. Available at http://www.ams.org/tran/2001-353-03/S0002-9947-00-02653-2/.

[202] J. Roberts. Mathematical aspects of local cohomology. In *Algèbres d'Opérateurs et Leurs Applications en Physique Mathématique*. Paris: CNRS, 1979, pp. 321–32.

[203] J. Roberts. Classical $6j$-symbols and the tetrahedron. *Geometry and Topology*, **3** (1999), 21–66.

[204] J. Roberts. Skein theory and Turaev–Viro invariants. *Topology*, **34** (1995), 771–87. Also available at http://math.ucsd.edu/~justin/papers.html.

[205] J. Rotman. *An Introduction to Homological Algebra*. New York: Academic Press, 1979.

[206] R. Rouquier. 2-Kac-Moody algebras. arXiv:0812.5023.

[207] C. Rovelli. Loop quantum gravity. *Living Reviews in Relativity*, **11**, (2008), 5. Available at http://relativity.livingreviews.org/Articles/lrr-2008-5/.

[208] C. Rovelli. *Quantum Gravity*. Cambridge: Cambridge University Press, 2004.

[209] C. Rovelli and L. Smolin. Loop space representation of quantum general relativity. *Nuclear Physics B*, **331** (1990), 80–152.

[210] C. Rovelli and L. Smolin. Spin networks and quantum gravity. *Physical Review*, **D52** (1995), 5743–59. arXiv:gr-qc/9505006.

[211] C. Schommer-Pries. *The Classification of Two-Dimensional Extended Topological Quantum Field Theories*. U. C. Berkeley: Ph.D. thesis, 2009. Also available at http://sites.google.com/site/chrisschommerpriesmath/.

[212] U. Schreiber. AQFT from $n$-functorial QFT. arXiv:0806.1079.

[213] A. Scorpan. *The Wild World of 4-Manifolds*. Providence, RI: AMS, 2005.

[214] G. B. Segal. The definition of conformal field theory. In *Topology, Geometry and Quantum Field Theory*. London Mathematical Society Lecture Note Series, **308**. Cambridge: Cambridge University Press, 2004, pp. 421–577.

[215] G. B. Segal. Topological structures in string theory. *Philosophical Transactions of the Royal Society of London*, **359**, No. 1784 (2001), 1389–98.

[216] N. Seiberg and E. Witten. Electric-magnetic duality, monopole condensation, and confinement in $N = 2$ supersymmetric Yang-Mills theory. *Nuclear Physics*, **B426** (1994), 19–52.

[217] N. Seiberg and E. Witten. Monopoles, duality and chiral symmetry breaking in $N = 2$ supersymmetric QCD. *Nuclear Physics*, **B431** (1994), 484–550.

[218] P. Selinger. Dagger compact closed categories and completely positive maps. In *Proceedings of the 3rd International Workshop on Quantum Programming Languages (QPL 2005). Electronic Notes in Theoretical Computer Science*, **170** (2007), 139–63. Also available at http://www.mscs.dal.ca/~selinger/papers.html\#dagger.

[219] M. -C. Shum. Tortile tensor categories. *Journal of Pure and Applied Algebra*, **93** (1994), 57–110.

[220] C. Simpson. A closed model structure for $n$-categories, internal Hom, $n$-stacks and generalized Seifert–Van Kampen. arXiv:alg-geom/9704006.

[221] C. Simpson. Limits in $n$-categories. arXiv:alg-geom/9708010.

[222] C. Simpson. Calculating maps between $n$-categories. arXiv:math/0009107.

[223] C. Simpson. On the Breen–Baez–Dolan stabilization hypothesis for Tamsamani's weak $n$-categories. arXiv:math/9810058.

[224] C. Simpson. Some properties of the theory of $n$-categories. arXiv:math/0110273.

[225] J. Stasheff. Homotopy associative $H$-spaces I, II. *Transactions of the AMS*, **108** (1963), 275–312.

[226] R. Street. The algebra of oriented simplexes. *Journal of Pure and Applied Algebra*, **49** (1987), 283–335.

[227] R. Street. Weak omega-categories. In Diagrammatic morphisms and applications. *Contemporary Mathematics*, **318**. Providence, RI: AMS, 2003, pp. 207–13.

[228] R. Street. An Australian conspectus of higher categories. Available at http://www.math.mq.edu.au/~street/.

[229] R. Street. The role of Michael Batanin's monoidal globular categories. In *Higher Category Theory*, eds. E. Getzler and M. Kapranov, *Contemporary Mathematics*, **230**. Providence, RI: AMS, 1998, pp. 99–116. Also available at http://citeseerx.ist.psu.edu/viewdoc/summary?doi=10.1.1.51.2074.

[230] Z. Tamsamani. Sur des notions de $n$-catégorie et $n$-groupoide non-strictes via des ensembles multi-simpliciaux. *K-Theory*, **16** (1999), 51–99. arXiv:alg-geom/9512006.

[231] Z. Tamsamani. Equivalence de la théorie homotopique des $n$-groupoides et celle des espaces topologiques $n$-tronqués. arXiv:alg-geom/9607010.

[232] V. G. Turaev. Shadow links and face models of stastistical mechanics. *Journal of Differential Geometry*, **36** (1992), 35–74.

[233] V. G. Turaev. *Quantum Invariants of Knots and 3-Manifolds* Berlin: de Gruyter, 1994.

[234] V. G. Turaev. Topology of shadows. In preparation.

[235] V. G. Turaev and O. Y. Viro. State sum invariants of 3-manifolds and quantum $6j$-symbols. *Topology*, **31** (1992), 865–902.

[236] M. Varagnolo and E. Vasserot. Canonical bases and Khovanov–Lauda algebras. arXiv:0901.3992.

[237] D. Verity. Complicial sets: Characterizing the simplicial nerves of strict $\omega$-categories. *Memoirs of the American Mathematical Society*, **905**, 2005. arXiv:math/0410412.

[238] D. Verity. Weak complicial sets, a simplicial weak $\omega$-category theory. Part I: Basic homotopy theory. arXiv:math/0604414.

[239] D. Verity. Weak complicial sets, a simplicial weak $\omega$-category theory. Part II: Nerves of complicial Gray-categories. arXiv:math/0604416.

[240] C. Weibel. *An Introduction to Homological Algebra*. Cambridge: Cambridge University Press, 1994.

[241] E. Wigner. On unitary representations of the inhomogeneous Lorentz group. *Annals of Mathematics*, **40** (1939), 149–204.

[242] F. Wilczek. Quantum mechanics of fractional-spin particles. *Physical Review Letters*, **49** (1982), 957–9.

[243] E. Witten. Topological quantum field theory. *Communications in Mathematical Physics*, **117** (1988), 353–86. Also available at http://projecteuclid.org/euclid.cmp/1104161738.

[244] E. Witten. Quantum field theory and the Jones polynomial. *Communications in Mathematical Physics*, **121** (1989), 351–99.

[245] C. N. Yang and R. Mills. Conservation of isotopic spin and isotopic gauge invariance. *Physical Review*, **96** (1954) 191–5.

[246] C. N. Yang and C. P. Yang. One-dimensional chain of anisotropic spin-spin interactions. I: Proof of Bethe's hypothesis for ground state in a finite system; II: Properties of the ground state energy per lattice site for an infinite system. *Physical Review*, **150** (1966), 321–7, 327–39.

[247] D. Yetter. *Functorial Knot Theory: Categories of Tangles, Categorical Deformations and Topological Invariants*. Singapore: World Scientific, 2001.

[248] D. Z. Zhang. C. N. Yang and contemporary mathematics. *Mathematical Intelligencer*, **14** (1993), 13–21.

[249] B. Zwiebach. *A First Course in String Theory*. Cambridge: Cambridge University Press, 2004.

CHAPTER 2

# A Universe of Processes and Some of Its Guises

Bob Coecke

## 2.1 Introduction

Our starting point is a particular "canvas" aimed to "draw" theories of physics, which has *symmetric monoidal categories* as its mathematical backbone. In this chapter, we consider the *conceptual foundations* for this canvas and how these can then be converted into mathematical structure.

With very little structural effort (i.e., in very abstract terms) and in a very short time span, the *categorical quantum mechanics* (CQM) research program, initiated by Abramsky and this author [6], has reproduced a surprisingly large fragment of quantum theory [3, 45, 48, 56, 60–62, 170, 179]. It also provides new insights both in *quantum foundations* and in *quantum information*—for example, in [49–51, 58, 59, 64, 79, 80]—and has even resulted in automated reasoning software called *quantomatic* [71–73], which exploits the deductive power of CQM, which is indeed a *categorical quantum logic* [77].

In this chapter, we complement the available material by not requiring prior knowledge of category theory and by pointing at connections to previous and current developments in the foundations of physics.

This research program is also in close synergy with developments elsewhere—for example, in representation theory [74], quantum algebra [176], knot theory [187], topological quantum field theory (TQFT) [132], and several other areas.

Philosophically speaking, this framework achieves the following:

- It shifts the conceptual focus from "material carriers," such as particles, fields, or other "material stuff," to "logical flows of information," mainly by encoding how things stand in *relation* to each other.
- Consequently, it privileges *processes* over states. The chief structural ingredient of the framework is the *interaction structure* on processes.
- In contrast to other ongoing *operational* approaches ([55] and references therein, [66], [109]), we do not take probabilities, nor properties, nor experiments as a priori, nor as generators of structure, but everything is encoded within the interaction of processes.

- In contrast to other ongoing *structural* approaches ([55] and references therein, [9], [10], [19], [76], [113], [117], and so forth); we do not start from a notion of system, systems now being "plugs" within a web of interacting processes. Hence, systems are organized within a structure for which compoundness is a player and not the structure of the system itself: a system is implicitly defined in terms of its relation(ship)/interaction with other systems.

So, for us, *composition of relation(ship)s* is the carrier of all structure—that is, how several relations make up one whole. For example, if $x_1, x_2, x_3, a$ are in relation(ship) $R_1$ and $y_1, y_2, y_3, a$ are in relation(ship) $R_2$, then this induces a relation(ship) between $x_1, x_2, x_3, y_1, y_2, y_3$:

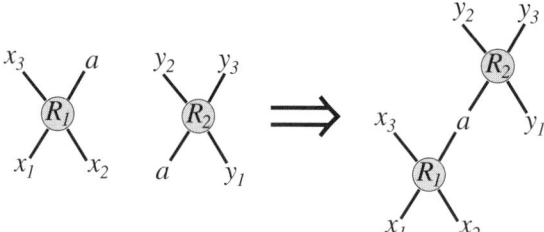

These relation(ship)s are much more general than the usual mathematical notion of a relation. A mathematical relation tells us only whether or not a thing relates to another thing, whereas for us also "the manner in which" a thing relates to another thing matters.

*Processes* are special kinds of relations that make up the actual "happenings." *Classicality* is an attribute of certain processes, and *measurements* are special kinds of processes, defined in terms of their capabilities to correlate other processes to these classical attributes.

So, rather than *quantization*, what we do is *classicization* within a universe of processes. For a certain theory, classicality and measurements may or may not exist because they are not a priori. For example, in analogy to "nonquantized field theories," one could consider *nonclassicized* theories within our setting.

Our attempt to spell out conceptual foundations is particularly timely given that other work in quantum foundations and ours are converging, most notably Hardy's recent work [107–109] and Chiribella, D'Ariano, and Perinotti's even more recent work [40,67]. Also, proponents of the "convex set approach" ([19], [20], and references therein) as well as those of the more traditional "Birkhoff–von Neumann style quantum logic" [27, 120, 144, 158] have meanwhile adopted an essential component of our framework [18, 22, 103, 111, 112, 115].

The mathematical flexibility of our framework allows one to craft hypothetical nonphysical universes, a practice that turns out to provide important insights in the theories of physics that we know and that recently gained popularity (e.g., [19], [23], [41], [174]). Such approaches provide an arena to explore how many physical phenomena arise within a theory from very few assumptions. Our approach has been particularly successful in this context, not only by producing many phenomena from little assumptions but also by casting theories that initially were defined within distinct mathematical frameworks within a single one. For example, it unifies quantum theory as well as Spekkens's toy theory [174] within a single framework [49], which enabled identification of the key differences [50] and also substantially simplified the presentation of the latter.

This chapter is structured as follows. Section 2.2 briefly sketches some earlier developments, be it because they provided ideas or because they exposed certain sources of failure. Section 2.3 introduces the primitives of our framework: systems, relations (and processes), and their composition. We show how these can be used to encode identical systems, symmetries and dynamics, variable causal structure, and an environment. Section 2.4 shows that in mathematical terms, these concepts give rise to symmetric monoidal categories. Next, in Section 2.5, we define classicality and measurement processes.

The author is not a professional philosopher but a "hell of a barfly," so the philosophical remarks throughout this chapter, of which there are plenty, should be taken with a grain of salt.

We are purposely somewhat vague on many fronts in order to leave several options available for the future; the reader is invited to fill in the blanks.

## 2.2 Some (Idiosyncratic) Lessons from the Past

In particular, we focus on the role of *operationalism* in quantum theory reconstructions, the *formal definition* of a physical property as proposed by the Geneva School (e.g., [122], [151]), the role of *processes* therein and forefront role of processes in quantum information, the manner in which algebraic quantum field theory (AQFT) [99, 100] retains the notion of a *system*, the modern logical view on the different guises of the *connective* "and" for systems, ideas of *relationalism* in physics [17, 167], the options of *discreteness* and *pointlessness* in quantum gravity, and the status of *foundations of mathematics* in all of this.

Although we make some reference to mathematical concepts in category theory, order theory, $C*$-algebra, quantum logic, linear logic, and quantum information, none of these is a prerequisite for the remainder of this chapter.

### 2.2.1 To Measure or Not to Measure

Although nature has not been created by us, the theories that describe it have been and, hence, unavoidably these will have to rely on concepts that make reference to our senses or some easy-to-grasp generalizations thereof. For example, as humans, we experience a three-dimensional space around us—hence, the important role of geometry in physics. Similarly, the symmetries that we observe around us have led to the importance of group theory in physics.

A fairly radical stance in this light is that of the typical *operationalist*. His or her take on quantum theory (and physics in general) is that measurement apparatuses constitute our window on nature. Different "schools" of operationalists each isolate an aspect of the measurement processes that they think causes the apparent nonclassicality of quantum theory, be it the structure of the space of probabilities or the structure of the verifiable propositions, and so forth.

This practice traces back to the early days of Hilbert space quantum mechanics. In *Mathematische Grundlagen der Quantenmechanik* [182], von Neumann stressed that the projectors make up self-adjoint operators that should be the fundamental ingredient of whatever formalism that describes the quantum world. Indeed, although he himself crafted Hilbert space quantum mechanics, he was also the first to denounce it in a letter to Birkhoff [26, 165]:

> I would like to make a confession which may seem immoral: I do not believe absolutely in Hilbert space any more.

This focus on projectors led to a sharp contrast with happenings in logic [27]:

> ... whereas for logicians the orthocomplementation properties of negation were the ones least able to withstand a critical analysis, the study of mechanics points to the distributive identities as the weakest link in the algebra of logic.

and ultimately resulted in Birkhoff–von Neumann quantum logic [27].

Via Mackey [143, 144], several structural paradigms emerged: the *Geneva School* [122, 158] inherited its lattice theoretic paradigm directly from Birkhoff and von Neumann; the *Ludwig School* [140, 141] associated the convex structure of state spaces attributed to experimental situations; and the *Foulis–Randall School* [88, 89, 185] considered the intersection structure of outcome spaces.

But this key role of the measurement process is rejected by many *realists* for whom physical properties of a system exist independent of any form of observation. For example, a star still obeys quantum theory even when not (directly) observed, and a red pencil does not stop being red when we are not observing it. More boldly put: Who measures the (entirely quantum) universe?[1]

The realist and operationalist views are typically seen as somewhat conflicting. But attributing properties to systems that are not being observed, while still subscribing to a clear operational meaning of basic concepts, was already explicitly realized within the Geneva School Mark II [11, 151]. Although its formal guise was still quite similar to Birkhoff–von Neumann quantum logic, the lattice structure is *derived* from an in-operational-terms precisely stated conception of "what it means for a system to possess a property."

The following example is from Aerts [11], and its pure classicality makes it intriguing in its own right. Consider a block of wood and the properties "floating" and "burning." If, with certainty, we want to know whether the block of wood possesses either of these properties, then we need to, respectively, throw it in the water and observe whether it floats, or set it on fire and observe whether it burns. Obviously, if we observed either, we altered the block of wood in such a manner that we will not be able anymore to observe the other. Still, it makes perfect sense for a block of wood to both be burnable and floatable.

In the Geneva School, one considers a system $A$ and the "yes/no" experiments $\{\alpha_i\}_i$ one can perform thereon. These experiments are related to each other in terms of a

---

[1] This utterance is regularly heard as a motivation for various histories interpretations [93, 98, 116], which, in turns, motivated the so-called topos approach to quantum theory [76, 113, 117]; we briefly discuss this approach at the end of this section.

preordering: for experiments $\alpha$ and $\beta$, we have that $\alpha \preceq \beta$ if and only if when we would perform $\alpha$ and obtain a "yes" answer with certainty, then we would also have obtained a "yes" answer with certainty when performing $\beta$. A *property* is then defined as an equivalence class for this preordering. The lattice structure on the induced partial ordering follows from the existence of certain product experiments.[2] Such a property is called *actual* if the physical system possesses it and *potential* otherwise.

### 2.2.2 Measurement among Other Processes

So, in the Geneva School Mark II, properties are a secondary notion emerging from considering experimental procedures on a given system $A$. The Geneva School Mark III emphasized the role of processes [42, 54, 63, 65, 86]. Faure, Moore, and Piron were able to derive unitarity of quantum evolution by cleverly exploiting the definition of a physical property.[3] Also, the (in)famous orthomodular law of quantum logic is about how properties propagate in measurement processes.[4] These results were a key motivation to organize physical processes within certain *categories*, which lift the operationally motivated lattice structure from systems to processes [54].

The crucial mathematical concept in the preceding is *Galois adjunctions*,[5] the order-theoretic counterpart to *adjoint functors* between categories [128]. These are by many category theoreticians considered the most important concept provided by category theory, in that almost all known mathematical constructions can be formulated in a very succinct manner in terms of these. Galois adjunctions were already implicitly present in the work by Pool in the late 1960s [160], which arguably was the first attempt to replace the quantum formalism by a formalism in which *processes* are the key players.[6]

From a more conceptual perspective, the idea that the structure of processes might help us to get a better understanding of nature was already present in the work of Whitehead in the 1950s [183] and the work of Bohr in the early 1960s [34]. It became

---

[2] The meet of a collection of properties arises from the experiment consisting of choosing among experiments that correspond to these properties [11, 151]. Because these are arbitrary meets, it also follows that the lattice has arbitrary joins (see, (e.g., [53].)

[3] Roughly, this argument goes as follows: if $\alpha_2$ is an experiment at time $t_2$ and $U$ is the unitary operation that describes how the system evolves from time $t_1$ to time $t_2$, then we can consider the experiment $\alpha_1$ at time $t_1$, which consists of first evolving the system according to $U$ and then performing $\alpha_2$. More generally, $U$ induces a mapping from experiments at time $t_2$ to experiments at time $t_1$, and one can show that from the definition of a property, it follows that this map must preserve all infima. Using the theory of Galois adjoints, it then follows that the map that describes how properties propagate during $U$ must preserve all suprema. The final purely technical step then involves using Wigner's theorem [184] and a modern category-theoretic account on projective geometry [85, 177].

[4] Explicitly, for $L$ the lattice of closed subspaces of a Hilbert space $\mathcal{H}$ and $P_a$ the projector on the subspace $a$ lifted to an operation on $L$, we have

$$[P_a : L \to L :: b \mapsto a \wedge (a^\perp \vee b) :] \dashv [(a \to_{\text{Sasaki}} -) : L \to L :: b \mapsto a^\perp \vee (a \wedge b)],$$

with $(- \to_{\text{Sasaki}} -)$ the (in)famous Sasaki hook [63]. In light of this argument, $P_a$ now plays the role of how properties propagate in quantum measurements, and $(a \to_{\text{Sasaki}} -)$ is the map that assigns to each property after the measurement one before the measurement.

[5] A survey in the light of the Geneva School approach is in [53].

[6] More details on this are in [152].

more prominent in the work of Bohm in the 1980s and later also in Hiley's work [31–33], who is still pursuing this line of research [114].

So why did neither Pool's work nor that by the Geneva School Mark III ever have any real impact? As discussed in great detail in [151], the entire Geneva School program only makes sense when considering "isolated systems" on which we can perform the experiments. This immediately makes it inappropriate to describe a system in interaction with another one. This is a notorious flaw of most quantum logic programs, which all drastically failed in providing a convincing abstract counterpart to the Hilbert space tensor product. In the approach outlined in this chapter, we consider an abstract counterpart to the Hilbert space tensor product as primitive. It encodes how systems interact with other systems; so, rather than explicitly given, its character is implicitly encoded in the structure on processes.

Today, the *measurement-based quantum computational model* (MBQC) poses a clear challenge for a theory of processes. MBQC is one of the most fascinating quantum computational architectures, and it relies on the dynamics of the measurement process for transforming the quantum state.[7] By modeling quantum process interaction in a *dagger compact closed category*, in [6, 43] Abramsky and this author trivialized computations within the Gottesman–Chuang *logic-gate teleportation* MBQC model [97]. The more sophisticated Raussendorf–Briegel *one-way* MBQC model [163, 164] was accounted for within a more refined categorical setting by Duncan, Perdrix, and this author [48, 62, 79, 80].

### 2.2.3 Systems from Processes

Less structurally adventurous than the Ludwig School, the Foulis–Randall School, and the Geneva School are the $C*$-algebra disciples, who prefer to stick somewhat closer to good old Hilbert space quantum mechanics. This path was again initiated by von Neumann after denouncing Birkhoff–von Neumann–style quantum logic.[8] A highlight of the $C*$-algebraic approach is AQFT [99, 100, 102], attributable mainly to Haag and Kastler. In contrast with most other presentations of quantum field theory, not only is AQFT mathematically solid, but it also has a clear conceptual foundation.

This approach takes as its starting point that every experiment takes place within some region of spacetime. Hence, to each spacetime region $R$[9], we assign the $C*$-algebra $A(R)$ of all observables associated to the experiments that potentially could take place in that region.[10] Although quantum field theory does not support the quantum mechanical notion of system attributable to the creation and annihilation of particles,

---

[7] Recently, Rau realized a reconstruction of Hilbert space based on a set of axioms that takes the fact that the one-way measurement-based quantum computational model can realize arbitrary evolutions as its key axiom [162] and proposes this dynamics-from-measurement processes as a new paradigm for quantum foundations.

[8] For a discussion of the what and the why of this, we refer the reader to Rédei [165].

[9] Which is typically restricted to open diamonds in Minkowski spacetime.

[10] It is a natural requirement that inclusion of regions $R \subseteq R'$ carries over to $C*$-algebra embeddings $A(R) \hookrightarrow A(R')$ because any experiment that can be performed within a certain region of spacetime can also be performed within a larger region of spacetime. The key axiom of AQFT is that spacelike separated regions correspond to commuting $C*$-algebras. All of these $C*$-algebras are then combined in a certain manner to form a giant $C*$-algebra $\mathcal{A}$. The connection with spacetime is retained by a mapping that sends each spacetime region

AQFT reintroduces by means of regions of spacetime and associated algebras of observables a meaningful notion of system $(R, A(R))$.[11]

In [41], $C*$-algebras also provided an arena for Clifton, Bub, and Halvorson to address Fuchs's and Brassard's challenge to reconstruct the quantum mechanical formalism in terms of information-theoretic constraints [38, 91, 92]. Meanwhile, it has been recognized by at least one of the authors that most of the work in this argument is done by the $C*$-algebra structure rather than by axioms [101]; hence, a more abstract mathematical arena is required.

### 2.2.4 The Logic of Interacting Processes

What does it mean to have two or more systems? That is, what is "$A$ and $B$":

1. I have a choice between $A$ and $B$.
2. I have both $A$ and $B$.
3. I have an unlimited availability of both $A$ and $B$.

Developments in logic have started to take account of these sorts of issues. In particular, Girard's *linear logic* [1, 8, 94, 178] (which originated in the late 1980s) makes the difference between either having the availability of one of two alternatives or having both alternatives available.[12] The first of the two conjunctions in linear logic, the *nonlinear* conjunction, is denoted by $\&$; the second one, the *linear* conjunction, is denoted by $\otimes$. The difference is

$$A \vdash A \& A \qquad A \& B \vdash A \qquad \text{while} \qquad A \not\vdash A \otimes A \qquad A \otimes B \not\vdash A.$$

That is, in words, from the fact that $A$ (resp. $A \& B$) holds, we can derive that also $A \& A$ (resp. $A$) holds; but, from the fact that $A$ (resp. $A \otimes B$) holds, we cannot derive that also $A \otimes A$ (resp. $A$) holds. Hence, the linear conjunction treats its arguments as genuine *resources*; that is, they cannot freely be copied or discarded. It is a *resource-sensitive* connective.

From a naive *truth-based* view on logic, where $A$ merely stands for the fact that this particular proposition holds, the failure of the last two entailments might look weird. However, a more modern view on logic is obtained in terms of *proof theory*. In this perspective, $A$ stands for the fact that one possesses a proof of $A$, $A \otimes B$ stands for the fact that one possesses a proof of $A$ and a proof of $B$, and $A \otimes A$ stands for the fact that one possesses two proofs of $A$.

---

on the corresponding sub-$C*$-algebra of $\mathcal{A}$, and the embeddings of $C*$-algebras now become themselves inclusions.

[11] Compact closed categories play a key role within AQFT [74, 75, 102], but their role in AQFT is conceptually totally different from this role in our framework. The natural manner to recast AQFT as a monoidal category, somewhat more in the spirit of the developments of this chapter, would be to replace the $C*$-algebra $\mathcal{A}$ by a monoidal category with the sub-$C*$-algebras of $\mathcal{A}$ as the objects, and completely positive maps as morphisms, subject to some technical issues to do with the nonuniqueness of the tensor product of $C*$-algebras.

[12] Since its birth, linear logic not only radically changed the area of logic, but it also has immediately played a very important role in computer science and still does [82]. The first occurrence of linear logic in the scientific literature was in Lambek's mathematical model for the grammar of natural languages [136] in the 1950s.

In proof theory, propositions mainly play a supporting role. What is of particular interest in proof theory is the *dynamics of proofs*: how to turn a long proof into a short one, how to eliminate lemmas, and so forth. In other words, the *derivation process* (i.e., proof) is the key player, and it is all about how proofs *compose* to make up another proof. The mathematical arena where all this takes place is that of *closed symmetric monoidal categories* (e.g., [169]).

One indeed can take the view that "states" stand to "systems" in physics as "proofs" stand to "propositions" in logic. "Physical processes" that turn a system into another in physics then correspond to "derivation processes" of one proposition into another in logic. In this view, systems serve mainly as things along which physical processes can be composed, a view that we shall adopt here.

### 2.2.5 Processes as Relations

Once one considers processes and their interactions as more fundamental than systems themselves, one enters the realm of *relationalism*.

One well-known recent example of relationalism is Barbour and Bertotti's take on relativity theory in terms of Mach's principle [17], which states that inertia of a material system is meaningful only in relation to its interaction with other material systems [142]. Rovelli's relational interpretation of quantum theory [167] considers all systems as equivalent—hence, not subscribing to a classical-quantum divide—and all information carried by systems as relative to other systems. Here, we will also adopt this relational view on physics.

One thing that relationalism provides is an alternative to the dominant "matter in spacetime" view on physical reality by taking spacetime to be a secondary construct. What it also does is relax the constrains imposed by no-go theorems on accounts of the measurement problem [95, 121, 131].[13] For example, if a system's character is defined by its relation to other systems, *contextuality*—rather than being something weird—becomes not just perfectly normal but also a *fundamental requirement for a theory not to be trivial*.[14]

The main problem with relationalism seems to be that although it is intuitively appealing, there is no clear formal conception. This is where category theory [83] provides a natural arena, in that it abstracts over the internal structure of *objects* (cf. the properties of a single physical system) and instead considers the structure of *morphisms* between systems (cf. how systems relate to each other). Monoidal categories [25, 146], moreover, come with an intrinsic notion of compound system. In their diagrammatic incarnation, these categories translate "being related" into the topological notion of "connectedness." The "nonfree" part of the structure then provides the modes in which things can be related. It seems to us that the *dagger compact* symmetric monoidal

---

[13] It is a common misconception that the Kochen–Specker theorem [131], (1967) would be in any way the first result of its kind. It is, in fact, a straightforward corollary of Gleason's theorem in 1957 [95], and a crisp, direct no-go theorem was already provided in 1963 by Jauch and Piron [121]. A discussion of this is in Belinfante's book [24].

[14] Obviously, this paragraph may be for many the most controversial, challenging, or interesting one in this chapter. They probably would have liked to see more on it. We expect to do this in future writings once we have obtained more formal support for our claims.

structure [7, 170] in particular provides a formal counterpart to the relational intuition. A more detailed and formal discussion of this issue is in Section 2.4.4.

### 2.2.6 Mathematical Rigor

One of the favorite activities of operationalists is to reconstruct quantum theory by imposing reasonable axioms on families of experimental situations. Some recent examples of such reconstructions are [66, 104, 162].

This tradition was initiated by Mackey [143] around 1957, with Piron's 1964 theorem as the first success [157]. The different attempts vary substantially in terms of their mathematical guise, in that some reconstructions start from the very foundations of mathematics (e.g., [157], [158], [173]), whereas others will take things like the real continuum as God-given in order to state the axioms in a very simple language (e.g., [104]). Quoting Lucien Hardy on this [104]:

> Various authors have set up axiomatic formulations of quantum theory, [...] The advantage of the present work is that there are a small number of simple axioms, [...] and the mathematical methods required to obtain quantum theory from these axioms are very straightforward (essentially just linear algebra).

Quoting Tom Yorke, singer of the Oxford-based band Radiohead [161], "Karma Police, arrest this man. He talks in Maths."

It is an undeniable fact that mathematical rigor is one of the key cornerstones of science. But, on the other hand, very important science has been developed long before there existed anything like a foundation of mathematics. Even in recent history, scientific progress was possible only by not subscribing to mathematical rigor, of which the problem of renormalization in quantum field theory is the most prominent witness, even leading to a Nobel Prize.

Ultimately, this boils down to the respect one gives to mathematics. Roughly put, is mathematics an a priori given thing that we can use to formulate our theories of physics, or is it something secondary that intends to organize our experiences—be it when reasoning, exploring nature, or whatever—and that should be adjusted to cope with our evolving spectrum of experiences? Simpler put, do we serve mathematics or does mathematics serves us?

Our approach is to assume a physical reality, with the things "out there" truly happening. We consider certain physical primitives—namely, relations and composition thereof. These primitives come with a notion of "sameness," which will play the role of equality—that is, it will tell us when compositions of relations are equal.[15] As a second step, we try to match these physical primitives with a mathematical structure—namely, particular kinds of categories. This, despite the great flexibility of category theory, will come at a certain cost.

---

[15] Let us mention that currently, even within the foundations of mathematics, we do not really know for what the sign "=" stands. In universal algebra, it is a binary predicate, but once one goes beyond classical logic, this breaks down. In first-order logic, equality is a distinguished binary relation. In higher-order logic, it is given by Leibniz identity, which identifies things with the same properties [137]. In Martin-Löf–type theory [150] and Bishop-style constructive mathematics [28], one uses yet again other notions of equality. In categorical logic [119], several options are still being explored. Credits for this concise summary go to Phil J. Scott.

In our view, Hilbert's proposal to *axiomatize physics*[16] is a very different ball game than *axiomatizing mathematics*,[17] something that also proved to be a far more delicate business than one imagined at first.

Our goal is also quite different from that of the reconstructionists. Rather than reproducing quantum theory with a set of reasonable axioms, our goal is to reproduce as much as possible physical phenomena with as little as possible "structural effort" or "axiomatic compromise," thereby providing a flexible setting that may be better adjusted to the theories of the future.

### 2.2.7 The Continuous or the Discrete?

In the light of future theories of quantum gravity, it has been argued that we may have to abandon our reliance on the continuum, be it either with respect to the structure of the space of states, spectra of observables, spacetime, or even probability valuation. Quoting Isham and Butterfield [118]:

> ...the success of [the edifice of physics] only shows the "instrumentalist utility" of the continuum—and not that physical quantities have real-number values...there is no good *a priori* reason why space should be a continuum; similarly, *mutatis mutandis* for time.

> ...limiting relative frequency interpretation seems problematic in the quantum gravity regime...for the other main interpretations of probability—subjective, logical or propensity—there seems to us to be no compelling *a priori* reason why it should be real numbers.

Once one abandons the continuum as a mathematical default, we need a paradigm or mechanism either to reproduce it or by which to replace it.

One option is "spaces without points," which has both a topological and geometric incarnation, respectively called *locales* and *frames* [123].[18] These spaces have been used both to model spectra as well as truth-values in the so-called topos approach to physical theories, which rose to prominence some ten years ago with the still ongoing work of Isham, collaborators, and followers.[19] Both the locales/frames as well as topos theory also provide a mathematical foundation for intuitionistic logic [124, 181]. In all of their guises, they have been particularly popular among computer scientists.

Also popular among computer scientists are discrete combinatorial spaces. In fact, computer scientists proposed various discrete spacetime structures [138, 156] well before physicists did so (e.g., Sorkin et al. [35]).

---

[16] Cf. Hilbert's sixth problem [166].

[17] Cf. Hilbert's second problem on the consistency of axiomatic arithmetic [166]. Gödel later showed that this issue cannot be settled within arithmetic itself [96].

[18] Locales and frames are a beautiful example of how the nature of a mathematical structure can change merely by changing the nature of its relation to other structures of the same kind rather than by changing the structure itself: in category theoretic terms, locales and frames are exactly the same objects, but they live in a different category; one obtains the category of locales simply by reversing the direction of the arrows in the category of frames.

[19] The first work in this area seems to be by Adelman and Corbett in 1993 [9, 10].

Our setting is flexible enough to accommodate both perspectives. For example, a topos gives rise to a so-called alegory of generalized relations [90], and similar categories arise when organizing combinatorial species [87, 125, 126].

In fact, even at a much more basic level, categories abstract over concrete well-pointed spaces, by abstracting over the actual structure of objects. They obviously also immediately provide a rich variety of combinatorial structures in that they themselves always form a graph.

## 2.3 Systems ← Relations ← Composition

We mentioned that operational approaches appeal to our everyday experiences or some easy-to-grasp generalization thereof. Also, here we make some reference to our perceptions but at a much more abstract level than in all of the aforementioned examples. Not measurement devices, or probabilities, or propositions, or classical mechanics concepts such as three-dimensional space, or concrete observables such as position, or the real continuum play any role.

We assume as primitive a flexible notion of *system*, a very general notion of *relation* between these, and two modes of *composition* of the latter, one that typically imposes dependencies between the processes that one composes and one that excludes dependencies. In graphical terms, these correspond with the primal topological distinction between "connected" and "disconnected," cf.:

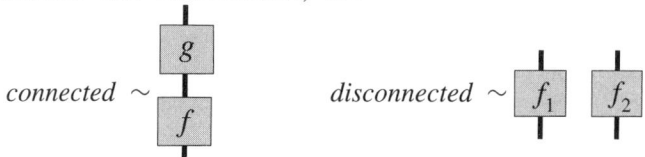

Within our approach, which models how relations compose to make up other relations, systems play the role of the "plugs" by means of which we can create dependencies between relations in one of the two modes of composition.

So it is in "bottom-up" order in which we introduce the basic concepts

$$\text{systems} \to \text{relations} \to \text{composition}$$

to appeal to the reader's intuition; the most important concept is composition. Relations are then those things that we can compose, and systems are the things along which we can compose these relations in a dependent manner.

This top-down view may seem to go in the opposite direction of a physicist's reductionist intuition. Nonetheless, it is something with which the physicist is well acquainted. For symmetry groups, it is not the elements of the group that are essential, but the way in which they multiply ($\sim$ compose) because the same set of elements may, in fact, carry many different group structures. In a similar manner that group structure conveys the shape of a space, the composition structure on relations will convey the "shape" of the "universe of processes."

In support of the reader's intuition, we refer to "properties" of a system when discussing concepts, but this has no defining status whatsoever. For this discussion, we inherit the "actual" versus "potential" terminology from the Geneva School, the first saying something about the state of the system, the second saying something about the system itself.

### 2.3.1 Systems

So the prime purpose of a notion of *system* is to support the notion of a relation, systems being those things along which relations can be composed.

More intuitively, by a system, we mean something identifiable about which we can pose questions and, hence, about which it makes sense to speak about "properties." It is the latter that are usually stated relative to our world of experiences. This, however, does not mean that a system is completely determined by what we consider to be its actual properties, nor does it mean that there necessarily exists an experiment by means of which we can verify these.[20] An example of a system that is not completely determined by its actual properties is one that is part of a larger system—that is, when considering "parts of a larger whole."

We denote systems by $A, B, C, \ldots$

**Example: Quantum Systems.** Quantum systems are the entities that we describe in Hilbert space quantum theory (e.g., position, momentum, or spin). Here, systems that are not completely determined by their actual properties are those described by density operators, which arise from a lack of knowledge as well as by tracing out part of a compound system. The need for a concept of system that is not characterized in terms of its actual properties becomes even more important in the case of quantum field theory, where we want to be able to consider what is relevant about the field for a certain region of spacetime.

**Example: AQFT and Beyond.** In AQFT, the systems are the $C*$-algebras associated to a region of spacetime [99, 102]. So a system is a pair $(R, A(R))$, where $R$ is a region of spacetime and $A(R)$ represents the observables attributed to that region. This idea of a pair consisting of a spacetime region and another mathematical object that encodes observables can be generalized to other manners of encoding observables—for example, in terms of *observable structures*—that is, special commutative dagger Frobenius algebras (see later) on an object in a dagger symmetric monoidal category (see later) as is done in categorical quantum mechanics [58, 60, 61]. These two perspectives are not that far apart, given that Vicary has shown [180] that finite dimensional $C*$-algebras are precisely the noncommutative generalizations of observable structures in the dagger symmetric monoidal category **FHilb** (see later).

By I, we denote the system that represents everything that we do not explicitly consider within our theory. One may refer to this as the *environment*—that is, what is not part of our domain of consideration. Intuitively put, it is the system that represents

---

[20] There are many things we can speak about without being able to set up an experiment, for example, simply because the technology is not (yet) available. One could consider speaking in terms of hypothetical or idealized experiments, but we do not know the technologies of the future yet. These will be based on theories of the future, and because crafting these theories of the future is exactly the purpose of this framework, guessing would lead to a circularity.

everything to which we do not attribute any properties whatsoever. Formally, it will play the role of the "trivial system" in that composing it with any other system $A$ will yield that system $A$ itself. Obviously, what is considered as $I$ may in part be a cognitive decision, or a technological constraint, or it may be a fundamental physical principle.[21] Here, these interpretational issues do not matter. What does matter is that there is a *domain of consideration* and that everything else falls under the umbrella of $I$.

**Example: Open Systems.** In open systems theory (e.g., [68], [133]), $I$ stands for the environment. In quantum theory, it is $I$ that is responsible for decoherence. Sections 2.4.5 and 2.5 elaborate on this issue in great detail.

Our account on systems as "a bag of things" may sound naive and, indeed, it is. A more realistic account that involves the notion of subsystem is discussed in Section 2.3.4. This will require that we first introduce some other concepts.

We denote "system $A$ *and* system $B$" by $A \otimes B$. The precise meaning of $A \otimes B$ will become clear later from what we mean by *composition of processes*. In particular, we will see that $A$ and $B$ in $A \otimes B$ will always be independent and, hence, distinct; that is, we cannot conjoin a system with itself.

The notation $A_1 \otimes A_2$ (wrongly) indicates that $A_1$ and $A_2$ are ordered. This is an unavoidable artifact of the one-dimensional linear notation that is employed in most natural languages as well as in most mathematical notation. Hence, $A_1 \otimes A_2$ is to be conceived as "a set of two systems" rather than as "an ordered pair of two systems."

**Example: AQFT and Beyond.** AQFT considers an inclusion order on diamond-shaped regions, which carries over on inclusion for $C*$-algebras. Intuitively, the joint system would consist of the union of the two regions and the corresponding union of $C*$-algebras, at least in the case that the regions are spacelike separated. But two regions do not make up a diamond anymore, so this naive notion of system $A$ and system $B$ would already take one beyond the AQFT framework. An article on this subject is in preparation by this author and Abramsky, Blute, Comeau, Porter, and Vicary [4].

### 2.3.2 Processes and Their Composition

*Processes* are relations that carry the "genuine physical substance" of a theory. They are those entities that we think of as actually "happening" or "taking place" (as opposed to; the *symmetry relations* discussed in Section 2.3.5). They arise by "orienting" a relation, that is, by assigning input/output roles to the systems it relates; that is, it is a relation that "happens" within a by-us perceived *causal structure* (cf. a partial ordering or, more generally, a directed graph).

Intuitively, a process embodies how properties of system $A$ are transformed into those of system $B$. The environment may play an important role in this.

---

[21] For example, the disciples of the so-called church of the larger Hilbert space seem to believe that system $I$ could always be eliminated from any situation in quantum theory.

The *type* of a process is the specification of the input system $A$ and the output system $B$, denoted as $A \to B$. We call $A$ the *input* and $B$ the *output* of the process. Processes themselves are denoted as $f : A \to B$.

**Example: Operations.** Processes can be the result of performing an operation on a system $A$ to produce system $B$ (e.g., measuring, imposing evolution, or any other kind of experimental setup). Our whole framework could be given a more radical operational connotation by restricting to processes arising from operations. It would then match Hardy's recent proposal [108].

**Example: Quantum Processes.** These processes include state preparations, evolutions, demolition and nondemolition measurement processes, and so forth.

**Example: Deinstrumentalizing Geneva School Mark II.** One can modify the Geneva School Mark II approach by replacing the experimental projects with any process $f$ that may cause a particular other process $f_{yes}$ to happen thereafter. Roughly put: a property of a system would then be an equivalence class of those processes that cause $f_{yes}$ to happen with certainty.

The trivial process from system $A$ to itself is denoted by $1_A : A \to A$. It "happens" in the sense that it asserts the existence of system $A$, and it trivially obeys causal order. These trivial processes are useful in that they provide a bridge between systems and processes by associating to each system a process.

For all other nontrivial processes, the input and the output are taken to be nonequal (i.e., if the type of a process is $A \to A$, then it is [equal to] $1_A$).

By a *state*, we mean a process of type $I \to A$; and by an *effect*, we mean a process of type $A \to I$. What is important for a state is indeed what it is, and not its origin, that can consequently be comprehended within I.

By a *weight*, we mean a process of type $I \to I$.

### 2.3.2.1 Sequential Composition

The *sequential* or *causal* or *dependent composition* of processes $f : A \to B$ and $g : B \to C$ is the process that relates input system $A$ to output system $C$. We denote it by:

$$g \circ f : A \to C.$$

We will also refer to $g \circ f$ as "$g$ after $f$" or as "first $f$ and then $g$."

**Example: Operations.** For processes resulting from operations, $g \circ f$ is the result of *first* performing operation $f$ *and then* performing operation $g$. The operations corresponding to states are preparation procedures.

**Example: Weights as Probabilities.** When we compose a state $\psi : I \to A$ and an effect $\pi : A \to I$, then the resulting weight $\pi \circ \psi : I \to I$ can be interpreted as the probability of the sequence "$\pi$ after $\psi$" to happen. That a projective measurement effect in quantum theory may be impossible for certain states and certain for others boils down to $\langle \phi | \circ | \psi \rangle = 0$ while $\langle \phi | \circ | \phi \rangle \neq 0$ for $|\psi\rangle \perp |\phi\rangle$. More generally, these weights can articulate likeliness of processes.

### 2.3.2.2 Separate Composition

The *separate* or *acausal* or *independent composition* of processes $f_1 : A_1 \to B_1$ and $f_2 : A_2 \to B_2$ is the process that relates input system $A_1 \otimes A_2$ to output system $B_1 \otimes B_2$. We denote it by:

$$f_1 \otimes f_2 : A_1 \otimes A_2 \to B_1 \otimes B_2.$$

The key distinction between sequential and separate composition in terms of "dependencies" between processes is imposed by the following constraint.

### 2.3.2.3 Independence Constraint on Separate Composition

A process is independent from any process to which it is not "connected via sequential composition," and the same holds for the systems that make up the types of these processes, with the exception of the environment I. In particular, within the compound process $f \otimes g$, the processes $f$ and $g$ are independent.

We precisely define what we mean by "connected via sequential composition" in Section 2.3.3 by relying on the topological notion of "connectedness." The spirit of this constraint is that causal connections can be established only via dependent composition, not by means of separate composition.

**Example: Operations.** When considering processes resulting from operations, $f_1$ and $f_2$ in $f_1 \otimes f_2$ are realized by two independent operations. This means that the setup in which one realizes one operation should be sufficiently isolated from the one that realizes the other operation.

The independence of $f_1$ and $f_2$ in $f_1 \otimes f_2$ imposes independence of $A_1$ and $A_2$ in $A_1 \otimes A_2$ and of $B_1$ and $B_2$ in $B_1 \otimes B_2$, which indeed forces $A_1$ and $A_2$ in $A_1 \otimes A_2$ and $B_1$ and $B_2$ in $B_1 \otimes B_2$ always to be distinct.

**Example: AQFT.** For intersecting regions $R$ and $R'$, the systems $(R, A(R))$ and $(R', A(R'))$ are obviously not independent. Neither are they for regions $R$ and $R'$ that are causally related.

In $f_1 \otimes f_2$, the two processes have to be independent, but this does not exclude that via causal composition with other processes dependencies can emerge.

**Example: Quantum Entanglement.** Although two quantum processes $f_1 : A_1 \to B_1$ and $f_2 : A_2 \to B_2$ are independent in $f_1 \otimes f_2$, it may, of course, be the case that because of common causes in the past measurements on their respective output systems, $B_1$ and $B_2$ may expose correlations. In that case, we are in fact considering $(f_1 \otimes f_2) \circ |\Psi\rangle$ where $|\Psi\rangle : I \to B_1 \otimes B_2$. In this case, as we shall see, $B_1$ and $B_2$ *are* connected via sequential composition.

Note that the exception of I in the independence constraint allows for

$$A \otimes I = A,$$

which affirms that the environment may always play a certain role in a process.

On a more philosophical note, within our setting, the independence constraint replaces the usual conception of *sufficient isolation* within the *scientific method* [151, 159]: we do not assume that systems or processes are sufficiently isolated, but our formal vehicle, which represents when we compose them implicitly, requires that they are independent, the environment excluded.

Now consider the four processes:

$$f_1 : A_1 \to B_1 \quad f_2 : A_2 \to B_2 \quad g_1 : B_1 \to C_1 \quad g_2 : B_2 \to C_2 \quad (2.1)$$

Note that causal composition of the processes $f_1 \otimes f_2$ and $g_1 \otimes g_2$ resulting from separate composition, which implies matching intermediate types, is well defined because $B_1$ and $B_2$ will always be taken to be distinct in $B_1 \otimes B_2$. But there is another manner in which we can compose these processes to make up a whole of type $A_1 \otimes A_2 \to C_1 \otimes C_2$—namely, by separate composition of $g_1 \circ f_1$ and $g_2 \circ f_2$. Although symbolically these two compounds are represented differently, *physically* they represent *the same* overall relation; hence the following.

### 2.3.2.4 Interaction Rule for Compositions

For processes (2.1), we have

$$(g_1 \circ f_1) \otimes (g_2 \circ f_2) = (g_1 \otimes g_2) \circ (f_1 \otimes f_2).$$

Similarly, for systems $A_1$ and $A_2$, we also have

$$1_{A_1} \otimes 1_{A_2} = 1_{A_1 \otimes A_2}.$$

### 2.3.2.5 Nonisolated Systems and Probabilistic Weights of States

The natural way to assert inclusion of *nonisolated* (or *open*) systems within a theory of processes is in terms of a particular kind of process

$$\top_A : A \to \mathrm{I}$$

that "feeds" a system into the environment I and hence explicitly realizes such a nonisolated system. Feeding a system $A$ into the environment can be achieved by taking a process $f : A \to B$ and by then "deciding" to consider $B$ as part of the environment I. As was the case for I, these feeding-into-the-environment processes may involve a cognitive component.

What characterizes such a *feed-into-environment process*? First, it is easily seen that we can always set

$$\top_{A \otimes B} := \top_A \otimes \top_B \quad \text{and} \quad \top_\mathrm{I} := 1_\mathrm{I}.$$

Second, it should be allowed to happen with certainty, independently on the state of the system, as opposed to, for example, the *projective measurement effects* in quantum theory already discussed. Denoting weights by $\mathbb{W}$, given a measure that assigns weights to each process, in particular to states $\mathbb{S}$,

$$|-| : \mathbb{S} \to \mathbb{W},$$

a feed-into-environment process $\top_A$; should be such that applying it leaves the weight of the state it is applied to invariant, that is, concretely,

$$\top_A \circ \psi = |\top_A \circ \psi| = |\psi| \qquad (2.2)$$

for all states $\psi : I \to A$, where the first equality merely says the weight of a weight is itself. But (2.2) can now also be dually interpreted: feed-into-environment processes are characterized in that they provide the measure for assigning weights to states, by postcomposing states with them.

**Example: Traces and Probabilities in Quantum Theory.** In quantum theory, the completely positive maps that trace out spaces play the role of the feed-into-environment processes:

$$tr_{\mathcal{H}} :: \rho_{\mathcal{H}} \mapsto \sum_i \langle i | \rho_{\mathcal{H}} | i \rangle.$$

These, indeed, stand for ignoring part of a system as well as for measuring the overall probability of a nonnormalized density matrix. Note in particular that these are the only completely positive maps that satisfy (2.2) for every possible state. In the language of Chiribella et al. [40], $tr_{\mathcal{H}}$ is the *unique deterministic effect* on $\mathcal{H}$.

### 2.3.3 Graphical Representation of Processes

The preceding specified data can be given a diagrammatic representation:

$$f \equiv \boxed{f} \qquad g \circ f \equiv \boxed{\genfrac{}{}{0pt}{}{g}{f}} \qquad f_1 \otimes f_2 \equiv \boxed{f_1}\ \boxed{f_2}$$

That is,

- a process is represented by a box with inputs and outputs
- $-\circ-$ is represented by connecting outputs to inputs
- $-\otimes-$ is represented by not connecting boxes

The object I will be represented by "no wire" and

$$\psi \equiv \triangledown_{\psi} : I \to A \qquad \pi \equiv \triangle^{\pi} : A \to I \qquad \omega \equiv \Diamond_{\omega} : I \to I.$$

What is particularly nice in this graphical representation is that the interaction rule automatically holds because translating both its left-hand side and its right-hand side into the graphical calculus both result in the same:

$$(g_1 \circ f_1) \otimes (g_2 \circ f_2) = (g_1 \otimes g_2) \circ (f_1 \otimes f_2) \quad \Leftrightarrow \quad \boxed{\genfrac{}{}{0pt}{}{g_1}{f_1}}\ \boxed{\genfrac{}{}{0pt}{}{g_2}{f_2}} = \boxed{\genfrac{}{}{0pt}{}{g_1}{f_1}}\ \boxed{\genfrac{}{}{0pt}{}{g_2}{f_2}}$$

# 146  A UNIVERSE OF PROCESSES AND SOME OF ITS GUISES

This shows also that the interaction rule is, in fact, nothing more than an artifact of one-dimensional symbolic notation!

Here is the definition we promised earlier:

**Definition.** Two processes are *connected via sequential composition* if, in the graphical representation, they are topologically connected.

There also is a direct translation of this graphical representation of processes to directed graphs and vice versa. The rules to do this include

- processes (i.e., boxes, triangles, diamonds, and so on) become the nodes
- systems (i.e., wires between boxes) become directed edges, with the direction pointing from what used to be an output to an input

Such a directed graph makes the underlying causal structure on processes explicit. The following is an example:

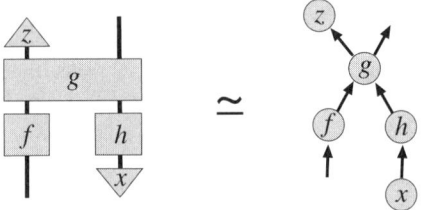

In this example, all processes are connected via sequential composition because the picture is, as a whole, connected.

Special processes can be give special notations, for example, a feed-into-environment process $\top_C : C \to I$ could be denoted as

$$\top_C \equiv \;\overset{\triangleq}{\top}\; \quad \text{so} \quad (1_B \otimes \top_C) \circ f \equiv \;\boxed{f}\;$$

for $f : A \to B \otimes C$.

In two-dimensional graphical language, as is the case for symbolic notation, systems appear in a certain order (cf. from left to right), which has no direct ontological counterpart. However, in the graphical notation, this order can be exploited to identify distinct systems in terms of their position within the order, hence, in part omitting the necessity to label the wires. More on this follows. One can, of course, also think of these pictures as living in three dimensions rather than in two dimensions or some even more abstract variation thereof. One calls graphs that exploit a third dimension *nonplanar* [171]. Planar graphs are subject to Kuratowski's characterization theorem [134].

### 2.3.4 Physical Scenarios, Snapshots, and Subsystems

A *physical scenario* is a collection of processes together with the *composition structure* in which they happen. By the *resolution* of a scenario, we mean the resulting overall process. A scenario comprises more information than its resolution in that it also comprehends the manner in which the overall process is decomposed in subprocesses. Obviously, the selection of a particular scenario that has a given process as its resolution has no actual physical content but is merely a subjective choice of what to consider as the parts of a whole.

Given such a physical scenario, one can consider a subset of the systems appearing within it, neither of which are causally related (e.g., those connected by the hand-drawn line in the following picture):

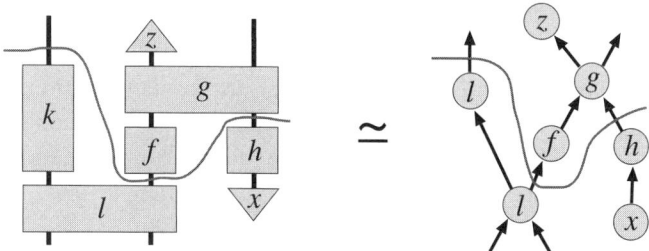

We call such a collection of systems appearing in a scenario a *snapshot*.

Note that it is not excluded that snapshots resulting from distinct scenarios are the same; for example, the hand-drawn line

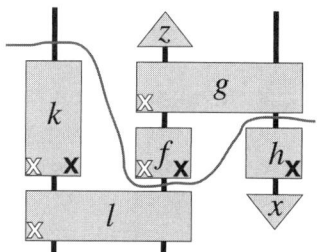

represents a snapshot both for the restriction of the boxes to those with a white cross as well as for the restriction to those with a black cross.

**Example: Relativistic Causal Histories.** These snapshots are Hardy's "systems" within his instrumental framework [108, 109]. In turns, these generalize Blute, Panangaden, and Ivanov's "locative slices" within their framework, which endows standard quantum mechanical operations with a causal ordering [30, 153]. A dual point of view was earlier put forward by Markopoulou [148].

These snapshots are indeed systems as much as any other system. But, as mentioned at the very beginning of this chapter, they are no longer the primal physical concept, but rather they are things along which we "decide" to decompose processes. It is the resolution of the snapshot that is physically the only primal concept.

Note that there exists a partial order on systems in terms of inclusion of snapshots. For example,

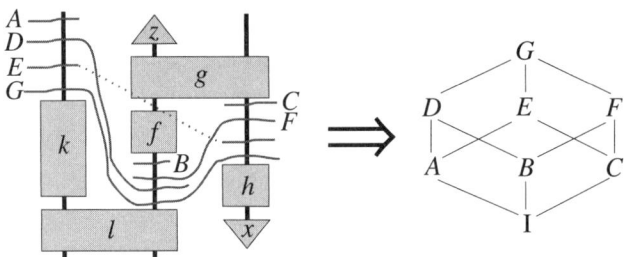

This in particular implies that "being nonequal" for systems—for example in $A \otimes B$—does not capture independence. In mathematical terms, it should be replaced by disjointness within the Boolean algebra structure arising from these snapshots. We are developing on formal account on this along with Lal [52].

### 2.3.5 Symmetry Relations

Unlike processes, *symmetry relations* do not represent actual "happenings," neither do they have to respect any by us perceived causal structure. Rather, they will enable one to express structural properties (i.e., symmetries) of the processes that make up the physical universe, in a manner similar to how the Galileo–Lorentz group conveys the shape of spacetime. But rather than being a structure on processes, in our approach they will interact with processes in the same manner as processes interact with each other, in terms of ∘ and ⊗, so that we can treat them as "virtual processes" within an "extended universe" that consists not only of processes but also of symmetry relations, as well as the relations arising when composing these. The interaction of processes and symmetry relations would, for example, embody how usually dynamics is derived in terms of representations of the Galileo–Lorentz group. Here, such a virtual process could, for example, be a Lorentz boost along a spacelike curve.

Intuitively, symmetry relations relate properties of one system to those of another system, and because the content carried by a process is how properties of system $A$ are transformed in those of system $B$, it indeed makes perfect sense to treat processes and symmetry relations on the same footing. Consequently, we can extend dependent composition to symmetry relations, but it obviously loses its causal connotation. Also, separate composition can evidently be extended to symmetry relations, separation now merely referring to some formal independence. Consequently, we can also still speak of scenarios and snapshots.

For some, the distinction between process and symmetry relation might seem somewhat artificial. But this would, in fact, advocate our framework even more.

**Example: Active and Passive Rotations in Classical Mechanics.** In classical mechanics, rotations of a rigid body are "processes" modeled in $SO(3)$, whereas the $SO(3)$ fragment of the Gallilei group consists of "symmetry relations," which assert the rotational symmetry of three-dimensional Euclidean space.

**Example: Inverses to Processes.** We define an *inverse* to a process $f : A \to B$ as the symmetry relation $f^{-1} : B \to A$, which satisfies

$$f^{-1} \circ f = 1_A \quad \text{and} \quad f \circ f^{-1} = 1_B. \tag{2.3}$$

It immediately follows that such an inverse, if it exists, is unique.

**Example: Identical Systems.** How can we describe distinct but "identical" systems? A pair of systems $A_1$ and $A_2$ is *identical* if it comes with a pair of mutually inverse relations $1_{A_1,A_2} : A_1 \to A_2$ and $1_{A_2,A_1} : A_2 \to A_1$. Explicitly,

$$1_{A_2,A_1} \circ 1_{A_1,A_2} = 1_{A_1} \quad \text{and} \quad 1_{A_1,A_2} \circ 1_{A_2,A_1} = 1_{A_2}.$$

Let $f : C \to A_1$ and $g : A_1 \to C$ be any processes. We set

$$F_{A_1,A_2} f := 1_{A_1,A_2} \circ f \quad \text{and} \quad G_{A_1,A_2} g := g \circ 1_{A_2,A_1}. \tag{2.4}$$

These can also be represented in a *commutative diagram* [145, 147]:

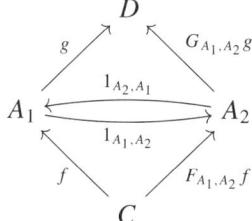

that is, a diagram in which any two paths that go from one system to another are equal. It also follows that

$$F_{A_2,A_1} F_{A_1,A_2} f = f \quad \text{and} \quad G_{A_2,A_1} G_{A_1,A_2} g = g.$$

Intuitively, the relations $1_{A_1,A_2}$ and $1_{A_2,A_1}$ identify the potential properties of systems $A_1$ and $A_2$ and do this in a mutually inverse manner because of $1_{A_1,A_2} \circ 1_{A_2,A_1} = 1_{A_2}$ and $1_{A_2,A_1} \circ 1_{A_1,A_2} = 1_{A_1}$. The assignments $F_{A_1,A_2}(-)$ (respectively $G_{A_1,A_2}(-)$) and $F_{A_2,A_1}(-)$ (respectively $G_{A_2,A_1}(-)$) identify processes involving $A_1$ and $A_2$ as output (respectively input) in a similar manner.

**Example: Identical Processes.** We leave it to the reader to combine the notion of inverse and that of identical systems into *identical processes*.

**Example: Bosonic States.** The *symmetric states*

$$\Psi : I \to A_1 \otimes \cdots \otimes A_n,$$

which describe nonisolated bosons, can now be defined. For any permutation,

$$\sigma : \{1, \ldots, n\} \to \{1, \ldots, n\}$$

we have

$$(1_{A_1, A_{\sigma(1)}} \otimes \cdots \otimes 1_{A_n, A_{\sigma(n)}}) \circ \Psi = \Psi; \quad (2.5)$$

that is, when permuting the roles of the (identical) systems that make up the joint system, then that state should remain invariant. Graphically, we represent the symmetry relation $1_{A_1, A_{\sigma(1)}} \otimes \cdots \otimes 1_{A_n, A_{\sigma(n)}}$ induced by the permutation $\sigma$ as "re-wiring" according to $\sigma$, for example,

$$1_{A_1, A_3} \otimes 1_{A_2, A_1} \otimes 1_{A_3, A_4} \otimes 1_{A_4, A_2} \equiv$$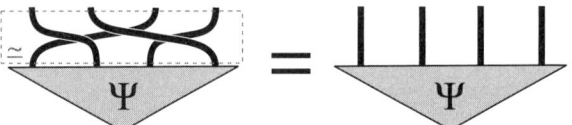

The dotted box and the "$\simeq$"-sign refer to the fact that the wires are different from those we have seen so far, which represented systems. Here, they encode a relation that changes systems. Equation (2.5) now becomes

We indeed now truly exploit the fact that in the graphical language, systems appear in a certain order, which can be used to identify systems. Symmetry relations that identify distinct identical systems now identify different positions within the order. More on this ordering and identity of systems is in Section 2.4.

We now combine symmetry relations representing identical systems with the notion of a process to derive the crucial notion of an evolution.

**Example: Evolutions.** By an *evolution*, we mean a scenario involving only causal composition and for which all maximal snapshots are identical in the preceding sense. Consider such a scenario with the process $f : A_0 \to B$ as its resolution, and let $A_\eta$ be a maximal snapshot distinct from $A_0$. Now consider the scenario that one obtains by restricting to those processes that happen before $A_\eta$, including $A_\eta$ itself; let the process $f_\eta : A_0 \to A_\eta$ be its resolution; and now consider the symmetry relation:

$$e_\eta := 1_{A_\eta, A} \circ f_\eta : A_0 \to A_0.$$

If the collection of all labels $\eta$ carries the structure of the real continuum, we obtain a generalization of the standard notion of an evolution in terms of a one-parameter family of "things"—here, symmetry relations—which, intuitively, relate potential properties of a system at time $\eta$—here, $A_0$ to those at time 0.

**Example: Symmetry Groups.** The maps $e_\eta : A_0 \to A_0$ in the previous example are special in that they relate a system to itself while typically not being identities. One could associate to each system a collection of such symmetry *endo*-relations that are closed under $\circ$ and each of which comes with an inverse; that is, for $f : A \to A$, there is $f^{-1} : A \to A$ such that $f \circ f^{-1} = f^{-1} \circ f = 1_A$. Such a collection plays the role of the symmetry groups in existing theories. It follows that a symmetry group of a system carries over to a symmetry group of an identical, and that evolutions respect symmetries.

**Example: Variable Causal Structure.** This example addresses a particular challenge posed by Lucien Hardy at a lecture in Barbados in the spring of 2008 [106].

Thus far, processes were required to respect some perceived causal structure. However, several authors argue that a framework that stands a chance to be of any use for describing quantum gravity should allow for variable causal structure (e.g., [105], [107]). Once we "solved" Einstein's equations in general relativity, then the causal structure is of course fixed, so varying causal structure does not boil down to merely dropping it but rather to allow for a variety of causal structures.

This is what we will establish here—namely, to introduce processes that have the potential to adopt many different causal incarnations while still maintaining the key role of composition within the theory. In other words, a certain causal incarnation becomes something like potential property.

For each system, we introduce two symmetry relations:

$$\cup_A : I \to A^* \otimes A \quad \text{and} \quad \cap_A : A \otimes A^* \to I,$$

to which we respectively refer to as *input–output reversal* and *output–input reversal*. These are subject to the following equations:

$$(\cap_A \otimes 1_A) \circ (1_A \otimes \cup_A) = 1_A \quad \text{and} \quad (1_{A^*} \otimes \cap_A) \circ (\cup_A \otimes 1_{A^*}) = 1_{A^*} \quad (2.6)$$

which state that reversing twice yields no reversal. For a process $f : C \otimes A \to B$ (resp. $g : A \to B \otimes C$), we can use reversal to produce a variation on it where the input $C$ (respectively, output $C$) has become an output (respectively input) $C^*$:

$$\tilde{f} = (1_{C^*} \otimes f) \circ (\cup_C \otimes 1_A) : A \to C^* \otimes B$$
$$\tilde{g} = (1_B \otimes \cap_C) \circ (g \otimes 1_{C^*}) : A \otimes C^* \to B.$$

Here, the "$*$" tells us that whereas $A$ was an input (respectively output) for process $f$ (respectively $g$), it is now converted into an output (respectively input). This is crucial when composing $f$ (respectively $g$) with other processes.

Putting this in pictures, we set

$$\cup_A \equiv \smile \quad \text{and} \quad \cap_A \equiv \frown$$

where the directions of the arrows represent the ∗s. The equations then become

$$\text{⌒⌣} = \uparrow \quad \text{and} \quad \text{⌣⌒} = \downarrow$$

and the converted processes depict as

$$\tilde{f} \equiv \boxed{f} \quad \text{and} \quad \tilde{g} \equiv \boxed{g}$$

By a *precausal process,* we mean a collection of potential processes that is closed under reversal of all inputs and outputs. For example, graphically

$$\left[\boxed{f}\right] := \left\{\boxed{f}, \boxed{f}, \boxed{f}, \boxed{f}\right\}$$

is such a precausal process. Similar to how composition of processes could be represented by directed graphs, one can show that composition of precausal processes can be represented by *un*directed graphs:

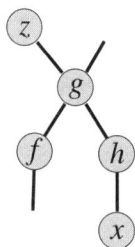

Indeed, by considering a node as representing a precausal process—that is, all of its potential causal incarnations—we obtain

$$\boxed{f \atop g} \simeq \boxed{f \atop g} = \boxed{g \atop f} = \boxed{\tilde{g} \atop \tilde{f}} \simeq \boxed{\tilde{g} \atop \tilde{f}}$$

Consequently, the directions on arrows carry no content.

Although the presentation of scenarios as nodes of undirected graphs is, of course, more concise than as collections of causal incarnations, the latter has the conceptual advantage that causal structure is attributed to processes. In our setup, these are the things that really "happen," whereas systems only play a supporting role; therefore, it is the processes that should carry the causal structure.

### 2.3.6 Vacuous Relations: Correcting Denotational Artifacts

Both processes and symmetry relations carry structural content of the theory under consideration. We mention one more kind of relation of which the sole purpose is to correct artifacts attributable to a particular choice of denotation. We have already pointed to the fact that when we denote separate composition either symbolically or diagrammatically, this unavoidably comes with some ordering because the points of a line are totally ordered.

To undo this, we need to state that all orderings are equivalent. Therefore, we introduce for each pair of systems $A_1$ and $A_2$ an invertible relation $\sigma_{A_1,A_2}$ which *exchanges* the order

$$\sigma_{A_1,A_2} : A_1 \otimes A_2 \to A_2 \otimes A_1 \quad \text{with} \quad \sigma_{A_2,A_1} \circ \sigma_{A_1,A_2} = 1_{A_1 \otimes A_2}.$$

These relations then generate arbitrary permutations, for example,[22]

$$: A_1 \otimes A_2 \otimes A_3 \otimes A_4 \to A_2 \otimes A_4 \otimes A_1 \otimes A_3.$$

To state that these exchanges of order are indeed vacuous, we have to assert that they do not affect the structural content of the theory—that is, the two compositions. First, separate compositions should be preserved:

$$\sigma_{B_1,B_2} \circ (f_1 \otimes f_2) = (f_2 \otimes f_1) \circ \sigma_{A_1,A_2}.$$

For example, for the preceding permutation of four systems, we have

That causal composition is also preserved then trivially follows:

$$\sigma_{C_1,C_2} \circ (g_1 \otimes g_2) \circ (f_1 \otimes f_2) = (g_1 \otimes g_2) \circ \sigma_{B_1,B_2} \circ (f_2 \otimes f_1)$$
$$= (g_1 \otimes g_2) \circ (f_2 \otimes f_1) \circ \sigma_{A_1,A_2}.$$

### 2.3.7 Summary of This Section

Within the proposed framework, a *physical theory* has the following ingredients:

- a collection of *relations* with two *compositions* $\circ$ and $\otimes$ thereon, subject to an *independence constraint*, as well as additional *equations* that specify for which scenarios the corresponding resolutions are equal;
- certain relations called *potential* (or *candidate*) *processes* that will act as the "actual physical substance" of the theory, according to

$$\frac{\text{actual process}}{\text{actual property}} \doteq \frac{\text{might happen}}{\text{might be}}$$

---

[22] Note the difference with the example of bosonic states earlier in this chapter in that now the wires relate a system with itself, just like identities do. They just shift the order. Hence,

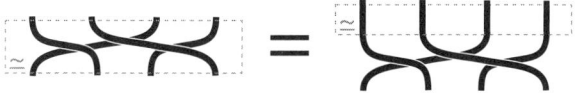

where the straight wires in the dotted box stand for change of system.

The following are examples of processes:
- states, effects, and weights
- processes resulting from performing an operation
- feed-in-environment processes witnessing nonisolation, and so on
- certain relations called *symmetry relations*, which carry additional structure of the theory; the following are examples of symmetry relations:
  - those that identify symmetries
  - those that identify identical systems
  - those that vary the causal structure, and so forth
- certain relations called *vacuous relations* that carry no physical content whatsoever but undo artifacts that are merely due to denotation

Subsequently, we identify some more ingredients, but first we see how we can cast these ingredients within standard mathematical structures.

## 2.4 The Mathematical Guise of Physical Theories

In set theory [37, 69], a *class* is a collection of which the members are defined by a predicate that they all obey. For example, the class of groups is defined as sets that come equipped with both a binary and a unary operation that obey the usual axioms of groups. By Russell's paradox, which can be restated as the fact that the collection of all sets itself does not form a set, it immediately follows that the collection of all groups together do not form a set; rather, they form a *proper class*.

### 2.4.1 Modeling Concession 1

The collection of all systems together forms a *class* and the collection of all relations of the same type forms a *set*.

This concession reflects standard mathematical practice[23] and, hence, is essential when trying to provide the "informal" ideas in the previous section with a more standard formal backbone, in terms of either axiomatics or more concrete models obeying this axiomatic.[24]

### 2.4.2 Axiomatics

The physical framework outlined in the preceding section, when subjected to the stated modeling concession 1, can be represented as a so-called strict symmetric monoidal category. For a more detailed discussion, we refer the reader to [57].

**Definition.** A *strict symmetric monoidal category* **C** consists of a class of *objects* $|\mathbf{C}|$, for each pair of objects $A, B \in |\mathbf{C}|$ a set of *morphisms* $\mathbf{C}(A, B)$,[25] a privileged *unit*

---

[23] There exist proposals to generalize this; e.g., the *universes* as in [36], Section 1.1.
[24] We refer the reader to [57] for a more detailed discussion of the sense in which we use *axiomatics* and *concrete models*, where rather than *axiomatics* we used the term *abstract*.
[25] Such a set $\mathbf{C}(A, B)$ of morphisms is usually referred to as a *homset*.

object $I \in |\mathbf{C}|$, for each object $A \in |\mathbf{C}|$ a privileged *identity* morphism $1_A \in \mathbf{C}(A, A)$, and the following operations and axioms:

- an associative binary operation $\otimes$ on $|\mathbf{C}|$ with unit $I$
- an associative binary operation $\otimes$ on $\bigcup_{A,B} \mathbf{C}(A, B)$ with unit $1_I$, and with $f_1 \otimes f_2 \in \mathbf{C}(A_1 \otimes A_2, B_1 \otimes B_2)$ for $f_1 \in \mathbf{C}(A_1, B_1)$ and $f_2 \in \mathbf{C}(A_2, B_2)$
- a partial associative binary operation $\circ$ on $\bigcup_{A,B} \mathbf{C}(A, B)$ restricted to pairs in $\mathbf{C}(B, C) \times \mathbf{C}(A, B)$ where $A, B, C \in |\mathbf{C}|$ are arbitrary, and for all $A, B \in |\mathbf{C}|$, all $f \in \mathbf{C}(A, B)$ have *right identity* $1_A$ and *left identity* $1_B$

Moreover, for all $A, B, A_1, B_1, C_1, A_2, B_2, C_2 \in |\mathbf{C}|$, $f_1 \in \mathbf{C}(A_1, B_1)$, $g_1 \in \mathbf{C}(B_1, C_1)$, $f_2 \in \mathbf{C}(A_2, B_2)$, and $g_2 \in \mathbf{C}(B_2, C_2)$, we have

$$(g_1 \circ f_1) \otimes (g_2 \circ f_2) = (g_1 \otimes g_2) \circ (f_1 \otimes f_2) \quad \text{and} \quad 1_{A_1} \otimes 1_{A_2} = 1_{A_1 \otimes A_2}.$$

Finally, for all $A_1, A_2 \in |\mathbf{C}|$, there is a privileged morphism

$$\sigma_{A_1, A_2} \in \mathbf{C}(A_1 \otimes A_2, A_2 \otimes A_1) \quad \text{with} \quad \sigma_{A_2, A_1} \circ \sigma_{A_1, A_2} = 1_{A_1 \otimes A_2}$$

such that for all $A_1, A_2, B_1, B_2 \in |\mathbf{C}|$, $f_1 \in \mathbf{C}(A_1, B_1)$, $f_2 \in \mathbf{C}(A_2, B_2)$, we have

$$\sigma_{B_1, B_2} \circ (f_1 \otimes f_2) = (f_2 \otimes f_1) \circ \sigma_{A_1, A_2}. \tag{2.7}$$

This is quite a mouthful, but there are very short, more elegant ways to say this that rely on higher-level category theory.[26] It is also a well-known fact that these strict monoidal categories are in exact correspondence with the kind of graphical calculi that we introduced to describe relations [127]. Although the use of this calculi traces back to Penrose's work in the early 1970s [155], it became a genuine formal discipline within the context of monoidal categories only with the work of Joyal and Street [127] in the 1990s. However, the first comprehensive detailed account on them was produced only in 2010 by Selinger [171], which provides an even nicer presentation. We say something more about these graphical presentations in Section 2.4.4.

We now show how the already discussed framework, subject to the modeling concession, can be interpreted in the language of strict symmetric monoidal categories. Recall here that an *isomorphism* in a category is a morphism $f : A \to B$, which has an inverse, precisely in the sense of Equation (2.3).

Systems are represented by objects of the symmetric monoidal category, relations by morphisms, and the compositions have been given matching notations. We discuss the role of some of the privileged morphisms:

- The *symmetry natural isomorphism*[27]

$$\{\sigma_{A_1, A_2} : A_1 \otimes A_2 \to A_2 \otimes A_1 \mid A_1, A_2 \in |\mathbf{C}|\}$$

  plays the role of the symmetry relation that undoes the unavoidable a priori ordering on systems when composing them with $\otimes$.

---

[26] For example, a (not-necessarily strict) symmetric monoidal category, which takes a lot more space to explicitly define than a strict one (see [8, 16, 147] for the usual definition and [57] for a discussion), is an *internal commutative monoid in the category of all categories*.

[27] The significance of the word *natural* here precisely boils down to validity of Equation (2.7).

- There may be several occurrences of the same object within a string of tensored objects (e.g., $A \otimes A$). To align this with the fact that all systems occurring in such an expression must be independent, we either

  c1 not assign any meaning to all objects and morphisms of the symmetric monoidal category but rather consider a subcategory of it with a partial tensor, an approach that is currently developed in [52]; or

  c2 represent distinct identical systems by the same object, which allows for the two $A$s in $A \otimes A$ to be interpreted as independent.

Here, c1 and c2 can also be seen as modeling concessions.

**Example: Compactness Models Variable Causal Structure.** A *compact (closed) category* [129, 130] is a symmetric monoidal category in which every object $A$ has a *dual* $A^*$; that is, there are morphisms $\cup_A : I \to A^* \otimes A$ and $\cap_A : A \otimes A^* \to I$ satisfying Equation (2.6). Equivalence classes of morphism then enable the model variable causal structure as already indicated.

**Example: Symmetry and Compactness in Communication Protocols.** In the preceding, the "symmetry" and "compact" structure represented relations that respectively undo the order on systems within scenarios or a causal structure. These morphisms can also play a more constructive role as special kinds of communication processes. It was this role that initially motivated the use of compact closed categories to model quantum protocols in [6]. To this end, we will treat the ordering of objects relative to the tensor as genuine locations in spacetime, represented by two agents, respectively named Ali and Bob. Then, the morphism $\sigma_{A_1, A_2} : A_1 \otimes A_2 \to A_2 \otimes A_1$ means that the agents *exchange* their physical systems. We can represent the agents by regions in the plane that extend vertically—that is, in the direction of causal composition:

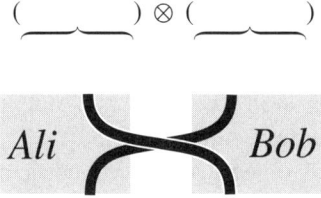

Moreover, if the category is compact closed, then by the axioms of compact closure we have the following equation between scenarios:

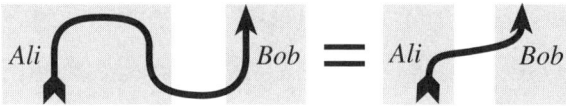

which can be interpreted as a correctness proof of *postselected quantum teleportation*. Here, $\cup_A : I \to A^* \otimes A$ represents a *Bellstate*, and $\cap_A : A \otimes A^* \to I$ represents a *postselected Belleffect*. A more detailed analysis as well as more sophisticated variations on the same theme that involve varying the entangled state and allowing for nondeterminism of the effects are in [6, 43, 47, 58, 59, 64].

**Example: Explicit Agents.** The previous example gives a concise presentation of protocols, but it is not completely consistent with our earlier interpretation of symmetry and compactness as symmetry relations. One possible manner to accommodate the use of these morphisms both as symmetry relations as well as processes is by explicitly introducing agents. To model agents, respectively named Ali and Bob, we take objects to be pairs consisting of an entry that represents the physical system together with an entry that represents the agent that possesses that system for that snapshot. Morphisms will be pairs consisting of the manner in which physical systems are processed, as well as specification of which agents possess it at the beginning and the end of the processes. We provide a rough idea of how naively this can formally be established, skimming over certain technical details. Take the *product category*

$$\mathbf{C} \times F_{\mathsf{SM}}\mathbf{Agents}$$

of the symmetric monoidal category **C** in which we model physical systems and the *free symmetric monoidal category over a category* **Agents**,[28] which has two objects *Ali* and *Bob* and only identities as morphisms:

$$\mathbf{Agents}(Ali, Ali) = \{1_{Ali}\} \qquad \mathbf{Agents}(Ali, Bob) = \emptyset$$

$$\mathbf{Agents}(Bob, Bob) = \{1_{Bob}\} \qquad \mathbf{Agents}(Bob, Ali) = \emptyset$$

The category $\mathbf{C} \times F_{\mathsf{SM}}\mathbf{Agents}$ inherits symmetric monoidal structure component-wise from **C** and $F_{\mathsf{SM}}\mathbf{Agents}$, with a symmetry morphism now of the form

$$\left(\sigma_{A_1, A_2}, \sigma_\pi\right) : (A_1, Ali) \otimes (A_2, Bob) \to (A_2, Bob) \otimes (A_1, Ali),$$

where $\pi$ is the (only) nontrivial permutation of two elements. This now represents the symmetry relation that undoes the ordering on objects. Conversely, the exchange process can now be differently represented by

$$\left(\sigma_{A_1, A_2}, 1_{Ali \otimes Bob}\right) : (A_1, Ali) \otimes (A_2, Bob) \to (A_2, Ali) \otimes (A_1, Bob).$$

So we have distinct morphisms representing both symmetry relations and processes, and the same can be done for postselected quantum teleportation.

### 2.4.3 Concrete Models

Thus far, we treated categories as a structure in their own right and, consequently, also the diagrammatic calculi. However, to realize existing theories such as quantum

---

[28] An overview of free constructions for the categories that we consider here is in [2]. The objects of the free symmetric monoidal category $F_{\mathsf{SM}}\mathbf{D}$ over a category **D** are finite lists of objects of **D**, and the morphisms are finite lists of morphisms of **D** together with a permutation of objects. Concretely, we can write these as $\sigma_\pi \circ (f : A_1 \to B_1, \ldots, A_n \to B_n) : (A_1, \ldots, A_n) \to (B_{\pi(1)}, \ldots, B_{\pi(n)})$ where $\pi : \{1, \ldots, n\} \to \{1, \ldots, n\}$ is a permutation. The permutation component alone provides the symmetry natural isomorphism.

theory, we need to consider *concrete models* of these. That is, the objects constitute some kind of mathematical strucure (e.g., Hilbert space), whereas the morphisms constitute mappings between these (e.g., linear maps). The monoidal tensor is then a binary construction on these.

But what we obtain in this manner are not strict symmetric monoidal categories but rather *symmetric monoidal categories*. In particular, we lose (strict) associativity and (strict) unitality of the tensor:

$$A \otimes (B \otimes C) \neq (A \otimes B) \otimes C \qquad I \otimes A \neq A \qquad A \otimes I \neq A$$

$$f \otimes (g \otimes h) \neq (f \otimes g) \otimes h \qquad 1_I \otimes f \neq f \qquad f \otimes 1_I \neq f.$$

This is a consequence of the fact that in set theory:

$$(x, (y, z)) \neq ((x, y), z) \qquad (*, x) \neq x \qquad (x, *) \neq x.$$

For a detailed discussion of this issue, we refer the reader to [57]. We mention here that the main consequence of this is the fact that in any standard textbook, the definition of a symmetric monoidal category may stretch many pages. The reason is that in one way or another, we need to articulate that $A \otimes (B \otimes C)$ and $(A \otimes B) \otimes C$ are in a special way related, similarly to how $A \otimes B$ and $B \otimes A$ relate was captured by the symmetry "natural isomorphisms."

The following are five examples of models of symmetric monoidal categories:

- (F)**Set**:=
  - Objects:= (finite) sets
  - Morphisms:= functions between these
  - Tensor:= the Cartesian product of sets
- (F)**Rel**:=
  - Objects:= (finite) sets
  - Morphisms:= (ordinary mathematical) relations between these
  - Tensor:= the Cartesian product of sets
- (F)**Hilb**:=
  - Objects:= (finite dimensional) Hilbert spaces
  - Morphisms:= linear maps between these
  - Tensor:= the Hilbert space tensor product
- WP(F)**Hilb**:=
  - Objects:= (finite dimensional) Hilbert spaces
  - Morphisms:= linear maps between these up to a global phase
  - Tensor:= the Hilbert space tensor product
- CP(F)**Hilb**:=
  - Objects:= (finite dimensional) Hilbert spaces
  - Morphisms:= completely positive maps between these[29]
  - Tensor:= the Hilbert space tensor product

The reason for restricting to finite sets/dimensions is explained in Section 2.4.4.

---

[29] These can, for example, be defined in terms of the Kraus representation $f :: \rho \mapsto \sum_i A_i^\dagger \rho A_i$ on the space of density matrices, where each $A_i$ is an $n \times n$-matrix [68, 133].

Mappings from one of these models, which take each object $A \in |\mathbf{C}|$ to an object $FA \in |\mathbf{C}|$ and which take each morphism $f \in \mathbf{C}(A, B)$ to a morphism $Ff \in \mathbf{C}(FA, FB)$, and which preserve the full symmetric monoidal structure, are called *strict monoidal functors*. If a strict monoidal functor is injective on homsets, it is called *faithful*. These strict monoidal functors allow one to relate different models to each other. For example, there are the identity-on-objects faithful strict monoidal functors[30]

$$F_{func} : \mathbf{Set} \hookrightarrow \mathbf{Rel} \qquad F_{pure} : \mathbf{WPFHilb} \hookrightarrow \mathbf{CPFHilb} \qquad (2.8)$$

as well as object-squaring (i.e., $\mathcal{H} \mapsto \mathcal{H} \otimes \mathcal{H}$) faithful strict monoidal functors[31]

$$F_{cp} : \mathbf{CPFHilb} \hookrightarrow \mathbf{FHilb} \qquad F_{cp} \circ F_{pure} : \mathbf{WPFHilb} \hookrightarrow \mathbf{FHilb}. \qquad (2.9)$$

But, in our view, the physical theories should primarily be formulated axiomatically rather than in terms of these models because it is at the axiomatic level that the conceptually meaningful entities live; hence, it is on those that structures should be imposed rather than providing concrete presentations of them, which typically would carry more information than necessary or meaningful. Ultimately, one would like to equip a strict symmetric monoidal category with enough structure so that we can derive all observable physical phenomena without the necessity to provide a concrete model.

Then, the choice of a particular model such as **WPFHilb** can be seen as a choice of *coordinate system,* which might enable one to solve a certain problem better than other coordinate systems. Hence, for us, the nonstrictness of the mathematical models is an unfortunate artifact, whereas the strictness that we took for granted when setting up the formalism, which is also implicitly present in the diagrammatic calculi, reflect the true state of affairs.

The category **FHilb** is the one that we typically have in mind in relation to quantum mechanics. But other models may provide the same features. These other models—in particular, those of a more combinatorial nature—might give some useful guidance toward, say, a theory of quantum gravity. Also, discrete models are also extremely useful for computer simulations.

### 2.4.4 Where Axioms and Models Meet: A Theorem

Consider the following four devices:

1. axiomatically described strict symmetric monoidal categories, possibly equipped with additional structure

---

[30] To see this for **WPFHilb** $\hookrightarrow$ **CPFHilb**, note that **WPFHilb** can be presented as a category with the same objects as **FHilb** but with maps of the form $f \otimes \bar{f} : \mathcal{H} \otimes \mathcal{H} \to \mathcal{H}' \otimes \mathcal{H}'$ (where $f : \mathcal{H} \to \mathcal{H}'$ is any linear map) as the morphisms in **WPFHilb**$(\mathcal{H}, \mathcal{H}')$ [45]. Similarly, also **CPFHilb** can be presented as a category with the homset **CPFHilb**$(\mathcal{H}, \mathcal{H}')$ containing maps of type $\mathcal{H} \otimes \mathcal{H} \to \mathcal{H}' \otimes \mathcal{H}'$, but now more general ones [170].
[31] This again relies on the presentations mentioned in the previous footnote.

2. axiomatically described symmetric monoidal categories (for which we refer to the many available textbooks and survey papers [8, 16, 147]) possibly equipped with additional structure
3. the diagrammatical calculus of strict symmetric monoidal categories (of which precise descriptions can be found in [127, 170, 171]) possibly equipped with additional graphical elements
4. the concrete category **FHilb**

If we establish an equation in one of these, what do we know about the validity of equations in one of the others?

**Theorem.** An equation between two scenarios in the language of symmetric monoidal categories follows from the axioms of symmetric monoidal categories, **if and only if** the corresponding equation between two scenarios in the language of strict symmetric monoidal categories follows from the axioms of strict symmetric monoidal categories, **if and only if** the corresponding equation in the graphical language follows from isomorphisms of diagrams.

The reader who wants to understand the nitty-gritty of this statement can consult Selinger [171]. The main point we wish to make here is that for all practical purposes, strict symmetric monoidal categories, general symmetric monoidal categories, and the corresponding graphical language are essentially one and the same thing! The first "if and only if" is referred to as either *MacLane's strictification theorem* or *coherence for symmetric monoidal categories* [57, 147].

So what about the concrete category **FHilb**? Because it is an example of a symmetric monoidal category, by the previously described result, whatever we prove about a strict one or within the diagrammatic language will automatically also hold for **FHilb**. Obviously, one would expect the converse not to hold because we are considering only a very particular symmetric monoidal category.

However, Selinger [172] recently elaborated on an existing result attributable to Hasegawa, Hofmann, and Plotkin [110] to show that there is, in fact, a converse statement, provided one adds some extra structure.

**Definition.** A *dagger compact (closed) category*[32] [6, 7, 170] is a compact (closed) category **C** together with a *dagger functor*; that is, for all $A, B \in |\mathbf{C}|$, a mapping

$$\dagger_{A,B} : \mathbf{C}(A, B) \to \mathbf{C}(B, A),$$

which is such that $\dagger_{A,B}$ and $\dagger_{B,A}$ are mutually inverse, and which moreover preserves the composition and the tensor structure, including units and identities.

---

[32] Strict dagger compact (closed) categories appeared in the work of Baez and Dolan [15] as a special case for $n = 1$ and $k = 3$ of k-tuply monoidal n-categories with duals.

Graphically, the dagger functor merely flips things upside-down [47, 170]:

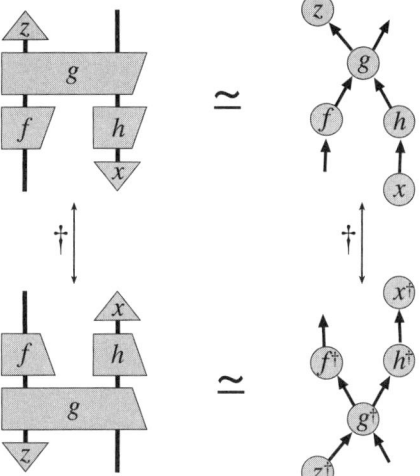

In the representation on the left, we made the boxes asymmetric to distinguish between a morphism and its dagger. In the one on the right, because nodes have no a priori orientation in the plane, we used an explicit involution on the symbols.

**Theorem.** An equation between two scenarios in the language of dagger compact categories follows from the axioms of dagger compact categories, **if and only if** the corresponding equation in the graphical language follows from isomorphisms of diagrams, **if and only if** an equation between two scenarios in the language of dagger compact (closed) categories holds in **FHilb**.

Because there are faithful strict monoidal functors that embed WP**FHilb** as well as CP**FHilb** within **FHilb**, the correspondence with the diagrammatic language also carries over to these models.

The question of whether we can carry this through for richer languages than that of dagger compact categories remains open. Still, the language of dagger compact categories already captures many important concepts: trace, transpose, conjugate, adjoint, inner product, unitary, and (complete) positivity [47].

Although admittedly the conceptual significance of the dagger is still being discussed,[33] besides *compactness*, the dagger is what truly gives a theory its *relational character*. In particular, it is a key property of the category **FRel** that the category **FSet** fails to admit: each relation has a converse relation. This is also the reason why **FRel** and **FHilb** are so alike in terms of their categorical structure, whereas **FRel** and **FSet** are very different in terms of categorical structure, despite the fact that **FRel** and **FSet** have the same objects and the same compositions, and that the morphisms of **FSet** are

---

[33] Also mathematically, there are some issues with the dagger to which some refer as "evil." The main problem is that the structure of the dagger—in particular, its strict action on objects—is not preserved under so-called categorical equivalences. This has been the subject of a recent long discussion on the categories mailing list involving all the "big shots" of the area.

a subset of those of **FRel**.[34] Intuitively, the reason for this is that linear maps (when conceived as matrices) can be seen as some kind of generalized relations in that they do not just encode whether two things relate but also in which manner that they relate, by means of a complex number. In contrast to the many who conceive quantum theory as a generalized probability theory, for us it is rather a theory of generalized relations, the latter now to be taken in its mathematical sense.

**Example: Spekkens's Toy Qubit Theory.** Spekkens [175] suggested dagger duality as an axiom for a class of theories that would generalize his toy qubit theory [174]. The concrete presentation of Spekkens's qubit theory as a dagger compact category **Spek** is in [49, 50, 81]. This presentation enabled a clear comparison with a dagger compact category **Stab** that encodes stabilizer qubit theory, from which it emerged that the only difference between the toy qubit theory and stabilizer qubit theory is the different group structure of the *phase groups* [50],[35] a concept introduced by Duncan and this author in [48].

We are also in a position to explain why we restricted to finite sets/dimensions. Whereas **Rel** is compact closed and has a dagger structure, **Hilb** neither is compact closed nor has a dagger.[36] Whereas some may take this as an objection to the dagger compact structure, we think that the fact that **Hilb** fails to be dagger compact may be an artifact of the Hilbert space structure rather than a feature of nature. Having said this, we do agree that dagger compactness surely is not the end of the story. In particular, we would like to conceive also the dagger as some kind of relation rather than as an operation on a category as a whole.[37]

### 2.4.5 Nonisolation in the von Neumann Quantum Model

To assert that a physical theory includes *nonisolated* (i.e., *open*) systems for every system $A \in |\mathbf{C}|$, we considered a designated process $\top_A : A \to \mathrm{I}$ with

$$\top_A \otimes \top_B = \top_{A \otimes B} \quad \text{and} \quad \top_\mathrm{I} = 1_\mathrm{I}, \qquad (2.10)$$

which are such that for all $A \in |\mathbf{C}|$, the mappings

$$\top_A \circ - : \mathbf{C}(\mathrm{I}, A) \to \mathbf{C}(\mathrm{I}, \mathrm{I})$$

assign the weights of each of these states. We now present a result that characterizes an additional condition that these processes have to satisfy relative to a collection of

---

[34] A detailed analysis of the similarities between **FRel** and **FHilb** and the differences between **FRel** and **FSet** is in [57]. To mention one difference: in **FSet**, the Cartesian product behaves like a nonlinear conjunction, whereas in **FRel** it behaves like a linear conjunction.

[35] These correspond to the two available four-element Abelian groups, the four-element cyclic group for stabilizer qubit theory, and the Klein four group for the toy qubit theory.

[36] We do obtain a dagger when restricting to bounded linear maps, and there are also category-theoretic technical tricks to have something very similar to compact structure [5].

[37] There are manners to do this, but we do not go into them here.

*isolated* (or *closed*) processes in order that:

$$\frac{\text{open processes}}{\text{closed processes}} \stackrel{.}{=} \frac{\text{mixed state quantum theory}}{\text{pure state quantum theory}} \stackrel{.}{=} \frac{\text{CPFHilb}}{\text{WPFHilb}}.$$

That is, in words, if we know that our theory of closed systems is ordinary quantum theory of closed systems, what do we have to require from the feed-into-environment processes such that the whole theory corresponds to quantum theory of open systems? This condition turns out to be nontrival.

Consider a symmetric monoidal category **C** with feed-into-environment processes; that is, for each $A \in |\mathbf{C}|$ a designated morphism $\top_A : A \to I$ satisfying Equation (2.10). Assume that it contains a subsymmetric monoidal category $\mathbf{C}_{pure}$, and we refer to the morphisms in it as *pure*. By a *purification* of a morphism of $f : A \to B$ in **C**, we mean a morphism $f_{pure} : A \to B \otimes C$ in $\mathbf{C}_{pure}$ such that:

$$(1_B \otimes \top_C) \circ f_{pure} = f. \tag{2.11}$$

We say that $\mathbf{C}_{pure}$ *generates* **C** whenever each morphism in **C** can be purified.

**Example: Purification in Probabilistic Theories.** The power of purification as a postulate is exploited by Chiribella, D'Ariano, and Perinotti [40].

By combining the results in [170] with those of [46], we obtain:

**Theorem.** If $\mathbf{C}_{pure} \simeq$ WP**FHilb** generates **C**, and if for the usual dagger functor on WP**FHilb** we have for all $f : C \to A$ and $g : C \to B$ in WP**FHilb**:

$$\top_A \circ f = \top_B \circ g \iff f^\dagger \circ f = g^\dagger \circ g, \tag{2.12}$$

then $\mathbf{C} \simeq$ CP**FHilb**.

More generally, for any pair **C** and $\mathbf{C}_{pure}$, the conditions in Equations (2.10, 2.11) and a slight generalization of Equation (2.12) together allow one to construct the whole category **C** from morphisms in $\mathbf{C}_{pure}$ by only using the dagger symmetric monoidal structure, together with a canonical inclusion of $\mathbf{C}_{pure}$ within **C**, a result that is obtained by combining the results in [170], [46], and [44]. If the category **C** is compact closed, as is the case for WP**FHilb**, then Equation (2.12) does suffice. In this case, following Selinger in [170], the open processes in the constructed category CP**C** all take the form

for some $f_{pure} : A \to B \otimes C$ in $\mathbf{C}_{pure}$, where the left-right reflection represents the composite of the dagger and *transposition*, explicitly:

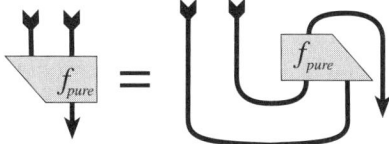

In WP**FHilb** this is nothing but complex conjugation (see [47] for more details on this). The subcategory of pure processes consists of those of the form:

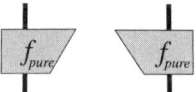

Graphically, condition (2.12) can then be rewritten as

and from it immediately follows, setting $g := 1_A$, that

Calling processes $f_{pure}$ that obey

$$f^\dagger_{pure} \circ f_{pure} = 1_A$$

*isometries*, it then follows that isometries are exactly those processes that leave the feed-into-environment processes invariant.

Condition (2.12) in the compact case can be equivalently presented as

which provides a direct translation between Selinger's presentation of WP**FHilb** and one that relies on the feed-into-environment processes. The more general form of the noncompact case mentioned previously is now obtained by "undoing" all compact morphisms, which requires introduction of symmetry morphisms:

Importantly, both (WP)**C** and CP**C** are symmetric monoidal (and compact) if **C** is, so CP**C** admits a graphical language in its own right without reference to the underlying symmetric monoidal category **C**.

### 2.4.6 Nonisolation and Causality

If we restrict to processes that "happen with certainty," then, as shown by Chiriballa, D'Ariano, and Perinotti, uniqueness of a deterministic effect enforces *causality* in the sense that states of compound systems have well-defined marginals [40]. In category-theoretic terms, this uniqueness means that I is *terminal*; that is, for each object $A$ there is a unique morphism of type $A \to I$, which will then play the role of $\top_A$. It then immediately follows that

$$\top_{A \otimes B} = \top_A \otimes \top_B,$$

and hence that there are no entangled effects. The manner in which

- connectedness in graphical calculus as expressing causal connections, and
- this notion of causality in terms of a terminal object

are related is currently being explored in collaboration with Lal [52].

## 2.5 Classicality and Measurement

For us, a classicality entity is one for which there are no limitations to be *shared* among many parties; that is, using quantum information terminology, which can be *broadcast* [21]. It is witnessed by a collection of processes that establish this sharing/broadcasting. To give an example, although an unknown quantum state cannot be cloned [70, 186], this scientific fact itself is, of course, available to every individual of the scientific community by means of writing an article about it and distributing copies of the journal in which it appears.

So our notion of classicality makes no direct reference to anything "material" but rather to the ability of a logical flow of information to admit "branching." In relational terms, it will be witnessed by the relation that identifies the branches as being identical. The power of this idea for describing quantum information tasks is discussed in an article with Simon Perdrix [62], where it is also discussed that *decoherence* can be seen as a material embodiment of this idea.

Our explorations have indeed made us realize that rather than starting from a classical theory that one subjects to a *quantization* procedure to produce a theory that can describe quantum systems, one obtains an elegant compositional mathematical framework when, starting from a "quantum" universe of processes, one identifies classicality in this manner. Put in slightly more mathematical terms, citing John Baez in TWF 268 [14] on our work:

> Mathematicians in particular are used to thinking of the quantum world as a mathematical structure resting on foundations of classical logic: first comes set theory, then Hilbert spaces on top of that. But what if it's really the other way around? What if classical mathematics is somehow sitting inside quantum theory? The world is quantum, after all.

This idea of "classical objects living in a quantum world governed by quantum rules" was introduced by Pavlovic and this author [60] and further elaborated in [56, 58, 61]. In terms of symmetric monoidal categories, we are speaking the language of certain kinds of so-called internal Frobenius algebras [39, 60, 61]. Interestingly, these

Frobenius algebras appeared first in the literature in Carboni and Walter's axiomatization of the category **Rel** [39].

### 2.5.1 Classicality

In what follows, when we (slightly abusively) denote several systems by the same symbol, we think of them as distinct identical systems and not as the same one.

The processes that establish the shareability that is characteristic for classicality implements an "equality" between the distinct instances of that entity, and we depict them in a "spider-like" manner:

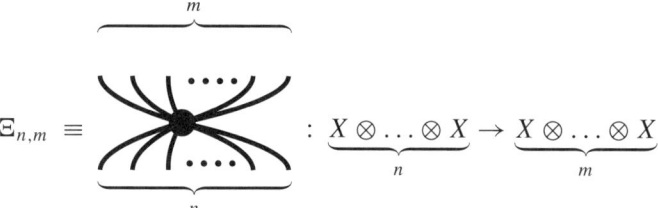

By transitivity of equality, it immediately follows that:

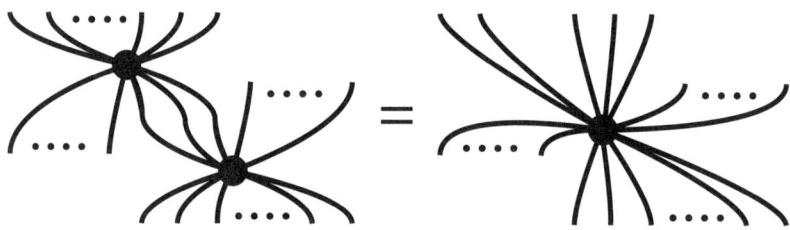

One may distinguish two kinds of classicality. The more restrictive first kind, called *controlled* (or *closed* or *pure*), requires sharing to be within the domain of consideration. This is explicitly realized by

$$\Xi_{1,1} = 1_X \qquad \text{i.e.,} \qquad \begin{array}{c}\bullet\\|\end{array} = \;|\;.$$

The second kind of classicality, called *uncontrolled* (or *open* or *mixed*) allows sharing to be outside our domain of consideration, which in the light of the previous composition rule is realized by:

$$\Xi^o_{1,0} = \top_X \qquad \text{i.e.,} \qquad \begin{array}{c}\circ\\|\end{array} = \;\triangleq\;,$$

because then we obtain:

$$(\top_X \otimes 1_{X \otimes \cdots \otimes X}) \circ \Xi^o_{n,m} = \Xi^o_{n,m-1} \qquad \text{i.e.}$$

that is, uncontrolled sharing is invariant under feed-into-environment processes.

These two forms of classicality may be naturally related to each other by introducing feed-into-environment processes within the closed spiders or, dually put, by considering closed spiders as purifications of the open ones

$$(\top_X \otimes 1_{X \otimes \cdots \otimes X}) \circ \Xi_{n,m} = \Xi^o_{n,m-1} \quad \text{i.e.}$$

which can, in fact, be summarized as the following two equations:

We can identify some special examples:

- erasing := $\Xi_{1,0} \equiv$
- cloning := $\Xi_{1,2} \equiv$
- correlating := $\Xi_{0,2} \equiv$
- comparing := $\Xi_{2,0} \equiv$
- matching := $\Xi_{2,1} \equiv$
- either := $\Xi_{0,1} \equiv$

Conceptually, it is more than fair to cast doubt on physical meaningfulness of closed spiders. To see this, it suffices to consider the erasing operation in light of Landauer's principle [139, 149]. Also, when thinking of a cloning operation, then one usually would assume some ancillary state onto which one clones, and this ancillary state by the very definition of state is an open process:

Still, it is useful to retain the closed spiders as an idealized concept given that their behavioral specifications exactly match the well-understood mathematical gadget of commutative Frobenius algebras (see Section 2.5.4).

A particularly relevant open spider is the purification of copying

- broadcasting := $\Xi^o_{1,1} \equiv$

Although it has the type of an identity, it is genuinely nontrivial.

**Example: Cloning and Broadcasting in Quantum Information.** A hint as to why the preceding may indeed characterize classicality on-the-nose comes from the *no-cloning theorem* [70, 186], which states that the only quantum states that can be *copied* by a single operation have to be orthogonal. Maximal sets of these jointly copyable states make up an orthonormal basis; that is, a (pure) classical "slice" of quantum theory. Similarly, the *no-broadcasting theorem* [21] states that the only quantum states that can be *broadcast* by a single operation correspond to a collection of density matrices that are diagonal in the same orthonormal basis. The following table summarizes cloneability/broadcastability:

|               | Pure classical | Mixed classical | Pure quantum | Mixed quantum |
|---------------|----------------|-----------------|--------------|---------------|
| Broadcastable: | Yes           | YES             | No           | No            |
| Cloneable:    | Yes            | NO              | No           | No            |

Conversely, for an orthonormal basis $\{|i\rangle\}_i$ of $\mathcal{H}$, the corresponding broadcast operation is the following completely positive map:

$$|i\rangle\langle j| \mapsto \delta_{ij}|i\rangle\langle i|.$$

Clearly, this completely positive map totally destroys coherence; hence, broadcasting is physically embodied by *decoherence*. Decoherence can indeed be seen as "sharing with (cf. coupling to) the environment."

Given that in quantum information both copying and broadcasting enable characterization an orthonormal basis, the question then remains to define candidate cloning/broadcasting operations in a manner that there is a one-to-one correspondence between such operations and orthonormal bases. That is exactly what we did, as the following theorem confirms.

**Theorem.** In (WP)**FHilb**, the previously defined families

$$\mathcal{X} = \{\Xi_{n,m} \mid n, m \in \mathbb{N}\}$$

of closed spiders are in bijective correspondence with orthogonal bases. Moreover, if we have that $\Xi_{n,m} = \Xi_{m,n}^{\dagger}$ for all $n, m$, then this basis is orthonormal.

To show this, we need to combine Lack's (highly abstract) account on spiders [135] (of which a more accessible direct presentation is in [56]) with a result obtained by Pavlovic, Vicary, and this author [61] (see Section 2.5.4).

This result states that all nondegenerate observables can indeed be bijectively represented by these sharing processes. Now we establish that spiders are also expressive enough to associate a corresponding "classical slice" of the universe of all processes to each family $\mathcal{X}$. We *assert classicality of a process* by imposing *invariance under broadcasting/decoherences* [58].[38] The copyability of pure classical data can be used to assert deterministic processes. We conveniently set $o_{\mathcal{X}} = \Xi_{1,1}^{o} \in \mathcal{X}$ and $\delta_{\mathcal{X}} = \Xi_{1,2} \in \mathcal{X}$.

---

[38] This is akin to Blume-Kohout and Zurek's *quantum Darwinism* [29, 188].

**Definition.** A *classical process* is a process of the form:

$$o_Y \circ f \circ o_X \equiv \boxed{f}$$

where $f : X \to Y$ can be an arbitrary process. Evidently, classical processes can be equivalently defined as processes $f : X \to Y$, which satisfy

$$o_Y \circ f \circ o_X = f \quad \text{that is} \quad \boxed{f} = \boxed{f}.$$

Such a classical process is *normalized* or *stochastic* if we have

$$\top_Y \circ f = \top_X \quad \text{that is} \quad \boxed{f} = \;\bigg|,$$

and it is deterministic moreover if we have that

$$\delta_Y \circ f = (f \otimes f) \circ \delta_X \quad \text{that is} \quad \boxed{f} = \boxed{f}\;\boxed{f}.$$

**Theorem.** In **CPFHilb**, normalized classical processes correspond exactly to the usual notion of stochastic maps—that is, matrices with positive real entries such that all columns add up to one—and deterministic processes correspond to functions—that is, matrices with exactly one entry in each column.

This result was shown by Paquette, Pavlovic, and this author in [58], where many other species of classical processes (doubly stochastic, partial processes, relations, ...) are defined in a similar manner.[39]

**Challenge.** Develop the above without any reference to closed spiders.

### 2.5.2 Measurement

Once we have identified these classical entities (provided they exist at all), we may wish to represent general processes relative to this entity. This is obviously what *observables* (or *measurements*) in quantum theory aim to do. One thing we know for a fact is that it

---

[39] We adopted Carboni and Walters's axiomatization of the category of relations [39], which also involved introducing the *Frobenius law*, to the probabilistic and the quantum case.

is not possible to represent the whole universe of processes by means of such an entity. So what is the best for which we could aim?

By a *probe*, we mean a process

$$(1_B \otimes o_\mathcal{X}) \circ m : A \to B \otimes X \quad \text{that is}$$

We call it a *nondemolition* probe if $A$ and $B$ are identical systems, which we denote by setting $A = B$, and we call it a *demolition* probe if $B$ is I. By a *von Neumann probe*, we mean a nondemolition one that is such that

$$(m \otimes 1_X) \circ m = (1_A \otimes \delta_\mathcal{X}) \circ m \quad \text{that is} \quad \phantom{xxx} = \phantom{xxx} \quad (2.13)$$

where we used grayscale to distinguish the systems $A$ and $X$ for the reader's convenience. There is a very straightforward interpretation to Equation (2.13): applying the same probe twice is equal to applying it once and then copying the output. More intuitively put, the $A$-output after the first application is such that the probe produces the same $X$-output (and also the same $A$-output) after a second application. This means that there is strict relationship between the $A$-output and the $X$-output for that probe:

Set $\frown_\mathcal{X} = \Xi_{2,0} \in \mathcal{X}$ and $e_\mathcal{X} = \Xi_{1,0} \in \mathcal{X}$.

**Theorem.** In (WP)**FHilb**, von Neumann probes exactly correspond with spectra of *mutually orthogonal idempotents* $\{P_i\}$; that is

$$P_i \circ P_j = \delta_{ij} \cdot P_i.$$

Moreover, if we have that this probe is *self-adjoint*

$$m^\dagger = (1_A \otimes \delta_\mathcal{X}) \circ (1_A \otimes \frown_\mathcal{X}) \quad \text{that is} \quad \phantom{xxx} = \phantom{xxx} \quad (2.14)$$

then these idempotents are *orthogonal* projectors; that is:

$$P_i^\dagger = P_i,$$

and if

$$(1_A \otimes e_\mathcal{X}) \circ m = 1_A, \quad \text{that is,} \quad \phantom{xxx} = \phantom{xxx} \quad (2.15)$$

then this spectrum is exhaustive, that is, $\sum_i P_i = 1_\mathcal{H}$.

This result was shown by Pavlovic and this author in [60].

**Challenge.** Develop the above without any reference to closed spiders.

## 2.5.3 Classicality in the von Neumann Quantum Model

We now discuss how classicality fits within the model of the environment (i.e., open systems) of Section 2.4.5. First note that classicality in this sense automatically yields *self-dual* (i.e., $X = X^*$) compactness:

$$\bigcap \cup = \bullet = |$$

In CPC, decoherences $o_X$ take the shape

$$\text{[diagram]} = \times$$

and, consequently, classical operations take the shape

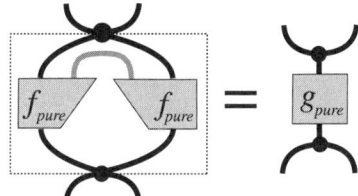

and for some process, $g : X \to Y$. Hence, we obtain

$$\frac{\text{classical}}{\text{nonclassical}} \;\dot{=}\; \frac{\text{one wire}}{\text{two wires}}.$$

This fact seems to be closely related to Hardy's axiom $K = N^2$ [104], which in turn is closely related to Barrett's *local tomography* assumption [23].

The interpretation of Equation (2.15) in light of Equations (2.13) and (2.14) is intriguing:

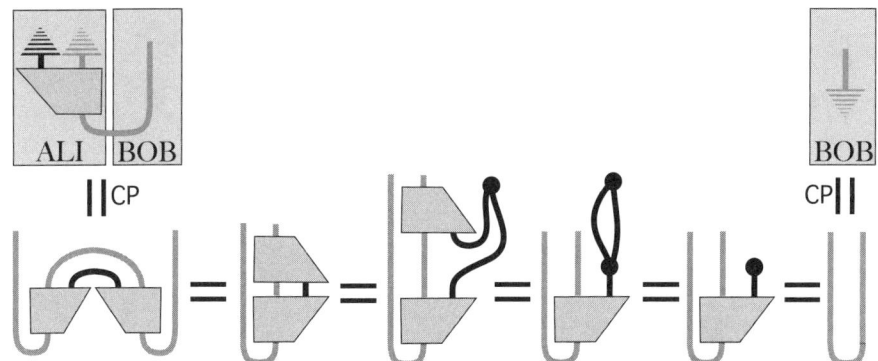

The left-upper picture articulates a protocol where Ali and Bob share a Bell state and Ali performs a measurement on it. We are interested in the resulting state at Bob's end; therefore, we feed Ali's outputs into the environment. Using Equations (2.13),

(2.14), and in particular also (2.15) within the CP-representation, it then follows that what Bob sees is the dagger of a feed-into-environment process—that is, a maximally mixed state. Hence, this protocol provides Bob with no knowledge whatsoever, hence not violating no-faster-than-light signaling.

**Challenge.** Develop the preceding without any reference to closed spiders. Then relate this to Hardy's and D'Arianio's research programs.

### 2.5.4 The Algebra of Classical Behaviors

The preceding seems to have little to do with the structures that we usually encounter in mathematics, and quantum theory in particular. We now relate it to "semifamiliar" mathematical structures, which are the ones that enable establishment of the relation with orthonormal bases already mentioned.

#### *2.5.4.1 Commutative Monoids and Commutative Comonoids*

A *commutative monoid* is a set $A$ with a binary map

$$-\bullet- : A \times A \to A$$

that is commutative, associative, and unital; that is,

$$(a \bullet b) \bullet c = a \bullet (b \bullet c) \qquad a \bullet b = b \bullet a \qquad a \bullet 1 = a.$$

In other words, which may appeal more to the physicist, it is a group without inverses. Note that we could also define a monoid as a one-object category. Slightly changing the $\bullet$-notation to

$$\mu : A \times A \to A$$

for which we now have

$$\mu(\mu(a, b), c) = \mu(a, \mu(b, c)) \qquad \mu(a, b) = \mu(b, a) \qquad \mu(a, 1) = a$$

enables us to write these conditions in a manner that makes no further reference to the elements $a, b, c \in A$; namely,

$$\mu \circ (\mu \times 1_A) = \mu \circ (1_A \times \mu) \qquad \mu = \mu \circ \sigma \qquad \mu \circ (1_A \times u) = 1_A$$

with

$$\sigma : A \times A \to A \times A :: (a, b) \mapsto (b, a) \qquad u : \{*\} \to A :: * \mapsto 1$$

where $\{*\}$ is any singleton. This perspective emphasizes how the map $\mu$ "interacts" with itself, as opposed to how it "acts" on elements, which clearly brings us closer to the process view advocated in this chapter.

This change of perspective also allows us to define a new concept merely by reversing the order of all compositions and types. Concretely, a *cocommutative comonoid* is a set $A$ with two maps:

$$\delta : A \to A \times A \qquad \text{and} \qquad e : A \to \{*\}$$

which is cocommutative, coassociative, and counital; that is,

$$(\delta \times 1_A) \circ \delta = (1_A \times \delta) \circ \delta \qquad \delta = \sigma \circ \delta \qquad (1_A \times e) \circ \delta = 1_A.$$

Obviously, there are many well-known examples of monoids, typically monoids with additional structure (e.g., groups). Another one is the two-element set {0, 1} equipped with the "and"-monoid:

$$\wedge : \{0, 1\} \times \{0, 1\} \to \{0, 1\} :: \begin{cases} (0,0) \mapsto 0 \\ (0,1) \mapsto 0 \\ (1,0) \mapsto 0 \\ (1,1) \mapsto 1 \end{cases} \qquad u_\wedge : \{*\} \to \{0, 1\} :: * \mapsto 1.$$

Given that from this perspective, monoids and comonoids are very similar things, why do we never encounter comonoids in a standard algebra textbook? Let us first look at an example of such a comonoid, just to show that such things do exist. Let $X$ be any set and

$$\delta : X \to X \times X :: x \mapsto (x, x) \qquad e : X \to \{*\} :: x \mapsto *.$$

The map $\delta$ *copies* the elements of $X$ and $e$ *erases* them. Here, coassociativity means that if we wish to obtain three copies of something, then after first making two copies, it does not matter which of these two we copy again. Cocommutativity tells us that after copying, we exchange the two copies we still have the same. Counitality tells us that if we first copy and then erase one of the copies, this is the same as doing nothing.

Now, the reason why you will not encounter any comonoids in a standard algebra textbook is simply because this example is the only example of a commutative comonoid; hence, it carries no real content (i.e., it freely arises from the underlying set). But the reason for the trivial nature of commutative comonoids is the fact of the following being functions:

$$\mu : A \times A \to A \qquad \delta : A \to A \times A \qquad u : \{*\} \to A \qquad e : A \to \{*\}.$$

In other words, $\mu$ and $\delta$ are morphisms in the category **FSet**. Although the concept of a commutative monoid is interesting in **FSet**, that of a cocommutative comonoid is not in **FSet**. However, if we put the preceding definition in the language of monoidal categories and pass to categories other than **FSet**, then the situation changes. In fact, if this category has a †-functor, then to each commutative monoid corresponds a cocommutative comonoid. This already happens when we relax the condition that $\mu$ and $\delta$ are functions to $\mu$ and $\delta$ being relations.

Let **C** be any symmetric monoidal category. A *commutative* **C**-*monoid* is an object $A \in |\mathbf{C}|$ with morphisms

$$\mu : A \otimes A \to A \qquad u : I \to A$$

which is commutative, associative, and unital; that is,

$$\mu \circ (\mu \otimes 1_A) = \mu \circ (1_A \otimes \mu) \qquad \mu = \mu \circ \sigma \qquad \mu \circ (1_A \otimes u) = 1_A.$$

Similarly, a *cocommutative* **C**-*comonoid* is an object $A$ with morphisms

$$\delta : A \to A \otimes A \qquad e : A \to I,$$

which is cocommutative, coassociative, and counital; that is,

$$(\delta \otimes 1_A) \circ \delta = (1_A \otimes \delta) \circ \delta \qquad \delta = \sigma \circ \delta \qquad (1_A \otimes e) \circ \delta = 1_A.$$

Now, putting all of this diagrammatically, a commutative **C**-monoid is a pair

$: A \otimes A \to A$ $\qquad$ $: I \to A$

satisfying

and a cocommutative **C**-comonoid is a pair

$: A \to A \otimes A$ $\qquad$ $: A \to I$

satisfying

Recall also that it is a general fact in algebra that if a binary operation $- \bullet -$ has both a left unit $1_l$ and right unit $1_r$, then these must be equal to:

$$1_l = 1_l \bullet 1_r = 1_r.$$

It then also follows that a commutative multiplication can only have one unit; that is, if it has a unit, then it is completely determined by the multiplication. This fact straightforwardly lifts to the more general kinds of monoids and comonoids that we discussed previously; therefore, we will on several occasions omit specification of the (co)unit.

Here is an example of commutative monoids and corresponding cocommutative comonoids in **FHilb** on a two-dimensional Hilbert space:

$$:: \begin{cases} |00\rangle, |01\rangle, |10\rangle \mapsto |0\rangle \\ |11\rangle \mapsto |1\rangle \end{cases} \quad \Big| \quad :: \begin{cases} |0\rangle \mapsto |00\rangle + |01\rangle + |10\rangle \\ |1\rangle \mapsto |11\rangle \end{cases}$$

The first monoid has the "and" operation applied to the $\{|0\rangle, |1\rangle\}$-basis as its multiplication. The comultiplication is the corresponding adjoint.

#### 2.5.4.2 Commutative Dagger Frobenius Algebras

Now consider the following three comultiplications:

$$\mu_Z = \begin{array}{c}\includegraphics[scale=0.5]{muz.png}\end{array} \ :: \ \begin{cases} |00\rangle \mapsto |0\rangle \\ |11\rangle \mapsto |1\rangle \end{cases} \quad \bigg| \quad \delta_Z = \begin{array}{c}\includegraphics[scale=0.5]{dz.png}\end{array} \ :: \ \begin{cases} |0\rangle \mapsto |00\rangle \\ |1\rangle \mapsto |11\rangle \end{cases}$$

$$\mu_X = \begin{array}{c}\end{array} \ :: \ \begin{cases} |++\rangle \mapsto |+\rangle \\ |--\rangle \mapsto |-\rangle \end{cases} \quad \bigg| \quad \delta_X = \begin{array}{c}\end{array} \ :: \ \begin{cases} |+\rangle \mapsto |++\rangle \\ |-\rangle \mapsto |--\rangle \end{cases}$$

$$\mu_Y = \begin{array}{c}\end{array} \ :: \ \begin{cases} |\sharp\sharp\rangle \mapsto |\sharp\rangle \\ |==\rangle \mapsto |=\rangle \end{cases} \quad \bigg| \quad \delta_Y = \begin{array}{c}\end{array} \ :: \ \begin{cases} |\sharp\rangle \mapsto |\sharp\sharp\rangle \\ |=\rangle \mapsto |==\rangle \end{cases}$$

Each of these is defined as a copying operation of some basis, respectively,

$$\mathcal{Z} = \{|0\rangle, |1\rangle\} \quad \mathcal{X} = \{|+\rangle = |0\rangle + |1\rangle, |-\rangle = |0\rangle - |1\rangle\}$$
$$\mathcal{Y} = \{|\sharp\rangle = |0\rangle + i|1\rangle, |=\rangle = |0\rangle - i|1\rangle\},$$

that is, the eigenstates for the usual Pauli operators. Each of these encodes a basis in the sense that we can recover the basis from the comultiplication as those vectors that satisfy

$$\delta(|\psi\rangle) = |\psi\rangle \otimes |\psi\rangle.$$

The fact that no other vector besides those that we, by definition, copy are in fact copied is a consequence of the previously mentioned no-cloning theorem [70, 186].

The corresponding multiplications are again their adjoints. These last examples embody the reason why we are interested in commutative comonoids. What is already remarkable at this stage is that each of these encodes an orthonormal basis in a language involving only composition and tensor. There is no reference whatsoever to either sums or scalar multiples, in contrast to the usual definition of an orthonormal basis $\{|i\rangle\}$ on a Hilbert space:

$$\forall |\psi\rangle \in \mathcal{H}, \exists (c_i)_i \in \mathbb{C}^n : |\psi\rangle = \sum_i c_i |i\rangle \quad \forall i, j : \langle i|j\rangle = \delta_{ij}.$$

But, there is more. In fact, one can endow these monoids and comonoids with some additional properties, expressible in a language involving only composition, tensor, and now also adjoint, such that they are in bijective correspondence with orthonormal bases.

**Definition.** A *commutative algebra* in a symmetric monoidal category is a pair consisting of a commutative monoid and a cocommutative comonoid on the same object. A *special commutative Frobenius algebra* is a commutative algebra that moreover is *special* and satisfies the *Frobenius law*, respectively:

A *special commutative dagger Frobenius algebra* or *classical structure* or *basis structure* in a dagger symmetric monoidal category is a *special commutative Frobenius algebra* for which the monoid is the dagger of the comonoid.

Think of an orthonormal basis $\mathcal{B} = \{|b_i\rangle\}_i$ for a Hilbert space $\mathcal{H}$ as the pair $(\mathcal{H}, \mathcal{B})$ consisting of the Hilbert space, which carries this basis as additional structure. Because the multiplication and the comultiplication are related by the dagger, and because having a unit is a property rather than a structure, we denote a special commutative dagger Frobenius algebra on an object $A$ as $(A, \delta)$.

**Theorem.** There is a bijective correspondence between orthogonal bases for finite-dimensional Hilbert spaces and special commutative Frobenius algebras in **FHilb**. This correspondence is realized by the mutually inverse mappings:

- Each special commutative Frobenius algebra $(\mathcal{H}, \delta)$ is mapped on $(\mathcal{H}, \mathcal{B}_\delta)$, where $\mathcal{B}_\delta$ consists of the set of vectors that are copied by $\delta$.
- Each orthonormal basis $(\mathcal{H}, \mathcal{B})$ is mapped on $(\mathcal{H}, \delta_\mathcal{B} : \mathcal{H} \to \mathcal{H} \otimes \mathcal{H})$, where $\delta_\mathcal{B}$ is the linear map that copies the vectors of $\mathcal{B}$.

Restricting to orthonormal bases corresponds to restricting to special commutative dagger Frobenius algebras. This result was shown by Pavlovic, Vicary, and this author in [61].

That classicality boils down to families of spiders is a consequence of the fact that special commutative dagger Frobenius algebras are in bijective correspondence with spiders, as our notation already indicated. This was shown by Lack [135], but in a manner so abstract that it may not be accessible to the reader. A more direct presentation of the proof is in [56].

### 2.5.4.3 Varying the Coordinate System

But what if we change the category? It turns out that this mathematical concept, when we look through coordinate systems other than **FHilb**, allows us to discover important quantum mechanical concepts in places where one does not expect it, most notably "complementarity" or "unbiasedness" [168].

Duncan and this author [48] defined *complementarity* in terms of special commutative dagger Frobenius algebras (i.e., still in terms of a language only involving composition, tensor, and adjoint) in a manner that yields the usual notion in **FHilb**. Concretely, it was shown that classical structures

$$\left(\mathcal{H}, \delta_G = \; \curlyvee \right) \quad \text{and} \quad \left(\mathcal{H}, \delta_R = \; \curlyvee \right)$$

in **FHilb** are complementary if and only if we have

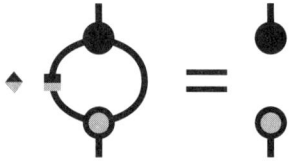

where ♠ is a normalizing scalar and ✠ is a so-called *dualizer* [48, 59], both of which are obtained by composing $\delta_g^\dagger$, $\delta_r$, $u_g$, and $u_r^\dagger$ in a certain manner.[40]

The quite astonishing fact discovered by Edwards and this author [49] was that even in **FRel**, one encounters such complementary classical structures, even already on the two-element set {0, 1}:

$$\begin{array}{c}\text{Z} \\[2pt] :: \begin{cases} (0,0) \mapsto 0 \\ (1,1) \mapsto 1 \end{cases} \quad\bigg|\quad \text{Z} :: \begin{cases} 0 \mapsto (0,0) \\ 1 \mapsto (1,1) \end{cases} \\[10pt] \oplus :: \begin{cases} (0,0), (1,1) \mapsto 0 \\ (0,1), (1,0) \mapsto 1 \end{cases} \quad\bigg|\quad \oplus :: \begin{cases} 0 \mapsto (0,0), (1,1) \\ 1 \mapsto (0,1), (1,0) \end{cases}\end{array}$$

Meanwhile, Pavlovic, Duncan, and Edwards and Evans et al. have classified all classical structures and complementarity situations in **FRel** [78, 84, 154].

**Example: Spekkens's Toy Qubit Theory.** It is a particular case of these complementarity situations in **FRel** that gives rise to Spekkens's toy qubit theory discussed herein and hence its striking resemblance to quantum theory. This exploration of **FRel** is still an unfinished story. Although, for example, Spekkens's toy theory is a local theory, we strongly suspect that we can discover nonlocality (in the sense of [50]) somewhere within **FRel**.[41]

## Acknowledgments

This chapter benefited from discussions with Samson Abramsky, John Baez, Rick Blute, Giulio Chiribella, Mauro D'Ariano, Andreas Döring, Chris Fuchs, Chris Isham, Keye Martin, David Moore, Prakash Panangaden, Simon Perdrix, Constantin Piron, Phil Scott, Rob Spekkens, Frank Valckenborgh, Jamie Vicary, and Alex Wilce at some point in the past; from discussions with my students Bill Edwards, Benjamin Jackson,

---

[40] Their explicit definition is not of importance here; it suffices to know that formally it witnesses the role played by complex conjugation in adjoints, in the sense that it becomes trivial (i.e., identity) when only real coefficients are involved, which is, for example, the case for the $Z$- and $X$-classical structures for which we have:

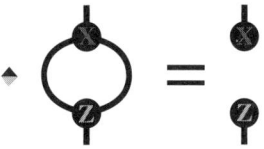

Those acquainted with the field of quantum algebra [176] might recognize here the defining equation of a Hopf algebra, with the dualizer playing the role of the antipode. The apparent nonsymmetrical left-hand-side picture becomes symmetric if we represent the bases in terms of the unitary operations that transform a chosen standard basis into them [64].

[41] For completeness, let us mention that in [13], Baez emphasizes structural similarities between **FHilb** and the category **2Cob** of one-dimensional closed manifolds and cobordisms between these, which play an important role in TQFT [12]. Whereas in **2Cob**, each object comes with a classical structure, there is never more than one, so in this coordinate system, there are no complementarity situations.

and Raymond Lal; and, in particular, from duels with Lucien Hardy, be it either on whatever they call beer in England, his obsession with the shape of tea bags, or his "appreciation" of mathematics, but in particular for one that can be seen at [106]. Credits go to Howard Barnum for recalling the Radiohead song lyric of *Karma Police* in an e-mail to Chris Fuchs, Marcus Appleby, and myself, just when I was about to finish this chapter. We also in particular like to thank Hans Halvorson and my students Ray Lal and Johan Paulsson for proofreading the manuscript. We acknowledge the Perimeter Institute for Theoretical Physics for a long-term visiting-scientist position. It was during my stay there that some of the ideas in this chapter were developed. This work is supported by the author's EPSRC Advanced Research Fellowship, by the EU FP6 STREP QICS, by a Foundational Questions Institute (FQXi) Large Grant, and by the U.S. Office of Naval Research.

# References

[1] S. Abramsky. Computational interpretations of linear logic. *Theoretical Computer Science*, **111** (1993), 3–57.

[2] S. Abramsky. Abstract scalars, loops, free traced and strongly compact closed categories. In *Algebra and Coalgebra in Computer Science, Proceedings of the First International Conference, CALCO 2005* (3–6 Sep 2005, Swansea, UK), *Lecture Notes in Computer Science*, vol. 3629, eds. J. L Fiadeiro, N. Harman, M. Roggenbach, and J. Rutten. Berlin-Heidelberg: Springer-Verlag, 2005, pp. 1–31. arXiv:0910.2931.

[3] S. Abramsky. No-cloning in categorical quantum mechanics. In *Semantic Techniques for Quantum Computation*, eds. I. Mackie and S. Gay. Cambridge: Cambridge University Press, 2009, pp. 1–28. arXiv:0910.2401.

[4] S. Abramsky, R. Blute, B. Coecke, M. Comeau, T. Porter, and J. Vicary. Towards compositional quantum relativity. Draft paper (2010).

[5] S. Abramsky, R. Blute, and P. Panangaden. Nuclear and trace ideals in tensored ∗-categories. *Journal of Pure and Applied Algebra*, **143** (1999), 3–47.

[6] S. Abramsky and B. Coecke. A categorical semantics of quantum protocols. *Proceedings of the 19th Annual IEEE Symposium on Logic in Computer Science (LICS 2004)*, 14–17 July 2004, Turku, Finland. IEEE, 2004, pp. 415–25. arXiv:quant-ph/0402130. Revised version: S. Abramsky and B. Coecke. Categorical quantum mechanics. In *Handbook of Quantum Logic and Quantum Structures*, eds. K. Engesser, D. M. Gabbay, and D. Lehmann. Amsterdam: North-Holland/Elsevier, 2009, pp. 261–323. arXiv:0808.1023.

[7] S. Abramsky and B. Coecke. Abstract physical traces. *Theory and Applications of Categories*, **14** (2005), 111–24. arXiv:0910.3144.

[8] S. Abramsky and N. Tzevelekos. Introduction to categories and categorical logic. In *New Structures for Physics*, Lecture Notes in Physics, vol. 813, eds. B. Coecke. Springer-Verlag, 2010, pp. 3–94.

[9] M. Adelman and J. V. Corbett. A sheaf model for intuitionistic quantum mechanics. *Applied Categorical Structures*, **3** (1995), 79–104.

[10] M. Adelman and J. V. Corbett. Quantum numbers viewed intuitionistically. In *Confronting the Infinite*, eds. A. Carey, W. J. Ellis, and P. A. Pearce. World Scientific Press, 1995.

[11] D. Aerts. The one and the many: Towards a unification of the quantum and the classical description of one and many physical entities. Doctoral dissertation, Free University of Brussels, 1981.

[12] M. Atiyah. Topological quantum field theories. *Publications Mathématique de l'Institut des Hautes Études Scientifiques*, **68** (1989), 175–86.

[13] J. C. Baez. Quantum quandaries: A category-theoretic perspective. In *The Structural Foundations of Quantum Gravity*, eds. D. Rickles, S. French, and J. T. Saatsi. Oxford: Oxford University Press, 2006, pp. 240–66. arXiv:quant-ph/0404040.

[14] J. C. Baez. This week's finds in mathematical physics, week 268 (2008). Available at http://math.ucr.edu/home/baez/week268.html.

[15] J. C. Baez and J. Dolan. Higher-dimensional algebra and topological quantum field theory. *Journal of Mathematical Physics*, **36** (1995), 6073–105. arXiv:q-alg/9503002.

[16] J. C. Baez and M. Stay. Physics, topology, logic and computation: A Rosetta Stone. In *New Structures for Physics*, Lecture Notes in Physics, vol. 813, ed. B. Coecke. Springer-Verlag, 2010, pp. 95–172. arXiv:0903.0340.

[17] J. B. Barbour and B. Bertotti. Mach's principle and the structure of dynamical theories. *Proceedings of the Royal Society of London A*, **382** (1982), 295–306.

[18] H. Barnum. Convex and categorical frameworks for information processing and physics. Lecture at: Categories, Quanta, Concepts (Perimeter Institute, Waterloo, Canada). Perimeter Institute Recorded Seminar Archive (PIRSA), PIRSA:09060028 (2009). Available at http://pirsa.org/09060028/.

[19] H. Barnum, J. Barrett, M. Leifer, and A. Wilce. Cloning and broadcasting in generic probabilistic theories (2006). arXiv:quant-ph/0611295.

[20] H. Barnum, J. Barrett, M. Leifer, and A. Wilce. A generalized no-broadcasting theorem. *Physical Review Letters*, **99** (2007), 240501. arXiv:0707.0620.

[21] H. Barnum, C. M. Caves, C. A. Fuchs, R. Jozsa, and B. Schumacher. Noncommuting mixed states cannot be broadcast. *Physical Review Letters*, **76** (1996), 2818–21. arXiv:quant-ph/9511010.

[22] H. Barnum and A. Wilce. Ordered linear spaces and categories as frameworks for information-processing characterizations of quantum and classical theory (2009). arXiv:0908.2354.

[23] J. Barrett. Information processing in general probabilistic theories. *Physical Review A*, **75** (2007), 032304. arXiv:quant-ph/0508211.

[24] F. J. Belinfante. *A Survey of Hidden-Variable Theories*. Oxford: Pergamon, 1973.

[25] J. Bénabou. Categories avec multiplication. *Comptes Rendus des Séances de l'Académie des Sciences Paris*, **256** (1963), 1887–90.

[26] G. Birkhoff. Von Neumann and lattice theory. *Bulletin of the American Mathematical Society*, **64** (1958), 50–6.

[27] G. Birkhoff and J. von Neumann. The logic of quantum mechanics. *Annals of Mathematics*, **37** (1936), 823–43.

[28] E. Bishop and D. Bridges. *Constructive Analysis*. New York: Springer-Verlag, 1985.

[29] R. Blume-Kohout and W. H. Zurek. Quantum Darwinism: Entanglement, branches, and the emergent classicality of redundantly stored quantum information. *Physical Review A*, **73** (2006), 062310. arXiv:quant-ph/050503.

[30] R. Blute, I. T. Ivanov, and P. Panangaden. Discrete quantum causal dynamics. *International Journal of Theoretical Physics*, **42** (2003), 2025–41. arXiv:gr-qc/0109053.

[31] D. J. Bohm. Time, the implicate order and pre-space. In *Physics and the Ultimate Significance of Time*, ed. D. R. Griffin. Albany, NY: SUNY Press, 1986, pp. 172–208.

[32] D. J. Bohm, P. G. Davies, and B. J. Hiley. Algebraic quantum mechanics and pre-geometry (1982). Draft paper. Published in *Quantum Theory: Reconsiderations of Foundations–3* (6–11 June 2005, Växjö, Sweden), AIP Conference Proceedings, vol. 810, eds. G. Adenier, A.-Y. Krennikov, and T. M. Nieuwenhuizen. Melville, NY: American Institute of Physics, 2006, pp. 314–24.

[33] D. J. Bohm and B. J. Hiley. Generalization of the twistor to Clifford algebras as a basis for geometry. *Revista Brasileira de Ensino de Fisica*, Volume Especial, Os 70 anos de Mario Schönberg,
[34] N. Bohr. *Atomic Physics and Human Knowledge*. New York: Science Editions, 1961.
[35] L. Bombelli, J. Lee, D. Meyer, and R. D. Sorkin. Space-time as a causal set. *Physical Review Letters*, **59** (1987), 521–4.
[36] F. Borceux. *Handbook of Categorical Algebra 1: Basic Category Theory*. Cambridge: Cambridge University Press, 1994.
[37] N. Bourbaki. *Elements of the History of Mathematics*. New York: Springer-Verlag, 1994.
[38] G. Brassard. Comments during a discussion at Quantum Foundations in the Light of Quantum Information and Cryptography, 17–19 May 2000, Montreal, Canada.
[39] A. Carboni and R. F. C. Walters. Cartesian bicategories I. *Journal of Pure and Applied Algebra*, **49** (1987), 11–32.
[40] G. Chiribella, G. M. D'Ariano, and P. Perinotti. Probabilistic theories with purification. *Physical Review A*, **81** (2010), 062348. arXiv:0908.1583.
[41] R. Clifton, J. Bub, and H. Halvorson. Characterizing quantum theory in terms of information-theoretic constraints. *Foundations of Physics*, **33** (2003), 1561–91. arXiv:quant-ph/0211089.
[42] B. Coecke. Quantum logic in intuitionistic perspective. *Studia Logica*, **70** (2001), 411–40. arXiv:math.L0/0011208.
[43] B. Coecke. Kindergarten quantum mechanics. In *Quantum Theory: Reconsiderations of Foundations–3* (6–11 June 2005, Växjö, Sweden), AIP Conference Proceedings, vol. 810, eds. G. Adenier, A.-Y. Krennikov, and T. M. Nieuwenhuizen. Melville, NY: American Institute of Physics, 2006, pp. 81–98. arXiv:quant-ph/0510032.
[44] B. Coecke. Complete positivity without positivity and without compactness. Oxford University Computing Laboratory Research Report PRG-RR-07-05, 2007. Available at http://www.comlab.ox.ac.uk/publications/publication54-abstract.html.
[45] B. Coecke. De-linearizing linearity: Projective quantum axiomatics from strong compact closure. *Electronic Notes in Theoretical Computer Science*, **170** (2007), 47–72. arXiv:quant-ph/0506134.
[46] B. Coecke. Axiomatic description of mixed states from Selinger's CPM-construction. *Electronic Notes in Theoretical Computer Science*, **210** (2008), 3–13.
[47] B. Coecke. Quantum picturalism. *Contemporary Physics*, **51** (2010), 59–83. arXiv:0908.1787.
[48] B. Coecke and R. W. Duncan. Interacting quantum observables. In *Proceedings of the 35th International Colloquium on Automata, Languages and Programming (ICALP)*, Lecture Notes in Computer Science, vol. 5126. New York: Springer-Verlag, 2008, pp. 298–310. Extended version: Interacting quantum observables: Categorical algebra and diagrammatics (2009). arXiv:0906.4725.
[49] B. Coecke and B. Edwards. Toy quantum categories, **270** (2011), *Electronic Notes in Theoretical Computer Science* 1:29–40. arXiv:0808.1037.
[50] B. Coecke, B. Edwards, and R. W. Spekkens. Phase groups and the origin of non-locality for qubits, *Electronic Notes in Theoretical Computer Science*, **270** (2011), 15–36. arXiv:1003.5005.
[51] B. Coecke and A. Kissinger. The compositional structure of multipartite quantum entanglement. In *Proceedings of the 37th International Colloquium on Automata, Languages and Programming (ICALP)*, Lecture Notes in Computer Science, vol. 6199. Springer-Verlag, 2010, pp. 297–308. arXiv:1002.2540.
[52] B. Coecke and R. Lal. Causal categories (2010). Proceedings of the 7th Workshop on Quantum Physics and Logic (QPL), eds. B. Coecke, P. Pannagaden, and P. Selinger. To appear in *Foundations of Physics*.

[53] B. Coecke and D. J. Moore. Operational Galois adjunctions. In *Current Research in Operational Quantum Logic: Algebras, Categories, Languages*, eds. B. Coecke, D. J. Moore, and A. Wilce. New York: Springer-Verlag, 2000, pp. 195–218. arXiv:quant-ph/0008021.

[54] B. Coecke, D. J. Moore, and I. Stubbe. Quantaloids describing causation and propagation for physical properties. *Foundations of Physics Letters*, **14** (2001), 133–45. arXiv:quant-ph/0009100.

[55] B. Coecke, D. J. Moore, and A. Wilce. Operational quantum logic: An overview. In *Current Research in Operational Quantum Logic: Algebras, Categories and Languages*, eds. B. Coecke, D. J. Moore, and A. Wilce. New York: Springer-Verlag, 2000, pp. 1–36. arXiv:quant-ph/0008019.

[56] B. Coecke and E. O. Paquette. POVMs and Naimark's theorem without sums. *Electronic Notes in Theoretical Computer Science*, **210** (2006), 131–52. arXiv:quant-ph/0608072.

[57] B. Coecke and E. O. Paquette. Categories for the practicing physicist. In *New Structures for Physics*, Lecture Notes in Physics, vol. 813, eds. B. Coecke. New York: Springer-Verlag, 2010, pp. 173–286. arXiv:0905.3010.

[58] B. Coecke, E. O. Paquette, and D. Pavlovic. Classical and quantum structuralism. In *Semantic Techniques for Quantum Computation*, eds. I. Mackie and S. Gay. Cambridge: Cambridge University Press, 2009, pp. 29–69. arXiv:0904.1997.

[59] B. Coecke, E. O. Paquette, and S. Perdrix. Bases in diagrammatic quantum protocols. *Electronic Notes in Theoretical Computer Science*, **218** (2008), 131–52. arXiv:0808.1037.

[60] B. Coecke and D. Pavlovic. Quantum measurements without sums. In *Mathematics of Quantum Computing and Technology*, eds. G. Chen, L. Kauffman, and S. Lamonaco. Oxford: Taylor & Francis, 2007, pp. 567–604. arXiv:quant-ph/0608035.

[61] B. Coecke, D. Pavlovic, and J. Vicary. A new description of orthogonal bases (2008). To appear in *Mathematical Structures in Computer Science*. arXiv:0810.0812.

[62] B. Coecke and S. Perdrix. Environment and classical channels in categorical quantum mechanics. In *Proceedings of the 19th EACSL Annual Conference on Computer Science Logic (CSL)*, Lecture Notes in Computer Science, vol. 6247. New York: Springer-Verlag, 2010. arXiv:1004.1598.

[63] B. Coecke and S. Smets. The Sasaki hook is not a [static] implicative connective induces a backward [in time] dynamic one that assigns causes. *International Journal of Theoretical Physics*, **43** (2004), 1705–36. arXiv:quant-ph/0111076.

[64] B. Coecke, B.-S. Wang, Q.-L. Wang, Y.-J. Wang, and Q.-Y. Zhang. Graphical calculus for quantum key distribution. *Electronic Notes in Theoretical Computer Science*, **270** (2011) 2:231–49.

[65] W. Daniel. Axiomatic description of irreversible and reversible evolution of a physical system. *Helvetica Physica Acta*, **62** (1989), 941–68.

[66] G. M. D'Ariano. Operational axioms for quantum mechanics. In *4th Conference on Foundations of Probability and Physics*, AIP Conference Proceedings. Melville, NY: American Institute of Physics, 2007, pp. 79–105. arXiv:quant-ph/0611094.

[67] G. M. D'Ariano. Candidates for principles of quantumness. Lecture at: Reconstructing Quantum Theory (Perimeter Institute, Waterloo, Canada). Perimeter Institute Recorded Seminar Archive (PIRSA), PIRSA:09080014 (2009). Available at http://pirsa.org/09080014/.

[68] E. B. Davies. *Quantum Theory of Open Systems*. New York: Academic Press, 1976.

[69] K. J. Devlin. *The Joy of Sets: Fundamentals of Contemporary Set Theory*. New York: Springer-Verlag, 1993.

[70] D. G. B. J. Dieks. Communication by EPR devices. *Physics Letters A*, **92** (1982), 271–2.

[71] L. Dixon. Demo of quantomatic software (2010). Available at http://www.comlab.ox.ac.uk/quantum/contetn/1005019/.

[72] L. Dixon and R. Duncan. Graphical reasoning in compact closed categories for quantum computation. *Annals of Mathematics and Artificial Intelligence*, **56** (2009), 23–42.

[73] L. Dixon, R. W. Duncan, A. Merry, and A. Kissinger. Quantomatic software (2010). Available at http://dream.inf.ed.ac.uk/projects/quantomatic/.

[74] S. Doplicher and J. E. Roberts. A new duality theory for compact groups. *Inventiones Mathematicae*, **98** (1989), 157–218.

[75] S. Doplicher and J. E. Roberts. Why there is a field algebra with a compact gauge group describing the superselection structure in particle physics. *Communications in Mathematical Physics*, **131** (1990), 51–107.

[76] A. Döring and C. J. Isham. A topos foundation for theories of physics: I. Formal languages for physics (2007). arXiv:quant-ph/0703060 (see also follow-up papers II, III, and IV).

[77] R. W. Duncan. Types for quantum computing. D.Phil. thesis, University of Oxford, 2006.

[78] R. Duncan and B. Edwards. Draft paper (2009); made redundant by slightly earlier appearance of [84].

[79] R. Duncan and S. Perdrix. Graph states and the necessity of Euler decomposition. In *Proceedings of Computability in Europe: Mathematical Theory and Computational Practice (CiE'09)*, Lecture Notes in Computer Science, vol. 5635, New York: Springer-Verlag, 2009, pp. 167–77. arXiv:0902.0500.

[80] R. Duncan and S. Perdrix. Rewriting measurement-based quantum computations with generalised flow. In *Proceedings of the 37th International Colloquium on Automata, Languages and Programming (ICALP)*, Lecture Notes in Computer Science, vol. 6199. New York: Springer-Verlag, 2010.

[81] B. Edwards. Non-locality in categorical quantum mechanics. D. Phil. thesis, University of Oxford, 2010.

[82] T. Ehrhard, J.-Y. Girard, P. Ruet, and P. J. Scott. *Linear Logic in Computer Science*, London Mathematical Society, Lecture Note Series, vol. 316. Cambridge University Press, 2004.

[83] S. Eilenberg and S. Mac Lane. General theory of natural equivalences. *Transactions of the American Mathematical Society*, **58** (1945), 231–94.

[84] J. Evans, R. Duncan, A. Lang, and P. Panangaden. Classifying all mutually unbiased bases in Rel (2009). arXiv:0909.4453.

[85] C.-A. Faure and A. Frölicher. *Modern Projective Geometry*. Amsterdam: Kluwer, 2000.

[86] C.-A. Faure, D. J. Moore, and C. Piron. Deterministic evolutions and Schrödinger flows. *Helvetica Physica Acta*, **68** (1995), 150–7.

[87] M. P. Fiore. Mathematical models of computational and combinatorial structures. In *Proceedings of the 8th International Conference on Foundations of Software Science and Computational Structures (FoSSaCS)*, Lecture Notes in Computer Science, vol. 3441, New York: Springer-Verlag, 2005, pp. 25–46.

[88] D. J. Foulis and C. H. Randall. Operational statistics. I. Basic concepts. *Journal of Mathematical Physics*, **13** (1972), 1667–75.

[89] D. J. Foulis and C. H. Randall. Operational statistics. II. Manuals of operations and their logics. *Journal of Mathematical Physics*, **14** (1973), 1472–80.

[90] P. J. Freyd and A. Scedrov. *Categories, Allegories*. Amsterdam: North-Holland, 1990.

[91] C. A. Fuchs. Information gain vs. state disturbance in quantum theory. *Fortschritte der Physik*, **46** (1998), 535–65. Reprinted in: *Quantum Computation: Where Do We Want to Go Tomorrow?*, ed. S. L. Braunstein. New York: Wiley-VCH Verlag, pp. 229–59. arXiv:quant-ph/9605014.

[92] C. A. Fuchs. Quantum foundations in the light of quantum information. In *Decoherence and Its Implications in Quantum Computation and Information Transfer: Proceedings of the NATO Advanced Research Workshop*, 25–30 June 2000, Mykonos, Greece. Eds. A. Gonis and P. E. A. Turchi. Amsterdam: IOS Press, 2001, pp. 38–82. arXiv:quant-ph/0106166.

[93] M. Gell-Mann and J. B. Hartle. Quantum mechanics in the light of quantum cosmology. In *Complexity, Entropy, and the Physics of Information*, ed. W. Zurek. Reading, MA: Addison-Wesley, 1990.

[94] J.-Y. Girard. Linear logic. *Theoretical Computer Science*, **50** (1987), 1–102.

[95] A. M. Gleason. Measures on the closed subspaces of a Hilbert space. *Journal of Mathematics and Mechanics*, **6** (1957), 885–93.

[96] K. Gödel. Über formal unentscheidbare sätze der Principia Mathematica und verwandter systeme I. *Monatshefte für Mathematik und Physik*, **38** (1931), 173–98.

[97] D. Gottesman and I. L. Chuang. Quantum teleportation is a universal computational primitive. *Nature*, **402** (1999), 390–3. arXiv:quant-ph/9908010.

[98] R. B. Griffiths. *Consistent Quantum Theory*. Cambridge: Cambridge University Press, 2003.

[99] R. Haag. *Local Quantum Physics: Fields, Particles, Algebras*. Berlin-Heidelberg: Springer-Verlag, 1992.

[100] R. Haag and D. Kastler. An algebraic approach to quantum field theory. *Journal of Mathematical Physics*, **5** (1964), 848–61.

[101] H. Halvorson. Private communication at Deep Beauty symposium, Princeton, New Jersey, 2007.

[102] H. Halvorson and M. Müger. Algebraic quantum field theory. In *Handbook of Philosophy of Physics*, eds. J. Butterfield and J. Earman. New York: Elsevier, 2006, pp. 731–922. arXiv:math-ph/0602036.

[103] J. Harding. A link between quantum logic and categorical quantum mechanics. *International Journal of Theoretical Physics*, **48** (2009), 769–802.

[104] L. Hardy. Quantum theory from five reasonable axioms (2001). arXiv:quant-ph/0101012.

[105] L. Hardy. Towards quantum gravity: A framework for probabilistic theories with non-fixed causal structure. *Journal of Physics A*, **40** (2007), 3081–99. arXiv:gr-qc/0608043.

[106] L. Hardy. The causaloid approach to quantum theory and quantum gravity. Lecture at: Logic, Physics and Quantum Information Theory, 2008 (Barbados). Available at http://categorieslogicphysics.wikidot.com/people$\#$lucienhardy.

[107] L. Hardy. Operational structures as a foundation for probabilistic theories. Lecture at: Categories, Quanta, Concepts (Perimeter Institute, Waterloo, Canada). Perimeter Institute Recorded Seminar Archive (PIRSA), PIRSA:09060015 (2009). Available at http://pirsa.org/09060015/.

[108] L. Hardy. Operational structures and natural postulates for quantum theory. Lecture at: Reconstructing Quantum Theory (Perimeter Institute, Waterloo, Canada). Perimeter Institute Recorded Seminar Archive (PIRSA), PIRSA:09080011 (2009). Available at http://pirsa.org/09080011/.

[109] L. Hardy. Foliable operational structures for general probabilistic theories. In this volume. arXiv:0912.4740 (2010).

[110] M. Hasegawa, M. Hofmann, and G. Plotkin. Finite dimensional vector spaces are complete for traced symmetric monoidal categories. In *Pillars of Computer Science*, Lecture Notes in Computer Science, vol. 4800, eds. A. Avron, N. Dershowitz, and A. Rabinovich. Berlin, Heidelberg: Springer-Verlag, 2008, pp. 367–85.

[111] C. Heunen. Categorical quantum models and logics. Ph.D. thesis, Radbout Universiteit Nijmegen. Amsterdam University Press, 2009.

[112] C. Heunen and B. Jacobs. Quantum logic in dagger kernel categories, *Electronic Notes in Theoretical Computer Science*. **270** (2011) 2:79–103. arXiv:0902.2355.

[113] C. Heunen, N. P. Landsman, and B. Spitters. A topos for algebraic quantum theory. *Communications in Mathematical Physics*, **291** (2009), 63–110. arXiv:0709.4364.

[114] B. J. Hiley. Process, distinction, groupoids and Clifford algebras: An alternative view of the quantum formalism. In *New Structures for Physics*, Lecture Notes in Physics, vol. 813, ed. B. Coecke. Berlin, Heidelberg: Springer-Verlag, 2010, pp. 705–52.

[115] P. Hines and S. L. Braunstein. The structure of partial isometries. In *Semantic Techniques for Quantum Computation*, eds. I. Mackie and S. Gay. Cambridge: Cambridge University Press, 2010, pp. 389–413.

[116] C. J. Isham. Topos theory and consistent histories: The internal logic of the set of all consistent sets. *International Journal of Theoretical Physics*, **36** (1997), 785–814. arXiv:gr-qc/9607069.

[117] C. J. Isham and J. Butterfield. Topos perspective on the Kochen-Specker theorem: I. Quantum states as generalized valuations. *International Journal of Theoretical Physics*, **37** (1998), 2669–733. arXiv:quant-ph/9803055 (see also follow-up papers II, III, and IV).

[118] C. J. Isham and J. Butterfield. Some possible roles for topos theory in quantum theory and quantum gravity. *Foundations of Physics*, **30** (2000), 1707–36. arXiv:gr-qc/9910005.

[119] B. Jacobs. *Categorical Logic and Type Theory*, Studies in Logic and the Foundations of Mathematics, vol. 141, eds. S. Abramsky, S. Artemov, R. A. Shore, and A. S. Troelstra. Amsterdam: Elsevier, 2000.

[120] J. M. Jauch. *Foundations of Quantum Mechanics*. Reading, MA: Addison-Wesley, 1968.

[121] J. M. Jauch and C. Piron. Can hidden variables be excluded in quantum mechanics? *Helvetica Physica Acta*, **36** (1963), 827–37.

[122] J. M. Jauch and C. Piron. On the structure of quantal proposition systems. *Helvetica Physica Acta*, **42** (1969), 842–8.

[123] P. T. Johnstone. *Stone Spaces*. Cambridge: Cambridge University Press, 1982.

[124] P. T. Johnstone. *Sketches of an Elephant: A Topos Theory Compendium*. Oxford: Oxford University Press, 2002.

[125] A. Joyal. Une théorie combinatoire des séries formelles. *Advances in Mathematics*, **42** (1981), 1–82.

[126] A. Joyal. Foncteurs analytiques et espèces de structures. In *Combinatoire énumérative*, Lecture Notes in Mathematics, vol. 1234, Springer-Verlag, 1986, pp. 126–59.

[127] A. Joyal and R. Street. The geometry of tensor calculus I. *Advances in Mathematics*, **88** (1991), 55–112.

[128] D. M. Kan. Adjoint functors. *Transactions of the American Mathematical Society*, **87** (1958), 294–329.

[129] G. M. Kelly. Many-variable functorial calculus. In *Coherence in Categories*, Lecture Notes in Mathematics, vol. 281, eds. G. M. Kelly, M. L. Laplaza, G. Lewis, and S. Mac Lane, Springer-Verlag, 1972, pp. 66–105.

[130] G. M. Kelly and M. L. Laplaza. Coherence for compact closed categories. *Journal of Pure and Applied Algebra*, **19** (1980), 193–213.

[131] S. Kochen and E. P. Specker. The problem of hidden variables in quantum mechanics. *Journal Mathathematics and Mechanics*, **17** (1967), 59–87.

[132] J. Kock. *Frobenius Algebras and 2D Topological Quantum Field Theories*. Cambridge: Cambridge University Press, 2003.

[133] K. Kraus. *States, Effects, and Operations*. New York: Springer-Verlag, 1983.

[134] K. Kuratowski. Sur le problème des courbes gauches en topologie. *Fundamenta Mathematicae*, **15** (1930), 271–83.

[135] S. Lack. Composing PROPs. *Theory and Applications of Categories*, **13** (2004), 147–63.

[136] J. Lambek. The mathematics of sentence structure. *American Mathematical Monthly*, **65** (1958), 154–69.

[137] J. Lambek and P. J. Scott. *Higher Order Categorical Logic*. Cambridge: Cambridge University Press, 1986.

[138] L. Lamport. Time, clocks, and the ordering of events in a distributed system. *Communications of the ACM*, **21** (1978), 558–65.

[139] R. Landauer. Irreversibility and heat generation in the computing process. *IBM Journal of Research and Development*, **5** (1961), 183–91.

[140] G. Ludwig. *An Axiomatic Basis of Quantum Mechanics. 1. Derivation of Hilbert Space*. Berlin: Springer-Verlag, 1985.

[141] G. Ludwig. *An Axiomatic Basis of Quantum Mechanics. 2. Quantum Mechanics and Macrosystems*. Berlin: Springer-Verlag, 1987.

[142] E. Mach. *The Science of Mechanics: A Critical and Historical Account of its Development*. LaSalle IL: Open Court, 1960.

[143] G. W. Mackey. Quantum mechanics and Hilbert space. *American Mathematical Monthly*, **64** (1957), 45–57.

[144] G. W. Mackey. *The Mathematical Foundations of Quantum Mechanics*. New York: W. A. Benjamin, 1963.

[145] S. Mac Lane. *Homology*. Die Grundlehren der Mathematischen Wissenschaften, vol. 114, Berlin: Springer-Verlag, 1963.

[146] S. Mac Lane. Natural associativity and commutativity. *Rice University Studies*, **49** (1963), 28–46.

[147] S. Mac Lane. *Categories for the Working Mathematician*, 2nd ed. New York: Springer-Verlag, 1971, 1998.

[148] F. Markopoulou. Quantum causal histories. *Classical and Quantum Gravity*, **17** (2000), 2059–77. arXiv:hep-th/9904009.

[149] O. J. E. Maroney. Information processing and thermodynamic entropy (2009). The Stanford Encyclopedia of Philosophy. Available at http://plato.stanford.edu/ entries/information-entropy/.

[150] P. Martin-Löf. *Intuitionistic Type Theory*. Napoli: Bibliopolis, 1984.

[151] D. J. Moore. On state spaces and property lattices. *Studies in History and Philosophy of Modern Physics*, **30** (1999), 61–83.

[152] D. J. Moore and F. Valckenborgh. Operational quantum logic: A survey and analysis. In *Handbook of Quantum Logic and Quantum Structures*, eds. K. Engesser, D. M. Gabbay, and D. Lehmann. Amsterdam: North-Holland/Elsevier, 2009, pp. 389–442.

[153] P. Panangaden. Discrete quantum causal dynamics. Lecture at: Categories, Quanta, Concepts (Perimeter Institute, Waterloo, Canada). Perimeter Institute Recorded Seminar Archive (PIRSA), PIRSA:09060029 (2009). Available at http://pirsa.org/09060029/.

[154] D. Pavlovic. Quantum and classical structures in nondeterminstic computation. In *Quantum Interaction*, Lecture Notes in Computer Science, vol. 5494, eds. P. Bruza, D. Sofge, W. Lawless, et al. New York: Springer, 2009, pp. 143–57, arXiv:0812.2266.

[155] R. Penrose. Applications of negative dimensional tensors. In *Combinatorial Mathematics and its Applications*, ed. D. Welsh. New York: Academic Press, 1971, pp. 221–4.

[156] C. A. Petri. State-transition structures in physics and in computation. *International Journal of Theoretical Physics*, **12** (1982), 979–92.

[157] C. Piron. Axiomatique quantique. *Helvetica Physica Acta*, **37** (1964), 439–68.

[158] C. Piron. *Foundations of Quantum Physics*. New York: W.A. Benjamin, 1976.

[159] H. Poincaré. *La Science et l'Hypothèse*. Paris: Flammarion, 1905.

[160] J. C. T. Pool. Baer *-semigroups and the logic of quantum mechanics. *Communications in Mathematical Physics*, **9** (1968), 118–41.

[161] Radiohead. Karma Police. *OK Computer*, 1997.

[162] J. Rau. Measurement-based quantum foundations. *Foundations of Physics* **41**(2011) 3:388–8. arXiv:0909.1036 (2010).

[163] R. Raussendorf, D. E. Browne, and H.-J. Briegel. Measurement-based quantum computation on cluster states. *Physical Review A*, **68** (2003), 022312. arXiv:quant-ph/0301052.

[164] R. Raussendorf and H.-J. Briegel. A one-way quantum computer. *Physical Review Letters*, **86** (2001), 5188.

[165] M. Rédei. Why John von Neumann did not like the Hilbert space formalism of quantum mechanics (and what he liked instead). *Studies in History and Philosophy of Modern Physics*, **27** (1997), 493–510.

[166] C. Reid. *Hilbert*. Berlin, New York: Springer-Verlag, 1970.

[167] C. Rovelli. Relational quantum mechanics. *International Journal of Theoretical Physics*, **35** (1996), 1637–78. arXiv:quant-ph/9609002.

[168] J. Schwinger. Unitary operator bases. *Proceedings of the National Academy of Sciences of the U.S.A.* **46** (1960), 570–9.

[169] R. A. G. Seely. Linear logic, ∗-autonomous categories and cofree algebras. *Contemporary Mathematics*, **92** (1998), 371–82.

[170] P. Selinger. Dagger compact closed categories and completely positive maps. *Electronic Notes in Theoretical Computer Science*, **170** (2007), 139–63.

[171] P. Selinger. A survey of graphical languages for monoidal categories. In *New Structures for Physics*, Lecture Notes in Physics, vol. 813, ed. B. Coecke. Springer-Verlag, 2010, pp. 289–356. arXiv:0908.3347.

[172] P. Selinger. Finite dimensional Hilbert spaces are complete for dagger compact closed categories. *Electronic Notes in Theoretical Computer Science*, **470**(2011): 113–19.

[173] M. P. Solèr. Characterization of Hilbert spaces by orthomodular spaces. *Communcations in Algebra*, **23** (1995), 219–43.

[174] R. W. Spekkens. Evidence for the epistemic view of quantum states: A toy theory. *Physical Review A*, **75** (2007), 032110. arXiv:quant-ph/0401052.

[175] R. W. Spekkens. Axiomatization through foil theories. Talk, 5 July 2007, University of Cambridge.

[176] R. Street. *Quantum Groups: A Path to Current Algebra*. Cambridge: Cambridge University Press, 2007.

[177] I. Stubbe and B. Van Steirteghem. Propositional systems, Hilbert lattices and generalized Hilbert spaces. In *Handbook of Quantum Logic and Quantum Structures*, eds. K. Engesser, D. M. Gabbay, and D. Lehmann. Amsterdam: Elsevier, 2007, pp. 477–524. arXiv:0710.2098.

[178] A. S. Troelstra. *Lectures on Linear Logic*. CSLI Publications, 1992.

[179] J. Vicary. A categorical framework for the quantum harmonic oscillator. *International Journal of Theoretical Physics*, **47** (2008), 3408–47. arXiv:0706.0711.

[180] J. Vicary. Categorical formulation of quantum algebras (2008). To appear in *Communications in Mathematical Physics*. arXiv:0805.0432.

[181] S. Vickers. *Topology via Logic*. Cambridge: Cambridge University Press, 1989.

[182] J. von Neumann. *Mathematische Grundlagen der Quantenmechanik*. Berlin: Springer-Verlag, 1932. English translation: *Mathematical Foundations of Quantum Mechanics*. Princeton, NJ: Princeton University Press, 1955.

[183] A. N. Whitehead. *Process and Reality*. New York: Harper & Row, 1957.

[184] E. P. Wigner. *Gruppentheorie*. Frederick Wieweg und Sohn, 1932. English translation: *Group Theory*. New York: Academic Press, Inc., 1959.

[185] A. Wilce. Test spaces and orthoalgebras. In *Current Research in Operational Quantum Logic: Algebras, Categories and Languages*, eds. B. Coecke, D. J. Moore, and A. Wilce. New York: Springer-Verlag, 2000, pp. 81–114. arXiv:quant-ph/0008019.

[186] W. K. Wootters and W. Zurek. A single quantum cannot be cloned. *Nature*, **299** (1982), 802–3.

[187] D. N. Yetter. *Functorial Knot Theory: Categories of Tangles, Coherence, Categorical Deformations, and Topological Invariants*. Singapore: World Scientific, 2001.

[188] W. Zurek. Quantum Darwinism. *Nature Physics*, **5** (2009), 181–8. arXiv:0903.5082.

CHAPTER 3

# Topos Methods in the Foundations of Physics[1]

Chris J. Isham

## 3.1 Introduction

More than forty years have passed since I first became interested in the problem of quantum gravity. During that time, there have been many diversions and, perhaps, some advances. Certainly, the naively optimistic approaches have long been laid to rest, and the schemes that remain have achieved some degree of stability. The original "canonical" program evolved into loop quantum gravity, which has become one of the two major approaches. The other, of course, is string theory—a scheme whose roots lie in the old Veneziano model of hadronic interactions, but whose true value became apparent only after it had been reconceived as a theory of quantum gravity.

However, notwithstanding these hard-won developments, certain issues in quantum gravity transcend any of the current schemes. These involve deep problems of both a mathematical and a philosophical kind and stem from a fundamental paradigm clash between general relativity—the apotheosis of classical physics—and quantum physics.

In general relativity, spacetime "itself" is modeled by a differentiable manifold $\mathcal{M}$, a set whose elements are interpreted as "spacetime points." The curvature tensor of the pseudo-Riemannian metric on $\mathcal{M}$ is then deemed to represent the gravitational field. As a classical theory, the underlying philosophical interpretation is realist: both the spacetime and its points truly "exist,"[2] as does the gravitational field.

Conversely standard quantum theory employs a background spacetime that is fixed ab initio in regard to both its differential structure and its metric/curvature. Furthermore, the conventional interpretation is thoroughly instrumentalist in nature, dealing as it does with counterfactual statements about what would happen (or, to be more precise, the probability of what would happen) *if* a measurement is made of some physical quantity.

---

[1] This chapter is based on joint work with Andreas Döring; see also the chapter he contributed to this volume.
[2] At least, that would be the view of unreconstructed, spacetime substantivalists. However, even purely within the realm of classical physics, this position has often been challenged, particularly by those who place emphasis on the relational features that are inherent in general relativity.

In regard to quantum gravity, the immediate question is:

> How can such a formalism be applied to space and time themselves?

Specifically, what could it mean to "measure" properties of space or time if the very act of measurement requires a spatiotemporal background within which it is made? And how can we meaningfully talk about the "probability" of the results of such measurements? A related question is what meaning, if any, can be ascribed to quantum superpositions of eigenstates of properties of space, time, or spacetime. Over the years, this issue has been much discussed by Roger Penrose, who concludes that the existing quantum formalism simply cannot be applied to space and time and that some new starting point is needed.

Another twist is provided by the subject of *quantum cosmology*, which aspires to apply quantum theory to the entire universe. There is no *prima facie* link between this aspiration and the subject of quantum gravity other than that we have become accustomed to discussing cosmology using various simple solutions to the classical equations of general relativity. However, irrespective of the link with quantum gravity proper, it is still thought-provoking to consider in general terms what it means to apply quantum theory to the "entire" universe. Of course, this might simply be a stupid thing to do, but if one does attempt it, then the problem of implementing instrumentalism becomes manifest: for where is the external observer if the system is the entire universe? In this situation, one can see clearly the attractions of finding a more realist interpretation of quantum theory.

The complexity of such questions is significantly enhanced by the lack of any experimental data that can be related unequivocally to the subject of quantum gravity. In normal theoretical physics, there is the (unholy) trinity of (1) real-world data, (2) mathematical framework, and (3) conceptual/philosophical framework. These three factors are closely interrelated in any real theory of the natural world: indeed, this tripartite structure underpins all theoretical physics.

However, in quantum gravity, the first factor is largely missing, and this raises some curious questions, the following in particular:

- Would we recognize the/a "correct" theory of quantum gravity even if it was handed to us on a plate? Certainly, the Popperian notion of refutation is hard to apply with such a sparsity of empirical data.
- What makes any particular idea a "good" one as far as the community is concerned? Relatedly, how is it that one particular research program becomes well established, whereas another falls by the wayside or, at best, gains only a relatively small following?

The second question is not just whimsical, especially for our younger colleagues, for it plays a key role in decisions about the award of research grants, postdoctoral positions, promotion, and the like.

In practice, many of the past and present research programs in quantum gravity have been developed by analogy with other theories and theoretical structures, particularly standard quantum field theory with gauge groups. And, as for why it is that certain ideas survive and others do not, the answer is partly that individual scientists indulge in their own philosophical prejudices and partly that they like using theoretical tools with which they are already familiar. This, after all, is the fastest route to writing new papers.

At this point, it seems appropriate to mention the ubiquitous, oft-maligned "Planck length." This fundamental unit of length comes from combining Planck's constant $\hbar$ (the "quantum" in quantum gravity) with Newton's constant $G$ (the "gravity" in quantum gravity) and the speed of light $c$ (which always lurks around) in the form

$$L_P := \left(\frac{G\hbar}{c^3}\right)^{1/2}, \qquad (3.1.1)$$

which has a value of approximately $10^{-35}$ meters; the corresponding Planck time (defined as $T_P := L_P/c$) has a value of approximately $10^{-42}$ seconds.

The general assumption is that something "dramatic" happens to the nature of space and time at these fundamental scales. Precisely what that dramatic change might be has been the source of endless speculation and conjecture. However, there is a fairly widespread anticipation that insofar as spatiotemporal concepts have any meaning at all in the "deep" quantum-gravity regime, the appropriate mathematical model will not be based on standard, continuum differential geometry. Indeed, it is not hard to convince oneself that from a physical perspective, the important ingredient in a spacetime model is not the "points" in that space but rather the "regions" in which physical entities can reside. In the context of a topological space, such regions are best modeled by open sets: the closed sets may be too "thin"[3] to contain a physical entity—the only physically meaningful closed sets are those with a nonempty interior. These reflections lead naturally to the subject of "pointless topology" and the theory of *locales*—a natural step along the road to topos theory.

However, another frequent conjecture is that what we normally call space and time (or *spacetime*) will only "emerge" from the correct quantum gravity formalism in some (classical?) limit. Thus, a fundamental theory of quantum gravity may (1) have no intrinsic reference at all to spatiotemporal concepts, and (2) be such that some of the spatiotemporal concepts that emerge in various limits are *nonstandard* and are modeled mathematically with something other than topology and differential geometry. All this leads naturally to the main question that lies behind the work reported in this chapter, namely:

> What is the status of (or justification for) using standard quantum theory when trying to construct a theory of quantum gravity?

It is notable that the main current programs in quantum gravity *do* all use essentially standard quantum theory. However, around fifteen years ago, I came to the conclusion that the use of standard quantum theory was fundamentally inconsistent, and I stopped working in quantum gravity proper. Instead, I began studying what, to me, were the central problems in quantum theory itself: a search that lead quickly to the use of topos theory.

---

[3] In a differentiable manifold, closed sets include points and lines, whereas physical entities "take up room."

Of the various fundamental issues that arise, I focus here on just two of them. The first is the problem mentioned already: viz., applying the standard instrumentalist interpretation of quantum theory to space and time. The second is what I claim is a category error in the use, a priori, of the real and complex numbers in the mathematical formulation of quantum theory when applied in a quantum-gravity context. However, to unpack this problem, it is first necessary to be more precise about the way that the real (and complex) numbers arise in quantum theory. This is the subject of the next section.

## 3.2 The Problem of Using the Real Numbers in Quantum Gravity

The real numbers arise in theories of physics in three different (but related) ways: (1) as the values of physical quantities, (2) as the values of probabilities, and (3) as a fundamental ingredient in models of space and time (especially in those based on differential geometry). All three are of direct concern vis-a-vis our worries about making unjustified, a priori assumptions in quantum theory. Let us consider them in turn.

One reason for assuming physical quantities to be real-valued is that traditionally (i.e., using methods existing in the predigital age), they are measured with rulers and pointers, or they are defined operationally in terms of such measurements. However, rulers and pointers are taken to be classical objects that exist in the physical space of classical physics, and this space is modeled using the reals. In this sense, there is a direct link between the space in which physical quantities take their values (what we call the "quantity-value space") and the nature of physical space or spacetime [8].

> Thus, assuming physical quantities to be real-valued is problematic in any theory in which space, or spacetime, is not modeled by a smooth manifold. This is a theoretical physics analogue of the philosophers' *category error*.

Of course, real numbers also arise as the value space for probabilities via the relative-frequency interpretation of probability. Thus, in principle, an experiment is to be repeated a large number, $N$, times, and the probability associated with a particular result is defined to be the ratio $N_i/N$, where $N_i$ is the number of experiments in which that particular result was obtained. The rational numbers $N_i/N$ necessarily lie between 0 and 1, and if the limit $N \to \infty$ is taken—as is appropriate for a hypothetical "infinite ensemble"—real numbers in the closed interval [0, 1] are obtained.

The relative-frequency interpretation of probability is natural in instrumentalist theories of physics, but it is inapplicable in the absence of any classical spatiotemporal background in which the necessary sequence of measurements can be made (as, for example, is the situation in quantum cosmology).

In the absence of a relativity-frequency interpretation, the concept of "probability" must be understood in a different way. One possibility involves the concept of "potentiality," or "latency." But, in this case, there is no compelling reason why the probability-value space should be a subset of the real numbers. The minimal requirement on this value-space is that it is an ordered set, so that one proposition can be said to be more or less probable than another. However, there is no prima facie reason why this set should be *totally* ordered.

> In fact, one of our goals is to dispense with probabilities altogether and to replace them with "generalized" truth values for propositions.

It follows that a key issue is the way in which the formalism of standard quantum theory is firmly grounded in the concepts of Newtonian space and time (or the relativistic extensions of these ideas), which are essentially assumed a priori. The big question is how this formalism can be modified/generalized/replaced so as to give a framework for physical theories that is

1. "Realist" in some meaning of that word; and
2. Not dependent a priori on the real and/or complex numbers.

For example, suppose we are told that there is a background spacetime $\mathcal{C}$, but it is modeled on something other than differential geometry—say, a causal set. Then, what is the quantum formalism that has the same relation to $\mathcal{C}$ as standard quantum theory does to Newtonian space and time?

This question is very nontrivial because the familiar Hilbert space formalism is rigid and does not lend itself to minor "fiddling." What is needed is something that is radically new and yet that can still embody the basic principles[4] of the quantum formalism, or beyond. To proceed, let us return to first principles and consider what can be said in general about the general structure of mathematical theories of a physical system.

## 3.3 Theories of a Physical System

### 3.3.1 The Realism of Classical Physics

> From the range of the basic questions of metaphysics we shall here ask this *one* question: What is a thing? The question is quite old. What remains ever new about it is merely that it must be asked again and again [7].
>
> —Martin Heidegger

Asking such questions is not a good way for a modern young philosopher to gain tenure. Plato was not ashamed to do so, and neither was Kant, but at some point in the

---

[4] Of course, the question of what precisely are these "basic principles" is much debatable. I have a fond memory of being in the audience for a seminar by John Wheeler at a conference on quantum gravity in the early 1970s. John was getting well into the swing of his usual enthusiastic lecturing style and made some forceful remark about the importance of the quantum principle. At that point, a hand was raised at the back of the lecture room, and a frail voice asked, "What *is* the quantum principle?" John Wheeler paused, looked thoughtfully at his interlocutor, and answered "Well, to be honest, I don't know." He paused again and then said, "Do you?" "No," came the reply. The questioner was Paul Dirac.

last century, the question became "Wittgensteined," and since then it is asked at one's peril. Fortunately, theoretical physicists are not confined in this way, and I address the issue front on. Indeed, why should a physicist be ashamed of this margaritiferous question? For is it not what we all strive to answer when we probe the physical world?

Heidegger's own response makes interesting reading:

> A thing is always something that has such and such properties, always something that is constituted in such and such a way. This something is the bearer of the properties; the something, as it were, that underlies the qualities.

Let us see how modern physics approaches this fundamental question about the beingness of Being.

In constructing a theory of any "normal"[5] branch of physics, the key ingredients are the mathematical representations of the following:

1. Space, time, or spacetime: the framework within which "things" are made manifest to us.
2. "States" of the system: the mathematical entities that determine "the way things are."
3. Physical quantities pertaining to the system.
4. "Properties" of the system: that is, propositions about the values of physical quantities. A state assigns truth values to these propositions.

In the case of standard quantum theory, the mathematical states are interpreted in a counterfactual, nonrealist sense as specifying only what would happen *if* certain measurements are made: so the phrase "the way things are" has to be broadened to include this view. This ontological concept must also apply to the "truth values," which are *intrinsically* probabilistic in nature.

In addition, "the way things are" is normally construed as referring to a specific "moment" of time, but this could be generalized to be compatible with whatever model of time/spacetime is being employed. This could even be extended to a "history theory," in which case "the way things are" means the way things are for all moments of time, or whatever is appropriate for the spacetime model being employed.

For theoretical physicists with a philosophical bent, a key issue is how such a mathematical framework implies, or encompasses, various possible philosophical positions. In particular, how does it interface with the position of "realism"?

This issue is made crystal clear in the case of classical physics, which is represented in the following way:

1. In Newtonian physics (which will suffice for illustrative purposes), space is represented by the three-dimensional Euclidean space, $\mathbb{R}^3$, and time by the one-dimensional space, $\mathbb{R}$.
2. To each system $S$ there is associated a set (actually, a symplectic differentiable manifold) $\mathcal{S}$ of states. At each moment of time, $t \in \mathbb{R}$, the system has a unique state $s_t \in \mathcal{S}$.
3. Any physical quantity, $A$, is represented by a function $\breve{A} : \mathcal{S} \to \mathbb{R}$. The associated interpretation is that if $s \in \mathcal{S}$ is a state, then the value of $A$ in that state is the real number

---

[5] That is, any branch of physics other than quantum gravity!

$\breve{A}(s)$. This is the precise sense in which the philosophical position of (naïve) realism is encoded in the framework of classical physics.

4. The basic propositions are of the form "$A \in \Delta$," which asserts that the value of the physical quantity $A$ lies in the (Borel) subset $\Delta$ of the real numbers. This proposition is represented mathematically by the subset $\breve{A}^{-1}(\Delta) \subseteq \mathcal{S}$—that is, the collection of all states, $s$, in $\mathcal{S}$ such that $\breve{A}(s) \in \Delta$.

This representation of propositions by subsets of the state space $\mathcal{S}$ has a fundamental implication for the *logical* structure of classical physics:

> The mathematical structure of set theory implies that *of necessity*, the propositions in classical physics have the logical structure of a *Boolean algebra*.

We note that classical physics is the paradigmatic implementation of Heidegger's view of a "thing" as the bearer of properties. However, in quantum theory, the situation is very different. There, the existence of any such realist interpretation is foiled by the famous Kochen–Specker theorem [12], which asserts that it is impossible to assign values to all physical quantities at once if this assignment is to satisfy the consistency condition that the value of a function of a physical quantity is that function of the value. For example, the value of "energy squared" is the square of the value of energy.

### 3.3.2 A Categorial Generalization of the Representation of Physical Quantities

The Kochen–Specker theorem implies that from Heidegger's worldview, there is *no* "way things are." To cope with this, physicists have historically fallen back on the instrumentalist interpretation of quantum theory with which we are all so familiar. However, as I keep emphasizing, in the context of quantum gravity, there are good reasons for wanting to achieve a more realist view, and the central question is how this can be done.

The problem is the great disparity between the mathematical formalism of classical physics—which is naturally realist—and the formalism of quantum physics—which is not. Let us summarize these different structures as they apply to a physical quantity $A$ in a system $S$ with associated propositions "$A \in \Delta$" where $\Delta \subseteq \mathbb{R}$.

**The classical theory of $S$:**

- The state space is a set $\mathcal{S}$.
- $A$ is represented by a function $\breve{A} : \mathcal{S} \to \mathbb{R}$.
- The proposition "$A \in \Delta$" is represented by the subset $\breve{A}^{-1}(\Delta) \subseteq \mathcal{S}$ of $\mathcal{S}$. The collection of all such subsets forms a Boolean lattice.

**The quantum theory of $S$:**

- The state space is a Hilbert space $\mathcal{H}$.
- $A$ is represented by a self-adjoint operator, $\hat{A}$, on $\mathcal{H}$.

- The proposition "$A \in \Delta$" is represented by the operator $\hat{E}[\hat{A} \in \Delta]$ that projects onto the subset $\Delta \cap \text{sp}(\hat{A})$ of the spectrum, $\text{sp}(\hat{A})$, of $\hat{A}$. The collection of all projection operators on $\mathcal{H}$ forms a *nondistributive* lattice.

The "nonrealism" of quantum theory is reflected in the fact that propositions are represented by elements of a nondistributive lattice, whereas in classical physics, a distributive lattice (Boolean algebra) is used.

So how can we find a formalism that goes beyond classical physics and yet retains some degree of realist interpretation? One possibility is to generalize the axioms of classical physics to a category, $\tau$, *other* than the category of sets. Such a representation of a physical system would include the following ingredients:

> **The $\tau$-category theory of $S$:**
> - There are two special objects, $\Sigma$, $\mathcal{R}$, in $\tau$ known respectively as the *state object* and *quantity-value object*.
> - A physical quantity, $A$, is represented by an arrow $\breve{A} : \Sigma \to \mathcal{R}$ in the category $\tau$.
> - Propositions about the physical world are represented by subobjects of the state object $\Sigma$. In standard physics, there must be some way of embedding into this structure propositions of the form "$A \in \Delta$" where $\Delta \subseteq \mathbb{R}$.

But does such "categorification" work? In particular, is there some category such that quantum theory can be rewritten in this way?

## 3.4 The Use of Topos Theory

### 3.4.1 The Nature of a Topos

The simple answer to the question of categorification is "no," not in general. The subobjects of an object in a general category do not have any logical structure, and I regard possessing such a structure to be a sine qua non for the propositions in a physical theory. However, there is a special type of category, known as a *topos*, in which the subobjects of any object *do* have a logical structure—a remark that underpins our entire research program.

Broadly speaking, a topos, $\tau$, is a category that behaves in certain critical respects just like the category, **Sets**, of sets. In particular, these include the following:

1. There is an *initial* object, $0_\tau$, and a *terminal object*, $1_\tau$. These are the analogues of, respectively, the empty set, $\emptyset$, and the singleton set $\{*\}$.

    A *global element* (or just *element*) of an object $X$ is defined to be an arrow $x : 1_\tau \to X$. This definition reflects the fact that in set theory, any element $x$ of a set $X$ can be associated with a unique arrow[6] $x : \{*\} \to X$ defined by[7] $x(*) := x$.

---

[6] The singleton, $\{*\}$, is not unique, of course. But any two singletons are isomorphic as sets, and it does not matter which one we choose.

[7] Note that, hopefully without confusion, we are using the same letter "$x$" for the element in $X$ and the associated function from $\{*\}$ to $X$.

The set of all global elements of an object $X$ is denoted[8] $\Gamma X$; that is, $\Gamma X := \text{Hom}_\tau(1_\tau, X)$.

2. One can form *products* and *co-products* of objects.[9] These are the analogue of the Cartesian product and disjoint union in set theory.

3. In set theory, the collection, $B^A$, of functions $f: A \to B$ between sets $A$ and $B$ is itself a *set*—that is, it is an object in the category of sets. In a topos, there is an analogous operation known as *exponentiation*. This associates to each pair of objects $A$, $B$ in $\tau$ an object $B^A$ with the characteristic property that

$$\text{Hom}_\tau(C, B^A) \simeq \text{Hom}_\tau(C \times A, B) \qquad (3.4.1)$$

for all objects $C$. In set theory, the relevant statement is that a parameterized family of functions $c \mapsto f_c : A \to B$, $c \in C$, is equivalent to a single function $F : C \times A \to B$ defined by $F(c, a) := f_c(a)$ for all $c \in C$, $a \in A$.

4. Each subset $A$ of a set $X$ is associated with a unique "characteristic function" $\chi_A : X \to \{0, 1\}$ defined by $\chi_A(x) := 1$ if $x \in A$, and $\chi_A(x) = 0$ if $x \notin A$. Here, 0 and 1 can be viewed as standing for "false" and "true," respectively, so that $\chi_A(x) = 1$ corresponds to the mathematical proposition "$x \in A$" being true.

This operation is mirrored in any topos. More precisely, there is a so-called subobject classifier, $\Omega_\tau$, that is the analogue of the set $\{0, 1\}$. Specifically, to each subobject $A$ of an object $X$ there is associated a *characteristic arrow* $\chi_A : X \to \Omega_\tau$; conversely, each arrow $\chi : X \to \Omega_\tau$ determines a unique subobject of $X$. Thus,

$$\text{Sub}(X) \simeq \text{Hom}_\tau(X, \Omega_\tau) \qquad (3.4.2)$$

where $\text{Sub}(X)$ denotes the collection of all subobjects of $X$.

5. The *power set*, $PX$, of any set $X$, is defined to be the set of all subsets of $X$. Each subset $A \subseteq X$ determines and is uniquely determined by its characteristic function $\chi_A : X \mapsto \{0, 1\}$. Thus, the set $PX$ is in bijective correspondence with the function space $\{0, 1\}^X$.

Analogously, in a general topos, $\tau$, we define the *power object* of an object $X$ to be $PX := \Omega_\tau^X$. It follows from Equation (3.4.1) that

$$\Gamma(PX) := \text{Hom}_\tau(1_\tau, PX) = \text{Hom}_\tau(1_\tau, \Omega_\tau^X)$$
$$\simeq \text{Hom}_\tau(1_\tau \times X, \Omega_\tau) \simeq \text{Hom}_\tau(X, \Omega_\tau)$$
$$\simeq \text{Sub}(X) \qquad (3.4.3)$$

$\text{Sub}(X)$ is always nonempty (because any object $X$ is always a subobject of itself), so it follows that for any $X$, the object $PX$ is nontrivial. As a matter of notation, if $A$ is a subobject of $X$, the corresponding arrow from $1_\tau$ to $PX$ is called the *name* of $A$ and is denoted $\ulcorner A \urcorner : 1_\tau \to PX$.

A key result for our purposes is that in any topos, the collection $\text{Sub}(X)$ of subobjects of any object $X$ forms a *Heyting algebra*, as does the set $\Gamma\Omega_\tau := \text{Hom}_\tau(1_\tau, \Omega_\tau)$ of global elements of the subobject classifier $\Omega_\tau$. A Heyting algebra is a distributive

---

[8] In general, $\text{Hom}_\tau(X, Y)$ denotes the collection of all arrows in $\tau$ from the object $X$ to the object $Y$.
[9] More generally, there are pull-backs and push-outs.

lattice, $\mathfrak{h}$, with top and bottom elements 1 and 0 and such that if $\alpha, \beta \in \mathfrak{h}$, there exists an element $\alpha \Rightarrow \beta$ in $\mathfrak{h}$ with the property

$$\gamma \preceq (\alpha \Rightarrow \beta) \quad \text{if and only if} \quad \gamma \wedge \alpha \preceq \beta. \tag{3.4.4}$$

Then, the "negation" of any $\alpha \in \mathfrak{h}$ is defined as

$$\neg \alpha := (\alpha \Rightarrow 0) \tag{3.4.5}$$

A Heyting algebra has all the properties of a Boolean algebra, except that the principle of *excluded middle* may not hold. That is, there may exist $\alpha \in \mathfrak{h}$ such that $\alpha \vee \neg \alpha \prec 1$; that is, $\alpha \vee \neg \alpha \neq 1$. Equivalently, there may be $\beta \in \mathfrak{h}$ such that $\beta \prec \neg\neg\beta$. Of course, in a Boolean algebra, we have $\beta = \neg\neg\beta$ for all $\beta$. We return to this feature shortly.

### 3.4.2 The Mathematics of "Neorealism"

Let us now consider the assignment of truth values to propositions in mathematics. In set theory, let $K \subseteq X$ be a subset of some set $X$, and let $x$ be an element of $X$. Then, the basic mathematical proposition "$x \in K$" is true if and only if $x$ is an element of the subset $K$. At the risk of seeming pedantic, the truth value, $[\![ x \in K ]\!]$, of this proposition can be written as

$$[\![ x \in K ]\!] := \begin{cases} 1 & \text{if } x \text{ belongs to } K; \\ 0 & \text{otherwise.} \end{cases} \tag{3.4.6}$$

In terms of the characteristic function $\chi_K : X \to \{0, 1\}$, we have

$$[\![ x \in K ]\!] = \chi_K \circ x \tag{3.4.7}$$

where $\chi_K \circ x : \{*\} \to \{0, 1\}$.[10]

The reason for writing the truth value Equation (3.4.6) as Equation (3.4.7) is that this is the form that generalizes to an arbitrary topos. More precisely, let $K \in \text{Sub}(X)$ with characteristic arrow $\chi_K : X \to \Omega_\tau$, and let $x$ be a global element of $X$ so that $x : 1_\tau \to X$. Then, the "generalized" truth value of the mathematical proposition "$x \in K$" is defined to be the arrow

$$[\![ x \in K ]\!] := \chi_K \circ x : 1_\tau \to \Omega_\tau. \tag{3.4.8}$$

Thus, $[\![ x \in K ]\!]$ belongs to the Heyting algebra $\Gamma\Omega_\tau$. The adjective *generalized* refers to the fact that in a generic topos, the Heyting algebra contains elements other than just 0 and 1. In this sense, a proposition in a topos can be only *partially* true: a concept that seems ideal for application to the fitful reality of a quantum system.

This brings us to our main contention, which is that it may be profitable to consider constructing theories of physics in a topos other than the familiar topos of sets [1–11]. The propositions in such a theory admit a "neo"-realist interpretation in the sense that there *is* a "way things are," but this is specified by generalized truth values that may not be just true (1) or false (0).

---

[10] For the notation to be completely consistent, the left-hand side of Equation (3.4.6) should really be written as $[\![ x \in K ]\!](*)$.

Such a theory has the following ingredients:

1. A physical quantity $A$ is represented by a $\tau$-arrow $\breve{A} : \Sigma \to \mathcal{R}$ from the state object $\Sigma$ to the quantity-value object $\mathcal{R}$.
2. Propositions about $S$ are represented by elements of the Heyting algebra, $\mathrm{Sub}(\Sigma)$, of subobjects of $\Sigma$. If $Q$ is such a proposition, we denote by $\delta(Q)$ the associated subobject of $\Sigma$; the "name" of $\delta(Q)$ is the arrow $\ulcorner \delta(Q) \urcorner : 1_\tau \to P\Sigma$.
3. The topos analogue of a state is a "pseudo-state," which is a particular type of subobject of $\Sigma$. Given this pseudo-state, each proposition can be assigned a truth value in the Heyting algebra $\Gamma \Omega_\tau$. Equivalently, we can use "truth objects" (see the following).

Conceptually, such a theory is neorealist in the sense that the propositions and their truth values belong to structures that are "almost" Boolean: in fact, they differ from Boolean algebras only insofar as the principle of excluded middle may not apply.

Thus, a theory expressed in this way "looks like" classical physics, except that classical physics always employs the topos **Sets**, whereas other theories—including, we claim, quantum theory—use a different topos. If the theory requires a background spacetime (or functional equivalent thereof), this would be represented by another special object, $\mathcal{M}$, in the topos. It would even be possible to mimic the actions of differential calculus if the topos is such as to support synthetic differential geometry.[11]

The presence of intrinsic logical structures in a topos has another striking implication. Namely, a topos can be used as a *foundation* for mathematics itself, just as set theory is used in the foundations of "normal" (or "classical") mathematics. Thus, classical physics is modeled in the topos of sets and thereby by standard mathematics. But a theory of physics modeled in a topos, $\tau$, other than **Sets** is being represented in an alternative mathematical universe! The absence of excluded middle means that proofs by contradiction cannot be used; but, apart from that, this so-called intuitionistic logic can be handled in the same way as classical logic.

A closely related feature is that each topos has an "internal language" that is functionally similar to the formal language on which set theory is based. It is this internal language that is used in formulating axioms for the mathematical universe associated with the topos. The same language is also used in constructing the neorealist interpretation of the physical theory.

### 3.4.3 The Idea of a Pseudo-State

In discussing the construction of truth values, it is important to distinguish clearly between truth values of *mathematical* propositions and truth values of *physical* propositions. A key step in constructing a physical theory is to translate the latter into the former. For example, in classical physics, the physical proposition "$A \in \Delta$" is represented by the subset $\breve{A}^{-1}(\Delta)$ of the state space, $\mathcal{S}$; that is,

$$\delta(A \, \varepsilon \, \Delta) := \breve{A}^{-1}(\Delta). \qquad (3.4.9)$$

---

[11] This approach to calculus is based on the existence of genuine infinitesimals in certain topoi. For example, see [13].

Then, for any state $s \in \mathcal{S}$, the truth value of the physical proposition "$A \in \Delta$" is defined to be the truth value of the *mathematical* proposition "$s \in \delta(A \, \varepsilon \, \Delta)$" (or, equivalently, the truth value of the mathematical proposition "$\breve{A}(s) \in \Delta$").

Given these ideas, in a topos theory one might expect to represent a physical state by a global element $s : 1_\tau \to \Sigma$ of the state object $\Sigma$. The truth value of a proposition, $Q$, represented by a subobject $\delta(Q)$ of $\Sigma$, would then be defined as the global element

$$v(Q; s) := [\![ s \in \delta(Q) ]\!] = \chi_{\delta(Q)} \circ s : 1_\tau \to \Omega_\tau \qquad (3.4.10)$$

of $\Omega_\tau$. However, in the topos version of quantum theory (see Section 3.5), it transpires that $\Sigma$ has *no global elements at all*: in fact, this turns out to be precisely equivalent to the Kochen–Specker theorem! This absence of global elements of the state object could well be a generic feature of topos-formulated physics, in which case we cannot use Equation (3.4.10) and need to proceed in a different way.

In set theory, there are two mathematical statements that are equivalent to "$s \in K$." These are (1):

$$K \in T^s, \qquad (3.4.11)$$

where the "truth object" $T^s$ is defined by[12]

$$T^s := \{ J \subseteq \mathcal{S} \mid s \in J \}, \qquad (3.4.13)$$

and (2):

$$\{s\} \subseteq K. \qquad (3.4.14)$$

Thus, in classical physics, "$A \in \Delta$" is true in a state $s$ if and only if (1) $\delta(A \, \varepsilon \, \Delta) \in T^s$ and (2) $\{s\} \subseteq \delta(A \, \varepsilon \, \Delta)$.

Let us consider in turn the topos analogue of these two options.

### 3.4.3.1 Option 1 (The Truth-Object Option)

Note that $T^s$ is a collection of subsets of $\mathcal{S}$: that is, $T^s \in \mathrm{Sub}(P\mathcal{S})$. The interesting thing about Equation (3.4.11) is that it is of the form "$x \in X$" and therefore has an immediate generalization to any topos. More precisely, in a general topos $\tau$, a truth object, $T$, would be a subobject of $P\Sigma$ (equivalently, a global element of $P(P\Sigma)$) with a characteristic arrow $\chi_T : P\Sigma \to \Omega_\tau$. Then, the physical proposition $Q$ has the topos truth value

$$v(Q; T) := [\![ \delta(Q) \in T ]\!] = \chi_T \circ \ulcorner \delta(Q) \urcorner : 1_\tau \to \Omega_\tau. \qquad (3.4.15)$$

The key remark is that although Equation (3.4.10) is inapplicable if there are no global elements of $\Sigma$, Equation (3.4.15) *can* be used because global elements of a power object (like $P(P\Sigma)$) always exist. Thus, in option 1, the analogue of a classical state $s \in \mathcal{S}$ is played by the truth object $T \in \mathrm{Sub}(P\Sigma)$.

---

[12] Note that $s$ can be recovered from $T^s$ via

$$\{s\} := \bigcap_{K \in T^s} K. \qquad (3.4.12)$$

### 3.4.3.2 Option 2 (The Pseudo-State Option)

We cannot use Equation (3.4.14) in a literal way in the topos theory because if there are no global elements, $s$, of $\Sigma$, trying to construct an analogue of $\{s\}$ is meaningless. However, although there are no global elements of $\Sigma$, there may nevertheless be certain subobjects, $\mathfrak{w}$, of $\Sigma$ (what we call "pseudo-states") that are, in some sense, as "close as we can get" to the (nonexistent) analogue of the singleton subsets, $\{s\}$, of $\mathcal{S}$. We must then consider the mathematical proposition "$\mathfrak{w} \subseteq \delta(Q)$," where $\mathfrak{w}$ and $\delta(Q)$ are both subobjects of $\Sigma$.

As we have seen in Equation (3.4.8), the truth value of the mathematical proposition "$x \in K$" is a global element of $\Omega_\tau$: as such, it may have a value other than 1 ("true") or 0 ("false"). In other words, "$x \in K$" can be only "partially true." What is important for us is that an analogous situation arises if $J$, $K$ are subobjects of some object $X$. Namely, a global element, $[\![ J \subseteq K ]\!]$, of $\Omega_\tau$ can be assigned to the proposition "$J \subseteq K$"; that is, there is a precise sense in which one subobject of $X$ can be only "partially" a subobject of another. In this scenario, a physical proposition $Q$ has the topos truth value

$$\nu(Q; \mathfrak{w}) := [\![ \mathfrak{w} \subseteq \delta(Q) ]\!] \qquad (3.4.16)$$

when the pseudo-state is $\mathfrak{w}$. The general definition of $[\![ J \subseteq K ]\!]$ is not important for our present purposes. In the quantum case, the explicit form is discussed in Chapter 4 in this volume by my collaborator, Andreas Döring.

To summarize what has been said so far, the key ingredients in formulating a theory of a physical system in a topos $\tau$ are the following:

1. There is a "state object," $\Sigma$, in $\tau$.
2. To each physical proposition, $Q$, there is associated a subobject, $\delta(Q)$, of $\Sigma$.
3. The analogue of a classical state is given by either (1) a truth object, $T$, or (2) a pseudo-state $\mathfrak{w}$. The topos truth value of the proposition $Q$ is then $[\![ \delta(Q) \in T ]\!]$, or $[\![ \mathfrak{w} \subseteq \delta(Q) ]\!]$, respectively.

Note that no mention has been made here of the quantity-value object $\mathcal{R}$. However, in practice, this also is expected to play a key role, not least in constructing the physical propositions. More precisely, we anticipate that each physical quantity $A$ will be represented by an arrow $\breve{A} : \Sigma \to \mathcal{R}$, and then a typical proposition will be of the form "$A \in \Xi$" where $\Xi$ is some subobject of $\mathcal{R}$.

If any "normal" physics is addressed in this way, physical quantities are expected to be *real*-valued, and the physical propositions are of the form "$A \in \Delta$" for some $\Delta \subseteq \mathbb{R}$. In this case, it is necessary to decide what subobject of $\mathcal{R}$ in the topos corresponds to the external quantity $\Delta \subseteq \mathbb{R}$.

There is subtlety here, however; although (in any topos we are likely to consider) there is a precise topos analogue of the real numbers,[13] it would be a mistake to assume that this is necessarily the quantity-value object; indeed, in our topos version of quantum theory, this is definitely *not* the case.

The question of relating $\Delta \subseteq \mathbb{R}$ to some subobject of the quantity-value object $\mathcal{R}$ in $\tau$ is just one aspect of the more general issue of distinguishing quantities that are

---

[13] Albeit defined using the analogue of Dedekind cuts, not Cauchy sequences.

*external* to the topos and those that are *internal*. Thus, for example, a subobject $\Xi$ of $\mathcal{R}$ in $\tau$ is an internal concept, whereas $\Delta \subseteq \mathbb{R}$ is external, referring as it does to something, $\mathbb{R}$, that lies outside $\tau$. Any reference to a background space, time, or spacetime would also be external. Ultimately, perhaps—or, at least, certainly in the context of quantum cosmology—one would want to have no external quantities at all. However, in any more "normal" branch of physics, it is natural that the propositions about the system refer to the external (to the theory) world to which theoretical physics is meant to apply. And, even in quantum cosmology, the actual collection of what counts as physical quantities is external to the formalism itself.

## 3.5 The Topos of Quantum Theory

### 3.5.1 The Kochen–Specker Theorem and Contextuality

To motivate our choice of topos for quantum theory, let us return again to the Kochen–Specker theorem, which asserts the impossibility of assigning values to all physical quantities while, at the same time, preserving the functional relations between them [12].

In a quantum theory, a physical quantity $A$ is represented by a self-adjoint operator $\hat{A}$ on the Hilbert space, $\mathcal{H}$, of the system. A "valuation" is defined to be a real-valued function $\lambda$ on the set of all bounded, self-adjoint operators, with the properties that (1) the value $\lambda(\hat{A})$ belongs to the spectrum of $\hat{A}$, and (2) the functional composition principle (or FUNC for short) holds:

$$\lambda(\hat{B}) = h(\lambda(\hat{A})) \tag{3.5.1}$$

for any pair of self-adjoint operators $\hat{A}$, $\hat{B}$ such that $\hat{B} = h(\hat{A})$ for some real-valued function $h$.

Several important results follow from this definition. For example, if $\hat{A}_1$ and $\hat{A}_2$ commute, it follows from the spectral theorem that there exists an operator $\hat{C}$ and functions $h_1$ and $h_2$ such that $\hat{A}_1 = h_1(\hat{C})$ and $\hat{A}_2 = h_2(\hat{C})$. It then follows from FUNC that

$$\lambda(\hat{A}_1 + \hat{A}_2) = \lambda(\hat{A}_1) + \lambda(\hat{A}_2) \tag{3.5.2}$$

and

$$\lambda(\hat{A}_1 \hat{A}_2) = \lambda(\hat{A}_1)\lambda(\hat{A}_2). \tag{3.5.3}$$

The Kochen–Specker theorem says that no valuations exist if $\dim(\mathcal{H}) > 2$. Conversely, Equations (3.5.2)–(3.5.3) show that if it existed, a valuation restricted to a commutative subalgebra of operators would be just an element of the *spectrum* of the algebra, and, of course, such elements *do* exist. Thus, valuations exist on any commutative subalgebra[14] of operators but not on the (noncommutative) algebra, $\mathcal{B}(\mathcal{H})$, of all bounded operators. We shall call such valuations "local."

Within the instrumentalist interpretation of quantum theory, the existence of local valuations is closely related to the possibility of making "simultaneous" measurements on commuting observables. However, the existence of local valuations also plays a key role in the so-called modal interpretations, in which values are given to the physical quantities that belong to some specific commuting set. The most famous such interpretation is that of David Bohm, where it is the configuration[15] variables in the system that are regarded as always "existing" (in the sense of possessing values).

The topos implication of these remarks stems from the following observations. First, let $V$, $W$ be a pair of commutative subalgebras with $V \subseteq W$. Then, any (local) valuation, $\lambda$, on $W$ restricts to give a valuation on the subalgebra $V$. More formally, if $\underline{\Sigma}_W$ denotes the set of all local valuations on $W$, there is a "restriction map" $r_{WV} : \underline{\Sigma}_W \to \underline{\Sigma}_V$ in which $r_{WV}(\lambda) := \lambda|_V$ for all $\lambda \in \underline{\Sigma}_W$. It is clear that if $U \subseteq V \subseteq W$, then

$$r_{VU}(r_{WV}(\lambda)) = r_{WU}(\lambda) \tag{3.5.4}$$

for all $\lambda \in \underline{\Sigma}_W$—that is, restricting from $W$ to $U$ is the same as going from $W$ to $V$ and then from $V$ to $U$.

Note that if one existed, a valuation, $\lambda$, on the noncommutative algebra of all operators on $\mathcal{H}$ would provide an association of a "local" valuation $\lambda_V := \lambda|_V$ to each commutative algebra $V$ such that, for all pairs $V$, $W$ with $V \subseteq W$, we have

$$\lambda_W|_V = \lambda_V. \tag{3.5.5}$$

The Kochen–Specker theorem asserts there are no such associations $V \mapsto \lambda_V \in \underline{\Sigma}_V$ if $\dim \mathcal{H} > 2$.

To explore this further, consider the situation where we have three commuting algebras $V$, $W_1$, $W_2$ with $V \subseteq W_1$ and $V \subseteq W_2$, and suppose that $\lambda_1 \in \underline{\Sigma}_{W_1}$ and $\lambda_2 \in \underline{\Sigma}_{W_2}$ are local valuations. If there is some commuting algebra $W$ so that (1) $W_1 \subseteq W$ and $W_2 \subseteq W$, and (2) there exists $\lambda \in \underline{\Sigma}_W$ such that $\lambda_1 = \lambda|_{W_1}$ and $\lambda_2 = \lambda|_{W_2}$, then Equation (3.5.4) implies that

$$\lambda_1|_V = \lambda_2|_V. \tag{3.5.6}$$

However, suppose now the elements of $W_1$ and $W_2$ do not all commute with each other—that is, there is no $W$ such that $W_1 \subseteq W$ and $W_2 \subseteq W$. Then, although valuations $\lambda_1 \in \underline{\Sigma}_{W_1}$ and $\lambda_2 \in \underline{\Sigma}_{W_2}$ certainly exist, there is no longer any guarantee that they can be chosen to satisfy the matching condition in Equation (3.5.6). Indeed, the Kochen–Specker theorem says precisely that it is impossible to construct a collection of local valuations, $\lambda_W$, for all commutative subalgebras $W$ such that all the matching conditions of the form of Equation (3.5.6) are satisfied.

---

[14] More precisely, because we want to include projection operators, we assume that the commutative algebras are von Neumann algebras. These algebras are defined over the complex numbers so that non–self-adjoint operators are included too.

[15] If taken literally, the word *configuration* implies that the state space of the underlying classical system must be a cotangent bundle $T^*Q$.

We note that triples $V, W_1, W_2$ of this type arise when there are noncommuting self-adjoint operators $\hat{A}_1, \hat{A}_2$ with a third operator $\hat{B}$ and functions $f_1, f_2 : \mathbb{R} \to \mathbb{R}$ such that $\hat{B} = f_1(\hat{A}_1) = f_2(\hat{A}_2)$. Of course, $[\hat{B}, \hat{A}_1] = 0$ and $[\hat{B}, \hat{A}_2] = 0$ so that, in physical parlance, $A_1$ and $B$ can be given "simultaneous values" (or can be measured simultaneously) as can $A_2$ and $B$. However, the implication of the preceding discussion is that the value ascribed to $B$ (respectively, the result of measuring $B$) depends on whether it is considered together with $A_1$ or together with $A_2$. In other words, the value of the physical quantity $B$ is *contextual*. This is often considered one of the most important implications of the Kochen–Specker theorem.

### 3.5.2 The Topos of Presheaves Sets$^{\mathcal{V}(\mathcal{H})^{\mathrm{op}}}$

To see how all this relates to topos theory, let us rewrite the preceding slightly. Thus, let $\mathcal{V}(\mathcal{H})$ denote the collection of all commutative subalgebras of operators on the Hilbert space $\mathcal{H}$. This is a partially ordered set with respect to subalgebra inclusion. Hence, it is also a category whose objects are just the commutative subalgebras of $\mathcal{B}(\mathcal{H})$; we shall call it the "category of contexts."

We view each commutative algebra as a context with which to view the quantum system in an essentially classical way in the sense that the physical quantities in any such algebra can be given consistent values, as in classical physics. Thus, each context is a "classical snapshot," or "worldview," or "window on reality." In any modal interpretation of quantum theory, only one context at a time is used,[16] but our intention is to use the collection of *all* contexts in one megastructure that will capture the entire quantum theory.

To do this, let us consider the association of the spectrum $\underline{\Sigma}_V$ (the set of local valuations on $V$) to each commutative subalgebra $V$. As explained, there are restriction maps $r_{WV} : \underline{\Sigma}_W \to \underline{\Sigma}_V$ for all pairs $V, W$ with $V \subseteq W$, and these maps satisfy the conditions in Equation (3.5.4). In the language of category theory, this means that the operation $V \mapsto \underline{\Sigma}_V$ defines the elements of a *contravariant functor*, $\underline{\Sigma}$, from the category $\mathcal{V}(\mathcal{H})$ to the category of sets; equivalently, it is a covariant functor from the opposite category, $\mathcal{V}(\mathcal{H})^{\mathrm{op}}$ to **Sets**.

Now, one of the basic results in topos theory is that for any category $\mathcal{C}$, the collection of covariant functors $F : \mathcal{C}^{\mathrm{op}} \to$ **Sets** is a topos, known as the *topos of presheaves* over $\mathcal{C}$, and is denoted **Sets**$^{\mathcal{C}^{\mathrm{op}}}$. In regard to quantum theory, our fundamental claim is that the theory can be reformulated so as to look like classical physics but in the topos **Sets**$^{\mathcal{V}(\mathcal{H})^{\mathrm{op}}}$. The object $\underline{\Sigma}$ is known as the *spectral presheaf* and plays a fundamental role in our theory. In purely mathematical terms, this has considerable interest as the foundation for a type of noncommutative spectral theory; from a physical perspective, we identify it as the state object in the topos.

The terminal object $1_{\mathbf{Sets}^{\mathcal{V}(\mathcal{H})^{\mathrm{op}}}}$ in the topos **Sets**$^{\mathcal{V}(\mathcal{H})^{\mathrm{op}}}$ is the presheaf that associates to each commutative algebra $V$ a singleton set $\{*\}_V$, and the restriction maps are the obvious ones. It is then easy to see that a global element $\lambda : 1_{\mathbf{Sets}^{\mathcal{V}(\mathcal{H})^{\mathrm{op}}}} \to \underline{\Sigma}$ of the spectral presheaf is an association to each $V$ of a spectral element $\lambda_V \in \underline{\Sigma}_V$ such that,

---

[16] In the standard instrumentalist interpretation of quantum theory, a context is selected by choosing to measure a particular set of commuting observables.

for all pairs $V$, $W$ with $V \subseteq W$, we have Equation (3.5.5). Thus, we have the following basic result:

> The Kochen–Specker theorem is equivalent to the statement that the spectral presheaf, $\underline{\Sigma}$, has no global elements.

Of course, identifying the topos and the state object is only the first step in constructing a topos formulation of quantum theory. The next key step is to associate a subobject of $\underline{\Sigma}$ with each physical proposition. In quantum theory, propositions are represented by projection operators, and so what we seek is a map

$$\delta : \mathcal{P}(\mathcal{H}) \to \mathrm{Sub}(\underline{\Sigma}) \tag{3.5.7}$$

where $\mathcal{P}(\mathcal{H})$ denotes the lattice of projection operators on $\mathcal{H}$. Thus, $\delta$ is a map from a nondistributive quantum logic to the (distributive) Heyting algebra, $\mathrm{Sub}(\underline{\Sigma})$, in the topos $\mathrm{Sets}^{\mathcal{V}(\mathcal{H})^{\mathrm{op}}}$.

The precise definition of $\delta(\hat{P})$, $\hat{P} \in \mathcal{P}(\mathcal{H})$, is described in Chapter 4 in this volume by Andreas Döring. Suffice it to say that at each context $V$, it involves approximating $\hat{P}$ with the "closest" projector to $\hat{P}$ that lies in $V$: an operator that we call "daseinisation" in honor of Heidegger's memorable existentialist perspective on ontology.

The next step is to construct the quantity-value object, $\underline{\mathcal{R}}$. We do this by applying the Gelfand spectral transformation in each context $V$. The most striking remark about the result is that $\underline{\mathcal{R}}$ is *not* the real-number object, $\underline{\mathbb{R}}$, in the topos, although the latter is a subobject of the former. One result of this construction is that we are able to associate an arrow $\breve{A} : \underline{\Sigma} \to \underline{\mathcal{R}}$ with each physical quantity $A$—that is, with each bounded, self-adjoint operator $\hat{A}$. This is a type of noncommutative spectral theory.

The final key ingredient is to construct the truth objects, or pseudostates; in particular, we wish to do this for each vector $|\psi\rangle$ in the Hilbert space $\mathcal{H}$. Again, the details can be found in Chapter 4 by Andreas Döring, but suffice it to say that the pseudostate, $\underline{\mathfrak{w}}^{|\psi\rangle}$ associated with $|\psi\rangle \in \mathcal{H}$, can be written in the simple form

$$\underline{\mathfrak{w}}^{|\psi\rangle} = \delta(|\psi\rangle\langle\psi|). \tag{3.5.8}$$

In other words, the pseudostate corresponding to the unit vector $|\psi\rangle$ is just the daseinisation of the projection operator onto $|\psi\rangle$.

## 3.6 Conclusions

In this chapter, I have revisited the oft-repeated statement about fundamental incompatibilities between quantum theory and general relativity. In particular, I have argued that the conventional quantum formalism is inadequate to the task of quantum gravity both in regard to (1) the nonrealist, instrumentalist interpretation; and (2) the a priori use of the real and complex numbers.

The alternative that my collaborator Andreas Döring and I suggest is to employ a mathematical formalism that "looks like" classical physics (so as to gain some degree of "realism") but in a topos, $\tau$, other than the topos of sets. The following are the ingredients in such a theory:

1. A state-object, $\Sigma$, and a quantity-value object, $\mathcal{R}$.
2. A map $\delta : \mathcal{P} \to \text{Sub}(\Sigma)$ from the set of propositions, $\mathcal{P}$, to the Heyting algebra, $\text{Sub}(\Sigma)$, of subobjects of $\Sigma$.
3. A set of pseudostates, $\mathfrak{w}$, or truth objects, $T$. Then, the truth value of the physical proposition $Q$ in the pseudostate $\mathfrak{w}$ (resp. the truth object $T$) is $[\![\, \mathfrak{w} \subseteq \delta(Q) \,]\!]$ (resp. $[\![\, \delta(Q) \in T \,]\!]$), which is a global element of the subobject classifier, $\Omega_\tau$, in $\tau$. The set $\Gamma \Omega_\tau$ of all such global elements is also a Heyting algebra.

In the case of quantum theory, in our published articles, we have shown in great detail how this program goes through with the topos being the category, $\mathbf{Sets}^{\mathcal{V}(\mathcal{H})^{op}}$, of presheaves over the category/poset $\mathcal{V}(\mathcal{H})$ of commutative subalgebras of the algebra of all bounded operators on the Hilbert space $\mathcal{H}$.

The next step in this program is to experiment with "generalizations" of quantum theory in which the context category is not $\mathcal{V}(\mathcal{H})$ but some category $\mathcal{C}$, which has no fundamental link with a Hilbert space, and therefore with the real or complex numbers. In a scheme of this type, the intrinsic contextuality of quantum theory is kept, but its domain of applicability could include spatiotemporal concepts that are radically different from those currently in use.

## Acknowledgments

I am most grateful to Andreas Döring for the stimulating collaboration that we have enjoyed in the last six years.

## References

[1] A. Döring and C. J. Isham. A topos foundation for theories of physics: I. Formal languages for physics. *Journal of Mathematical Physics*, **49** (2008), 053515.

[2] A. Döring and C. J. Isham. A topos foundation for theories of physics: II. Daseinisation and the liberation of quantum theory. *Journal of Mathematical Physics*, **49** (2008), 053516.

[3] A. Döring and C. J. Isham. A topos foundation for theories of physics: III. The representation of physical quantities with arrows $\breve{A} : \underline{\Sigma} \to \underline{\mathbb{R}^{\succeq}}$. *Journal of Mathematical Physics*, **49** (2008), 053517.

[4] A. Döring and C. J. Isham. A topos foundation for theories of physics: IV. Categories of systems. *Journal of Mathematical Physics*, **49** (2008), 053518.

[5] A. Döring and C. J. Isham. "What is a thing?": Topos theory in the foundations of physics. To appear in *New Structures in Physics*, Lecture Notes in Physics, vol. 813, ed. B. Coecke. New York: Springer, 2010 (in press).

[6] J. Hamilton, C. J. Isham, and J. Butterfield. A topos perspective on the Kochen–Specker theorem: III. Von Neumann algebras as the base category. *International Journal of Theoretical Physics*, **39** (2000), 1413–36.

[7] M. Heidegger. *What Is a Thing?* South Bend, IN: Regenery/Gateway, 1967.

[8] C. J. Isham. Some reflections on the status of conventional quantum theory when applied to quantum gravity. In *Proceedings of the Conference in Honour of Stephen Hawking's 60th Birthday*, ed. G. Gibbons. Cambridge: Cambridge University Press, 2003.

[9] C. J. Isham and J. Butterfield. A topos perspective on the Kochen–Specker theorem: I. Quantum states as generalised valuations. *International Journal of Theoretical Physics*, **37** (1998), 2669–733.

[10] C. J. Isham and J. Butterfield. A topos perspective on the Kochen–Specker theorem: II. Conceptual aspects, and classical analogues. *International Journal of Theoretical Physics*, **38** (1999), 827–59.

[11] C. J. Isham and J. Butterfield. A topos perspective on the Kochen–Specker theorem: IV. Interval valuations. *International Journal of Theoretical Physics*, **41** (2002), 613–39.

[12] S. Kochen and E. P. Specker. The problem of hidden variables in quantum mechanics. *Journal of Mathematics and Mechanics*, **17** (1967), 59–87.

[13] I. Moerdijk and G. E. Reyes. *Models for Smooth Infinitesimal Analysis*. New York: Springer-Verlag, 1991.

# CHAPTER 4

# The Physical Interpretation of Daseinisation

Andreas Döring

## 4.1 Introduction

This chapter provides a conceptual discussion and physical interpretation of some of the quite abstract constructions in the topos approach to physics. In particular, the daseinisation process for projection operators and for self-adjoint operators is motivated and explained from a physical point of view. Daseinisation provides the bridge between the standard Hilbert space formalism of quantum theory and the new topos-based approach to quantum theory. As an illustration, I show all constructions explicitly for a three-dimensional Hilbert space and the spin-$z$ operator of a spin-1 particle. Throughout, I refer to joint work with Chris Isham, and this chapter is intended to serve as a companion to the one he contributed to this volume.

### 4.1.1 The Topos Approach

The topos approach to quantum theory was initiated by Isham [21] and Butterfield and Isham [19, 22–24]. It was developed and broadened into an approach to the formulation of physical theories in general by Isham and by this author [12–15]. The long article [16] gives a more or less exhaustive[1] and coherent overview of the approach. More recent developments are the description of arbitrary states by probability measures [9] and further developments [10] concerning the new form of quantum logic that constitutes a central part of the topos approach. For background, motivation, and the main ideas, see also Isham's Chapter 3 in this volume.

Most of the work so far has been done on standard nonrelativistic quantum theory. A quantum system is described by its *algebra of physical quantities*. Often, this can be assumed to be $\mathcal{B}(\mathcal{H})$, the algebra of all bounded operators on the (separable) Hilbert space $\mathcal{H}$ of the system. More generally, one can use a suitable operator algebra. For conceptual and pragmatic reasons, we assume that the algebra of physical quantities is

---

[1] and probably exhausting.

a *von Neumann algebra* $\mathcal{N}$ (see, for example, [25, 29]). For our purposes, this poses no additional technical difficulty. Physically, it allows the description of quantum systems with symmetry or superselection rules. The reader unfamiliar with von Neumann algebras can always assume that $\mathcal{N} = \mathcal{B}(\mathcal{H})$.

Quantum theory usually is identified with the Hilbert space formalism together with some interpretation. This works fine in a vast number of applications. Moreover, the Hilbert space formalism is very rigid. One cannot just change some part of the structure because this typically brings down the whole edifice. Yet there are serious conceptual problems with the usual instrumentalist interpretations of quantum theory that become even more severe when one tries to apply quantum theory to gravity and cosmology. For a discussion of some of these conceptual problems, see Isham's Chapter 3 in this volume.

The topos approach provides not merely another interpretation of the Hilbert space formalism but also a mathematical reformulation of quantum theory based on structural and conceptual considerations. The resulting formalism is *not* a Hilbert space formalism. The fact that such a reformulation is possible at all is somewhat surprising. (Needless to say, many open questions remain.)

### 4.1.2 Daseinisation

This chapter focuses mainly on how the new topos formalism relates to the standard Hilbert space formalism. The main ingredient is the process that we coined *daseinisation*. It will be shown how daseinisation relates familiar structures in quantum theory to structures within the topos associated with a quantum system. Hence, daseinisation gives the "translation" from the ordinary Hilbert space formalism to the topos formalism.

The presentation here keeps the use of topos theory to an absolute minimum, and we do not assume any familiarity with category theory beyond the basics. Some notions and results from functional analysis (e.g., the spectral theorem) are used. As an illustration, we will show what all constructions look like concretely for the algebra $\mathcal{N} = \mathcal{B}(\mathbb{C}^3)$ and the spin-$z$ operator, $\hat{S}_z \in \mathcal{B}(\mathbb{C}^3)$, of a spin-1 particle.

In fact, there are two processes called daseinisation. The first is *daseinisation of projections*, which maps projection operators to certain subobjects of the state object. (The state object and its subobjects are defined herein.) This provides the bridge from ordinary Birkhoff–von Neumann quantum logic to a new form of quantum logic that is based on the internal logic of a topos associated with the quantum system. The second is *daseinisation of self-adjoint operators*, which maps each self-adjoint operator to an arrow in the topos. Mathematically, the two forms of daseinisation are related, but conceptually they are quite different.

The topos approach emphasizes the role of *classical perspectives* on a quantum system. A classical perspective or *context* is nothing but a set of commuting physical quantities or, more precisely, the abelian von Neumann algebra generated by such a set. One of the main ideas is that *all* classical perspectives should be taken into account simultaneously. But given, for example, a projection operator, which represents a proposition about the value of a physical quantity in standard quantum theory, one is immediately faced with the problem that the projection operator is contained in some contexts but not in all. The idea is to approximate the projection in all of those contexts that do not contain it. Likewise, self-adjoint operators, which represent physical quantities, must be approximated suitably to all contexts. Daseinisation is nothing but a process of systematic approximation to all classical contexts.

This chapter is organized as follows: in Section 4.2, we discuss propositions and their representation in classical physics and standard quantum theory. Section 4.3 presents some basic structures in the topos approach to quantum theory. In Section 4.4, the representation of propositions in the topos approach is discussed—this is via daseinisation of projections. Section 4.5, which is the longest and most technical, shows how physical quantities are represented in the topos approach; here, we present daseinisation of self-adjoint operators. Section 4.6 concludes the chapter.

## 4.2 Propositions and Their Representation in Classical Physics and Standard Quantum Theory

**Propositions.** Let $S$ be some physical system, and let $A$ denote a physical quantity pertaining to the system $S$ (e.g., position, momentum, energy, angular momentum, spin). We are concerned with propositions of the form "the physical quantity $A$ has a value in the set $\Delta$ of real numbers," for which we use the shorthand notation "$A \varepsilon \Delta$." In all applications, $\Delta$ will be a Borel subset of the real numbers $\mathbb{R}$.

Arguably, physics is fundamentally about what we can say about the truth values of propositions of the form "$A \varepsilon \Delta$" (where $A$ varies over the physical quantities of the system and $\Delta$ varies over the Borel subsets of $\mathbb{R}$) when the system is in a given state. Of course, the state may change in time, and hence the truth values also will change in time.[2]

Speaking about propositions like "$A \varepsilon \Delta$" requires a certain amount of conceptualizing. We accept that it is sensible to talk about physical systems as suitably separated entities, that each such physical system is characterized by its physical quantities, and that the range of values of a physical quantity $A$ is a subset of the real numbers. If we want to assign truth values to propositions, then we need the concept of a state of a physical system $S$, and the truth values of propositions will depend on the state. If we regard these concepts as natural and basic (and potentially as prerequisites for doing physics at all), then a proposition "$A \varepsilon \Delta$" *refers to the world* in the most direct conceivable sense.[3] We will deviate from common practice, though, by insisting that also in quantum theory, a proposition is about "how things are" and is not to be understood as a counterfactual statement; that is, it is not merely about what we would obtain as a measurement result if we were to perform a measurement.

---

[2] One could also think of propositions about *histories* of a physical system $S$ such that propositions refer to multiple times, and the assignment of truth values to such propositions (in a given, evolving state). For some very interesting recent results on a topos formulation of histories, see Flori's recent article [18].

[3] Of course, as explained by Isham in Chapter 3, it is one of the central motivations for the whole topos program to find a mathematical framework for physical theories that is not depending fundamentally on the real numbers. In particular, the premise that physical quantities have real values is doubtful in this light. How this seeming dilemma is solved in the topos approach to quantum theory is clarified in Section 4.5.

### 4.2.1 Classical Physics, State Spaces, and Realism

In classical physics, a state of the system $S$ is represented by an element of a set; namely, by a point of the state space $\mathcal{S}$ of the system.[4] Each physical quantity $A$ is represented by a real-valued function $f_A : \mathcal{S} \to \mathbb{R}$ on the state space. A proposition "$A \, \varepsilon \, \Delta$" is represented by a certain subset of the state space; namely the set $f_A^{-1}(\Delta)$. We assume that $f_A$ is (at least) measurable. Because $\Delta$ is a Borel set, the subset $f_A^{-1}(\Delta)$ of state space representing "$A \, \varepsilon \, \Delta$" is a Borel set too.

In any state, each physical quantity $A$ has a value, which is simply given by evaluation of the function $f_A$ representing $A$ at the point $s \in \mathcal{S}$ of state space representing the state—that is, the value is $f_A(s)$. Every proposition "$A \, \varepsilon \, \Delta$" has a Boolean truth value in each given state $s \in \mathcal{S}$:

$$v(\text{``}A \, \varepsilon \, \Delta\text{''}; s) = \begin{cases} \textit{true} & \text{if } s \in f_A^{-1}(\Delta) \\ \textit{false} & \text{if } s \notin f_A^{-1}(\Delta). \end{cases} \qquad (4.1)$$

Let "$A \, \varepsilon \, \Delta$" and "$B \, \varepsilon \, \Gamma$" be two different propositions, represented by the Borel subsets $f_A^{-1}(\Delta)$ and $f_B^{-1}(\Gamma)$ of $\mathcal{S}$, respectively. The union $f_A^{-1}(\Delta) \cup f_B^{-1}(\Gamma)$ of the two Borel subsets is another Borel subset, and this subset represents the proposition "$A \, \varepsilon \, \Delta$ or $B \, \varepsilon \, \Gamma$" (disjunction). Similarly, the intersection $f_A^{-1}(\Delta) \cap f_B^{-1}(\Gamma)$ is a Borel subset of $\mathcal{S}$, and this subset represents the proposition "$A \, \varepsilon \, \Delta$ and $B \, \varepsilon \, \Gamma$" (conjunction). Both conjunction and disjunction can be extended to arbitrary countable families of Borel sets representing propositions. The set-theoretic operations of taking unions and intersections distribute over each other. Moreover, the negation of a proposition "$A \, \varepsilon \, \Delta$" is represented by the complement $\mathcal{S} \setminus f_A^{-1}(\Delta)$ of the Borel set $f_A^{-1}(\Delta)$. Clearly, the empty subset $\emptyset$ of $\mathcal{S}$ represents the trivially false proposition, and the maximal Borel subset, $\mathcal{S}$ itself, represents the trivially true proposition. The set $\mathcal{B}(\mathcal{S})$ of Borel subsets of the state space $\mathcal{S}$ of a classical system thus is a Boolean $\sigma$-algebra—that is, a $\sigma$-complete distributive lattice with complement. Stone's theorem shows that every Boolean algebra is isomorphic to the algebra of subsets of a suitable space, so Boolean logic is closely tied to the use of sets.

The fact that in a given state $s$ each physical quantity has a value and each proposition has a truth value makes classical physics a *realist* theory.[5]

### 4.2.2 Representation of Propositions in Standard Quantum Theory

In contrast to that, in quantum physics it is not possible to assign real values to all physical quantities at once. This is the content of the Kochen–Specker theorem [26] (for the generalization to von Neumann algebras, which is also needed for this chapter, see [7]). In the standard Hilbert space formulation of quantum theory, each physical

---

[4] We avoid the usual synonym *phase space*, which seems to be a historical misnomer. In any case, there are no phases in a phase space.

[5] Here, we could enter into an interesting but potentially never-ending discussion on physical reality, ontology, epistemology, and so forth. We avoid that and posit that for the purpose of this chapter, a realist theory is one that in any given state allows us to assign truth values to all propositions of the form "$A \, \varepsilon \, \Delta$." Moreover, we require that there is a suitable logical structure, in particular a deductive system, in which we can argue about (representatives of) propositions.

quantity $A$ is represented by a self-adjoint operator $\hat{A}$ on some Hilbert space $\mathcal{H}$. The range of possible real values that $A$ can take is given by the spectrum $\mathrm{sp}(\hat{A})$ of the operator $\hat{A}$. Of course, there is a notion of states in quantum theory: in the simplest version, they are given by unit vectors in the Hilbert space $\mathcal{H}$. There is a particular mapping taking self-adjoint operators and unit vectors to real numbers; namely, the evaluation

$$(\hat{A}, |\psi\rangle) \longmapsto \langle\psi| \hat{A} |\psi\rangle. \tag{4.2}$$

In general (unless $|\psi\rangle$ is an eigenstate of $\hat{A}$), this real value is *not* the value of the physical quantity $A$ in the state described by $\psi$. It rather is the expectation value, which is a statistical and instrumentalist notion. The physical interpretation of the mathematical formalism of quantum theory fundamentally depends on measurements and observers.

According to the spectral theorem, propositions like "$A \, \varepsilon \, \Delta$" are represented by projection operators $\hat{P} = \hat{E}[A \, \varepsilon \, \Delta]$. Each projection $\hat{P}$ corresponds to a unique closed subspace $U_{\hat{P}}$ of the Hilbert space $\mathcal{H}$ of the system and vice versa. The intersection of two closed subspaces is a closed subspace, which can be taken as the definition of a conjunction. The closure of the subspace generated by two closed subspaces is a candidate for the disjunction, and the function sending a projection $\hat{P}$ to $\hat{1} - \hat{P}$ is an orthocomplement. Birkhoff and von Neumann suggested in their seminal paper [1] interpreting these mathematical operations as providing a logic for quantum systems.

At first sight, this looks similar to the classical case: the Hilbert space $\mathcal{H}$ now takes the role of a state space, while its closed subspaces represent propositions. But there is an immediate, severe problem: if the Hilbert space $\mathcal{H}$ is at least two-dimensional, then the lattice $\mathcal{L}(\mathcal{H})$ of closed subspaces is *nondistributive* (and so is the isomorphic lattice of projections). This makes it very hard to find a proper semantics of quantum logic.[6]

## 4.3 Basic Structures in the Topos Approach to Quantum Theory

### 4.3.1 Contexts

As argued by Isham in this volume, the topos formulation of quantum theory is based on the idea that one takes the collection of *all* classical perspectives on a quantum system. No single classical perspective can deliver a complete picture of the quantum system, but the collection of all of them may well do. As we will see, it is also of great importance how the classical perspectives relate to each other.

These ideas are formalized in the following way: we consider a quantum system as being described by a *von Neumann algebra* $\mathcal{N}$ (see, for example, [25]). Such an algebra is always given as a subalgebra of $\mathcal{B}(\mathcal{H})$, the algebra of all bounded operators on a suitable Hilbert space $\mathcal{H}$. We can always assume that the identity in $\mathcal{N}$ is the identity

---

[6] We cannot discuss the merits and shortcomings of quantum logic here. Suffice it to say that there are more conceptual and interpretational problems and a large number of developments and abstractions from standard quantum logic (i.e., the lattice of closed subspaces of Hilbert space). An excellent review is in [2].

operator $\hat{1}$ on $\mathcal{H}$. Because von Neumann algebras are much more general than just the algebras of the form $\mathcal{B}(\mathcal{H})$ and all our constructions work for arbitrary von Neumann algebras, we present the results in this general form. For a very good introduction to the use of operator algebras in quantum theory, see [17]. Von Neumann algebras can be used to describe quantum systems with symmetry or superselection rules. Throughout, we use the very simple example of $\mathcal{N} = \mathcal{B}(\mathbb{C}^3)$ to illustrate the constructions.

We assume that the self-adjoint operators in the von Neumann algebra $\mathcal{N}$ associated with our quantum system represent the *physical quantities* of the system. A quantum system has noncommuting physical quantities, so the algebra $\mathcal{N}$ is *nonabelian*. The $\mathbb{R}$-vector space of self-adjoint operators in $\mathcal{N}$ is denoted as $\mathcal{N}_{sa}$.

A classical perspective is given by a collection of commuting physical quantities. Such a collection determines an *abelian subalgebra*, typically denoted as $V$, of the nonabelian von Neumann algebra $\mathcal{N}$. An abelian subalgebra is often called a *context*, and we will use the notions classical perspective, context, and abelian subalgebra synonymously. We consider only abelian subalgebras that

- are von Neumann algebras—that is, they are closed in the weak topology; the technical advantage is that the spectral theorem holds in both $\mathcal{N}$ and its abelian von Neumann subalgebras, and that the lattices of projections in $\mathcal{N}$ and its abelian von Neumann subalgebras are complete
- contain the identity operator $\hat{1}$

Given a von Neumann algebra $\mathcal{N}$, let $\mathcal{V}(\mathcal{N})$ be the set of all its abelian von Neumann subalgebras that contain $\hat{1}$. By convention, the trivial abelian subalgebra $V_0 = \mathbb{C}\hat{1}$ is not contained in $\mathcal{V}(\mathcal{N})$. If some subalgebra $V' \in \mathcal{V}(\mathcal{N})$ is contained in a larger subalgebra $V \in \mathcal{V}(\mathcal{N})$, then we denote the inclusion as $i_{V'V} : V' \to V$. Clearly, $\mathcal{V}(\mathcal{N})$ is a partially ordered set under inclusion, and as such is a simple kind of *category* (see, for example, [27]). The objects in this category are the abelian von Neumann subalgebras, and the arrows are the inclusions. Clearly, there is at most one arrow between any two objects. We call $\mathcal{V}(\mathcal{N})$ the *category of contexts* of the quantum system described by the von Neumann algebra $\mathcal{N}$.

Considering the abelian parts of a nonabelian structure may seem trivial, yet in fact it is not because the context category $\mathcal{V}(\mathcal{N})$ keeps track of the relations between contexts: whenever two abelian subalgebras $V, \tilde{V}$ have a nontrivial intersection, then there are inclusion arrows

$$V \longleftarrow V \cap \tilde{V} \longrightarrow \tilde{V} \tag{4.3}$$

in $\mathcal{V}(\mathcal{N})$. Every self-adjoint operator $\hat{A} \in V \cap \tilde{V}$ can be written as $g(\hat{B})$ for some $\hat{B} \in V_{sa}$, where $g : \mathbb{R} \to \mathbb{R}$ is a Borel function, and, at the same time, as another $h(\hat{C})$ for some $\hat{C} \in \tilde{V}_{sa}$ and some other Borel function $h$.[7] The point is that whereas $\hat{A}$ commutes with $\hat{B}$ and $\hat{A}$ commutes with $\hat{C}$, the operators $\hat{B}$ and $\hat{C}$ do not necessarily commute. In that way, the context category $\mathcal{V}(\mathcal{N})$ encodes a lot of information about

---

[7] This follows from the fact that each abelian von Neumann algebra $V$ is generated by a single self-adjoint operator; see, for example, Prop. 1.21 in [29]. A Borel function $g : \mathbb{R} \to \mathbb{C}$ takes a self-adjoint operator $\hat{A}$ in a von Neumann algebra to another operator $g(\hat{A})$ in the same algebra. If $g$ is real-valued, then $g(\hat{A})$ is self-adjoint. For details, see [25, 29].

the algebraic structure of $\mathcal{N}$, not just between commuting operators, and about the relations between contexts.

If $V' \subset V$, then the context $V'$ contains fewer self-adjoint operators and fewer projections than the context $V$, so one can describe less physics from the perspective of $V'$ than from the perspective of $V$. The step from $V$ to $V'$ hence involves a suitable kind of *coarse-graining*. We see later how daseinisation of projections (resp. self-adjoint operators) implements this informal idea of coarse-graining.

**Example 1:** Let $\mathcal{H} = \mathbb{C}^3$, and let $\mathcal{N} = \mathcal{B}(\mathbb{C}^3)$, the algebra of all bounded linear operators on $\mathbb{C}^3$. $\mathcal{B}(\mathbb{C}^3)$ is the algebra $M_3(\mathbb{C})$ of $3 \times 3$ matrices with complex entries, acting as linear transformations on $\mathbb{C}^3$. Let $(\psi_1, \psi_2, \psi_3)$ be an orthonormal basis of $\mathbb{C}^3$, and let $(\hat{P}_1, \hat{P}_2, \hat{P}_3)$ be the three projections onto the one-dimensional subspaces $\mathbb{C}\psi_1$, $\mathbb{C}\psi_2$, and $\mathbb{C}\psi_3$, respectively. Clearly, the projections $\hat{P}_1, \hat{P}_2, \hat{P}_3$ are pairwise orthogonal—that is, $\hat{P}_i \hat{P}_j = \delta_{ij} \hat{P}_i$.

There is an abelian subalgebra $V$ of $\mathcal{B}(\mathbb{C}^3)$ generated by the three projections. One can use von Neumann's double commutant construction and define $V = \{\hat{P}_1, \hat{P}_2, \hat{P}_3\}''$ (see [25]). More concretely, $V = \text{lin}_\mathbb{C}(\hat{P}_1, \hat{P}_2, \hat{P}_3)$. Even more explicitly, one can pick the matrix representation of the projections $\hat{P}_i$ such that

$$\hat{P}_1 = \begin{pmatrix} 1 & 0 & 0 \\ 0 & 0 & 0 \\ 0 & 0 & 0 \end{pmatrix}, \quad \hat{P}_2 = \begin{pmatrix} 0 & 0 & 0 \\ 0 & 1 & 0 \\ 0 & 0 & 0 \end{pmatrix}, \quad \hat{P}_3 = \begin{pmatrix} 0 & 0 & 0 \\ 0 & 0 & 0 \\ 0 & 0 & 1 \end{pmatrix}. \quad (4.4)$$

The abelian algebra $V$ generated by these projections then consists of all diagonal $3 \times 3$ matrices (with complex entries on the diagonal). Clearly, there is no larger abelian subalgebra of $\mathcal{B}(\mathbb{C}^3)$ that contains $V$, so $V$ is maximal abelian.

Every orthonormal basis $(\tilde{\psi}_1, \tilde{\psi}_2, \tilde{\psi}_3)$ determines a maximal abelian subalgebra $\tilde{V}$ of $\mathcal{B}(\mathbb{C}^3)$. There is some redundancy, though: if two bases differ only by a permutation or phase factors of the basis vectors, then they generate the same maximal abelian subalgebra. Evidently, there are uncountably many maximal abelian subalgebras of $\mathcal{B}(\mathbb{C}^3)$.

There also are nonmaximal abelian subalgebras. Consider, for example, the algebra generated by $\hat{P}_1$ and $\hat{P}_2 + \hat{P}_3$. This algebra is denoted as $V_{\hat{P}_1}$ and is given as $V_{\hat{P}_1} = \text{lin}_\mathbb{C}(\hat{P}_1, \hat{P}_2 + \hat{P}_3) = \mathbb{C}\hat{P}_1 + \mathbb{C}(\hat{1} - \hat{P}_1)$. Here, we use that $\hat{P}_2 + \hat{P}_3 = \hat{1} - \hat{P}_1$. There are uncountably many nonmaximal subalgebras of $\mathcal{B}(\mathbb{C}^3)$.

The trivial projections $\hat{0}$ and $\hat{1}$ are contained in every abelian subalgebra. An algebra of the form $V_{\hat{P}_1}$ contains two nontrivial projections: $\hat{P}_1$ and $\hat{1} - \hat{P}_1$. There is the trivial abelian subalgebra $V_0 = \mathbb{C}\hat{1}$ consisting of multiples of the identity operator only. By convention, we will not consider the trivial algebra (it is not included in the partially ordered set $\mathcal{V}(\mathcal{N})$).

The set $\mathcal{V}(\mathcal{N}) = \mathcal{V}(\mathcal{B}(\mathbb{C}^3)) =: \mathcal{V}(\mathbb{C}^3)$ of abelian subalgebras of $\mathcal{B}(\mathbb{C}^3)$ hence can be divided into two subsets: the maximal abelian subalgebras, corresponding to orthonormal bases as described, and the ones of the form $V_{\hat{P}_1}$ that are generated by a single projection.[8] Clearly, given two different maximal abelian subalgebras $V$ and $\tilde{V}$, neither contains the other. Similarly, if we have two nonmaximal

---

[8] In other (larger) algebras than $\mathcal{B}(\mathbb{C}^3)$, the situation is more complicated, of course.

algebras $V_{\hat{P}_1}$ and $V_{\hat{P}_2}$, neither will contain the other, with one exception: if $\hat{P}_2 = \hat{1} - \hat{P}_1$, then $V_{\hat{P}_1} = V_{\hat{P}_2}$. This comes from the more general fact that if a unital operator algebra contains an operator $\hat{A}$, then it also contains the operator $\hat{1} - \hat{A}$.

Of course, nonmaximal subalgebras can be contained in maximal ones. Let $V$ be the maximal abelian subalgebra generated by three pairwise orthogonal projections $\hat{P}_1, \hat{P}_2, \hat{P}_3$. Then, $V$ contains three nonmaximal abelian subalgebras—namely, $V_{\hat{P}_1}, V_{\hat{P}_2}$ and $V_{\hat{P}_3}$. Importantly, each nonmaximal abelian subalgebra is contained in many different maximal ones: consider two projections $\hat{Q}_2, \hat{Q}_3$ onto one-dimensional subspaces in $\mathbb{C}^3$ such that $\hat{P}_1, \hat{Q}_2, \hat{Q}_3$ are pairwise orthogonal. We assume that $\hat{Q}_2, \hat{Q}_3$ are such that the maximal abelian subalgebra $\tilde{V}$ generated by $\hat{P}_1, \hat{Q}_2, \hat{Q}_3$ is different from the algebra $V$ generated by the three pairwise orthogonal projections $\hat{P}_1, \hat{P}_2, \hat{P}_3$. Then, both $V$ and $\tilde{V}$ contain $V_{\hat{P}_1}$ as a subalgebra. This argument makes clear that $V_{\hat{P}_1}$ is contained in every maximal abelian subalgebra that contains the projection $\hat{P}_1$.

We take inclusion of smaller into larger abelian subalgebras (here of nonmaximal into maximal ones) as a partial order on the set $\mathcal{V}(\mathbb{C}^3)$.

Because $\mathbb{C}^3$ is finite-dimensional, the algebra $\mathcal{B}(\mathbb{C}^3)$ is both a von Neumann and a $C^*$-algebra. The same holds for the abelian subalgebras. We do not have to worry about weak closedness. On infinite-dimensional Hilbert spaces, these questions become important. In general, an abelian von Neumann subalgebra $V$ is generated by the collection $\mathcal{P}(V)$ of its projections (which all commute with each other, of course) via the double commutant construction.

### 4.3.2 Gelfand Spectra and the Spectral Presheaf

#### *4.3.2.1 Introduction*

We now make use of the fact that each abelian ($C^*$-) algebra $V$ of operators can be seen as an algebra of continuous functions on a suitable topological space, the *Gelfand spectrum* of $V$. At least locally, for each context, we thus obtain a mathematical formulation of quantum theory that is similar to classical physics, with a state space and physical quantities as real-valued functions on this space. The "local state spaces" are then combined into one global structure, the *spectral presheaf*, which serves as a state space analogue for quantum theory.

#### *4.3.2.2 Gelfand Spectra*

Let $V \in \mathcal{V}(\mathcal{N})$ be an abelian subalgebra of $\mathcal{N}$, and let $\Sigma_V$ denote its Gelfand spectrum. $\Sigma_V$ consists of the multiplicative states on $V$—that is, the positive linear functionals $\lambda : V \longrightarrow \mathbb{C}$ of norm 1 such that

$$\forall \hat{A}, \hat{B} \in V : \lambda(\hat{A}\hat{B}) = \lambda(\hat{A})\lambda(\hat{B}). \tag{4.5}$$

The elements $\lambda$ of $\Sigma_V$ also are pure states of $V$ and algebra homomorphisms from $V$ to $\mathbb{C}$. (It is useful to have these different aspects in mind.) The set $\Sigma_V$ is a compact Hausdorff space in the weak* topology [25, 29].

Let $\lambda \in \Sigma_V$ be given. It can be shown that for any self-adjoint operator $\hat{A} \in V_{\text{sa}}$, it holds that

$$\lambda(\hat{A}) \in \text{sp}(\hat{A}), \qquad (4.6)$$

and every element $s \in \text{sp}(\hat{A})$ is given as $s = \tilde{\lambda}(\hat{A})$ for some $\tilde{\lambda} \in \Sigma_V$. Moreover, if $g : \mathbb{R} \to \mathbb{R}$ is a continuous function, then

$$\lambda(g(\hat{A})) = g(\lambda(\hat{A})). \qquad (4.7)$$

This implies that an element $\lambda$ of the Gelfand spectrum $\Sigma_V$ can be seen as a *valuation* on $V_{\text{sa}}$—that is, a function sending each self-adjoint operator in $V$ to some element of its spectrum in such a way that Equation (4.7), the FUNC principle, holds. Within each context $V$, all physical quantities in $V$ (and only those!) can be assigned values at once, and the assignment of values commutes with taking (Borel) functions of the operators. Every element $\lambda$ of the Gelfand spectrum gives a different valuation on $V_{\text{sa}}$.

The well-known *Gelfand representation theorem* shows that the abelian von Neumann algebra $V$ is isometrically $*$-isomorphic (i.e., isomorphic as a $C^*$-algebra) to the algebra of continuous, complex-valued functions on its Gelfand spectrum $\Sigma_V$.[9] The isomorphism is given by

$$V \longrightarrow C(\Sigma_V) \qquad (4.8)$$
$$\hat{A} \longmapsto \overline{A}, \qquad (4.9)$$

where $\overline{A}(\lambda) := \lambda(\hat{A})$ for all $\lambda \in \Sigma_V$. The function $\overline{A}$ is called the *Gelfand transform* of the operator $\hat{A}$. If $\hat{A}$ is a self-adjoint operator, then $\overline{A}$ is real valued, and $\overline{A}(\Sigma_V) = \text{sp}(\hat{A})$.

The fact that the self-adjoint operators in an abelian (sub)algebra $V$ can be written as real-valued functions on the compact Hausdorff space $\Sigma_V$ means that the Gelfand spectrum $\Sigma_V$ plays a role exactly like the state space of a classical system. Of course, because $V$ is abelian, not all self-adjoint operators $\hat{A} \in \mathcal{N}_{\text{sa}}$ representing physical quantities of the quantum system are contained in $V$. We interpret the Gelfand spectrum $\Sigma_V$ as a *local state space* at $V$. *Local* here means "at the abelian part $V$ of the global nonabelian algebra $\mathcal{N}$."

For a projection $\hat{P} \in \mathcal{P}(V)$, we have

$$\lambda(\hat{P}) = \lambda(\hat{P}^2) = \lambda(\hat{P})\lambda(\hat{P}), \qquad (4.10)$$

so $\lambda(\hat{P}) \in \{0, 1\}$ for all projections. This implies that the Gelfand transform $\overline{P}$ of $\hat{P}$ is the characteristic function of some subset of $\Sigma_V$.[10] Let $A$ be a physical quantity that is represented by some self-adjoint operator in $V$, and let "$A \, \varepsilon \, \Delta$" be a proposition about the value of $A$. Then, we know from the spectral theorem that there is a projection $\hat{E}[A \, \varepsilon \, \Delta]$ in $V$ that represents the proposition. We saw that $\lambda(\hat{E}[A \, \varepsilon \, \Delta]) \in \{0, 1\}$, and by interpreting 1 as "true" and 0 as "false," we see that a valuation $\lambda \in \Sigma_V$ assigns a Boolean truth-value to each proposition "$A \, \varepsilon \, \Delta$."

---

[9] The Gelfand representation theorem holds for abelian $C^*$-algebras. Because every von Neumann algebra is a $C^*$-algebra, the theorem applies to our situation.

[10] The fact that a characteristic function can be continuous shows that the Gelfand topology on $\Sigma_V$ is pretty wild: the Gelfand spectrum of an abelian von Neumann algebra is *extremely disconnected*—that is, the closure of each open set is open.

The precise relation between projections in $V$ and subsets of the Gelfand spectrum of $V$ is as follows: let $\hat{P} \in \mathcal{P}(V)$, and define

$$S_{\hat{P}} := \{\lambda \in \Sigma_V \mid \lambda(\hat{P}) = 1\}. \tag{4.11}$$

One can show that $S_{\hat{P}} \subseteq \Sigma_V$ is *clopen*—that is, closed and open in $\Sigma_V$. Conversely, every clopen subset $S$ of $\Sigma_V$ determines a unique projection $\hat{P}_S \in \mathcal{P}(V)$, given as the inverse Gelfand transform of the characteristic function of $S$. There is a lattice isomorphism

$$\alpha : \mathcal{P}(V) \longrightarrow \mathcal{C}l(\Sigma_V) \tag{4.12}$$
$$\hat{P} \longmapsto S_{\hat{P}}$$

between the lattice of projections in $V$ and the lattice of clopen subsets of $\Sigma_V$. Thus, starting from a proposition "$A \,\varepsilon\, \Delta$," we obtain a clopen subset $S_{\hat{E}[A\,\varepsilon\,\Delta]}$ of $\Sigma_V$. This subset consists of all valuations—that is, pure states $\lambda$ of $V$ such that $\lambda(\hat{E}[A \,\varepsilon\, \Delta]) = 1$ — which means that the proposition "$A \,\varepsilon\, \Delta$" is true in the state/under the valuation $\lambda$. This strengthens the interpretation of $\Sigma_V$ as a local state space.

We saw that each context $V$ provides us with a local state space, and one of the main ideas in the work by Isham and Butterfield [19, 22–24] was to form a single global object from all of these local state spaces.

To keep track of the inclusion relations between the abelian subalgebras, we must relate the Gelfand spectra of larger and smaller subalgebras in a natural way. Let $V, V' \in \mathcal{V}(\mathcal{N})$ such that $V' \subseteq V$. Then, there is a mapping

$$r : \Sigma_V \longrightarrow \Sigma_{V'} \tag{4.13}$$
$$\lambda \longmapsto \lambda|_{V'},$$

sending each element $\lambda$ of the Gelfand spectrum of the larger algebra $V$ to its restriction $\lambda|_{V'}$ to the smaller algebra $V'$. It is well known that the mapping $r : \Sigma_V \to \Sigma_{V'}$ is continuous with respect to the Gelfand topologies and surjective. Physically, this means that every valuation $\lambda'$ on the smaller algebra is given as the restriction of a valuation on the larger algebra.

### 4.3.2.3 The Spectral Presheaf

We can now define the central object in the topos approach to quantum theory: the *spectral presheaf* $\underline{\Sigma}$. A *presheaf* is a contravariant, **Sets**-valued functor; see, for example, [27].[11] Our base category, the domain of this functor, is the context category $\mathcal{V}(\mathcal{N})$. To each object in $\mathcal{V}(\mathcal{N})$—that is, to each context $V$—we assign its Gelfand spectrum $\underline{\Sigma}_V := \Sigma_V$; and to each arrow in $\mathcal{V}(\mathcal{N})$—that is, each inclusion $i_{V'V}$—we assign a function from the set $\underline{\Sigma}_V$ to the set $\underline{\Sigma}_{V'}$. This function is $\underline{\Sigma}(i_{V'V}) := r$ (see Equation 4.13). It implements the concept of coarse-graining on the level of the local state spaces (Gelfand spectra).

---

[11] I make only minimal use of category theory in this chapter. In particular, the following definition should be understandable in itself, without further knowledge of functors, and so forth.

The spectral presheaf $\underline{\Sigma}$ associated with the von Neumann algebra $\mathcal{N}$ of a quantum system is the analogue of the state space of a classical system. The spectral presheaf is not a set and, hence, not a space in the usual sense. It rather is a particular, **Sets**-valued functor, built from the Gelfand spectra of the abelian subalgebras of $\mathcal{N}$ and the canonical restriction functions between them.

We come back to our example, $\mathcal{N} = \mathcal{B}(\mathbb{C}^3)$, and describe its spectral presheaf.

**Example 2:** Let $\mathcal{N} = \mathcal{B}(\mathbb{C}^3)$ as in Example 1, and let $V \in \mathcal{V}(\mathbb{C}^3)$ be a maximal abelian subalgebra. As discussed, $V$ is of the form $V = \{\hat{P}_1, \hat{P}_2, \hat{P}_3\}''$ for three pairwise orthogonal rank-1 projections $\hat{P}_1, \hat{P}_2, \hat{P}_3$. The Gelfand spectrum $\Sigma_V$ of $V$ has three elements (and is equipped with the discrete topology, of course). The spectral elements are given as

$$\lambda_i(\hat{P}_j) = \delta_{ij} \quad (i = 1, 2, 3). \tag{4.14}$$

Let $\hat{A} \in V$ be an arbitrary operator. Then, $\hat{A} = \sum_{i=1}^{3} a_i \hat{P}_i$ for some (unique) complex coefficients $a_i$. We have $\lambda_i(\hat{A}) = a_i$. If $\hat{A}$ is self-adjoint, then the $a_i$ are real and are the eigenvalues of $\hat{A}$.

The nonmaximal abelian subalgebra $V_{\hat{P}_1}$ has two elements in its Gelfand spectrum. Let us call them $\lambda_1'$ and $\lambda_2'$, where $\lambda_1'(\hat{P}) = 1$ and $\lambda_1'(\hat{1} - \hat{P}) = 0$, while $\lambda_2'(\hat{P}) = 0$ and $\lambda_2'(\hat{1} - \hat{P}) = 1$. The restriction mapping $r : \Sigma_V \to \Sigma_{V'}$ sends $\lambda_1$ to $\lambda_1'$. Both $\lambda_2$ and $\lambda_3$ are mapped to $\lambda_2'$.

Analogous relations hold for the other nonmaximal abelian subalgebras $V_{\hat{P}_2}$ and $V_{\hat{P}_3}$ of $V$. Because the spectral presheaf is given by assigning to each abelian subalgebra its Gelfand spectrum, and to each inclusion the corresponding restriction function between the Gelfand spectra, we have a complete, explicit description of the spectral presheaf $\underline{\Sigma}$ of the von Neumann algebra $\mathcal{B}(\mathbb{C}^3)$.

The generalization to higher dimensions is straightforward. In particular, if $\dim(\mathcal{H}) = n$ and $V$ is a maximal abelian subalgebra of $\mathcal{B}(\mathbb{C}^n)$, then Gelfand spectrum consists of $n$ elements. Topologically, the Gelfand spectrum $\Sigma_V$ is a discrete space. (For infinite-dimensional Hilbert spaces, where more complicated von Neumann algebras than $\mathcal{B}(\mathbb{C}^n)$ exist, this is not true in general.)

## 4.4 Representation of Propositions in the Topos Approach—Daseinisation of Projections

We now consider the representation of propositions like "$A \varepsilon \Delta$" in our topos scheme. We already saw that the spectral presheaf $\underline{\Sigma}$ is an analogue of the state space $\mathcal{S}$ of a classical system. In analogy to classical physics, where propositions are represented by (Borel) subsets of state space, in the topos approach, propositions will be represented by suitable subobjects of the spectral presheaf.

### 4.4.1 Coarse-Graining of Propositions

Consider a proposition "$A \varepsilon \Delta$," where $A$ is some physical quantity of the quantum system under consideration. We assume that there is a self-adjoint operator $\hat{A}$ in the von

Neumann algebra $\mathcal{N}$ of the system that represents $A$. From the spectral theorem, we know that there is a projection operator $\hat{P} := \hat{E}[A \, \varepsilon \, \Delta]$ that represents the proposition "$A \, \varepsilon \, \Delta$." As is well known, the mapping from propositions to projections is many-to-one. One can form equivalence classes of propositions to obtain a bijection. In a slight abuse of language, we will usually refer to *the* proposition represented by a projection. Because $\mathcal{N}$ is a von Neumann algebra, the spectral theorem holds. In particular, for all propositions "$A \, \varepsilon \, \Delta$," we have $\hat{P} = \hat{E}[A \, \varepsilon \, \Delta] \in \mathcal{P}(\mathcal{N})$. The projections representing propositions are all contained in the von Neumann algebra. If we had chosen a more general $C^*$-algebra instead, this would not have been the case.

We stipulate that in the topos formulation of quantum theory, all classical perspectives/contexts must be taken into account at the same time, so we have to adapt the proposition "$A \, \varepsilon \, \Delta$" resp. its representing projection $\hat{P}$ to all contexts. The idea is very simple: in every context $V$, we pick the strongest proposition implied by "$A \, \varepsilon \, \Delta$" that can be made from the perspective of this context. On the level of projections, this amounts to taking the *smallest* projection in any context $V$ that is larger than or equal to $\hat{P}$.[12] We write, for all $V \in \mathcal{V}(\mathcal{N})$,

$$\delta^o(\hat{P})_V := \bigwedge \{\hat{Q} \in \mathcal{P}(V) \mid \hat{Q} \geq \hat{P}\} \tag{4.15}$$

for this approximation of $\hat{P}$ to $V$. If $\hat{P} \in \mathcal{P}(V)$, then clearly the approximation will give $\hat{P}$ itself—that is, $\delta^o(\hat{P})_V = \hat{P}$. If $\hat{P} \notin \mathcal{P}(V)$, then $\delta^o(\hat{P})_V > \hat{P}$. For many contexts $V$, it holds that $\delta^o(\hat{P})_V = \hat{1}$.

Because the projection $\delta^o(\hat{P})_V$ lies in $V$, it corresponds to a proposition about some physical quantity described by a self-adjoint operator *in* $V$. If $\delta^o(\hat{P})_V$ happens to be a spectral projection of the operator $\hat{A}$, then the proposition corresponding to $\delta^o(\hat{P})_V$ is of the form "$A \, \varepsilon \, \Gamma$" for some Borel set $\Gamma$ of real numbers that is larger than the set $\Delta$ in the original proposition "$A \, \varepsilon \, \Delta$."[13] The fact that $\Gamma \supset \Delta$ shows that the mapping

$$\delta^o_V : \mathcal{P}(\mathcal{N}) \longrightarrow \mathcal{P}(V) \tag{4.16}$$
$$\hat{P} \longmapsto \delta^o(\hat{P})_V$$

implements the idea of coarse-graining on the level of projections (resp. the corresponding propositions).

If $\delta^o(\hat{P})_V$ is not a spectral projection of $\hat{A}$, it still holds that it corresponds to a proposition "$B \, \varepsilon \, \Gamma$" about the value of some physical quantity $B$ represented by a self-adjoint operator $\hat{B}$ in $V$ and, because $\hat{P} < \delta^o(\hat{P})_V$, this proposition can still be seen as a coarse-graining of the original proposition "$A \, \varepsilon \, \Delta$" corresponding to $\hat{P}$ (even though $A \neq B$).

Every projection is contained in some abelian subalgebra, so there is always some $V$ such that $\delta^o(\hat{P})_V = \hat{P}$. From the perspective of this context, no coarse-graining takes place.

---

[12] Here, we use the interpretation that $\hat{P} < \hat{Q}$ for two projections $\hat{P}, \hat{Q} \in \mathcal{P}(\mathcal{N})$ means that the proposition represented by $\hat{P}$ implies the proposition represented by $\hat{Q}$. This is customary in quantum logic, and we use it to motivate our construction, although the partial order on projections will *not* be the implication relation for the form of quantum logic resulting from our scheme.

[13] We remark that in order to have $\delta^o(\hat{P})_V \in \mathcal{P}(V)$, it is sufficient, but *not* necessary, that $\hat{A} \in V_{\text{sa}}$.

We call the original proposition "$A \, \varepsilon \, \Delta$" that we want to represent the *global* proposition, and a proposition "$B \, \varepsilon \, \Gamma$" corresponding to the projection $\delta^o(\hat{P})_V$ is called a *local* proposition. *Local* here again means "at the abelian part $V$" and has no spatiotemporal connotation.

We can collect all the local approximations into a mapping $\hat{P} \mapsto (\delta^o(\hat{P})_V)_{V \in \mathcal{V}(\mathcal{N})}$ whose component at $V$ is given by Equation (4.15). From every global proposition, we thus obtain one coarse-grained local proposition for each context $V$. In the next step, we consider the whole collection of coarse-grained local propositions, relate it to a suitable subobject of the spectral presheaf, and regard this object as the representative of the global proposition.

### 4.4.2 Daseinisation of Projections

Let $V$ be a context. Because $V$ is an abelian von Neumann algebra, the lattice $\mathcal{P}(V)$ of projections in $V$ is a distributive lattice. Moreover, $\mathcal{P}(V)$ is complete and orthocomplemented. We saw in Equation (4.12) that there is a lattice isomorphism between $\mathcal{P}(V)$ and $\mathcal{Cl}(\underline{\Sigma}_V)$, the lattice of clopen subsets of $\underline{\Sigma}_V$.

We already have constructed a family $(\delta^o(\hat{P})_V)_{V \in \mathcal{V}(\mathcal{N})}$ of projections from a single projection $\hat{P} \in \mathcal{P}(\mathcal{N})$ representing a global proposition. For each context $V$, we now consider the clopen subset $\alpha(\delta^o(\hat{P})_V) = S_{\delta^o(\hat{P})_V} \subseteq \underline{\Sigma}_V$. We thus obtain a family $(S_{\delta^o(\hat{P})_V})_{V \in \mathcal{V}(\mathcal{N})}$ of clopen subsets, one for each context. One can show that whenever $V' \subset V$, then

$$S_{\delta^o(\hat{P})_V}|_{V'} = \{\lambda|_{V'} \mid \lambda \in S_{\delta^o(\hat{P})_V}\} = S_{\delta^o(\hat{P})_{V'}}, \qquad (4.17)$$

that is, the clopen subsets in the family $(S_{\delta^o(\hat{P})_V})_{V \in \mathcal{V}(\mathcal{N})}$ "fit together" under the restriction mappings of the spectral presheaf $\underline{\Sigma}$ (see Thm. 3.1 in [13]). This means that the family $(S_{\delta^o(\hat{P})_V})_{V \in \mathcal{V}(\mathcal{N})}$ of clopen subsets, together with the restriction mappings between them, forms a *subobject* of the spectral presheaf. This particular subobject will be denoted as $\underline{\delta}(\hat{P})$ and is called the *daseinisation of* $\hat{P}$.

A subobject of the spectral presheaf (or any other presheaf) is the analogue of a subset of a space. Concretely, for each $V \in \mathcal{V}(\mathcal{N})$, the set $S_{\delta^o(\hat{P})_V}$ is a subset of the Gelfand spectrum of $V$. Moreover, the subsets for different $V$ are not arbitrary but rather fit together under the restriction mappings of the spectral presheaf.

The collection of all subobjects of the spectral presheaf is a complete *Heyting algebra* and is denoted as Sub($\underline{\Sigma}$). It is the analogue of the power set of the state space of a classical system.

Let $\underline{S}$ be a subobject of the spectral presheaf $\underline{\Sigma}$ such that the component $S_V \subseteq \underline{\Sigma}_V$ is a clopen subset for all $V \in \mathcal{V}(\mathcal{N})$. We call such a subobject a *clopen* subobject. All subobjects obtained from daseinisation of projections are clopen. One can show that the clopen subobjects form a complete Heyting algebra Sub$_{cl}(\underline{\Sigma})$ (see Thm. 2.5 in [13]). This complete Heyting algebra can be seen as the analogue of the $\sigma$-complete Boolean algebra of Borel subsets of the state space of a classical system. In our constructions, the Heyting algebra Sub$_{cl}(\underline{\Sigma})$ will be more important than the bigger Heyting algebra Sub($\underline{\Sigma}$), just as in classical physics, where the Boolean algebra of measurable subsets of state space is technically more important than the algebra of all subsets.

Starting from a global proposition "$A \varepsilon \Delta$" with corresponding projection $\hat{P}$, we have constructed the clopen subobject $\underline{\delta(\hat{P})}$ of the spectral presheaf. This subobject, the analogue of the measurable subset $f_A^{-1}(\Delta)$ of the state space of a classical system, is the representative of the global proposition "$A \varepsilon \Delta$." The following mapping is called *daseinisation of projections*:

$$\underline{\delta} : \mathcal{P}(\mathcal{N}) \longrightarrow \text{Sub}_{\text{cl}}(\underline{\Sigma}) \qquad (4.18)$$
$$\hat{P} \longmapsto \underline{\delta(\hat{P})}.$$

It has the following properties:

1. If $\hat{P} < \hat{Q}$, then $\underline{\delta(\hat{P})} < \underline{\delta(\hat{Q})}$—that is, daseinisation is order preserving.
2. The mapping $\underline{\delta} : \mathcal{P}(\mathcal{N}) \to \text{Sub}_{\text{cl}}(\underline{\Sigma})$ is injective—that is, two (inequivalent) propositions correspond to two different subobjects.
3. $\underline{\delta(\hat{0})} = \underline{0}$, the empty subobject, and $\underline{\delta(\hat{1})} = \underline{\Sigma}$. The trivially false proposition is represented by the empty subobject, and the trivially true proposition is represented by the whole of $\underline{\Sigma}$.
4. For all $\hat{P}, \hat{Q} \in \mathcal{P}(\mathcal{N})$, it holds that $\underline{\delta(\hat{P} \vee \hat{Q})} = \underline{\delta(\hat{P})} \vee \underline{\delta(\hat{Q})}$—that is, daseinisation preserves the disjunction (Or) of propositions.
5. For all $\hat{P}, \hat{Q} \in \mathcal{P}(\mathcal{N})$, it holds that $\underline{\delta(\hat{P} \wedge \hat{Q})} \leq \underline{\delta(\hat{P})} \wedge \underline{\delta(\hat{Q})}$—that is, daseinisation does not preserve the conjunction (And) of propositions.
6. In general, $\underline{\delta(\hat{P})} \wedge \underline{\delta(\hat{Q})}$ is not of the form $\underline{\delta(\hat{R})}$ for a projection $\hat{R} \in \mathcal{P}(\mathcal{N})$, and daseinisation is not surjective.

The domain of the mapping $\underline{\delta}$, the lattice $\mathcal{P}(\mathcal{N})$ of projections in the von Neumann algebra $\mathcal{N}$ of physical quantities, is the lattice that Birkhoff and von Neumann suggested as the algebraic representative of propositional quantum logic [1] (in fact, they considered the case $\mathcal{N} = \mathcal{B}(\mathcal{H})$). Thus, daseinisation of projections can be seen as a "translation" mapping between ordinary, Birkhoff–von Neumann quantum logic, which is based on the nondistributive lattice of projections $\mathcal{P}(\mathcal{N})$, and the topos form of propositional quantum logic, which is based on the distributive lattice $\text{Sub}_{\text{cl}}(\underline{\Sigma})$. The latter more precisely is a Heyting algebra.

The quantum logic aspects of the topos formalism and the physical and conceptual interpretation of these relations are discussed in some detail in [10], building on [13, 16, 22]. The resulting new form of quantum logic has many attractive features and avoids a number of the well-known conceptual problems of standard quantum logic [2].

**Example 3:** Let $\mathcal{H} = \mathbb{C}^3$, and let $S_z$ be the physical quantity "spin in $z$-direction." We consider the proposition "$S_z \varepsilon (-0.1, 0.1)$"—that is, "the spin in $z$-direction has a value between $-0.1$ and $0.1$." (Because the eigenvalues are $-\frac{1}{\sqrt{2}}, 0,$ and $\frac{1}{\sqrt{2}}$, this amounts to saying that the spin in $z$-direction is 0.) The self-adjoint operator representing $S_z$ is

$$\hat{S}_z = \frac{1}{\sqrt{2}} \begin{pmatrix} 1 & 0 & 0 \\ 0 & 0 & 0 \\ 0 & 0 & -1 \end{pmatrix}, \qquad (4.19)$$

and the eigenvector corresponding to the eigenvalue 0 is $|\psi\rangle = (0, 1, 0)$. The projection $\hat{P} := \hat{E}[S_z \varepsilon (-0.1, 0.1)] = |\psi\rangle\langle\psi|$ onto this eigenvector hence is

$$\hat{P} = \begin{pmatrix} 0 & 0 & 0 \\ 0 & 1 & 0 \\ 0 & 0 & 0 \end{pmatrix}. \quad (4.20)$$

There is exactly one context $V \in \mathcal{V}(\mathbb{C}^3)$ that contains the operator $\hat{S}_z$—namely, the (maximal) context $V_{\hat{S}_z} = \{\hat{P}_1, \hat{P}_2, \hat{P}_3\}''$ generated by the projections

$$\hat{P}_1 = \begin{pmatrix} 1 & 0 & 0 \\ 0 & 0 & 0 \\ 0 & 0 & 0 \end{pmatrix}, \ \hat{P}_2 = \hat{P} = \begin{pmatrix} 0 & 0 & 0 \\ 0 & 1 & 0 \\ 0 & 0 & 0 \end{pmatrix}, \ \hat{P}_3 = \begin{pmatrix} 0 & 0 & 0 \\ 0 & 0 & 0 \\ 0 & 0 & 1 \end{pmatrix}. \quad (4.21)$$

In the following, we consider different kinds of contexts $V$ and the approximations $\delta^o(\hat{P})_V$ of $\hat{P} = \hat{P}_2$ to these contexts.

1. $V_{\hat{S}_z}$ **and its subcontexts:** Of course, $\delta^o(\hat{P})_{V_{\hat{S}_z}} = \hat{P}$. Let $V_{\hat{P}_1} = \{\hat{P}_1, \hat{P} + \hat{P}_3\}''$, then $\delta^o(\hat{P})_{V_{\hat{P}_1}} = \hat{P} + \hat{P}_3$. Analogously, $\delta^o(\hat{P})_{V_{\hat{P}_3}} = \hat{P} + \hat{P}_1$, but $\delta^o(\hat{P})_{V_{\hat{P}}} = \hat{P}$, because $V_{\hat{P}}$ contains the projection $\hat{P}$.

2. **Maximal contexts that share the projection $\hat{P}$ with $V_{\hat{S}_z}$ and their subcontexts:** Let $\hat{Q}_1, \hat{Q}_3$ be two orthogonal rank-1 projections that are both orthogonal to $\hat{P}$ (e.g., projections obtained from $\hat{P}_1, \hat{P}_3$ by a rotation about the axis determined by $\hat{P}$). Then, $V = \{\hat{Q}_1, \hat{P}, \hat{Q}_3\}''$ is a maximal algebra that contains $\hat{P}$, so $\delta^o(\hat{P})_V = \hat{P}$. Clearly, there are uncountably many contexts $V$ of this form.

    For projections $\hat{Q}_1, \hat{P}, \hat{Q}_3$ as described previously, we have $\delta^o(\hat{P})_{V_{\hat{Q}_1}} = \hat{P} + \hat{Q}_3, \delta^o(\hat{P})_{V_{\hat{Q}_3}} = \hat{Q}_1 + \hat{P}$ (and $\delta^o(\hat{P})_{V_{\hat{P}}} = \hat{P}$, as mentioned before).

3. **Contexts that contain a rank-2 projection $\hat{Q}$ such that $\hat{Q} > \hat{P}$:** We first note that there are uncountably many projections $\hat{Q}$ of this form.
    a. The trivial cases of contexts containing $\hat{Q} > \hat{P}$ are (1) maximal contexts containing $\hat{P}$, and (2) the context $V_{\hat{Q}} = V_{\hat{1}-\hat{Q}}$.
    b. There are also maximal contexts that do not contain $\hat{P}$ but do contain $\hat{Q}$: let $\hat{Q}_1, \hat{Q}_2$ be two orthogonal rank-1 projections such that $\hat{Q}_1 + \hat{Q}_2 = \hat{Q}$, where $\hat{Q}_1, \hat{Q}_2 \neq \hat{P}$. Let $\hat{Q}_3 := \hat{1} - (\hat{Q}_1 + \hat{Q}_2)$. For any given $\hat{Q}$, there are uncountably many contexts of the form $V = \{\hat{Q}_1, \hat{Q}_2, \hat{Q}_3\}$ (because one can rotate around the axis determined by $\hat{Q}_3$).

    For all of these contexts, $\delta^o(\hat{P})_V = \hat{Q}$, of course.

4. **Other contexts:** All other contexts contain neither $\hat{P}$ nor a projection $\hat{Q} > \hat{P}$ that is not the identity $\hat{1}$. This means that the smallest projection larger than $\hat{P}$ contained in such a context is the identity, so $\delta^o(\hat{P})_V = 1$.

    We have constructed $\delta^o(\hat{P})_V$ for all contexts. Recalling that for each context $V$ there is a lattice isomorphism $\alpha$ from the projections to the clopen subsets of the Gelfand spectrum $\Sigma_V$ (see Equation 4.12), we can easily write down the clopen subsets $S_{\delta^o(\hat{P})_V}$ ($V \in \mathcal{V}(\mathbb{C}^3)$) corresponding to the projections $\delta^o(\hat{P})_V$ ($V \in \mathcal{V}(\mathbb{C}^3)$). The daseinisation $\delta(\hat{P})$ of the projection $\hat{P}$ is nothing but the collection $(S_{\delta^o(\hat{P})_V})_{V \in \mathcal{V}(\mathbb{C}^3)}$ (together with the canonical restriction mappings described in Equation [4.13] and Example 2).

## 4.5 Representation of Physical Quantities—Daseinisation of Self-Adjoint Operators

### 4.5.1 Physical Quantities as Arrows

In classical physics, a physical quantity $A$ is represented by a function $f_A$ from the state space $\mathcal{S}$ of the system to the real numbers. Both the state space and the real numbers are sets (with extra structure), so they are objects in the topos **Sets** of sets and functions.[14] The function $f_A$ representing a physical quantity is an arrow in the topos **Sets**.

The topos reformulation of quantum physics uses structures in the topos $\mathbf{Sets}^{\mathcal{V}(\mathcal{N})^{op}}$, the topos of presheaves over the context category $\mathcal{V}(\mathcal{N})$. The objects in the topos, called *presheaves*, are the analogues of sets, and the arrows between them, called *natural transformations*, are the analogues of functions. We now represent a physical quantity $A$ by a suitable arrow, denoted $\breve{\delta}(\hat{A})$, from the spectral presheaf $\underline{\Sigma}$ to some other presheaf related to the real numbers. Indeed, there is a real-number object $\underline{\mathbb{R}}$ in $\mathbf{Sets}^{\mathcal{V}(\mathcal{N})^{op}}$ that is very much like the familiar real numbers: for every $V$, the component $\underline{\mathbb{R}}_V$ simply is the ordinary real numbers, and all restriction functions are just the identity. But the Kochen–Specker theorem [7, 26] tells us that it is impossible to assign (sharp) real values to all physical quantities, which suggests that the presheaf $\underline{\mathbb{R}}$ is *not* the right codomain for our arrows $\breve{\delta}(\hat{A})$ representing physical quantities.

Instead, we will be using a presheaf, denoted as $\underline{\mathbb{R}}^{\leftrightarrow}$, that takes into account that (1) one cannot assign sharp values to all physical quantities in any given state in quantum theory, and (2) we have coarse-graining. We first explain how coarse-graining will show up: if we want to assign a "value" to a physical quantity $A$, then we again have to find some global expression involving all contexts $V \in \mathcal{V}(\mathcal{N})$. Let $\hat{A} \in \mathcal{N}_{sa}$ be the self-adjoint operator representing $A$. For those contexts $V$ that contain $\hat{A}$, there is no problem, but if $\hat{A} \notin V$, we will have to approximate $\hat{A}$ by a self-adjoint operator *in V*. Actually, we will use two approximations: one self-adjoint operator in $V$ approximating $\hat{A}$ from below (in a suitable order, specified herein), and one operator approximating $\hat{A}$ from above. Daseinisation of self-adjoint operators is nothing but the technical device achieving this approximation.

We will show that the presheaf $\underline{\mathbb{R}}^{\leftrightarrow}$ basically consists of real intervals, to be interpreted as "unsharp values." Upon coarse-graining, the intervals can only get bigger. Assume we have a given physical quantity $A$, some state of the system, and two contexts $V, V'$ such that $V' \subset V$. We use the state to assign some interval $[a, b]$ as an "unsharp value" to $A$ at $V$. Then, at $V'$, we will have a bigger interval $[c, d] \supseteq [a, b]$, being an even more unsharp value of $A$. It is important to note, though, that every self-adjoint operator is contained in some context $V$. If the state is an eigenstate of $\hat{A}$ with eigenvalue $a$, then at $V$, we will assign the "interval" $[a, a]$ to $A$. In this sense, the eigenvector-eigenvalue link is preserved, regardless of the unsharpness in values introduced by coarse-graining. As always in the topos scheme, it is of central importance that the interpretationally relevant structures—here, the "values" of physical

---

[14] We do not give the technical definition of a *topos* here. The idea is that a topos is a category that is structurally similar to **Sets**, the category of small sets and functions. Of course, **Sets** itself is a topos.

quantities—are global in nature. One has to consider all contexts at once and not just argue locally at some context $V$, which will necessarily only give partial information.

#### 4.5.1.1 Approximation in the Spectral Order

Let $\hat{A} \in \mathcal{N}_{sa}$ be a self-adjoint operator, and let $V \in \mathcal{V}(\mathcal{N})$ be a context. We assume that $\hat{A} \notin V_{sa}$, so we have to approximate $\hat{A}$ in $V$. The idea is very simple: we take a pair of self-adjoint operators, consisting of the largest operator in $V_{sa}$ smaller than $\hat{A}$ and the smallest operator in $V_{sa}$ larger than $\hat{A}$. The only nontrivial point is the question of which order on self-adjoint operators to use.

The most commonly used order is the *linear order*: $\hat{A} \leq \hat{B}$ if and only if $\langle \psi | \hat{A} | \psi \rangle \leq \langle \psi | \hat{B} | \psi \rangle$ for all vector states $\langle \psi | \_ | \psi \rangle : \mathcal{N} \to \mathbb{C}$. Yet we will approximate in another order on the self-adjoint operators the so-called *spectral order* [3, 28]. This has the advantage that the spectra of the approximated operators are subsets of the spectrum of the original operator, which is not the case for the linear order in general.

The spectral order is defined in the following way: let $\hat{A}, \hat{B} \in \mathcal{N}_{sa}$ be two self-adjoint operators in a von Neumann algebra $\mathcal{N}$, and let $(\hat{E}_r^{\hat{A}})_{r \in \mathbb{R}}, (\hat{E}_r^{\hat{B}})_{r \in \mathbb{R}}$ be their spectral families (see, for example, [25]). Then,

$$\hat{A} \leq_s \hat{B} \quad \text{if and only if} \quad \forall r \in \mathbb{R} : \hat{E}_r^{\hat{A}} \geq \hat{E}_r^{\hat{B}}. \tag{4.22}$$

Equipped with the spectral order, $\mathcal{N}_{sa}$ becomes a boundedly complete lattice. The spectral order is coarser than the linear order—that is, $\hat{A} \leq_s \hat{B}$ implies $\hat{A} \leq \hat{B}$. The two orders coincide on projections and for commuting operators.

Consider the set $\{\hat{B} \in V_{sa} \mid \hat{B} \leq_s \hat{A}\}$—that is, those self-adjoint operators in $V$ that are spectrally smaller than $\hat{A}$. Because $V$ is a von Neumann algebra, its self-adjoint operators form a boundedly complete lattice under the spectral order, so the previous set has a well-defined maximum with respect to the spectral order. It is denoted as

$$\delta^i(\hat{A})_V := \bigvee \{\hat{B} \in V_{sa} \mid \hat{B} \leq_s \hat{A}\}. \tag{4.23}$$

Similarly, we define

$$\delta^o(\hat{A})_V := \bigwedge \{\hat{B} \in V_{sa} \mid \hat{B} \geq_s \hat{A}\}, \tag{4.24}$$

which is the smallest self-adjoint operator in $V$ that is spectrally larger than $\hat{A}$.

Using the definition in Equation (4.22) of the spectral order, it is easy to describe the spectral families of the operators $\delta^i(\hat{A})_V$ and $\delta^o(\hat{A})_V$:

$$\forall r \in \mathbb{R} : \hat{E}_r^{\delta^o(\hat{A})_V} = \delta^i(\hat{E}_r^{\hat{A}})_V \tag{4.25}$$

and

$$\forall r \in \mathbb{R} : \hat{E}_r^{\delta^i(\hat{A})_V} = \delta^o(\hat{E}_r^{\hat{A}})_V. \tag{4.26}$$

It turns out (see [6, 13]) that the spectral family defined by the former definition is right continuous. The latter expression must be amended slightly if we want a right-continuous spectral family: we simply enforce right continuity by setting

$$\forall r \in \mathbb{R} : \hat{E}_r^{\delta^i(\hat{A})_V} = \bigwedge_{s > r} \delta^o(\hat{E}_s^{\hat{A}})_V. \tag{4.27}$$

These equations show the close mathematical link between the approximations in the spectral order of projections and self-adjoint operators.

The self-adjoint operators $\delta^i(\hat{A})_V$, $\delta^o(\hat{A})_V$ are the best approximations to $\hat{A}$ in $V_{sa}$ with respect to the spectral order. $\delta^i(\hat{A})_V$ approximates $\hat{A}$ from below, $\delta^o(\hat{A})_V$ from above. Because $V$ is an abelian $C^*$-algebra, we can consider the Gelfand transforms of these operators to obtain a pair

$$(\overline{\delta^i(\hat{A})_V}, \overline{\delta^o(\hat{A})_V}) \tag{4.28}$$

of continuous functions from the Gelfand spectrum $\Sigma_V$ of $V$ to the real numbers. One can show that [5, 14]

$$\mathrm{sp}(\delta^i(\hat{A})_V) \subseteq \mathrm{sp}(\hat{A}), \quad \mathrm{sp}(\delta^o(\hat{A})_V) \subseteq \mathrm{sp}(\hat{A}), \tag{4.29}$$

so the Gelfand transforms actually map into the spectrum $\mathrm{sp}(\hat{A})$ of the original operator $\hat{A}$. Let $\lambda \in \Sigma_V$ be a pure state of the algebra $V$.[15] Then, we can evaluate $(\overline{\delta^i(\hat{A})_V}(\lambda), \overline{\delta^o(\hat{A})_V}(\lambda))$ to obtain a pair of real numbers in the spectrum of $\hat{A}$. The first number is smaller than the second because the first operator $\delta^i(\hat{A})_V$ approximates $\hat{A}$ from below in the spectral order (and $\delta^i(\hat{A})_V \leq_s \hat{A}$ implies $\delta^i(\hat{A})_V \leq \hat{A}$), while the second operator $\delta^o(\hat{A})_V$ approximates $\hat{A}$ from above. The pair $(\overline{\delta^i(\hat{A})_V}(\lambda), \overline{\delta^o(\hat{A})_V}(\lambda))$ of real numbers is identified with the interval $[\overline{\delta^i(\hat{A})_V}(\lambda), \overline{\delta^o(\hat{A})_V}(\lambda)]$ and interpreted as the component at $V$ of the unsharp value of $\hat{A}$ in the (local) state $\lambda$.

#### 4.5.1.2 The Presheaf of "Values"

We now see the way in which intervals show up as "values." If we want to define a presheaf encoding this, we encounter a certain difficulty. It is no problem to assign to each context $V$ the collection $\mathbb{IR}_V$ of real intervals, but if $V' \subset V$, then we need a restriction mapping from $\mathbb{IR}_V$ to $\mathbb{IR}_{V'}$. Both sets are the same, so the naive guess is to take an interval in $\mathbb{IR}_V$ and to map it to the same interval in $\mathbb{IR}_{V'}$. But we already know that the "values" of our physical quantities come from approximations of a self-adjoint operator $\hat{A}$ to all the contexts. If $V' \subset V$, then in general we have $\delta^i(\hat{A})_{V'} <_s \delta^i(\hat{A})_V$[16], which implies $\overline{\delta^i(\hat{A})_{V'}}(\lambda|_{V'}) \leq \overline{\delta^i(\hat{A})_V}(\lambda)$ for $\lambda \in \Sigma_V$. Similarly, $\overline{\delta^o(\hat{A})_{V'}}(\lambda|_{V'}) \geq \overline{\delta^o(\hat{A})_V}(\lambda)$, so the intervals that we obtain get bigger when we go from larger contexts to smaller ones as a result of coarse-graining. Our restriction mapping from the intervals at $V$ to the intervals at $V'$ must take this into account.

The intervals at different contexts clearly also depend on the state $\lambda \in \Sigma_V$ that one picks. Moreover, we want a presheaf that is not tied to the operator $\hat{A}$ but can provide "values" for all physical quantities and their corresponding self-adjoint operators.

The idea is to define a presheaf such that at each context $V$ we have all intervals (including those of the form $[a, a]$), plus all possible restrictions to the same or larger intervals at smaller contexts $V' \subset V$. Although this sounds daunting, there is a very simple way to encode this: let $\downarrow V := \{V' \in \mathcal{V}(\mathcal{N}) \mid V' \subseteq V\}$. This is a

---

[15] We remark that, in general, this cannot be identified with a state of the nonabelian algebra $\mathcal{N}$, so we still have a local argument here.

[16] This, of course, is nothing but coarse-graining on the level of self-adjoint operators.

partially ordered set. We now consider a function $\mu : \downarrow V \to \mathbb{R}$ that *preserves the order*; that is, $V' \subset V$ implies $\mu(V') \leq \nu(V)$. Analogously, let $\nu : \downarrow V \to \mathbb{R}$ denote an *order-reversing* function; that is, $V' \subset V$ implies $\nu(V') \geq \nu(V)$. Additionally, we assume that for all $V' \subseteq V$, it holds that $\mu(V') \leq \nu(V')$. The pair $(\mu, \nu)$ thus gives one interval $(\mu(V'), \nu(V'))$ for each context $V' \subseteq V$, and "going down the line" from $V$ to smaller subalgebras $V', V'', \ldots$, these intervals can only get bigger (or stay the same). Of course, each pair $(\mu, \nu)$ gives a specific sequence of intervals (for each given $\lambda$). To have all possible intervals and sequences built from them, we simply consider the collection of all pairs of order-preserving and order-reversing functions.

This is formalized in the following way: let $\underline{\mathbb{R}^{\leftrightarrow}}$ be the presheaf given

(a) on objects: for all $V \in \mathcal{V}(\mathcal{N})$, let

$$\underline{\mathbb{R}^{\leftrightarrow}}_V := \{(\mu, \nu) \mid \mu, \nu : \downarrow V \to \mathbb{R}, \mu \text{ order-preserving}, \nu \text{ order-reversing}, \mu \leq \nu\}; \quad (4.30)$$

(b) on arrows: for all inclusions $i_{V'V} : V' \to V$, let

$$\underline{\mathbb{R}^{\leftrightarrow}}(i_{V'V}) : \underline{\mathbb{R}^{\leftrightarrow}}_V \longrightarrow \underline{\mathbb{R}^{\leftrightarrow}}_{V'} \quad (4.31)$$
$$(\mu, \nu) \longmapsto (\mu|_{V'}, \nu|_{V'}).$$

The presheaf $\underline{\mathbb{R}^{\leftrightarrow}}$ is where physical quantities take their "values" in the topos version of quantum theory. It is the analogue of the set of real numbers, where physical quantities in classical physics take their values.

#### 4.5.1.3 Daseinisation of a Self-Adjoint Operator

We now start to assemble the approximations to the self-adjoint operator $\hat{A}$ into an arrow from the spectral presheaf $\underline{\Sigma}$ to the presheaf $\underline{\mathbb{R}^{\leftrightarrow}}$ of values. Such an arrow is a so-called natural transformation (see, e.g., [27]) because these are the arrows in the topos **Sets**$^{\mathcal{V}(\mathcal{N})^{op}}$. For each context $V \in \mathcal{V}(\mathcal{N})$, we define a function, which we will denote as $\breve{\delta}(\hat{A})_V$, from the Gelfand spectrum $\underline{\Sigma}_V$ of $V$ to the collection $\underline{\mathbb{R}^{\leftrightarrow}}_V$ of pairs of order-preserving and order-reversing functions from $\downarrow V$ to $\mathbb{R}$. In a second step, we see that these functions fit together in the appropriate sense to form a natural transformation.

The function $\breve{\delta}(\hat{A})_V$ is constructed as follows: let $\lambda \in \underline{\Sigma}_V$. In each $V' \in \downarrow V$, we have one approximation $\delta^i(\hat{A})_{V'}$ to $\hat{A}$. If $V'' \subset V'$, then $\delta^i(\hat{A})_{V''} \leq_s \delta^i(\hat{A})_{V'}$, which implies $\delta^i(\hat{A})_{V''} \leq \delta^i(\hat{A})_{V'}$. One can evaluate $\lambda$ at $\delta^i(\hat{A})_{V'}$ for each $V'$, giving an order-preserving function:

$$\mu_\lambda : \downarrow V \longrightarrow \mathbb{R} \quad (4.32)$$
$$V' \longmapsto \lambda|_{V'}(\delta^i(\hat{A})_{V'}) = \lambda(\delta^i(\hat{A})_{V'}).$$

Similarly, using the previous approximations $\delta^o(\hat{A})_{V'}$ to $\hat{A}$, we obtain an order-reversing function:

$$\nu_\lambda : \downarrow V \longrightarrow \mathbb{R} \quad (4.33)$$
$$V' \longmapsto \lambda|_{V'}(\delta^o(\hat{A})_{V'}) = \lambda(\delta^o(\hat{A})_{V'}).$$

This allows us to define the desired function:

$$\breve{\delta}(\hat{A})_V : \underline{\Sigma}_V \longrightarrow \underline{\mathbb{R}^{\leftrightarrow}}_V \quad (4.34)$$

$$\lambda \longmapsto (\mu_\lambda, \nu_\lambda).$$

Obviously, we obtain one function $\breve{\delta}(\hat{A})_V$ for each context $V \in \mathcal{V}(\mathcal{N})$. The condition for these functions to form a natural transformation is the following: whenever $V' \subset V$, one must have $\breve{\delta}(\hat{A})_V(\lambda)|_{V'} = \breve{\delta}(\hat{A})_{V'}(\lambda|_{V'})$. In fact, this equality follows trivially from the definition. We thus have arrived at an arrow

$$\breve{\delta}(\hat{A}) : \underline{\Sigma} \longrightarrow \underline{\mathbb{R}^{\leftrightarrow}} \quad (4.35)$$

representing a physical quantity $A$. The arrow $\breve{\delta}(\hat{A})$ is called the *daseinisation of $\hat{A}$*. This arrow is the analogue of the function $f_A : \mathcal{S} \to \mathbb{R}$ from the state space to the real numbers representing the physical quantity $A$ in classical physics.

### 4.5.1.4 Pure States and Assignment of "Values" to Physical Quantities

Up to now, we have always argued with elements $\lambda \in \underline{\Sigma}_V$. This is a local argument, and we clearly need a global representation of states. Let $\psi$ be a unit vector in Hilbert space, identified with the pure state $\langle\psi| \_ |\psi\rangle : \mathcal{N} \to \mathbb{C}$ as usual. In the topos approach, such a pure state is represented by the subobject $\underline{\delta(\hat{P}_\psi)}$ of the spectral presheaf $\underline{\Sigma}$—that is, the daseinisation of the projection $\hat{P}_\psi$ onto the line $\mathbb{C}\psi$ in Hilbert space. Details can be found in [16].

The subobject $\underline{\mathfrak{w}^\psi} := \underline{\delta(\hat{P}_\psi)}$ is called the *pseudostate* associated to $\psi$. It is the analogue of a point $s \in \mathcal{S}$ of the state space of a classical system. Importantly, the pseudostate is not a global element of the presheaf $\underline{\Sigma}$. Such a global element would be the closest analogue of a point in a set, but as Isham and Butterfield observed [22], the spectral presheaf $\underline{\Sigma}$ has no global elements at all—a fact that is equivalent to the Kochen–Specker theorem. Nonetheless, subobjects of $\underline{\Sigma}$ of the form $\underline{\mathfrak{w}^\psi} = \underline{\delta(\hat{P}_\psi)}$ are minimal subobjects in an appropriate sense and, as such, are as close to points as possible (see [9, 16]).

The "value" of a physical quantity $A$, represented by an arrow $\breve{\delta}(\hat{A})$, in a state described by $\underline{\mathfrak{w}^\psi}$ is then

$$\breve{\delta}(\hat{A})(\underline{\mathfrak{w}^\psi}). \quad (4.36)$$

This, of course, is the analogue of the expression $f_A(s)$ in classical physics, where $f_A$ is the real-valued function on the state space $\mathcal{S}$ of the system representing the physical quantity $A$, and $s \in \mathcal{S}$ is the state of the system.

The expression $\breve{\delta}(\hat{A})(\underline{\mathfrak{w}^\psi})$ turns out to describe a considerably more complicated object than the classical expression $f_A(s)$. One can easily show that $\breve{\delta}(\hat{A})(\underline{\mathfrak{w}^\psi})$ is a subobject of the presheaf $\underline{\mathbb{R}^{\leftrightarrow}}$ of "values" (see Section 8.5 in Döring and Isham [16]). Concretely, it is given at each context $V \in \mathcal{V}(\mathcal{N})$ by

$$(\breve{\delta}(\hat{A})(\underline{\mathfrak{w}^\psi}))_V = \breve{\delta}(\hat{A})_V(\underline{\mathfrak{w}^\psi}_V) = \{\breve{\delta}(\hat{A})_V(\lambda) \mid \lambda \in \underline{\mathfrak{w}^\psi}_V\}. \quad (4.37)$$

As we saw in the preceding, each $\breve{\delta}(\hat{A})_V(\lambda)$ is a pair of functions $(\mu_\lambda, \nu_\lambda)$ from the set $\downarrow V$ to the reals and, more specifically, to the spectrum $sp(\hat{A})$ of the operator $\hat{A}$. The function

$\mu_\lambda$ is order-preserving, so it takes smaller and smaller values as one goes down from $V$ to smaller subalgebras $V', V'', \ldots$, whereas $\nu_\lambda$ is order-reversing and takes larger and larger values. Together, they determine one real interval $[\mu_\lambda(V'), \nu_\lambda(V')]$ for every context $V' \in \downarrow V$. Every $\lambda \in \underline{\mathfrak{w}}^\psi{}_V$ determines one such sequence of nested intervals. The "value" of the physical quantity $A$ in the given state is the collection of all of these sequences of nested intervals for all the different contexts $V \in \mathcal{V}(\mathcal{N})$.

We conjecture that the usual expectation value $\langle \psi | \hat{A} | \psi \rangle$ of the physical quantity $A$—which, of course, is a single real number—can be calculated from the "value" $\breve{\delta}(\hat{A})(\underline{\mathfrak{w}}^\psi)$. It is easy to see that each real interval showing up in (the local expressions of) $\breve{\delta}(\hat{A})(\underline{\mathfrak{w}}^\psi)$ contains the expectation value $\langle \psi | \hat{A} | \psi \rangle$.

In Döring [9], it was shown that arbitrary quantum states (not just pure ones) can be represented by probability measures on the spectral presheaf. It was also shown how the expectation values of physical quantities can be calculated using these measures.

Instead of closing this section by showing what the relevant structures look like concretely for our standard example $\mathcal{N} = \mathcal{B}(\mathbb{C}^3)$, we introduce an efficient way of calculating the approximations $\delta^i(\hat{A})_V$ and $\delta^o(\hat{A})_V$ first, which works uniformly for all contexts $V \in \mathcal{V}(\mathcal{N})$. This will simplify the calculation of the arrow $\breve{\delta}(\hat{A})$ significantly.

### 4.5.2 Antonymous and Observable Functions

Our aim is to define two functions, $g_{\hat{A}}, f_{\hat{A}}$, from which we can calculate the approximations $\delta^i(\hat{A})_V$ resp. $\delta^o(\hat{A})_V$ to $\hat{A}$ for all contexts $V \in \mathcal{V}(\mathcal{N})$. More precisely, we describe the Gelfand transforms of the operators $\delta^i(\hat{A})_V$ and $\delta^o(\hat{A})_V$ (for all $V$). These are functions from the Gelfand spectrum $\Sigma_V$ to the reals. The main idea, from de Groote, is to use the fact that each $\lambda \in \Sigma_V$ corresponds to a maximal *filter* in the projection lattice $\mathcal{P}(V)$ of $V$. Hence, we consider the Gelfand transforms as functions on filters; by "extending" each filter in $\mathcal{P}(V)$ to a filter in $\mathcal{P}(\mathcal{N})$ (nonmaximal in general), we arrive at functions on filters in $\mathcal{P}(\mathcal{N})$, the projection lattice of the nonabelian von Neumann algebra $\mathcal{N}$. First, we collect a few basic definitions and facts about filters.

#### 4.5.2.1 Filters

Let $\mathbb{L}$ be a lattice with zero element 0. A subset $F$ of elements of $\mathbb{L}$ is a *(proper) filter* (or *[proper] dual ideal*) if (1) $0 \notin F$; (2) $a, b \in F$ implies $a \wedge b \in F$; and (3) $a \in F$ and $b \geq a$ imply $b \in F$. When we speak of filters, we always mean *proper* filters. We write $\mathcal{F}(\mathbb{L})$ for the set of filters in a lattice $\mathbb{L}$. If $\mathcal{N}$ is a von Neumann algebra, then we write $\mathcal{F}(\mathcal{N})$ for $\mathcal{F}(\mathcal{P}(\mathcal{N}))$.

A subset $B$ of elements of $\mathbb{L}$ is a *filter base* if (1) $0 \notin B$, and (2) for all $a, b \in B$ there exists a $c \in B$ such that $c \leq a \wedge b$.

Maximal filters and maximal filter bases in a lattice $\mathbb{L}$ with 0 exist by Zorn's lemma, and each maximal filter is a maximal filter base and vice versa. The maximal filters in the projection lattice of a von Neumann algebra $\mathcal{N}$ are denoted by $\mathcal{Q}(\mathcal{N})$. Following de Groote [4], one can equip this set with a topology and consider its relations to well-known constructions (in particular, Gelfand spectra of abelian von Neumann algebras). Here, we do not consider the topological aspects.

If $V$ is an abelian subalgebra of a von Neumann algebra $\mathcal{N}$, then the projections in $V$ form a distributive lattice $\mathcal{P}(V)$; that is, for all $\hat{P}, \hat{Q}, \hat{R} \in \mathcal{P}(V)$, we have

$$\hat{P} \wedge (\hat{Q} \vee \hat{R}) = (\hat{P} \wedge \hat{Q}) \vee (\hat{P} \wedge \hat{R}), \tag{4.38}$$

$$\hat{P} \vee (\hat{Q} \wedge \hat{R}) = (\hat{P} \vee \hat{Q}) \wedge (\hat{P} \vee \hat{R}). \tag{4.39}$$

The projection lattice of a von Neumann algebra is distributive if and only if the algebra is abelian.

A maximal filter in a complemented, distributive lattice $\mathbb{L}$ is called an *ultrafilter*; hence, $\mathcal{Q}(\mathbb{L})$ is the space of ultrafilters in $\mathbb{L}$. We do not use the notion of "ultrafilter" for maximal filters in *non*distributive lattices like the projection lattices of nonabelian von Neumann algebras.

The characterizing property of an ultrafilter $U$ in a distributive, complemented lattice $\mathbb{L}$ is that for each element $a \in \mathbb{L}$, either $a \in U$ or $a^c \in U$, where $a^c$ denotes the complement of $a$. This can easily be seen: a complemented lattice has a maximal element 1, and we have $a \vee a^c = 1$ by definition. Let us assume that $U$ is an ultrafilter, but $a \notin U$ and $a^c \notin U$. In particular, $a \notin U$ means that there is some $b \in U$ such that $b \wedge a = 0$. Using distributivity of the lattice $\mathbb{L}$, we get

$$b = b \wedge (a \vee a^c) = (b \wedge a) \vee (b \wedge a^c) = b \wedge a^c, \tag{4.40}$$

so $b \leq a^c$. Because $b \in U$ and $U$ is a filter, this implies $a^c \in U$, contradicting our assumption. If $\mathbb{L}$ is the projection lattice $\mathcal{P}(V)$ of an abelian von Neumann algebra $V$, then the maximal element is the identity operator $\hat{1}$ and the complement of a projection is given as $\hat{P}^c := \hat{1} - \hat{P}$. Each ultrafilter $q \in \mathcal{Q}(V)$ hence contains either $\hat{P}$ or $\hat{1} - \hat{P}$ for all $\hat{P} \in \mathcal{P}(V)$.

#### 4.5.2.2 Elements of the Gelfand Spectrum and Filters

We already saw that for all projections $\hat{P} \in \mathcal{P}(V)$ and all elements $\lambda$ of the Gelfand spectrum $\Sigma_V$ of $V$, it holds that $\lambda(\hat{P}) \in \{0, 1\}$. Given $\lambda \in \Sigma_V$, it is easy to construct a maximal filter $F_\lambda$ in the projection lattice $\mathcal{P}(V)$ of $V$: let

$$F_\lambda := \{\hat{P} \in \mathcal{P}(V) \mid \lambda(\hat{P}) = 1\}. \tag{4.41}$$

It is clear that $\hat{P} \in F_\lambda$ implies $\hat{Q} \in F_\lambda$ for all projections $\hat{Q} \geq \hat{P}$. Let $\hat{P}_1, \hat{P}_2 \in F_\lambda$. From the fact that $\lambda$ is a multiplicative state of $V$, we obtain $\lambda(\hat{P}_1 \wedge \hat{P}_2) = \lambda(\hat{P}_1 \hat{P}_2) = \lambda(\hat{P}_1)\lambda(\hat{P}_2) = 1$, so $\hat{P}_1 \wedge \hat{P}_2 \in F_\lambda$. Finally, we have $1 = \lambda(\hat{1}) = \lambda(\hat{P} + \hat{1} - \hat{P}) = \lambda(\hat{P}) + \lambda(\hat{1} - \hat{P})$ for all $\hat{P} \in \mathcal{P}(V)$, so either $\hat{P} \in F_\lambda$ or $\hat{1} - \hat{P} \in F_\lambda$ for all $\hat{P}$, which shows that $F_\lambda$ is a maximal filter in $\mathcal{P}(V)$, indeed. It is straightforward to see that the mapping

$$\beta : \Sigma_V \longrightarrow \mathcal{Q}(V) \tag{4.42}$$
$$\lambda \longmapsto F_\lambda$$

from the Gelfand spectrum of $V$ to the set $\mathcal{Q}(V)$ of maximal filters in $\mathcal{P}(V)$ is injective. This is the first step in the construction of a homeomorphism between these two spaces. De Groote has shown the existence of such a homeomorphism in Thm. 3.2 of [4].

Given a filter $F$ in the projection lattice $\mathcal{P}(V)$ of some context $V$, one can "extend" it to a filter in the projection lattice $\mathcal{P}(\mathcal{N})$ of the full nonabelian algebra $\mathcal{N}$ of physical quantities. One simply defines the $\mathcal{N}$-*cone over* $F$ as

$$\mathcal{C}_{\mathcal{N}}(F) := \uparrow F = \{\hat{Q} \in \mathcal{P}(\mathcal{N}) \mid \exists \hat{P} \in F : \hat{P} \leq \hat{Q}\}. \tag{4.43}$$

This is the smallest filter in $\mathcal{P}(\mathcal{N})$ containing $F$.

We need the following technical but elementary lemma:

**Lemma 4.** Let $\mathcal{N}$ be a von Neumann algebra, and let $\mathcal{T}$ be a von Neumann subalgebra such that $\hat{1}_{\mathcal{T}} = \hat{1}_{\mathcal{N}}$. Let

$$\delta_{\mathcal{T}}^i : \mathcal{P}(\mathcal{N}) \longrightarrow \mathcal{P}(\mathcal{T}) \tag{4.44}$$

$$\hat{P} \longmapsto \delta^i(\hat{P})_{\mathcal{T}} := \bigvee \{\hat{Q} \in \mathcal{P}(\mathcal{T}) \mid \hat{Q} \leq \hat{P}\}.$$

Then, for all filters $F \in \mathcal{F}(\mathcal{T})$,

$$(\delta_{\mathcal{T}}^i)^{-1}(F) = \mathcal{C}_{\mathcal{N}}(F). \tag{4.45}$$

**PROOF.** If $\hat{Q} \in F \subset \mathcal{P}(\mathcal{T})$, then $(\delta_{\mathcal{T}}^i)^{-1}(\hat{Q}) = \{\hat{P} \in \mathcal{P}(\mathcal{N}) \mid \delta^i(\hat{P})_{\mathcal{T}} = \hat{Q}\}$. Let $\hat{P} \in \mathcal{P}(\mathcal{N})$ be such that there is a $\hat{Q} \in F$ with $\hat{Q} \leq \hat{P}$—that is, $\hat{P} \in \mathcal{C}_{\mathcal{N}}(F)$. Then, $\delta^i(\hat{P})_{\mathcal{T}} \geq \hat{Q}$, which implies $\delta^i(\hat{P})_{\mathcal{T}} \in F$ because $F$ is a filter in $\mathcal{P}(\mathcal{T})$. This shows that $\mathcal{C}_{\mathcal{N}}(F) \subseteq (\delta_{\mathcal{T}}^i)^{-1}(F)$. Now, let $\hat{P} \in \mathcal{P}(\mathcal{N})$ be such that there is no $\hat{Q} \in F$ with $\hat{Q} \leq \hat{P}$. Because $\delta^i(\hat{P})_{\mathcal{T}} \leq \hat{P}$, there also is no $\hat{Q} \in F$ with $\hat{Q} \leq \delta^i(\hat{P})_{\mathcal{T}}$, so $\hat{P} \notin (\delta_{\mathcal{T}}^i)^{-1}(F)$. This shows that $(\delta_{\mathcal{T}}^i)^{-1}(F) \subseteq \mathcal{C}_{\mathcal{N}}(F)$. ∎

**Definition of *antonymous* and *observable* functions.** Let $\mathcal{N}$ be a von Neumann algebra, and let $\hat{A}$ be a self-adjoint operator in $\mathcal{N}_{sa}$. As before, we denote the spectral family of $\hat{A}$ as $(\hat{E}_r^{\hat{A}})_{r \in \mathbb{R}} = \hat{E}^{\hat{A}}$.

**Definition 5.** The *antonymous function of* $\hat{A}$ is defined as

$$g_{\hat{A}} : \mathcal{F}(\mathcal{N}) \longrightarrow \mathrm{sp}(\hat{A}) \tag{4.46}$$

$$F \longmapsto \sup\{r \in \mathbb{R} \mid \hat{1} - \hat{E}_r^{\hat{A}} \in F\}. \tag{4.47}$$

The *observable function of* $\hat{A}$ is

$$f_{\hat{A}} : \mathcal{F}(\mathcal{N}) \longrightarrow \mathrm{sp}(\hat{A}) \tag{4.48}$$

$$F \longmapsto \inf\{r \in \mathbb{R} \mid \hat{E}_r^{\hat{A}} \in F\}. \tag{4.49}$$

It is straightforward to see that, indeed, the range of these functions is the spectrum $\mathrm{sp}(\hat{A})$ of $\hat{A}$.

We observe that the mapping described in Equation (4.23) can be generalized: when approximating a self-adjoint operator $\hat{A}$ with respect to the spectral order in a von Neumann subalgebra of $\mathcal{N}$, then one can use an arbitrary, not necessarily abelian, subalgebra $\mathcal{T}$ of the nonabelian algebra $\mathcal{N}$. The only condition is that the unit elements in $\mathcal{N}$ and $\mathcal{T}$ coincide. We define

$$\delta_{\mathcal{T}}^i : \mathcal{N}_{sa} \longrightarrow \mathcal{T}_{sa} \qquad (4.50)$$
$$\hat{A} \longmapsto \delta^i(\hat{A})_{\mathcal{T}} = \bigvee \{\hat{B} \in \mathcal{T}_{sa} \mid \hat{B} \leq_s \hat{A}\}.$$

Analogously, generalizing Equation (4.24)

$$\delta_{\mathcal{T}}^o : \mathcal{N}_{sa} \longrightarrow \mathcal{T}_{sa} \qquad (4.51)$$
$$\hat{A} \longmapsto \delta^o(\hat{A})_{\mathcal{T}} = \bigwedge \{\hat{B} \in \mathcal{T}_{sa} \mid \hat{B} \geq_s \hat{A}\}.$$

The following proposition is central:

**Proposition 6.** Let $\hat{A} \in \mathcal{N}_{sa}$. For all von Neumann subalgebras $\mathcal{T} \subseteq \mathcal{N}$ such that $\hat{1}_{\mathcal{T}} = \hat{1}_{\mathcal{N}}$ and all filters $F \in \mathcal{F}(\mathcal{T})$, we have

$$g_{\delta^i(\hat{A})_{\mathcal{T}}}(F) = g_{\hat{A}}(\mathcal{C}_{\mathcal{N}}(F)). \qquad (4.52)$$

**PROOF.** We have

$$\begin{aligned}
g_{\delta^i(\hat{A})_{\mathcal{T}}}(F) &= \sup\{r \in \mathbb{R} \mid \hat{1} - \hat{E}_r^{\delta^i(\hat{A})_{\mathcal{T}}} \in F\} \\
&\stackrel{(4.27)}{=} \sup\{r \in \mathbb{R} \mid \hat{1} - \bigwedge_{\mu > r} \delta^o(\hat{E}_\mu^{\hat{A}})_{\mathcal{T}} \in F\} \\
&= \sup\{r \in \mathbb{R} \mid \hat{1} - \delta^o(\hat{E}_r^{\hat{A}})_{\mathcal{T}} \in F\} \\
&= \sup\{r \in \mathbb{R} \mid \delta^i(\hat{1} - \hat{E}_r^{\hat{A}})_{\mathcal{T}} \in F\} \\
&= \sup\{r \in \mathbb{R} \mid \hat{1} - \hat{E}_r^{\hat{A}} \in (\delta_{\mathcal{T}}^i)^{-1}(F)\} \\
&\stackrel{\text{Lemma 4}}{=} \sup\{r \in \mathbb{R} \mid \hat{1} - \hat{E}_r^{\hat{A}} \in \mathcal{C}_{\mathcal{N}}(F)\} \\
&= g_{\hat{A}}(\mathcal{C}_{\mathcal{N}}(F)).
\end{aligned}$$

∎

This shows that the antonymous function $g_{\hat{A}} : \mathcal{F}(\mathcal{N}) \to \text{sp}(\hat{A})$ of $\hat{A}$ encodes in a simple way *all* the antonymous functions $g_{\delta^i(\hat{A})_{\mathcal{T}}} : \mathcal{F}(\mathcal{T}) \to \text{sp}(\delta^i(\hat{A}))_{\mathcal{T}}$ corresponding to the approximations $\delta^i(\hat{A})_{\mathcal{T}}$ (from the spectral order) to $\hat{A}$ to von Neumann subalgebras $\mathcal{T}$ of $\mathcal{N}$. If, in particular, $\mathcal{T}$ is an abelian subalgebra, then the set $\mathcal{Q}(\mathcal{T})$ of *maximal* filters in $\mathcal{D}(\mathcal{T})$ can be identified with the Gelfand spectrum of $\mathcal{T}$ (see [5]), and $g_{\delta^i(\hat{A})_{\mathcal{T}}}|_{\mathcal{Q}(\mathcal{T})}$ can be identified with the Gelfand transform of $\delta^i(\hat{A})_{\mathcal{T}}$. Thus, we get:

**Corollary 7.** The antonymous function $g_{\hat{A}} : \mathcal{F}(\mathcal{N}) \to \text{sp}(\hat{A})$ encodes all the Gelfand transforms of the inner daseinised operators of the form $\delta^i(\hat{A})_V$, where $V \in \mathcal{V}(\mathcal{N})$ is an abelian von Neumann subalgebra of $\mathcal{N}$ (such that $\hat{1}_V = \hat{1}_{\mathcal{N}}$). Concretely, if $\mathcal{Q}(V)$ is the space of maximal filters in $\mathcal{P}(V)$ and $\Sigma_V$ is the Gelfand spectrum of $V$, then we have an identification $\beta : \underline{\Sigma}_V \to \mathcal{Q}(V)$, $\lambda \mapsto F_\lambda$ (see Equation 4.42). Let $\overline{\delta^i(\hat{A})_V} : \Sigma_V \to \text{sp}(\delta^i(\hat{A})_V)$ be the Gelfand transform of $\delta^i(\hat{A})_V$. We can identify $\overline{\delta^i(\hat{A})_V}$ with $g_{\delta^i(\hat{A})_V}|_{\mathcal{Q}(V)}$ and, hence, from Proposition 6,

$$\overline{\delta^i(\hat{A})_V}(\lambda) = g_{\delta^i(\hat{A})_V}(F_\lambda) = g_{\hat{A}}(\mathcal{C}_{\mathcal{N}}(F_\lambda)). \qquad (4.53)$$

Not surprisingly, there is a similar result for observable functions and outer daseinisation. This result was first proved in a similar form by de Groote [6].

**Proposition 8.** Let $\hat{A} \in \mathcal{N}_{sa}$. For all von Neumann subalgebras $\mathcal{T} \subseteq \mathcal{N}$ such that $\hat{1}_\mathcal{T} = \hat{1}_\mathcal{N}$ and all filters $F \in \mathcal{F}(\mathcal{T})$, we have

$$f_{\delta^o(\hat{A})_\mathcal{T}}(F) = f_{\hat{A}}(\mathcal{C}_\mathcal{N}(F)). \tag{4.54}$$

**PROOF.** We have

$$\begin{align}
f_{\delta^o(\hat{A})_\mathcal{T}}(F) &:= \inf\{r \in \mathbb{R} \mid \hat{E}_r^{\delta^o(\hat{A})_\mathcal{T}} \in F\} \tag{4.55}\\
&\stackrel{(4.25)}{=} \inf\{r \in \mathbb{R} \mid \delta^i(\hat{E}_r^{\hat{A}})_\mathcal{T} \in F\} \tag{4.56}\\
&= \inf\{r \in \mathbb{R} \mid \hat{E}_r^{\hat{A}} \in (\delta_\mathcal{T}^i)^{-1}(F)\} \tag{4.57}\\
&\stackrel{\text{Lemma 4}}{=} \inf\{r \in \mathbb{R} \mid \hat{E}_r^{\hat{A}} \in \mathcal{C}_\mathcal{N}(F)\} \tag{4.58}\\
&= f_{\hat{A}}(\mathcal{C}_\mathcal{N}(F)). \tag{4.59}
\end{align}$$

∎

This shows that the observable function $f_{\hat{A}} : \mathcal{F}(\mathcal{N}) \to \text{sp}(\hat{A})$ of $\hat{A}$ encodes in a simple way *all* the observable functions $f_{\delta^o(\hat{A})_\mathcal{T}} : \mathcal{F}(\mathcal{N}) \to \text{sp}(\delta^o(\hat{A})_\mathcal{T})$ corresponding to approximations $\delta^o(\hat{A})_\mathcal{T}$ (from the previous spectral order) to $\hat{A}$ to von Neumann subalgebras $\mathcal{T}$ of $\mathcal{N}$. If, in particular, $\mathcal{T} = V$ is an abelian subalgebra, then we get:

**Corollary 9.** The observable function $f_{\hat{A}} : \mathcal{F}(\mathcal{N}) \to \text{sp}(\hat{A})$ encodes all the Gelfand transforms of the outer daseinised operators of the form $\delta^o(\hat{A})_V$, where $V \in \mathcal{V}(\mathcal{N})$ is an abelian von Neumann subalgebra of $\mathcal{N}$ (such that $\hat{1}_V = \hat{1}_\mathcal{N}$). Using the identification $\beta : \underline{\Sigma}_V \to \mathcal{Q}(V)$, $\lambda \mapsto F_\lambda$ between the Gelfand and the space of maximal filters (see Equation 4.42), we can identify the Gelfand transform $\overline{\delta^o(\hat{A})_V} : \underline{\Sigma}_V \to \text{sp}(\delta^o(\hat{A})_V)$ of $\delta^o(\hat{A})_V$ with $f_{\delta^o(\hat{A})_V}|_{\mathcal{Q}(\mathcal{T})}$ and, hence, from Proposition 8,

$$\overline{\delta^o(\hat{A})_V}(\lambda) = f_{\delta^o(\hat{A})_V}(F_\lambda) = f_{\hat{A}}(\mathcal{C}_\mathcal{N}(F_\lambda)). \tag{4.60}$$

#### 4.5.2.3 Vector States from Elements of Gelfand Spectra

As emphasized in Section 4.5.1, an element of $\lambda$ of the Gelfand spectrum $\underline{\Sigma}_V$ of some context $V$ is only a local state: it is a pure state of the abelian subalgebra $V$, but it is not a state of the full nonabelian algebra $\mathcal{N}$ of physical quantities. We will now show how certain local states $\lambda$ can be extended to global states (i.e., states of $\mathcal{N}$).

Let $V$ be a context that contains at least one rank-1 projection.[17] Let $\hat{P}$ be such a projection, and let $\psi \in \mathcal{H}$ be a unit vector such that $\hat{P}$ is the projection onto the ray $\mathbb{C}\psi$. ($\psi$ is fixed up to a phase). We write $\hat{P} = \hat{P}_\psi$. Then,

$$F_{\hat{P}_\psi} := \{\hat{Q} \in \mathcal{P}(V) \mid \hat{Q} \geq \hat{P}_\psi\} \tag{4.61}$$

---

[17] Depending on whether the von Neumann algebra $\mathcal{N}$ contains rank-1 projections, there may or may not be such contexts. For the case $\mathcal{N} = \mathcal{B}(\mathcal{H})$, there, of course, are many rank-1 projections and, hence, contexts $V$ of the required form.

clearly is an ultrafilter in $\mathcal{P}(V)$. It thus corresponds to some element $\lambda_\psi$ of the Gelfand spectrum $\underline{\Sigma}_V$ of $V$. Being an element of the Gelfand spectrum, $\lambda_\psi$ is a pure state of $V$. The cone

$$\mathcal{C}_\mathcal{N}(F_{\hat{P}_\psi}) = \{\hat{Q} \in \mathcal{P}(\mathcal{N}) \mid \hat{Q} \geq \hat{P}_\psi\} \tag{4.62}$$

is a maximal filter in $\mathcal{P}(\mathcal{N})$. It contains all projections in $\mathcal{N}$ that represent propositions "$A \, \varepsilon \, \Delta$" that are (totally) true in the vector state $\psi$ on $\mathcal{N}$.[18] This vector state is uniquely determined by the cone $\mathcal{C}_\mathcal{N}(F_{\hat{P}_\psi})$. The "local state" $\lambda_\psi$ on the context $V \subset \mathcal{N}$ can thus be extended to a "global" vector state. On the level of filters, this extension corresponds to the cone construction.

For a finite-dimensional Hilbert space $\mathcal{H}$, one necessarily has $\mathcal{N} = \mathcal{B}(\mathcal{H})$. Let $V$ be an arbitrary maximal context, and let $\lambda \in \underline{\Sigma}_V$. Then, $\lambda$ is of the form $\lambda = \lambda_{\hat{P}_\psi}$ for some unit vector $\psi \in \mathcal{H}$ and corresponding rank-1 projection $\hat{P}_\psi$. Hence, every such $\lambda_{\hat{P}_\psi}$ can be extended to a vector state $\psi$ on the whole of $\mathcal{B}(\mathcal{H})$.

### 4.5.2.4 The Eigenstate-Eigenvalue Link

We now show that the arrow $\breve{\delta}(\hat{A})$ constructed from a self-adjoint operator $\hat{A} \in \mathcal{N}_{\text{sa}}$ preserves the eigenstate-eigenvalue link in a suitable sense. We employ the relation between ultrafilters in $\mathcal{P}(V)$ of the form $F_{\hat{P}_\psi}$ (see Equation 4.61) and maximal filters in $\mathcal{P}(\mathcal{N})$ established in the previous paragraph, but "read it backwards."

Let $\hat{A}$ be some self-adjoint operator, and let $\psi$ be an eigenstate of $\hat{A}$ with eigenvalue $a$. Let $V$ be a context that contains $\hat{A}$, and let $\hat{P}_\psi$ be the projection determined by $\psi$— that is, the projection onto the one-dimensional subspace (ray) $\mathbb{C}\psi$ of Hilbert space. Then, $\hat{P}_\psi \in V$ from the spectral theorem. Consider the maximal filter

$$F := \{\hat{Q} \in \mathcal{P}(\mathcal{N}) \mid \hat{Q} \geq \hat{P}_\psi\} \tag{4.63}$$

in $\mathcal{P}(\mathcal{N})$ determined by $\hat{P}_\psi$. Clearly, $F \cap \mathcal{P}(V) = \{\hat{Q} \in \mathcal{P}(V) \mid \hat{Q} \geq \hat{P}_\psi\}$ is an ultrafilter in $\mathcal{P}(V)$—namely, the ultrafilter $F_{\hat{P}_\psi}$ from Equation (4.61), and $F = \mathcal{C}_\mathcal{N}(F_{\hat{P}_\psi})$. Let $\lambda_\psi$ be the element of $\underline{\Sigma}_V$ (i.e., local state of $V$) determined by $F_{\hat{P}_\psi}$. The state $\lambda_\psi$ is nothing but the restriction of the vector state $\langle\psi| \_ |\psi\rangle$ (on $\mathcal{N}$) to the context $V$; hence,

$$\langle\psi| \hat{A} |\psi\rangle = \lambda_\psi(\hat{A}) = a. \tag{4.64}$$

Because $\hat{A} \in V_{\text{sa}}$, we have $\delta^i(\hat{A})_V = \delta^o(\hat{A})_V = \hat{A}$. This implies that

$$\begin{aligned}
\breve{\delta}(\hat{A})_V(\lambda_\psi)(V) &= [g_{\delta^i(\hat{A})_V}(\lambda_\psi), f_{\delta^o(\hat{A})_V}(\lambda_\psi)] \\
&= [g_{\hat{A}}(\lambda_\psi), f_{\hat{A}}(\lambda_\psi)] \\
&= [\underline{A}(\lambda_\psi), \overline{A}(\lambda_\psi)] \\
&= [\lambda_\psi(\hat{A}), \lambda_\psi(\hat{A})] \\
&= [a, a].
\end{aligned} \tag{4.65}$$

That is, the arrow $\breve{\delta}(\hat{A})$ delivers (the interval only containing) the eigenvalue $a$ for an eigenstate $\psi$ (resp. the corresponding $\lambda_\psi$) at a context $V$ that actually contains $\hat{A}$. In

---

[18] Here, we again use the usual identification between unit vectors and vector states, $\psi \mapsto \langle\psi| \_ |\psi\rangle$.

this sense, the eigenstate-eigenvalue link is preserved, and locally at $V$ the value of $\hat{A}$ actually becomes a single real number as expected.

**Example 10:** We return to our example, the spin-1 system. It is described by the algebra $\mathcal{B}(\mathbb{C}^3)$ of bounded operators on the three-dimensional Hilbert space $\mathbb{C}^3$. In particular, the spin-$z$ operator is given by

$$\hat{S}_z = \frac{1}{\sqrt{2}} \begin{pmatrix} 1 & 0 & 0 \\ 0 & 0 & 0 \\ 0 & 0 & -1 \end{pmatrix}. \tag{4.66}$$

This is the matrix expression for $\hat{S}_z$ with respect to the basis $e_1, e_2, e_3$ of eigenvectors of $\hat{S}_z$. Let $\hat{P}_1, \hat{P}_2, \hat{P}_3$ be the corresponding rank-1 projections onto the eigenspaces. The algebra $V_{\hat{S}_z} := \{\hat{P}_1, \hat{P}_2, \hat{P}_3\}''$ generated by these projections is a maximal abelian subalgebra of $\mathcal{B}(\mathbb{C}^3)$, and it is the only maximal abelian subalgebra containing $\hat{S}_z$.

Let us now consider what the approximations $\delta^i(\hat{S}_z)_V$ and $\delta^o(\hat{S}_z)_V$ of $\hat{S}_z$ to other contexts $V$ look like. To do so, we first determine the antonymous and the observable function of $\hat{S}_z$. The spectral family of $\hat{S}_z$ is given by

$$\hat{E}_\lambda^{\hat{S}_z} = \begin{cases} \hat{0} & \text{if } \lambda < -\frac{1}{\sqrt{2}} \\ \hat{P}_3 & \text{if } -\frac{1}{\sqrt{2}} \leq \lambda < 0 \\ \hat{P}_3 + \hat{P}_2 & \text{if } 0 \leq \lambda < \frac{1}{\sqrt{2}} \\ \hat{1} & \text{if } \lambda \geq \frac{1}{\sqrt{2}}. \end{cases} \tag{4.67}$$

The antonymous function $g_{\hat{S}_z}$ of $\hat{S}_z$ is

$$g_{\hat{S}_z} : \mathcal{F}(\mathcal{B}(\mathbb{C}^3)) \longrightarrow \text{sp}(\hat{S}_z) \tag{4.68}$$

$$F \longmapsto \sup\{r \in \mathbb{R} \mid \hat{1} - \hat{E}_r^{\hat{S}_z} \in F\},$$

and the observable function is

$$f_{\hat{S}_z} : \mathcal{F}(\mathcal{B}(\mathbb{C}^3)) \longrightarrow \text{sp}(\hat{S}_z) \tag{4.69}$$

$$F \longmapsto \inf\{r \in \mathbb{R} \mid \hat{E}_r^{\hat{S}_z} \in F\}.$$

Let $\hat{Q}_1, \hat{Q}_2, \hat{Q}_3$ be three pairwise orthogonal rank-1 projections, and let $V = \{\hat{Q}_1, \hat{Q}_2, \hat{Q}_3\}'' \in \mathcal{V}(\mathcal{B}(\mathbb{C}^3))$ be the maximal context determined by them. Then, the Gelfand spectrum $\underline{\Sigma}_V$ has three elements; let us denote them $\lambda_1, \lambda_2$, and $\lambda_3$ (where $\lambda_i(\hat{Q}_j) = \delta_{ij}$). Clearly, $\hat{Q}_i$ is the smallest projection in $V$ such that $\lambda_i(\hat{Q}_i) = 1$ ($i = 1, 2, 3$). Each $\lambda_i$ defines a maximal filter in the projection lattice $\mathcal{P}(V)$ of $V$:

$$F_{\lambda_i} = \{\hat{Q} \in \mathcal{P}(V) \mid \hat{Q} \geq \hat{Q}_i\}. \tag{4.70}$$

The Gelfand transform $\overline{\delta^i(\hat{S}_z)_V}$ of $\delta^i(\hat{S}_z)_V$ is a function from the Gelfand spectrum $\underline{\Sigma}_V$ of $V$ to the spectrum of $\hat{S}_z$. Equation (4.53) shows how to calculate this function from the antonymous function $g_{\hat{S}_z}$:

$$\overline{\delta^i(\hat{S}_z)_V}(\lambda_i) = g_{\hat{S}_z}(\mathcal{C}_{\mathcal{B}(\mathbb{C}^3)}(F_{\lambda_i})), \tag{4.71}$$

where $\mathcal{C}_{\mathcal{B}(\mathbb{C}^3)}(F_{\lambda_i})$ is the cone over the filter $F_{\lambda_i}$—that is,

$$\mathcal{C}_{\mathcal{B}(\mathbb{C}^3)}(F_{\lambda_i}) = \{\hat{R} \in \mathcal{P}(\mathcal{B}(\mathbb{C}^3)) \mid \exists \hat{Q} \in F_{\lambda_i} : \hat{R} \geq \hat{Q}\}$$
$$= \{\hat{R} \in \mathcal{P}(\mathcal{B}(\mathbb{C}^3)) \mid \hat{R} \geq \hat{Q}_i\}. \qquad (4.72)$$

Now, it is easy to actually calculate $\overline{\delta^i(\hat{S}_z)_V}$: for all $\lambda_i \in \underline{\Sigma}_V$,

$$\overline{\delta^i(\hat{S}_z)_V}(\lambda_i) = \sup\{r \in \mathbb{R} \mid \hat{1} - \hat{E}_r^{\hat{S}_z} \in \mathcal{C}_{\mathcal{B}(\mathbb{C}^3)}(F_{\lambda_i})\}$$
$$= \sup\{r \in \mathbb{R} \mid \hat{1} - \hat{E}_r^{\hat{S}_z} \geq \hat{Q}_i\}. \qquad (4.73)$$

Analogously, we obtain $\overline{\delta^o(\hat{S}_z)_V}$: for all $\lambda_i$,

$$\overline{\delta^o(\hat{S}_z)_V}(\lambda_i) = \inf\{r \in \mathbb{R} \mid \hat{E}_r^{\hat{S}_z} \in \mathcal{C}_{\mathcal{B}(\mathbb{C}^3)}(F_{\lambda_i})\}$$
$$= \inf\{r \in \mathbb{R} \mid \hat{E}_r^{\hat{S}_z} \geq \hat{Q}_i\}. \qquad (4.74)$$

Using the expression Equation (4.67) for the spectral family of $\hat{S}_z$, we can see directly that the values $\overline{\delta^i(\hat{S}_z)_V}(\lambda_i), \overline{\delta^o(\hat{S}_z)_V}(\lambda_i)$ lie in the spectrum of $\hat{S}_z$ as expected. A little less obviously, $\overline{\delta^i(\hat{S}_z)_V}(\lambda_i) \leq \overline{\delta^o(\hat{S}_z)_V}(\lambda_i)$, so we can think of the pair of values as an interval $[\overline{\delta^i(\hat{S}_z)_V}(\lambda_i), \overline{\delta^o(\hat{S}_z)_V}(\lambda_i)]$.

We had assumed that $V = \{\hat{Q}_1, \hat{Q}_2, \hat{Q}_2\}''$ is a *maximal* abelian subalgebra, but this is no actual restriction. A nonmaximal subalgebra $V'$ has a Gelfand spectrum $\underline{\Sigma}_{V'}$ consisting of two elements. All arguments work analogously. The important point is that for each element $\lambda$ of the Gelfand spectrum, there is a unique projection $\hat{Q}$ in $V'$ corresponding to $\lambda$, given as the smallest projection in $V'$ such that $\lambda(\hat{Q}) = 1$.

This means that we can now write down explicitly the natural transformation $\breve{\delta}(S_z) : \underline{\Sigma} \to \underline{\mathbb{R}^{\leftrightarrow}}$, the arrow in the presheaf topos representing the physical quantity "spin in $z$-direction," initially given by the self-adjoint operator $\hat{S}_z$. For each context $V \in \mathcal{V}(\mathcal{B}(\mathbb{C}^3))$, we have a function

$$\breve{\delta}(S_z)_V : \underline{\Sigma}_V \longrightarrow \underline{\mathbb{R}^{\leftrightarrow}}_V \qquad (4.75)$$
$$\lambda \longmapsto (\mu_\lambda, \nu_\lambda);$$

compare Equation (4.34). According to Equation (4.32), $\mu_\lambda : \downarrow V \to \mathbb{R}$ is given as

$$\mu_\lambda : \downarrow V \longrightarrow \mathbb{R} \qquad (4.76)$$
$$V' \longmapsto \lambda(\delta^i(\hat{S}_z)_{V'}) = \overline{\delta^i(\hat{S}_z)_{V'}}(\lambda).$$

Let $\hat{Q}_{V'} \in \mathcal{P}(V')$ be the projection in $V'$ corresponding to $\lambda$—that is, the smallest projection in $V'$ such that $\lambda(\hat{Q}_{V'}) = 1$. Note that this projection $\hat{Q}_{V'}$ depends on $V'$. For each $V' \in \downarrow V$, the element $\lambda$ (originally an element of the Gelfand spectrum $\underline{\Sigma}_V$) is considered as an element of $\underline{\Sigma}_{V'}$, given by the restriction $\lambda|_{V'}$ of the original $\lambda$ to the smaller algebra $V'$. Then, Equation (4.73) implies, for all $V' \in \downarrow V$,

$$\mu_\lambda(V') = \sup\{r \in \mathbb{R} \mid \hat{1} - \hat{E}_r^{\hat{S}_z} \geq \hat{Q}_{V'}\}. \qquad (4.77)$$

From Equation (4.33), we obtain

$$\nu_\lambda : \downarrow V \longrightarrow \mathbb{R} \tag{4.78}$$

$$V' \longmapsto \lambda(\delta^o(\hat{S}_z)_{V'}) = \overline{\delta^o(\hat{S}_z)_{V'}}(\lambda),$$

and with Equation (4.74), we get for all $V' \in \downarrow V$,

$$\nu_\lambda(V') = \inf\{r \in \mathbb{R} \mid \hat{E}^{\hat{S}_z} \geq \hat{Q}_{V'}\}. \tag{4.79}$$

We finally want to calculate the "value" $\breve{\delta}(\hat{S}_z)(\underline{\mathfrak{w}}^\psi)$ of the physical quantity "spin in $z$-direction" in the (pseudo)state $\underline{\mathfrak{w}}^\psi$. As described in Section 4.5.1, $\psi$ is a unit vector in the Hilbert space (resp. a vector state) and $\underline{\mathfrak{w}}^\psi = \delta(\hat{P}_\psi)$ is the corresponding pseudostate, a subobject of $\underline{\Sigma}$. The "value" $\breve{\delta}(\hat{S}_z)(\underline{\mathfrak{w}}^\psi)$ is given at $V \in \mathcal{V}(\mathcal{B}(\mathbb{C}^3))$ as

$$(\breve{\delta}(\hat{S}_z)(\underline{\mathfrak{w}}^\psi))_V = \breve{\delta}(\hat{S}_z)_V(\underline{\mathfrak{w}}^\psi{}_V) = \{\breve{\delta}(\hat{S}_z)_V(\lambda) \mid \lambda \in \underline{\mathfrak{w}}^\psi{}_V\}. \tag{4.80}$$

Here, $\underline{\mathfrak{w}}^\psi{}_V = \alpha(\delta^o(\hat{P}_\psi)_V) = S_{\delta^o(\hat{P}_\psi)_V}$ is a (clopen) subset of $\underline{\Sigma}_V$ (cf. Equation 4.12). This means that for each context $V$, the state determines a collection $\underline{\mathfrak{w}}^\psi{}_V$ of elements of the Gelfand spectrum $\underline{\Sigma}_V$ of $V$. We then evaluate the component $\breve{\delta}(\hat{S}_z)_V$ of the arrow/natural transformation representing spin-$z$, given by Equation (4.75) (and subsequent equations), on all the $\lambda \in \underline{\mathfrak{w}}^\psi{}_V$ to obtain the component at $V$ of the "value" $\breve{\delta}(\hat{S}_z)(\underline{\mathfrak{w}}^\psi)$. Each $\lambda \in \underline{\mathfrak{w}}^\psi$ gives a sequence of intervals; one interval for each $V' \in \downarrow V$ such that if $V'' \subset V'$, the interval at $V''$ contains the interval at $V'$.

The example can easily be generalized to other operators and higher dimensions. Other finite-dimensional Hilbert spaces present no further conceptual difficulty at all. Of course, infinite-dimensional Hilbert spaces bring a host of new technical challenges, but the main tools used in the calculation—the antonymous and observable functions ($g_{\hat{A}}$ resp. $f_{\hat{A}}$)—are still available. Because they encode the approximations in the spectral order of an operator $\hat{A}$ to *all* contexts $V \in \mathcal{V}(\mathcal{N})$, the natural transformation $\breve{\delta}(\hat{A})$ corresponding to a self-adjoint operator $\hat{A}$ can be written down efficiently, without the need to actually calculate the approximations to all contexts separately.

## 4.6 Conclusion

We have shown how daseinisation relates central aspects of the standard Hilbert space formalism of quantum theory to the topos formalism. Daseinisation of projections gives subobjects of the spectral presheaf $\underline{\Sigma}$. These subobjects form a Heyting algebra, and every pure state allows us to assign truth-values to all propositions. The resulting new form of quantum logic is contextual, multivalued, and intuitionistic. For more details on the logical aspects, see [8, 11, 16].

The daseinisation of self-adjoint operators gives arrows from the state object $\underline{\Sigma}$ to the presheaf $\underline{\mathbb{R}^\leftrightarrow}$ of "values." Of course, the "value" $\breve{\delta}(\hat{A})(\underline{\mathfrak{w}}^\psi)$ of a physical quantity in a pseudostate $\underline{\mathfrak{w}}^\psi$ is considerably more complicated than the value of a physical quantity in classical physics, which is just a real number. The main point, though, is that in the topos approach all physical quantities *do* have (generalized) values in any given state—something that clearly is not the case in ordinary quantum mechanics. Moreover, we have shown in Example 10 how to calculate the "value" $\breve{\delta}(\hat{S}_z)(\underline{w^\psi})$.

Recently, Heunen, Landsman, and Spitters suggested a closely related scheme using topoi in quantum theory (see [20], as well as their contribution to this volume and references therein). All the basic ingredients are the same: a quantum system is described by an algebra of physical quantities, in their case a Rickart $C^*$-algebra; associated with this algebra is a spectral object whose subobjects represent propositions; and physical quantities are represented by arrows in a topos associated with the quantum system. The choice of topos is very similar to ours: as the base category, one considers all abelian subalgebras (i.e., contexts) of the algebra of physical quantities and orders them partially under inclusion. In our scheme, we choose the topos to be contravariant, **Sets**-valued functors (called *presheaves*) over the context category, whereas Heunen et al. choose *co*-variant functors. The use of covariant functors allows the construction of the spectral object as the topos-internal Gelfand spectrum of a topos-internal abelian $C^*$-algebra canonically defined from the external nonabelian algebra of physical quantities.

Despite the similarities, there are some important conceptual and interpretational differences between the original contravariant approach and the covariant approach. These differences and their physical consequences will be discussed in a forthcoming article [11] see also [30].

Summing up, the topos approach provides a reformulation of quantum theory in a way that seemed impossible up to now because of powerful no-go theorems like the Kochen–Specker theorem. Despite the Kochen–Specker theorem, there is a suitable notion of state "space" for a quantum system in analogy to classical physics: the spectral presheaf.

## Acknowledgments

I thank Chris Isham for the great collaboration and his constant support. I am very grateful to Hans Halvorson for organizing the enjoyable October 2007 *Deep Beauty* symposium, which led to the development of this volume.

## References

[1] G. Birkhoff and J. von Neumann. The logic of quantum mechanics. *Annals of Mathematics*, **37** (1936), 823–43.

[2] M. L. Dalla Chiara and R. Giuntini. Quantum logics. In *Handbook of Philosophical Logic*, vol. VI, eds. G. Gabbay and F. Guenthner. Dordrecht: Kluwer, 2002, pp. 129–228.

[3] H. F. de Groote. On a canonical lattice structure on the effect algebra of a von Neumann algebra. arXiv:math-ph/0410018v2 (2004).

[4] H. F. de Groote. Observables I: Stone spectra. arXiv:math-ph/0509020 (2005).

[5] H. F. de Groote. Observables II: Quantum observables. arXiv:math-ph/0509075 (2005).

[6] H. F. de Groote. Observables IV: The presheaf perspective. arXiv:0708.0677 [math-ph] (2007).

## REFERENCES

[7] A. Döring. Kochen–Specker theorem for von Neumann algebras. *International Journal Theoretical Physics*, **44** (2005), 139–60. arXiv:quant-ph/0408106.

[8] A. Döring. Topos theory and "neo-realist" quantum theory. In *Quantum Field Theory: Competitive Models*, eds. B. Fauser, J. Tolksdorf, and E. Zeidler. Boston, Berlin: Birkhäuser Basel, 2009. pp. 25–48. arXiv:0712.4003.

[9] A. Döring. Quantum states and measures on the spectral presheaf. *Advanced Science Letters*, **2** (2009), 291–301; special issue on "Quantum Gravity, Cosmology and Black Holes," ed. M. Bojowald. arXiv:0809.4847.

[10] A. Döring. Topos quantum logic and mixed states. *Proceedings of the 6th Workshop on Quantum Physics and Logic*, QPL VI (8–9 April 2009, Oxford, United Kingdom). To appear in *Electronic Notes in Theoretical Computer Science*, eds. B. Coecke, P. Panangaden, and P. Selinger.

[11] A. Döring. Algebraic quantum theory in a topos: A Comparison. In preparation (2010).

[12] A. Döring and C. J. Isham. A topos foundation for theories of physics: I. Formal languages for physics. *Journal of Mathematical Physics*, **49** (2008), 053515. arXiv:quant-ph/0703060.

[13] A. Döring and C. J. Isham. A topos foundation for theories of physics: II. Daseinisation and the liberation of quantum theory. *Journal of Mathematical Physics*, **49** (2008), 053516. arXiv:quant-ph/0703062.

[14] A. Döring and C. J. Isham. A topos foundation for theories of physics: III. The representation of physical quantities with arrows $\breve{\delta}(A) : \underline{\Sigma} \to \underline{\mathbb{R}^{\succeq}}$. *Journal of Mathematical Physics*, **49** (2008), 053517. arXiv:quant-ph/0703064.

[15] A. Döring and C. J. Isham. A topos foundation for theories of physics: IV. Categories of systems. *Journal of Mathematical Physics*, **49** (2008), 053518. arXiv:quant-ph/0703066.

[16] A. Döring and C. J. Isham. "What is a thing?": Topos theory in the foundations of physics. In *New Structures in Physics*, Lecture Notes in Physics, vol. 813, ed. B. Coecke. Berlin, Heidelberg: Springer 2011. arXiv:0803.0417.

[17] G. G. Emch. *Mathematical and Conceptual Foundations of 20th-Century Physics*. Amsterdam: North Holland, 1984.

[18] C. Flori. A topos formulation of history quantum theory. arXiv:0812.1290. *Journal of Mathematical Physics*, **51** (2010), 053527.

[19] J. Hamilton, C. J. Isham, and J. Butterfield. A topos perspective on the Kochen–Specker theorem: III. Von Neumann algebras as the base category. *International Journal of Theoretical Physics*, **39** (2000), 1413–36. arXiv:quant-ph/9911020.

[20] C. Heunen, N. P. Landsman, and B. Spitters. A topos for algebraic quantum theory. *Communications in Mathematical Physics*, **291** (2009), 63–110. arXiv:0709.4364v3 [quant-ph].

[21] C. J. Isham. Topos theory and consistent histories: The internal logic of the set of all consistent sets. *International Journal of Mathematical Physics*, **36** (1997), 785–814. arXiv:gr-qc/9607069.

[22] C. J. Isham and J. Butterfield. A topos perspective on the Kochen–Specker theorem: I. Quantum states as generalised valuations. *International Journal of Theoretical Physics*, **37** (1998), 2669–733. arXiv:quant-ph/9803055v4.

[23] C. J. Isham and J. Butterfield. A topos perspective on the Kochen–Specker theorem: II. Conceptual aspects, and classical analogues. *International Journal of Theoretical Physics*, **38** (1999), 827–59. arXiv:quant-ph/9808067v2.

[24] C. J. Isham and J. Butterfield. A topos perspective on the Kochen–Specker theorem: IV. Interval valuations. *International Journal of Theoretical Physics*, **41** (2002), 613–39. arXiv:quant-ph/0107123.

[25] R. V. Kadison and J. R. Ringrose. *Fundamentals of the Theory of Operator Algebras, Volume 1: Elementary Theory*. New York: Academic Press, 1983.

[26] S. Kochen and E. P. Specker. The problem of hidden variables in quantum mechanics. *Journal of Mathematical and Mechanics*, **17** (1967), 59–87.

[27] S. Mac Lane. *Categories for the Working Mathematician,* 2nd ed. New York, Berlin, Heidelberg: Springer, 1998.
[28] M. P. Olson. The self-adjoint operators of a von Neumann algebra form a conditionally complete lattice. *Proceedings of the AMS,* **28** (1971), 537–44.
[29] M. Takesaki. *Theory of Operator Algebras I.* Berlin, Heidelberg, New York: Springer, 1979.
[30] S. A. M. Wolters, "Contravariant vs Convariant Quantum Logic: A comparison of Two Topos-Theoretic Approaches to Quantum Theory," preprint, arXIV:1010.2031v1(2010).

CHAPTER 5

# Classical and Quantum Observables

Hans F. de Groote[1]

"Neue Blicke durch die alten Löcher"

– Georg Christoph Lichtenberg, Aphorismen

In classical mechanics, an *observable* is represented by a real valued (smooth or continuous or measurable) function on an appropriate phase space, whereas in von Neumann's axiomatic approach to quantum physics [14], an observable is represented by a bounded self-adjoint operator $A$ acting on a Hilbert space $\mathcal{H}$.

Here, a natural question arises: Is the structural difference between classical and quantum observables fundamental, or is there some background structure showing that classical and quantum observables are on the same footing? Indeed, such a background structure exists, and I describe in this chapter some of its features.

## 5.1 The Stone Spectrum of a Lattice

Let $\mathbb{L}$ be a lattice. We usually assume that $\mathbb{L}$ contains a smallest element 0 and a greatest element 1. We can weaken these assumptions: if $\mathbb{L}$ does not contain a smallest or a greatest element, it can be adjoined without altering the structure of $\mathbb{L}$ essentially. Moreover, it suffices for the following that the minimum of two elements is always defined—that is, that $\mathbb{L}$ is a $\wedge$-semilattice. Our general references to lattice theory are [1] and [17], but all we need can also be found in [5].

A presheaf of sets on $\mathbb{L}$ is a contravariant functor $\mathcal{S} : \mathbb{L} \to \mathbf{Set}$, where we consider $\mathbb{L}$ as a category in the obvious way. We want to define an associated sheaf—not on $\mathbb{L}$ but on a topological space closely connected to $\mathbb{L}$. If $\mathcal{S}$ is a presheaf on a topological space $X$, the associated sheaf is a local homeomorphism $\pi : \mathcal{E}(\mathcal{S}) \to X$, where the

---

[1] Sadly, Professor de Groote passed away shortly after submitting his manuscript for inclusion in this book. The chapter presented here has been edited for matters of style and to correct typographical errors but otherwise reflects Professor de Groote's original manuscript.

*etale space* $\mathcal{E}(\mathcal{S})$ is the disjoint union of the *stalks*,

$$\mathcal{S}_x = \varinjlim_{U \in \mathfrak{U}(x)} \mathcal{S}(U).$$

Here, $\mathfrak{U}(x)$ denotes the system of open neighborhoods of $x \in X$. We can mimic this construction for the more general case of an arbitrary lattice $\mathbb{L}$. $\mathfrak{U}(x)$ is a filter base in the lattice $\mathcal{T}(X)$ of all open subsets of $X$; that is:

**Definition 5.1.1.** A nonvoid subset $B$ of a lattice $\mathbb{L}$ is called *a filter base* in $\mathbb{L}$ if

1. $0 \notin B$.
2. If $a, b \in B$, there is an element $c \in B$ such that $c \leq a, b$.

Of course, it makes no sense to define stalks for arbitrary filter bases. But there is an obvious class of distinguished filter bases: the maximal ones. It is easy to see that a maximal filter base in a lattice $\mathbb{L}$ is nothing else but a maximal dual ideal in $\mathbb{L}$:

**Definition 5.1.2.** A (proper) *dual ideal* in $\mathbb{L}$ is a subset $\mathcal{J} \subseteq \mathbb{L}$ such that

1. $0 \notin \mathcal{J}$.
2. If $a, b \in \mathcal{J}$, then $a \wedge b \in \mathcal{J}$.
3. If $a, b \in \mathbb{L}$, $a \in \mathcal{J}$ and $a \leq b$, then $b \in \mathbb{L}$.

We denote the set of all dual ideals in $\mathbb{L}$ by $\mathcal{D}(\mathbb{L})$ and the set of all maximal dual ideals in $\mathbb{L}$ by $\mathcal{Q}(\mathbb{L})$. Elements of $\mathcal{Q}(\mathbb{L})$ are also called *quasipoints* because they serve as a surrogate for points in the process of sheafification of a presheaf.

$\mathcal{D}(\mathbb{L})$ bears a natural topology: it is generated by the sets

$$\mathcal{D}_a(\mathbb{L}) := \{\mathcal{J} \in \mathcal{D}(\mathbb{L}) \mid a \in \mathcal{J}\} \quad (a \in \mathbb{L}).$$

Because

$$\mathcal{D}_0(\mathbb{L}) = \emptyset, \ \mathcal{D}_1(\mathbb{L}) = \mathcal{D}(\mathbb{L}), \ \text{and} \ \mathcal{D}_a(\mathbb{L}) \cap \mathcal{D}_b(\mathbb{L}) = \mathcal{D}_{a \wedge b}(\mathbb{L}) \ \text{for all} \ a, b \in \mathbb{L},$$

$\{\mathcal{D}_a(\mathbb{L}) \mid a \in \mathbb{L}\}$ is a basis for this topology. Its restriction to $\mathcal{Q}(\mathbb{L})$ is a Hausdorff space in which the sets $\mathcal{Q}_a(\mathbb{L})$ are clopen.[2] But note that this is not true for $\mathcal{D}(\mathbb{L})$ and all $\mathcal{D}_a(\mathbb{L})$! M. H. Stone introduced this topology in his famous article [21] for the special case $\mathcal{Q}(\mathbb{L})$, where $\mathbb{L}$ is a Boolean algebra. We therefore call it the *Stone topology* on $\mathcal{D}(\mathbb{L})$. Later, we prove that if $\mathbb{L}$ is the projection lattice of an abelian von Neumann algebra $\mathcal{R}$, then $\mathcal{Q}(\mathbb{L})$, equipped with the restriction of the Stone topology, is homeomorphic to the Gelfand spectrum of $\mathcal{R}$.

**Definition 5.1.3.** Let $\mathbb{L}$ be an arbitrary lattice and let $\mathcal{Q}(\mathbb{L})$ be the set of its maximal dual ideals. Then $\mathcal{Q}(\mathbb{L})$, together with (restriction of) the Stone topology, is called the *Stone spectrum of* $\mathbb{L}$.

The Stone spectrum $\mathcal{Q}(\mathbb{L})$ reflects some properties of the lattice $\mathbb{L}$. We restrict our discussion to *orthomodular* lattices. For their proofs, we refer to [5]. If $a, b$ are elements

---

[2] We use the common abbreviation *clopen* for "closed and open."

of an orthomodular lattice $\mathbb{L}$, then $\mathcal{Q}_a(\mathbb{L}) \subseteq \mathcal{Q}_b(\mathbb{L})$ implies $a \leq b$. In particular,

$$\mathcal{Q}_a(\mathbb{L}) = \mathcal{Q}_b(\mathbb{L}) \iff a = b.$$

Distributivity of an orthomodular lattice $\mathbb{L}$ can be characterized by topological properties of $\mathcal{Q}(\mathbb{L})$:

**Proposition 5.1.1.** *The following properties of an orthomodular lattice $\mathbb{L}$ are equivalent:*

1. *$\mathbb{L}$ is distributive.*
2. *$\mathcal{Q}_a(\mathbb{L}) \cup \mathcal{Q}_b(\mathbb{L}) = \mathcal{Q}_{a \vee b}(\mathbb{L})$ for all $a, b \in \mathbb{L}$.*
3. *$\mathcal{Q}_a(\mathbb{L}) \cup \mathcal{Q}_{a^\perp}(\mathbb{L}) = \mathcal{Q}(\mathbb{L})$ for all $a \in \mathbb{L}$.*
4. *The only clopen subsets of $\mathcal{Q}(\mathbb{L})$ are the sets $\mathcal{Q}_a(\mathbb{L})$ $(a \in \mathbb{L})$.*

**Definition 5.1.4.** A lattice $\mathbb{L}$ is called of *finite type* if

$$\overline{\bigcup_{k \in \mathbb{K}} \mathcal{Q}_{a_k}(\mathbb{L})} = \mathcal{Q}_{\bigvee_{k \in \mathbb{K}} a_k}(\mathbb{L})$$

holds for all *increasing* families $(a_k)_{k \in \mathbb{K}}$ in $\mathbb{L}$.

**Lemma 5.1.2.** *An orthomodular lattice $\mathbb{L}$ is of finite type if and only if*

$$a \wedge (\bigvee_{k \in \mathbb{K}} a_k) = \bigvee_{k \in \mathbb{K}} (a \wedge a_k)$$

*for all $a \in \mathbb{L}$ and all increasing families $(a_k)_{k \in \mathbb{K}}$ in $\mathbb{L}$.*

Finiteness of a von Neumann algebra $\mathcal{R}$ can also be characterized by topological properties of its Stone spectrum:

**Theorem 5.1.1.** *The projection lattice $\mathcal{P}(\mathcal{R})$ of a von Neumann algebra $\mathcal{R}$ is of finite type if and only if $\mathcal{R}$ is of finite type.*

## 5.2 Stone Spectra of von Neumann Algebras

Let $\mathcal{R}$ be an abelian von Neumann algebra acting on a Hilbert space $\mathcal{H}$. We give a new proof for the known result that $\beta \mapsto \tau_\beta$ is a homeomorphism from $\mathcal{Q}(\mathcal{R})$ onto the Gelfand spectrum of $\mathcal{R}$ using the generalized Gleason theorem. This proof is much shorter than the old one [5], but it is less elementary.

**Theorem 5.2.1.** *Let $\mathcal{R}$ be an abelian von Neumann algebra. Then, the Gelfand spectrum of $\mathcal{R}$ is homeomorphic to the Stone spectrum of $\mathcal{R}$.*

**PROOF.** Let $\beta \in \mathcal{Q}(\mathcal{R})$. We define a function $\tau_\beta : \mathcal{P}(\mathcal{R}) \to \{0, 1\}$ by

$$\tau_\beta(P) := \begin{cases} 1 & \text{if } P \in \beta \\ 0 & \text{if } P \notin \beta. \end{cases}$$

(Note that because $\mathcal{P}(\mathcal{R})$ is distributive, $P \notin \beta$ is equivalent to $I - P \in \beta$.) Note further that $\tau_\beta$ is multiplicative on $\mathcal{P}(\mathcal{R})$:

$$\forall\, P, Q \in \mathcal{P}(\mathcal{R}): \quad \tau_\beta(PQ) = \tau_\beta(P)\tau_\beta(Q).$$

It follows now directly from elementary properties of quasipoints in distributive lattice that

$$\tau_\beta\left(\sum_{k=1}^{n} P_k\right) = \sum_{k=1}^{n} \tau_\beta(P_k)$$

for all pairwise orthogonal $P_1, \ldots, P_n \in \mathcal{P}(\mathcal{R})$; that is, $\tau_\beta$ is a probability measure on $\mathcal{P}(\mathcal{R})$. By the generalization of Gleason's theorem (from Christensen, Yeadon et al. [18]), $\tau_\beta$ extends to a unique state of $\mathcal{R}$. We denote this state again by $\tau_\beta$. It is easy to prove that this state is multiplicative: by the spectral theorem, the algebra $lin_\mathbb{C}\mathcal{P}(\mathcal{R})$ of all finite linear combinations of projections from $\mathcal{R}$ is dense in $\mathcal{R}$, and a direct calculation shows that $\tau_\beta$ is multiplicative on $lin_\mathbb{C}\mathcal{P}(\mathcal{R})$.

Conversely, let $\tau$ be a character of $\mathcal{R}$. Then, it is easy to see that

$$\beta_\tau := \{P \in \mathcal{P}(\mathcal{R}) \mid \tau(P) = 1\}$$

is a quasipoint in $\mathcal{R}$ and that

$$\tau_{\beta_\tau} = \tau, \qquad \beta_{\tau_\beta} = \beta$$

holds. Therefore, the mapping $\tau \mapsto \beta_\tau$ is a bijection of the Gelfand spectrum $\Omega(\mathcal{R})$ of $\mathcal{R}$ onto the Stone spectrum $\mathcal{Q}(\mathcal{R})$ of $\mathcal{R}$. To prove that this mapping is a homeomorphism, we have only to show that it is continuous because $\Omega(\mathcal{R})$ and $\mathcal{Q}(\mathcal{R})$ are compact Hausdorff spaces.

Let $\tau_0 \in \Omega(\mathcal{R})$, $0 < \varepsilon < 1$, and let $P \in \mathcal{P}(\mathcal{R})$ such that $\tau_0(P) = 1$. Then,

$$N_w(\tau_0) := \{\tau \in \Omega(\mathcal{R}) \mid |\tau(P) - \tau_0(P)| < \varepsilon\}$$

is an open neighborhood of $\tau_0$, and from $\varepsilon < 1$, we conclude

$$\begin{aligned} \tau \in N_w(\tau_0) &\iff \tau(P) = \tau_0(P) \\ &\iff P \in \beta_\tau \\ &\iff \beta_\tau \in \mathcal{Q}_P(\mathcal{R}). \end{aligned}$$

This means that $N_w(\tau_0)$ is mapped bijectively onto the open neighborhood $\mathcal{Q}_P(\mathcal{R})$ of $\beta_{\tau_0}$. The $\mathcal{Q}_P(\mathcal{R})$ with $P \in \beta_{\tau_0}$ form a neighborhood base of $\beta_{\tau_0}$ in the Stone topology of $\mathcal{Q}(\mathcal{R})$. Hence, $\tau \mapsto \beta_\tau$ is continuous. ∎

In general, it seems a hard problem to determine the Stone spectrum of a von Neumann algebra, acting on an infinite-dimensional Hilbert space $\mathcal{H}$, explicitly. Even for $\mathcal{L}(\mathcal{H})$, only the following simple result is known:

**Proposition 5.2.1 [5].** If $\mathfrak{B}$ is a quasipoint of the projection-lattice $\mathcal{P}(\mathcal{L}(\mathcal{H}))$, then either

1. $\mathfrak{B}$ contains a projection of finite rank and, in this case, it is of the form $\mathfrak{B} = \mathfrak{B}_{\mathbb{C}x} := \{P \in \mathcal{P}(\mathcal{L}(\mathcal{H})) \mid P_{\mathbb{C}x} \leq P\}$; or
2. $\mathfrak{B}$ contains all projections of finite corank.

Moreover, if $\mathcal{H}$ is infinite-dimensional, $\mathcal{Q}(\mathcal{H})$ is not locally compact.

But even if $\mathcal{Q}(\mathcal{H})$ were known much better, this would not help much for general von Neumann subalgebras $\mathcal{R}$ of $\mathcal{L}(\mathcal{H})$ because the mapping

$$\zeta : \mathcal{Q}(\mathcal{H}) \to \mathcal{D}(\mathcal{R})$$
$$\mathfrak{B} \mapsto \mathfrak{B} \cap \mathcal{R}$$

does not map into $\mathcal{Q}(\mathcal{R})$! This follows from the following proposition.

**Proposition 5.2.2 [8].** Let $\mathcal{R}$ be a von Neumann algebra with center $\mathcal{C}$ and let $\mathcal{A}$ be a von Neumann subalgebra of $\mathcal{C}$. Then, the mapping

$$\zeta_\mathcal{A} : \mathfrak{B} \mapsto \mathfrak{B} \cap \mathcal{A}$$

is an open continuous—and, therefore, identifying—mapping from $\mathcal{Q}(\mathcal{R})$ onto $\mathcal{Q}(\mathcal{A})$. Moreover,

$$\zeta_\mathcal{A}(\mathfrak{B}) = \{s_\mathcal{A}(P) \mid P \in \mathfrak{B}\}$$

for all $\mathfrak{B} \in \mathcal{Q}(\mathcal{R})$, where

$$s_\mathcal{A}(P) := \bigwedge \{Q \in \mathcal{P}(\mathcal{A}) \mid P \leq Q\}$$

is the $\mathcal{A}$-support of $P \in \mathcal{P}(\mathcal{R})$.

Conversely, if $\mathcal{M}$ is a von Neumann subalgebra of $\mathcal{R}$ such that $\mathfrak{B} \cap \mathcal{M} \in \mathcal{Q}(\mathcal{M})$ for all $\mathfrak{B} \in \mathcal{Q}(\mathcal{R})$, then $\mathcal{M}$ is contained in the center of $\mathcal{R}$.

Nevertheless, we have a complete description of the Stone spectrum of finite von Neumann algebras of type I. We just state the main result:

**Theorem 5.2.2 [8].** Let $\mathcal{R}$ be a von Neumann algebra of type $I_n$ ($n \in \mathbb{N}$) with center $\mathcal{C}$. Then, the Stone spectrum $\mathcal{Q}(\mathcal{R})$ of $\mathcal{R}$ is a locally compact space, the projection mapping $\zeta_\mathcal{C} : \mathcal{Q}(\mathcal{R}) \to \mathcal{Q}(\mathcal{C})$ is a local homeomorphism and, therefore, has discrete fibers. The unitary group $\mathcal{U}(\mathcal{R})$ of $\mathcal{R}$ acts transitively on each fiber of $\zeta_\mathcal{C}$. Therefore, each fiber $\zeta_\mathcal{C}^{-1}(\beta)$ can be represented as a homogeneous space $\mathcal{U}(\mathcal{R})/\mathcal{U}(\mathcal{R})_\mathfrak{B}$, where the isotropy group $\mathcal{U}(\mathcal{R})_\mathfrak{B}$ of $\mathfrak{B} \in \zeta_\mathcal{C}(\beta)$ is given by

$$\mathcal{U}(\mathcal{R})_\mathfrak{B} = \{T \in \mathcal{U}(\mathcal{R}) \mid [T]_\beta [E]_\beta = [E]_\beta [T]_\beta \text{ for all abelian } E \in \mathfrak{B}\}.$$

Here, $[A]_\beta$ denotes the equivalence class of $A \in \mathcal{R}$ modulo the equivalence relation $\sim_\beta$ over $\beta \in \mathcal{Q}(\mathcal{C})$, defined by

$$\forall A, B \in \mathcal{R}: \ A \sim_\beta B :\iff \exists \, p \in \beta : \ pA = pB.$$

## 5.3 Observable Functions

If $A$ is a bounded self-adjoint operator acting on a Hilbert space $\mathcal{H}$, then $A$ has a *spectral decomposition*

$$A = \int_{-\infty}^{\infty} \lambda \, dE_\lambda,$$

where $(E_\lambda)_{\lambda \in \mathbb{R}}$ is the *spectral family* of $A$. It is a family in the projection lattice $\mathcal{P}(\mathcal{L}(\mathcal{H}))$ of $\mathcal{H}$ with the following properties:

1. $E_\lambda \leq E_\mu$ if $\lambda \leq \mu$.
2. $E_\lambda = \bigwedge_{\mu > \lambda} E_\mu$ for all $\lambda \in \mathbb{R}$.
3. There are $a, b \in \mathbb{R}$ such that $E_\lambda = 0$ for $\lambda < a$ and $E_\lambda = I$ for $\lambda \geq b$.

The integral is meant in the sense of norm-converging Riemann–Stieltjes sums. The spectral family $E$ is uniquely determined by $A$ and, if $\mathcal{R}$ is a von Neumann algebra acting on $\mathcal{H}$, then $A$ belongs to $\mathcal{R}$ if and only if all the $E_\lambda$ do. If $A$ is not bounded, condition (3) must be replaced by the more general

4. $\bigwedge_{\lambda \in \mathbb{R}} E_\lambda = 0$ and $\bigvee_{\lambda \in \mathbb{R}} E_\lambda = I$.

Of course, one can define spectral families in a complete lattice in quite the same manner:[3]

**Definition 5.3.1.** Let $\mathbb{L}$ be a lattice. A family $E = (E_\lambda)_{\lambda \in \mathbb{R}}$ in $\mathbb{L}$ is called a *spectral family* if it satisfies the following three properties:

1. $E_\lambda \leq E_\mu$ for $\lambda \leq \mu$.
2. $E_\lambda = \bigwedge_{\mu > \lambda} E_\mu$ for all $\lambda \in \mathbb{R}$.
3. $\bigwedge_{\lambda \in \mathbb{R}} E_\lambda = 0$ and $\bigvee_{\lambda \in \mathbb{R}} E_\lambda = 1$.

$E$ is called bounded if there are $a, b \in \mathbb{R}$ such that $E_\lambda = 0$ for $\lambda < a$ and $E_\lambda = 1$ for $E_\lambda \geq b$.

We restrict in the following our discussion to *bounded* spectral families in $\mathbb{L}$. There are interesting and important examples outside the range of functional analysis:

**Example 5.3.1.** The set of all open subsets of a topological space $M$ is a complete lattice $\mathcal{T}(M)$, where

$$\bigvee_{k \in \mathbb{K}} U_k = \bigcup_{k \in \mathbb{K}} U_k, \quad \bigwedge_{k \in \mathbb{K}} U_k = int(\bigcap_{k \in \mathbb{K}} U_k), \quad 0 = \emptyset, \quad 1 = M$$

for all families $(U_k)_{k \in \mathbb{K}}$ in $M$. Let $f : M \to \mathbb{R}$ be a bounded continuous function on a Hausdorff space $M$. Then,

$$\forall \lambda \in \mathbb{R} : \quad E_\lambda := int(f^{-1}(]-\infty, \lambda]))$$

defines a spectral family $E : \mathbb{R} \to \mathcal{T}(M)$. The natural guess for defining a spectral family corresponding to $f$ would be

$$\lambda \mapsto f^{-1}(]-\infty, \lambda[).$$

In general, this is only a *prespectral family*: it satisfies all properties of a spectral family, except continuity from the right. This is cured by *spectralization*—that

---

[3] If $\mathbb{L}$ is an arbitrary lattice, then the existence of the suprema and infima in the following definition should be regarded as a part of the requirements. But note that for the existence of these suprema and infima, we need only that $\mathbb{L}$ is $\sigma$-complete because a spectral family is increasing and the rational numbers are dense in $\mathbb{R}$.

is, by the switch to

$$\lambda \mapsto \bigwedge_{\mu > \lambda} f^{-1}(]-\infty, \mu[).$$

But

$$\bigwedge_{\mu > \lambda} f^{-1}(]-\infty, \mu[) = int(\bigcap_{\mu > \lambda} f^{-1}(]-\infty, \mu[)) = int(f^{-1}(]-\infty, \lambda])),$$

which shows that our original definition is the natural one.

We show in Section 5.4 that

$$\forall \, x \in M : \; f(x) = \inf\{\lambda \mid x \in E_\lambda\},$$

so one can recover the function $f$ from its spectral family $E$.

The following definition is fundamental:

**Definition 5.3.2.** Let $E = (E_\lambda)_{\lambda \in \mathbb{R}}$ be a bounded spectral family in a complete lattice $\mathbb{L}$. Then,

$$\begin{aligned} f_E : \mathcal{D}(\mathbb{L}) &\to \mathbb{R} \\ \mathcal{J} &\mapsto \inf\{\lambda \in \mathbb{R} \mid E_\lambda \in \mathcal{J}\} \end{aligned}$$

is called the *observable function of E*.

We are mainly interested in the restriction of $f_E$ to the Stone spectrum $\mathcal{Q}(\mathbb{L})$ of $\mathbb{L}$. We denote this restriction also by $f_E$ and call it again the *observable function of E*.

**Proposition 5.3.1.** The mapping $E \mapsto f_E$ from the set $\mathcal{E}_b(\mathbb{L})$ of all bounded spectral families in the $\sigma$-complete lattice $\mathbb{L}$ to the set of functions $\mathcal{Q}(\mathbb{L}) \to \mathbb{R}$ is injective.

**PROOF.** Let $E$, $F$ be bounded spectral families in $\mathbb{L}$ such that $f_E = f_F$. Assume that there is some $\lambda \in \mathbb{R}$ such that $E_\lambda \wedge (E_\lambda \wedge F_\lambda)^\perp \ne 0$. If $\mathfrak{B}$ is any quasipoint in $\mathbb{L}$ that contains $E_\lambda \wedge (E_\lambda \wedge F_\lambda)^\perp$, then also $E_\lambda \in \mathfrak{B}$ and, therefore, $f_E(\mathfrak{B}) \le \lambda$. But $f_E(\mathfrak{B}) < \lambda$ would imply $f_F(\mathfrak{B}) < \lambda$; hence, $F_\lambda \in \mathfrak{B}$, a contradiction. Thus, $f_F(\mathfrak{B}) = f_E(\mathfrak{B}) = \lambda$ for all $\mathfrak{B} \in \mathcal{Q}_{E_\lambda \wedge (E_\lambda \wedge F_\lambda)^\perp}(\mathbb{L})$, so

$$\forall \, \mu > \lambda : \; F_\mu \in \bigcap \mathcal{Q}_{E_\lambda \wedge (E_\lambda \wedge F_\lambda)^\perp}(\mathbb{L}).$$

But $\bigcap \mathcal{Q}_{E_\lambda \wedge (E_\lambda \wedge F_\lambda)^\perp}(\mathbb{L}) = H_{E_\lambda \wedge (E_\lambda \wedge F_\lambda)^\perp}$, the principal dual ideal generated by $E_\lambda \wedge (E_\lambda \wedge F_\lambda)^\perp$, so $F_\mu \ge E_\lambda \wedge (E_\lambda \wedge F_\lambda)^\perp$ for all $\mu > \lambda$. Hence, also $F_\lambda \ge E_\lambda \wedge (E_\lambda \wedge F_\lambda)^\perp$, a contradiction again. This shows $E_\lambda = E_\lambda \wedge F_\lambda$; that is, $E_\lambda \le F_\lambda$ for all $\lambda \in \mathbb{R}$. Because the argument is symmetric in $E$ and $F$, we have $E = F$. ∎

Moreover, it is surprisingly easy to see that the function associated to a bounded spectral family in a $\sigma$-complete orthomodular lattice is continuous:

**Proposition 5.3.2.** Let $E$ be a bounded spectral family in a $\sigma$-complete orthomodular lattice $\mathbb{L}$. Then, the function $f_E : \mathcal{Q}(\mathbb{L}) \to \mathbb{R}$ associated to $E$ is continuous.

**PROOF.** If $f_E(\mathfrak{B}_0) = \lambda$ and $\varepsilon > 0$, take $\mathfrak{B} \in \mathcal{Q}_{E_{\lambda+\varepsilon}}(\mathbb{L}) \setminus \mathcal{Q}_{E_{\lambda-\varepsilon}}(\mathbb{L})$. Then,

$$\lambda - \varepsilon \leq f_E(\mathfrak{B}) \leq \lambda + \varepsilon,$$

and because $\mathcal{Q}_{E_{\lambda+\varepsilon}}(\mathbb{L}) \setminus \mathcal{Q}_{E_{\lambda-\varepsilon}}(\mathbb{L})$ is an open neighborhood of $\mathfrak{B}_0$, we see that $f_E$ is continuous. ∎

It is easy to see that this proof is also valid for the lattice $\mathcal{T}(M)$ of all open subsets of a Hausdorff space $M$.

Note that in contrast to the foregoing result, the function $f_E : \mathcal{D}(\mathbb{L}) \to \mathbb{R}$ is only upper semicontinuous. We give a simple example:

**Example 5.3.2.** Let $\mathcal{R}$ be a von Neumann algebra. The observable function $f_{I-P} : \mathcal{D}(\mathcal{R}) \to \mathbb{R}$ for a projection $P \in \mathcal{R}$ is given by

$$f_{I-P} = 1 - \chi_{\mathcal{D}_P(\mathcal{R})}.$$

$f_{I-P}$ is continuous if and only if $\mathcal{D}_P(\mathcal{R})$ is open (which is true by definition) and closed. Now

$$\mathcal{J} \in \overline{\mathcal{D}_P(\mathcal{R})} \iff \forall Q \in \mathcal{J} : \mathcal{D}_Q(\mathcal{R}) \cap \mathcal{D}_P(\mathcal{R}) \neq \emptyset$$
$$\iff \forall Q \in \mathcal{J} : \mathcal{D}_{P \wedge Q}(\mathcal{R}) \neq \emptyset$$
$$\iff \forall Q \in \mathcal{J} : P \wedge Q \neq 0$$

and, therefore, $\overline{\mathcal{D}_P(\mathcal{R})} = \mathcal{D}_P(\mathcal{R})$ if and only if

$$\forall \mathcal{J} \in \overline{\mathcal{D}_P(\mathcal{R})} : ((\forall Q \in \mathcal{J} : P \wedge Q \neq 0) \implies P \in \mathcal{J}).$$

This leads to the following example. Let $P, P_1 \in \mathcal{P}(\mathcal{R})$ such that $0 \neq P < P_1$. Then, $Q \wedge P = P$ for all $Q \in H_{P_1}$ but $P \notin H_{P_1}$. Hence, $H_{P_1} \in \overline{\mathcal{D}_P(\mathcal{R})} \setminus \mathcal{D}_P(\mathcal{R})$.

We now show that observable functions are a generalization of the Gelfand transforms of self-adjoint elements of an abelian von Neumann algebra. It is easy to see that the observable function of $A = \sum_{k=1}^{n} a_k P_k$, where $a_1, \ldots, a_n \in \mathbb{R}$ and $P_1, \ldots, P_n \in \mathcal{P}(\mathcal{R})$ are pairwise orthogonal, is given by

$$f_A = \sum_{k=1}^{n} a_k \chi_{\mathcal{Q}_{P_k}(\mathcal{R})}.$$

This is also the *Gelfand transform* $\hat{A}$ of $A$, according to the fact that the Gelfand spectrum of $\mathcal{R}$ is homeomorphic to the Stone spectrum of $\mathcal{R}$ and a quasipoint of $\mathcal{R}$ is contained in at most one of the sets $\mathcal{Q}_{P_k}(\mathcal{R})$ $(k \leq n)$. Now, let $A$ be an arbitrary self-adjoint element of $\mathcal{R}$, let $m := \min sp(A)$, $M := \max sp(A)$, and, given $\varepsilon > 0$, we choose $\lambda_1, \ldots, \lambda_n \in \mathbb{R}$ such that $0 < \lambda_{k+1} - \lambda_k < \varepsilon$, $\lambda_n > M$, $\lambda_1 < m$. Moreover, let

$$A_\varepsilon := \sum_{k=1}^{n-1} \lambda_k (E_{\lambda_{k+1}} - E_{\lambda_k}).$$

Then, for any $\beta \in \mathcal{Q}(\mathcal{R})$,

$$\exists! k \in \{1, \ldots, k-1\} : \beta \in \mathcal{Q}_{E_{\lambda_{k+1}}}(\mathcal{R}) \setminus \mathcal{Q}_{E_{\lambda_k}}(\mathcal{R});$$

hence,
$$| f_A - f_{A_\varepsilon} |_\infty \leq \varepsilon.$$
Similarly,
$$| \hat{A} - \hat{A}_\varepsilon |_\infty \leq \varepsilon.$$
Therefore,
$$| f_A - \hat{A} |_\infty \leq | f_A - f_{A_\varepsilon} |_\infty + | \hat{A}_\varepsilon - \hat{A} |_\infty \leq 2\varepsilon$$
for all $\varepsilon > 0$, so
$$f_A = \hat{A}.$$
Thus, we have proved:

**Theorem 5.3.1.** If $\mathcal{R}$ is an abelian von Neumann algebra, and if the Gelfand spectrum of $\mathcal{R}$ is identified with the Stone spectrum of $\mathcal{R}$ via the homeomorphism $\tau \mapsto \beta_\tau$, then for all $A \in \mathcal{R}_{sa}$, the space of self-adjoint elements of $\mathcal{R}$, the observable function of $A$ is the Gelfand transform of $A$.

In Section 5.5, we give a topological characterization of observable functions among the continuous functions on the Stone spectrum of the complete orthomodular lattice $\mathbb{L}$. Here, we will give a lattice-theoretical characterization of observable functions among all real-valued functions on $\mathcal{D}(\mathbb{L})$. As an important consequence, we show that observable functions can be described by functions on $\mathbb{L} \setminus \{0\}$.

We define the spectrum of a spectral family in a lattice by generalizing a characteristic property of the spectrum of a self-adjoint operator:

**Definition 5.3.3.** Let $E$ be a bounded spectral family in a complete orthomodular lattice $\mathbb{L}$. The set $re(E)$ of all $\lambda \in \mathbb{R}$ such that $E$ is constant on a neighborhood of $\lambda$ is called the resolvent set of $E$. Its complement $sp(E) := \mathbb{R} \setminus re(E)$ is called the *spectrum of* $E$.

Clearly, $sp(E)$ is a compact subset of $\mathbb{R}$.

**Proposition 5.3.3 [6].** Let $f_E : \mathcal{Q}(\mathbb{L}) \to \mathbb{R}$ be the observable function of a bounded spectral family in $\mathbb{L}$. Then,
$$f_E(\mathcal{Q}(\mathbb{L})) = sp(E).$$

There are two fundamental properties of the function $f_E : \mathcal{D}(\mathbb{L}) \to \mathbb{R}$. The first one is expressed in the following.

**Proposition 5.3.4.** Let $(\mathcal{J}_j)_{j \in J}$ be a family in $\mathcal{D}(\mathbb{L})$. Then,
$$f_E(\bigcap_{j \in J} \mathcal{J}_j) = \sup_{j \in J} f_E(\mathcal{J}_j).$$

The other follows.

**Proposition 5.3.5.** $f_E : \mathcal{D}(\mathbb{L}) \to \mathbb{R}$ is upper semicontinuous; that is,
$$\forall \mathcal{J}_0 \in \mathcal{D}(\mathcal{R}) \ \forall \varepsilon > 0 \ \exists a \in \mathcal{J}_0 \ \forall \mathcal{J} \in \mathcal{D}_a(\mathbb{L}) : \ f_E(\mathcal{J}) < f_E(\mathcal{J}_0) + \varepsilon.$$

The proofs are simple and can be found in [6]. These two properties are fundamental for the abstract characterization of observable functions. Note that the property that $f_E$ is decreasing and upper semicontinuous is equivalent to the property

$$\forall \mathcal{J} \in \mathcal{D}(\mathbb{L}) : f_E(\mathcal{J}) = \inf\{f_E(H_a) | \, a \in \mathcal{J}\}.$$

**Definition 5.3.4.** A function $f : \mathcal{D}(\mathbb{L}) \to \mathbb{R}$ is called an *abstract observable function* if it is upper semicontinuous and satisfies the intersection condition

$$f(\bigcap_{j \in J} \mathcal{J}_j) = \sup_{j \in J} f(\mathcal{J}_j)$$

for all families $(\mathcal{J}_j)_{j \in J}$ in $\mathcal{D}(\mathbb{L})$.

A direct consequence of the intersection condition is the following:

**Remark 5.3.1.** Let $\lambda \in imf$. Then, the inverse image $f^{-1}(\lambda) \subseteq \mathcal{D}(\mathbb{L})$ has a minimal element $\mathcal{J}_\lambda$, which is simply given by

$$\mathcal{J}_\lambda = \bigcap\{\mathcal{J} \in \mathcal{D}(\mathbb{L}) | \, f(\mathcal{J}) = \lambda\}.$$

Moreover,

$$E_\lambda = \inf \mathcal{J}_\lambda.$$

**Theorem 5.3.2 [6].** Let $\mathbb{L}$ be a complete orthomodular lattice and $f : \mathcal{D}(\mathbb{L}) \to \mathbb{R}$ an abstract observable function. Then, there is a unique spectral family $E$ in $\mathbb{L}$ such that $f = f_E$.

The *proof* proceeds in three steps. In the first step, we construct from the abstract observable function $f$ an increasing family $(E_\lambda)_{\lambda \in imf}$ in $\mathcal{P}(\mathcal{R})$ and show in a second step that this family can be extended to a spectral family in $\mathcal{R}$. Finally, in the third step, we show that the self-adjoint operator $A \in \mathcal{R}$ corresponding to that spectral family has observable function $f_A = f$ and that $A$ is uniquely determined by $f$. We present here only the first two steps and omit the third because it is rather technical.

**Step 1.** Let $\lambda \in imf$ and let $\mathcal{J}_\lambda \in \mathcal{D}(\mathcal{R})$ be the smallest dual ideal such that $f(\mathcal{J}_\lambda) = \lambda$. In view of Remark 5.3.1, we are forced to define

$$E_\lambda := \inf \mathcal{J}_\lambda.$$

**Lemma 5.3.1.** The family $(E_\lambda)_{\lambda \in imf}$ is increasing.

**PROOF.** Let $\lambda, \mu \in imf$, $\lambda < \mu$. Then,

$$\begin{aligned} f(\mathcal{J}_\mu) &= \mu \\ &= max(\lambda, \mu) \\ &= max(f(\mathcal{J}_\lambda), f(\mathcal{J}_\mu)) \\ &= f(\mathcal{J}_\lambda \cap \mathcal{J}_\mu). \end{aligned}$$

Hence, by the minimality of $\mathcal{J}_\mu$,

$$\mathcal{J}_\mu \subseteq \mathcal{J}_\lambda \cap \mathcal{J}_\mu \subseteq \mathcal{J}_\lambda$$

and therefore $E_\lambda \leq E_\mu$. ∎

**Lemma 5.3.2.** $f$ is monotonely continuous—that is, if $(\mathcal{J}_j)_{j \in J}$ is an increasing net in $\mathcal{D}(\mathbb{L})$, then

$$f(\bigcup_{j \in J} \mathcal{J}_j) = \lim_j f(\mathcal{J}_j).$$

**PROOF.** Obviously, $\mathcal{J} := \bigcup_{j \in J} \mathcal{J}_j \in \mathcal{D}(\mathbb{L})$. As $f$ is decreasing, $f(\mathcal{J}) \leq f(\mathcal{J}_j)$ for all $j \in J$ and $(f(\mathcal{J}_j))_{j \in J}$ is a decreasing net of real numbers. Hence,

$$f(\mathcal{J}) \leq \lim_j f(\mathcal{J}_j).$$

Let $\varepsilon > 0$. Because $f$ is upper semicontinuous, there is an $a \in \mathcal{J}$ such that $f(\mathcal{I}) < f(\mathcal{J}) + \varepsilon$ for all $\mathcal{I} \in \mathcal{D}_a(\mathbb{L})$. Now, $a \in \mathcal{J}_k$ for some $k \in J$ and therefore

$$\lim_j f(\mathcal{J}_j) \leq f(\mathcal{J}_k) < f(\mathcal{J}) + \varepsilon,$$

which shows that also $\lim_j f(\mathcal{J}_j) \leq f(\mathcal{J})$ holds. ∎

**Corollary 5.3.1.** *The image of an abstract observable function is compact.*

**PROOF.** Because $\{I\} \subseteq \mathcal{J}$ for all $\mathcal{J} \in \mathcal{D}(\mathcal{R})$, we have $f \leq f(\{I\})$ on $\mathcal{D}(\mathbb{L})$. If $\lambda, \mu \in imf$ and $\lambda < \mu$, then $\mathcal{J}_\mu \subseteq \mathcal{J}_\lambda$; hence, $\bigcup_{\lambda \in imf} \mathcal{J}_\lambda$ is a dual ideal and therefore contained in a maximal dual ideal $\mathfrak{B} \in \mathcal{D}(\mathbb{L})$. This shows $f(\mathfrak{B}) \leq f$ on $\mathcal{D}(\mathbb{L})$ and, consequently, $imf$ is bounded. Let $\lambda \in \overline{imf}$. Then, there is an increasing sequence $(\mu_n)_{n \in \mathbb{N}}$ in $imf$ converging to $\lambda$, or there is a decreasing sequence $(\mu_n)_{n \in \mathbb{N}}$ in $imf$ converging to $\lambda$. In the first case, we have $\mathcal{J}_{\mu_{n+1}} \subseteq \mathcal{J}_{\mu_n}$ for all $n \in \mathbb{N}$ and therefore for $\mathcal{J} := \bigcap_n \mathcal{J}_{\mu_n} \in \mathcal{D}(\mathbb{L})$

$$f(\mathcal{J}) = \sup_n f(\mathcal{J}_{\mu_n}) = \sup_n \mu_n = \lambda.$$

In the second case, we have $\mathcal{J}_{\mu_n} \subseteq \mathcal{J}_{\mu_{n+1}}$ for all $n \in \mathbb{N}$ and therefore $\mathcal{J} := \bigcup_n \mathcal{J}_{\mu_n} \in \mathcal{D}(\mathbb{L})$. Hence,

$$f(\mathcal{J}) = \lim_n f(\mathcal{J}_{\mu_n}) = \lim_n \mu_n = \lambda.$$

Therefore, $\lambda \in imf$ in both cases; that is, $imf$ is also closed. ∎

**Step 2.** We now extend $(E_\lambda)_{\lambda \in imf}$ to a spectral family $E^f := (E_\lambda)_{\lambda \in \mathbb{R}}$. In defining $E^f$, we have in mind, of course, that the spectrum of $E^f$ should coincide with $imf$. This forces us to define $E_\lambda$ for $\lambda \notin imf$ in the following way. For $\lambda \notin imf$, let

$$S_\lambda := \{\mu \in imf | \mu < \lambda\}.$$

Then, we define

$$E_\lambda := \begin{cases} 0 & \text{if } S_\lambda = \emptyset \\ E_{\sup S_\lambda} & \text{otherwise.} \end{cases}$$

Note that $f(\{I\}) = \max imf$ and that $\mathcal{J}_{f(\{I\})} = \{I\}$.

**Proposition 5.3.6.** $E^f$ is a spectral family.

**PROOF.** The only remaining point to prove is that $E^f$ is continuous from the right; that is, that $E_\lambda = \bigwedge_{\mu > \lambda} E_\mu$ for all $\lambda \in \mathbb{R}$. This is obvious if $\lambda \notin imf$ or if there is some $\delta > 0$ such that $]\lambda, \lambda + \delta[ \cap imf = \emptyset$. Therefore, we are left with the case that there is a strictly decreasing sequence $(\mu_n)_{n \in \mathbb{N}}$ in $imf$ converging to $\lambda$. For all $n \in \mathbb{N}$, we have $f(\mathcal{J}_{\mu_n}) > f(\mathcal{J}_\lambda)$ and therefore $\mathcal{J}_{\mu_n} \subseteq \mathcal{J}_\lambda$. Hence, $\bigcup_n \mathcal{J}_{\mu_n} \subseteq \mathcal{J}_\lambda$ and

$$f(\bigcup_n \mathcal{J}_{\mu_n}) = \lim_n f(\mathcal{J}_{\mu_n}) = \lambda$$

implies $\bigcup_n \mathcal{J}_{\mu_n} = \mathcal{J}_\lambda$ by the minimality of $\mathcal{J}_\lambda$. If $a \in \mathcal{J}_\lambda$, then $a \in \mathcal{J}_{\mu_n}$ for some $n$ and therefore $E_{\mu_n} \leq a$. This shows $\bigwedge_{\mu > \lambda} E_\mu \leq a$. Because $a \in \mathcal{J}_\lambda$ is arbitrary, we can conclude that $\bigwedge_{\mu > \lambda} E_\mu \leq E_\lambda$. The reverse inequality is obvious. ■

The theorem confirms that there is no difference between "abstract" and "concrete" observable functions; therefore, we will speak generally of observable functions.

Let $f : \mathcal{D}(\mathbb{L}) \to \mathbb{R}$ be an observable function. Then, $f$ is determined by its restriction to the set of principle dual ideals in $\mathbb{L}$; that is, by the function

$$\begin{aligned} r_f : \mathbb{L} \setminus \{0\} &\to \mathbb{R} \\ a &\mapsto f(H_a). \end{aligned}$$

Because $\bigcap_{k \in \mathbb{K}} H_{a_k} = H_{\bigvee_{k \in \mathbb{K}} a_k}$, the function $r : \mathbb{L} \setminus \{0\} \to \mathbb{R}$ is *completely increasing*; that is, it is bounded and satisfies

$$r_f(\bigvee_{k \in \mathbb{K}} a_k) = \sup_{k \in \mathbb{K}} r_f(a_k)$$

for each family $(a_k)_{k \in \mathbb{K}}$ in $\mathbb{L} \setminus \{0\}$.

Conversely, let $r : \mathbb{L} \setminus \{0\} \to \mathbb{R}$ be any completely increasing function. We can extend $r$ to a function $f_r : \mathcal{D}(\mathbb{L}) \to \mathbb{R}$ by defining

$$\forall \mathcal{J} \in \mathcal{D}(\mathbb{L}) : \quad f_r(\mathcal{J}) := \inf\{r(a) \mid H_a \subseteq \mathcal{J}\}.$$

**Proposition 5.3.7.** The function $f_r : \mathcal{D}(\mathbb{L}) \to \mathbb{R}$ induced by the completely increasing function $r : \mathbb{L} \setminus \{0\} \to \mathbb{R}$ is an observable function.

**PROOF.** In view of Theorem 5.3.2, we have to show that $f_r$ satisfies

$$f_r(\bigcap_{k \in \mathbb{K}} \mathcal{J}_k) = \sup_{k \in \mathbb{K}} f_r(\mathcal{J}_k)$$

for all families $(\mathcal{J}_k)_{k \in \mathbb{K}}$ in $\mathcal{D}(\mathcal{R})$. Because $f_r$ is decreasing, we have

$$f_r(\bigcap_{k \in \mathbb{K}} \mathcal{J}_k) \geq \sup_{k \in \mathbb{K}} f_r(\mathcal{J}_k).$$

Let $\varepsilon > 0$ and choose $a_k \in \mathcal{J}_k$ ($k \in \mathbb{K}$) such that $r(a_k) < f_r(\mathcal{J}_k) + \varepsilon$. Now, $\bigcap_k H_{a_k} \subseteq \bigcap_k \mathcal{J}_k$, $f_r$ is decreasing, and $r$ is completely increasing; hence,

$$f_r(\bigcap_k \mathcal{J}_k) \leq f_r(\bigcap_k H_{a_k}) = r(\bigvee_k a_k) = \sup_k r(a_k) \leq \sup_k f_r(\mathcal{J}_k) + \varepsilon$$

and therefore

$$f_r(\bigcap_{k \in \mathbb{K}} \mathcal{J}_k) \leq \sup_{k \in \mathbb{K}} f_r(\mathcal{J}_k). \qquad \blacksquare$$

## 5.4 Classical Observables

### 5.4.1 Continuous Functions

In the previous subsection, we saw that self-adjoint elements $A$ of a von Neumann algebra $\mathcal{R}$ correspond to certain bounded continuous real valued functions $f_A$ on the Stone spectrum $\mathcal{Q}(\mathcal{R})$ of $\mathcal{P}(\mathcal{R})$.

Now, we show that continuous real valued functions on a Hausdorff space $M$ can be described by spectral families with values in the complete lattice $\mathcal{T}(M)$ of open subsets of $M$. These spectral families $E : \mathbb{R} \to \mathcal{T}(M)$ can be characterized abstractly by a certain property of the mapping $E$. Thus, also a classical observable has a "quantum mechanical" description. Similar results hold for functions on a set $M$ that are measurable with respect to a $\sigma$-algebra of subsets of $M$. Because of the limited size of this chapter, we confine ourselves to the statement of the main results.

We begin with some simple examples:

**Example 5.4.1.** The following settings define spectral families in $\mathcal{T}(\mathbb{R})$:

$$E_\lambda^{id} := ]-\infty, \lambda[, \tag{5.1}$$

$$E_\lambda^{abs} := ]-\lambda, \lambda[ \tag{5.2}$$

$$E_\lambda^{ln} := ]-\exp(\lambda), \exp(\lambda)[ \tag{5.3}$$

$$E_\lambda^{step} := ]-\infty, \lfloor\lambda\rfloor[ \tag{5.4}$$

where $\lfloor\lambda\rfloor$ denotes the "floor of $\lambda \in \mathbb{R}$":

$$\lfloor\lambda\rfloor = \max\{n \in \mathbb{Z} \mid n \leq \lambda\}.$$

The names of these spectral families sound somewhat crazy at the moment, but we justify them soon.

In close analogy to the case of spectral families in the lattice $\mathbb{L}(\mathcal{H})$, each spectral family in $\mathcal{T}(M)$ induces a function on a subset of $M$.

**Definition 5.4.1.** Let $E : \mathbb{R} \to \mathcal{T}(M)$ be a spectral family in $\mathcal{T}(M)$. Then,

$$\mathcal{D}(E) := \{x \in M \mid \exists \lambda \in \mathbb{R} : x \notin E_\lambda\}$$

is called the *admissible domain of $E$*.

Note that

$$\mathcal{D}(E) = M \setminus \bigcap_{\lambda \in \mathbb{R}} E_\lambda.$$

**Remark 5.4.1.** The admissible domain $\mathcal{D}(E)$ of a spectral family $E$ in $\mathcal{T}(M)$ is dense in $M$.

Conversely, it may happen that $\mathcal{D}(E) \neq M$. The spectral family $E^{ln}$ is a simple example:

$$\forall \lambda \in \mathbb{R} : 0 \in E_\lambda^{ln}.$$

Every spectral family $E$ in $\mathcal{T}(M)$ induces a function $f_E : \mathcal{D}(E) \to \mathbb{R}$:

**Definition 5.4.2.** Let $E$ be a spectral family in $\mathcal{T}(M)$ with admissible domain $\mathcal{D}(E)$. Then, the function $f_E : \mathcal{D}(E) \to \mathbb{R}$, defined by

$$\forall x \in \mathcal{D}(E) : f_E(x) := \inf\{\lambda \in \mathbb{R} \mid x \in E_\lambda\},$$

is called the *function induced* by $E$.

The functions induced by our foregoing examples are

$$f_{E^{id}}(x) = x \tag{5.5}$$

$$f_{E^{abs}}(x) = |x| \tag{5.6}$$

$$f_{E^{ln}}(x) = \ln|x| \quad \text{and} \quad \mathcal{D}(E^{ln}) = \mathbb{R} \setminus \{0\} \tag{5.7}$$

$$f_{E^{step}} = \sum_{n \in \mathbb{Z}} n \chi_{[n,n+1]}. \tag{5.8}$$

There is a fundamental difference between the spectral families $E^{id}$, $E^{abs}$, $E^{ln}$ on the one side and $E^{step}$ on the other. The function induced by $E^{step}$ is not continuous. This fact is mirrored in the spectral families: the first three spectral families have the property

$$\forall \lambda < \mu : \overline{E_\lambda} \subseteq E_\mu.$$

Obviously, $E^{step}$ fails to have this property.

**Definition 5.4.3.** A spectral family $E$ in $\mathcal{T}(M)$ is called *regular* if

$$\forall \lambda < \mu : \overline{E_\lambda} \subseteq E_\mu$$

holds.

Because $\mathcal{T}(M)$ is Heyting algebra with pseudocomplement $U^c := int(M \setminus U)$ ($= M \setminus \overline{U}$) ($U \in \mathcal{T}(M)$, the regularity of $E$ can be expressed by

$$E_\lambda^c \cup E_\mu = M \quad \text{for all} \quad \lambda < \mu.$$

**Remark 5.4.2.** The admissible domain $\mathcal{D}(E)$ of a regular spectral family $E$ in $\mathcal{T}(M)$ is an open (and dense) subset of $M$.

**Remark 5.4.3.** If $E$ in $\mathcal{T}(M)$ is a regular spectral family, then for all $\lambda \in \mathbb{R}$, $E_\lambda$ is a regular open set; that is, $E_\lambda$ is the interior of its closure.

The importance of regular spectral families becomes manifest in the following:

**Theorem 5.4.1 [7].** Let $M$ be a Hausdorff space. Then, every continuous function $f : M \to \mathbb{R}$ induces a regular spectral family $E^f$ in $\mathcal{T}(M)$ by

$$\forall \lambda \in \mathbb{R}: \; E^f_\lambda := int(f^{-1}(]-\infty, \lambda])).$$

The admissible domain $\mathcal{D}(E^f)$ equals $M$ and the function $f_{E^f} : M \to \mathbb{R}$ induced by $E^f$ is $f$. Conversely, if $E$ is a regular spectral family in $\mathcal{T}(M)$, then the function

$$f_E : \mathcal{D}(E) \to \mathbb{R}$$

induced by $E$ is continuous and the induced spectral family $E^{f_E}$ in $\mathcal{T}(\mathcal{D}(E))$ is the restriction of $E$ to the admissible domain $\mathcal{D}(E)$:

$$\forall \lambda \in \mathbb{R}: \; E^{f_E}_\lambda = E_\lambda \cap \mathcal{D}(E).$$

One may wonder why we have defined the function that is induced by a spectral family $E$ on $M$ and not on the Stone spectrum $\mathcal{Q}(\mathcal{T}(M))$. A quasipoint $\mathfrak{B} \in \mathcal{Q}(\mathcal{T}(M))$ is called *finite* if $\bigcap_{U \in \mathfrak{B}} \overline{U} \neq \emptyset$. If $\mathfrak{B}$ is finite, then this intersection consists of a single element $x_\mathfrak{B} \in M$, and we call $\mathfrak{B}$ a quasipoint *over* $x_\mathfrak{B}$. Note that for a compact space $M$, all quasipoints are finite. Moreover, one can show that for compact $M$, the mapping $pt : \mathfrak{B} \mapsto x_\mathfrak{B}$ from $\mathcal{Q}(\mathcal{T}(M))$ onto $M$ is continuous and *identifying*.

**Remark 5.4.4 [7].** Let $E$ be a regular spectral family in $\mathcal{T}(M)$ and let $x \in \mathcal{D}(E)$. Then, for all quasipoints $\mathfrak{B}_x \in \mathcal{Q}(\mathcal{T}(M))$ over $x$, we have

$$f_E(\mathfrak{B}_x) = f_E(x).$$

Therefore, if $M$ is compact, it makes no difference whether we define $f_E$ in $M$ or in $\mathcal{Q}(\mathcal{T}(M))$.

### 5.4.2 Measurable Functions

If $M$ is a nonempty set and $\mathfrak{A}(M)$ is a $\sigma$-algebra of subsets[4] of $M$, then every $\mathfrak{A}(M)$-measurable function $f : M \to \mathbb{R}$ induces a spectral family $E^f$ in $\mathfrak{A}(M)$ by

$$E^f_\lambda := f^{-1}(]-\infty, \lambda]).$$

Conversely, every spectral family $E$ in $\mathfrak{A}(M)$ defines a $\mathfrak{A}(M)$-measurable function $f_E : M \to \mathbb{R}$ by

$$f_E(x) := \inf\{\lambda \in \mathbb{R} \mid x \in E_\lambda\},$$

and it is easy to see that these constructions are inverse to each other:

$$E^{f_E} = E \quad \text{and} \quad f_{E^f} = f.$$

---

[4] This means that the lattice operations are the ordinary set-theoretical operations.

This result suggests the following definition:

**Definition 5.4.4.** Let $\mathfrak{A}$ be an arbitrary $\sigma$-algebra. A spectral family in $\mathfrak{A}$ is called a *generalized $\mathfrak{A}$-measurable function*.

To fill this definition with content, we use the theorem of Loomis and Sikorski [20] that an abstract $\sigma$-algebra $\mathfrak{A}$ is representable as a quotient $\mathfrak{A}(M)/\mathcal{I}$ of a $\sigma$-algebra $\mathfrak{A}(M)$ of sets and a $\sigma$-ideal $\mathcal{I} \subseteq \mathfrak{A}(M)$. It is easy to show [5] that the Stone spectrum of the quotient lattice $\mathfrak{A}(M)/\mathcal{I}$ is homeomorphic to the compact subspace

$$\mathcal{Q}^{\mathcal{I}}(\mathfrak{A}(M)) := \{\mathfrak{B} \in \mathcal{Q}(\mathfrak{A}(M)) \mid \mathfrak{B} \cap \mathcal{I} = \emptyset\}$$

of $\mathcal{Q}(\mathfrak{A}(M))$. We proved in [5] the following theorem:

**Theorem 5.4.3.** Let $M$ be a nonempty set, $\mathfrak{A}(M)$ a $\sigma$-algebra of subsets of $M$, and $\mathcal{I}$ a $\sigma$-ideal in $\mathfrak{A}(M)$. Furthermore, let $\mathcal{F}_{\mathfrak{A}(M)}(M, \mathbb{C})$ be the abelian algebra of all bounded $\mathfrak{A}(M)$-measurable functions $M \to \mathbb{C}$. $\mathcal{F}_{\mathfrak{A}(M)}(M, \mathbb{C})$ is a $C^*$-algebra with respect to the supremum-norm and the set $\mathcal{F}(\mathcal{I})$ of all $f \in \mathcal{F}_{\mathfrak{A}(M)}(M, \mathbb{C})$ that vanish outside some set $A \in \mathcal{I}$ is a closed ideal in $\mathcal{F}_{\mathfrak{A}(M)}(M, \mathbb{C})$. Then, the Gelfand spectrum of the quotient $C^*$-algebra $\mathcal{F}_{\mathfrak{A}(M)}(M, \mathbb{C})/\mathcal{F}(\mathcal{I})$ is homeomorphic to the Stone spectrum of the $\sigma$-algebra $\mathfrak{A}(M)/\mathcal{I}$.

Note that $\mathcal{F}_{\mathfrak{A}(M)}(M, \mathbb{C})/\mathcal{F}(\mathcal{I})$ may fail to be a von Neumann algebra [16, 5.7.21(iv)].

**Theorem 5.4.3 [7].** Let $\mathfrak{A}$ be a $\sigma$-algebra, represented as a quotient $\mathfrak{A}(M)/\mathcal{I}$ of a $\sigma$-algebra $\mathfrak{A}(M)$ of subsets of a set $M$ modulo a $\sigma$-ideal $\mathcal{I}$ in $\mathfrak{A}(M)$. Then, the Gelfand transformation of the $C^*$-algebra $\mathcal{F}_{\mathfrak{A}(M)}(M, \mathbb{C})/\mathcal{F}(\mathcal{I})$, restricted to the self-adjoint part $\mathcal{F}_{\mathfrak{A}(M)}(M, \mathbb{R})/\mathcal{F}_{\mathbb{R}}(\mathcal{I})$, is given by

$$\Gamma : \mathcal{F}_{\mathfrak{A}(M)}(M, \mathbb{R})/\mathcal{F}_{\mathbb{R}}(\mathcal{I}) \to C(\mathcal{Q}(\mathfrak{A}), \mathbb{R})$$
$$[\varphi] \mapsto f_{E^{[\varphi]}}.$$

Moreover, if $\mathcal{Q}(\mathfrak{A})$ is identified with $\mathcal{Q}^{\mathcal{I}}(\mathfrak{A}(M))$, we can write

$$f_{E^{[\varphi]}} = f_{E^{\varphi}}|_{\mathcal{Q}(\mathfrak{A})},$$

where $\varphi \mapsto f_{E^{\varphi}}$ is the Gelfand transformation on $\mathcal{F}_{\mathfrak{A}(M)}(M, \mathbb{R})$.

We have already mentioned that for an abstract $\sigma$-algebra $\mathfrak{A}$, spectral families in $\mathfrak{A}$ are the adequate substitutes for $\mathfrak{A}$-measurable functions. If $\mathfrak{A}$ is represented as a quotient $\mathfrak{A}(M)/\mathcal{I}$ with $\mathfrak{A}(M)$ and $\mathcal{I}$ as in the preceding, we shall show that each *bounded* spectral family in $\mathfrak{A}$ is the quotient modulo $\mathcal{I}$ of a spectral family in $\mathfrak{A}(M)$. This means that every bounded spectral family in $\mathfrak{A}$ can be lifted to the spectral family of a bounded $\mathfrak{A}(M)$-measurable function. The proof rests on the foregoing theorem.

**Corollary 5.4.1 [7].** Let $E$ be a bounded spectral family in a $\sigma$-algebra $\mathfrak{A}$ and let $\mathfrak{A} = \mathfrak{A}(M)/\mathcal{I}$, where $\mathfrak{A}(M)$ is a $\sigma$-algebra of subsets of some set $M$ and $\mathcal{I}$ is a $\sigma$-ideal in $\mathfrak{A}(M)$. Then, there is some $\varphi \in \mathcal{F}_{\mathfrak{A}(M)}(M, \mathbb{R})$ such that

$$\forall \lambda \in \mathbb{R} : E_\lambda = [E_\lambda^\varphi].$$

$\varphi$ is unique up to equivalence modulo $\mathcal{F}_{\mathbb{R}}(\mathcal{I})$.

## 5.5 Topological Characterization of Observable Functions

Let $\mathbb{L}$ be an arbitrary complete[5] orthomodular lattice. Consider the mapping

$$\mathcal{Q} : \mathbb{L} \to \mathcal{T}(\mathcal{Q}(\mathbb{L}))$$
$$a \mapsto \mathcal{Q}_a(\mathbb{L}).$$

The mapping $\mathcal{Q}$ satisfies

$$\mathcal{Q}(a \wedge b) = \mathcal{Q}(a) \cap \mathcal{Q}(b),$$

but not

$$\mathcal{Q}(a \vee b) = \mathcal{Q}(a) \cup \mathcal{Q}(b)$$

for all $a, b \in \mathbb{L}$, unless $\mathbb{L}$ is distributive. Nevertheless, it follows that the cone $C_{\mathcal{T}(\mathcal{Q}(\mathbb{L}))}(\mathcal{Q}(\mathcal{J}))$ of the image $\mathcal{Q}(\mathcal{J})$ of any dual ideal $\mathcal{J}$ in $\mathbb{L}$ is a dual ideal in $\mathcal{T}(\mathcal{Q}(\mathbb{L}))$. Next, we show that $\mathcal{Q}$ maps spectral families to spectral families.

**Lemma 5.5.1.** Let $E = (E_\lambda)_{\lambda \in \mathbb{R}}$ be a (bounded) spectral family in $\mathbb{L}$. Then, $\mathcal{Q}(E) := (\mathcal{Q}_{E_\lambda}(\mathbb{L}))_{\lambda \in \mathbb{R}}$ is a spectral family in $\mathcal{T}(\mathcal{Q}(\mathbb{L}))$.

**PROOF.** The only nontrivial property is the right continuity of $\mathcal{Q}(E)$. Let $\lambda \in \mathbb{R}$. It follows directly from $E_\lambda = \bigwedge_{\mu > \lambda} E_\mu$ that $\mathcal{Q}_{E_\lambda}(\mathbb{L}) \subseteq int \bigcap_{\mu > \lambda} \mathcal{Q}_{E_\mu}(\mathbb{L})$ holds. For any $\mathfrak{B} \in int \bigcap_{\mu > \lambda} \mathcal{Q}_{E_\mu}(\mathbb{L})$, there is a nonzero $a \in \mathbb{L}$ such that $\mathcal{Q}_a(\mathbb{L}) \subseteq \bigcap_{\mu > \lambda} \mathcal{Q}_{E_\mu}(\mathbb{L})$. This means that $a \leq E_\mu$ for all $\mu > \lambda$; hence, $a \leq \bigwedge_{\mu > \lambda} E_\mu = E_\lambda$ and, therefore, $\mathcal{Q}_a(\mathbb{L}) \subseteq \mathcal{Q}_{E_\lambda}(\mathbb{L})$. ∎

**Proposition 5.5.1.** Let $E$ be a bounded spectral family in $\mathbb{L}$ and let $f_E, f_{\mathcal{Q}(E)} : \mathcal{Q}(\mathbb{L}) \to \mathbb{R}$ be the continuous functions induced by the bounded spectral families $E$ and $\mathcal{Q}(E)$, respectively. Then,

$$f_{\mathcal{Q}(E)} = f_E.$$

**PROOF.** Because each $\mathcal{Q}_{E_\lambda}(\mathbb{L})$ is clopen, $\mathcal{Q}(E)$ is regular; hence, $f_{\mathcal{Q}(E)}$ is continuous. Now, for any quasipoint $\mathfrak{B} \in \mathcal{Q}(\mathbb{L})$, we obtain

$$f_E(\mathfrak{B}) = \inf\{\lambda \in \mathbb{R} \mid E_\lambda \in \mathfrak{B}\}$$
$$= \inf\{\lambda \in \mathbb{R} \mid \mathfrak{B} \in \mathcal{Q}_{E_\lambda}(\mathbb{L})\}$$
$$= f_{\mathcal{Q}(E)}(\mathfrak{B}).$$
∎

This result leads to a characterization of those continuous functions that are induced by spectral families in $\mathbb{L}$.

**Theorem 5.5.1.** A bounded continuous function $f : \mathcal{Q}(\mathbb{L}) \to \mathbb{R}$ is induced by a bounded spectral family $E$ in $\mathbb{L}$ if and only if

$$\forall \lambda \in \mathbb{R} \, \exists \, a_\lambda \in \mathbb{L} : \, int \, f^{-1}(]-\infty, \lambda]) = \mathcal{Q}_{a_\lambda}(\mathbb{L}).$$

**PROOF.** Because $f_E = f_{\mathcal{Q}(E)}$ and $int \, f^{-1}(]-\infty, \lambda]) = \mathcal{Q}_{E_\lambda}(\mathbb{L})$, the condition is satisfied for observable functions. Conversely, if the condition is satisfied,

---
[5] Or at least $\sigma$-complete.

the elements $a_\lambda \in \mathbb{L}$ are uniquely determined by $f$ because the lattice $\mathbb{L}$ is orthomodular. All we have to show is that the family $(a_\lambda)_{\lambda \in \mathbb{R}}$ is a spectral family. For all $\lambda \in \mathbb{R}$, we have

$$\mathcal{Q}_{a_\lambda}(\mathbb{L}) = int \bigcap_{\mu > \lambda} \mathcal{Q}_{a_\mu}(\mathbb{L}),$$

which implies $a_\lambda \leq a_\mu$ for all $\mu > \lambda$; hence, $a_\lambda \leq \bigwedge_{\mu > \lambda} a_\mu$. If $a \leq a_\mu$ for all $\mu > \lambda$, then $\mathcal{Q}_a(\mathbb{L}) \subseteq int \bigcap_{\mu > \lambda} \mathcal{Q}_{a_\mu}(\mathbb{L}) = \mathcal{Q}_{a_\lambda}(\mathbb{L})$; hence, $a \leq a_\lambda$. Therefore, $a_\lambda = \bigwedge_{\mu > \lambda} a_\mu$. ∎

Now, let $\mathcal{R}$ be a von Neumann algebra. We obtain from the foregoing results that $\mathcal{R}$ is abelian if and only if $\mathcal{O}(\mathcal{R}) = C_b(\mathcal{Q}(\mathcal{R}))$:

**Corollary 5.5.1.** A von Neumann algebra $\mathcal{R}$ is abelian if and only if $\mathcal{O}(\mathcal{R}) = C_b(\mathcal{Q}(\mathcal{R}))$.

**PROOF.** If $\mathcal{R}$ is abelian, $\mathcal{Q}(\mathcal{R})$ is a Stonean space; hence, the interior of any closed set is clopen. Moreover, $\mathcal{P}(\mathcal{R})$ is distributive; hence, all clopen subsets of $\mathcal{Q}(\mathcal{R})$ are of the form $\mathcal{Q}_P(\mathcal{R})$ for some $P \in \mathcal{P}_0(\mathcal{R})$. Conversely, if $\mathcal{O}(\mathcal{R}) = C_b(\mathcal{Q}(\mathcal{R}))$, and $W$ is a clopen subset of $\mathcal{Q}(\mathcal{R})$, $f := 1 - \chi_W$ is an observable function and $W = f^{-1}(]-\infty, 0])$; hence, $W = \mathcal{Q}_P(\mathcal{R})$ for some $P \in \mathcal{P}(\mathcal{R})$. This implies that $\mathcal{R}$ is abelian. ∎

## 5.6 The Presheaf Perspective

### 5.6.1 The Spatial Presheaf of Spectral Families

Let $\mathbb{L}$ be a complete orthomodular lattice. For $p \in \mathbb{L} \setminus \{0\}$, we denote by $\mathcal{E}(p)$ the set of all bounded spectral families in the ideal $\mathcal{I} := \{a \in \mathbb{L} \mid a \leq p\}$. Boundedness of $E \in \mathcal{E}(p)$ means here that there are $a, b \in \mathbb{R}$ such that $E_\lambda = 0$ for all $\lambda < a$ and $E_\lambda = p$ for all $\lambda \geq b$. For $p, q \in \mathbb{L} \setminus \{0\}$, $p \leq q$, we define a restriction map

$$\varrho_p^q := \mathcal{E}(q) \to \mathcal{E}(p)$$
$$E \mapsto E \wedge p,$$

where

$$\forall \lambda \in \mathbb{R} : (E \wedge p)_\lambda := E_\lambda \wedge p.$$

Obviously, $\varrho_p^p = id_{\mathcal{E}(p)}$ for all $p \in \mathbb{L} \setminus \{0\}$ and

$$\forall p, q, r \in \mathbb{L} \setminus \{0\} : (p \leq q \leq r \implies \varrho_p^q \circ \varrho_q^r = \varrho_p^r).$$

Therefore:

**Definition 5.6.1.** The collection of sets $\mathcal{E}(p)$, $p \in \mathbb{L} \setminus \{0\}$, together with the restriction maps $\varrho_p^q$, $p \leq q$, forms a presheaf $\mathcal{PE}_\mathbb{L}$, which we call the spatial presheaf of spectral families in $\mathbb{L}$.

**Remark 5.6.1.** It is not difficult to show that the spatial presheaf is complete[6] if the lattice is distributive.

We like to motivate the term *spatial* in this definition.

Let $\mathbb{L} := \mathcal{P}(\mathcal{R})$, where $\mathcal{R}$ is a von Neumann algebra acting on a Hilbert space $\mathcal{H}$. If $E$ is a bounded spectral family in $\mathcal{P}(\mathcal{R})$ and $P \in \mathcal{P}_0(\mathcal{R})$, then $E \wedge P$ is a bounded spectral family in the von Neumann algebra $P\mathcal{R}P$ acting on the closed subspace $P\mathcal{H}$ of $\mathcal{H}$. If $A \in \mathcal{R}_{sa}$ is the self-adjoint operator corresponding to $E$, then the self-adjoint operator $A_P$ corresponding to $E \wedge P$ is called the *spatial restriction of A to PRP*. Note that, in general, $A_P \neq PAP$.

Now, we want to describe the *sheaf* $\mathcal{SE}_\mathbb{L}$ associated to $\mathcal{PE}_\mathbb{L}$.

If $q, r \in \mathbb{L} \setminus \{0\}$, $E \in \mathcal{E}(q)$, and $F \in \mathcal{E}(r)$, we say that $E$ is equivalent to $F$ at the quasipoint $\mathfrak{B} \in \mathcal{Q}(\mathbb{L})$, if there is a $p \in \mathfrak{B}$ such that $p \leq q, r$ and

$$E \wedge p = F \wedge p.$$

Note that if $E$ and $F$ are equivalent at $\mathfrak{B}$, then necessarily $q, r \in \mathfrak{B}$. We denote by $[E]_\mathfrak{B}$ the equivalence class of $E$ at $\mathfrak{B}$. It is the *germ* of $E$ at $\mathfrak{B}$.

Now, let $\mathcal{J} \in \mathcal{D}(\mathbb{L})$, and let $E \in \mathcal{E}(q)$, $F \in \mathcal{E}(r)$ such that there is a $p \in \mathcal{J}$ with $p \leq q, r$ and $E \wedge p = F \wedge p$. $p \in \mathcal{J}$ implies

$$\forall \lambda \in \mathbb{R} : E_\lambda \in \mathcal{J} \iff E_\lambda \wedge p \in \mathcal{J}$$

and, hence,

$$f_E(\mathcal{J}) = f_{E \wedge p}(\mathcal{J})$$

for all $\mathcal{J} \in \mathcal{D}_p(\mathbb{L})$. Therefore, because $E \wedge p = F \wedge p$, we get

$$f_E = f_F \text{ on } \mathcal{D}_p(\mathbb{L}).$$

Conversely, assume that $f_E = f_F$ on $\mathcal{Q}_p(\mathbb{L})$ holds for some $p \in \mathbb{L} \setminus \{0\}$ with $p \leq q, r$. Then, for all $\lambda \in \mathbb{R}$ with $E_\lambda \wedge p \neq 0$:

$$f_E(H_{E_\lambda \wedge p}) = f_E(\bigcap \mathcal{Q}_{E_\lambda \wedge p}(\mathbb{L})) = \sup_{\mathfrak{B} \in \mathcal{Q}_{E_\lambda \wedge p}(\mathbb{L})} f_E(\mathfrak{B}) = \sup_{\mathfrak{B} \in \mathcal{Q}_{E_\lambda \wedge p}(\mathbb{L})} f_F(\mathfrak{B}) = f_F(H_{E_\lambda \wedge p}).$$

Hence, $E_\lambda \wedge p \leq F_\lambda \wedge p$ and, by a symmetrical argument, $F_\lambda \wedge p \leq E_\lambda \wedge p$. Thus, we have proved the following:

**Proposition 5.6.1.** Two bounded spectral families $E$ and $F$ in a complete orthomodular lattice $\mathbb{L}$ are equivalent at $\mathfrak{B} \in \mathcal{Q}(\mathbb{L})$ if and only if there is a $p \in \mathfrak{B}$ such that $f_E = f_F$ on $\mathcal{Q}_p(\mathbb{L})$.

**Lemma 5.6.1.** Let $p \in \mathbb{L} \setminus \{0\}$. Then, $\mathcal{Q}(\mathbb{L}_p)$ is homeomorphic to $\mathcal{Q}_p(\mathbb{L})$.

PROOF. The homeomorphism is given by the mapping

$$soc_p : \mathcal{Q}_p(\mathbb{L}) \to \mathcal{Q}(\mathbb{L}_p)$$
$$\mathfrak{B} \mapsto \mathfrak{B}_p,$$

---

[6] Completeness of a presheaf over a complete lattice is defined in the same way as for presheaves over a topological space: just replace unions by joins and intersections by meets.

where $\mathfrak{B}_p := \{a \in \mathfrak{B} \mid a \le p\}$ denotes the "socle" of $\mathfrak{B}$ under $p \in \mathfrak{B}$. It is easy to see that each socle of a quasipoint determines that quasipoint uniquely. ∎

Using this homeomorphism, we can (and do) regard $\mathcal{Q}(\mathbb{L}_p)$ as a subset of $\mathcal{Q}(\mathbb{L}_q)$ if $p \le q$.

For $p \in \mathbb{L} \setminus \{0\}$, let $\mathcal{O}(\mathbb{L}_p)$ be the set of observable functions of spectral families in $\mathbb{L}_p$. If $p, q \in \mathbb{L} \setminus \{0\}$, $p \le q$, we have a natural restriction map

$$\mathcal{O}_p^q : \mathcal{O}(\mathbb{L}_q) \to \mathcal{O}(\mathbb{L}_p)$$
$$f \mapsto f \wedge p,$$

where

$$f \wedge p := f_{|\mathcal{Q}(\mathbb{L}_p)}.$$

Then, the collection of all $\mathcal{O}(\mathbb{L}_p)$, together with the restriction maps $\mathcal{O}_p^q$, forms a presheaf $\mathcal{PO}_\mathbb{L}^s$. The associated sheaf, $\mathcal{O}_\mathbb{L}^s$, is isomorphic to the sheaf $\mathcal{SE}_\mathbb{L}$:

**Proposition 5.6.2.** *The sheaves $\mathcal{SE}_\mathbb{L}$ and $\mathcal{O}_\mathbb{L}^s$ are isomorphic. The isomorphism is simply given by the mapping $[E]_\mathfrak{B} \mapsto [f_E]_\mathfrak{B}$, where $[f_E]_\mathfrak{B}$ denotes the germ of the observable function $f_E$ at the quasipoint $\mathfrak{B}$.*

This result follows immediately from our foregoing discussion.

### 5.6.2 The Algebraic Presheaf of Spectral Families

Again, $\mathbb{L}$ denotes a complete orthomodular lattice with minimal element 0 and maximal element 1. Moreover, let $\mathcal{B}_\mathbb{L}$ be the set of all complete Boolean sublattices $\mathcal{B}$ of $\mathbb{L}$ with $0, 1 \in \mathcal{B}$. The main example that we have in mind here is the projection lattice of a von Neumann algebra $\mathcal{R}$. Switching from spectral families to self-adjoint operators means that we consider abelian von Neumann subalgebras of $\mathcal{R}$ instead of complete Boolean sublattices of $\mathbb{L}$. Another possible application is the lattice of causally closed subsets of a spacetime [2].

#### 5.6.2.1 Algebraic Restriction of Spectral Families

Let $\mathcal{E}(\mathbb{M})$ be the set of all bounded spectral families in the complete orthomodular sublattice $\mathbb{M}$ of $\mathbb{L}$. (We always assume that complete sublattices contain 0 and 1.)

**Lemma 5.6.2.** *Let $E, F$ be spectral families in $\mathbb{L}$ and let $f_E, f_F : \mathbb{L} \setminus \{0\} \to \mathbb{R}$ be their corresponding observable functions. Then,*

$$f_E \le f_F \iff \forall \lambda \in \mathbb{R} : F_\lambda \le E_\lambda.$$

**PROOF.** If $f_E \le f_F$ on $\mathbb{L} \setminus \{0\}$, and if $a \in \mathbb{L} \setminus \{0\}$, then

$$\inf\{\mu \in \mathbb{R} \mid a \le E_\mu\} \le \inf\{\mu \in \mathbb{R} \mid a \le F_\mu\}.$$

Hence, if $F_\lambda \ne 0$, then $\inf\{\mu | F_\lambda \le F_\mu\} \le \lambda$ and therefore

$$\lambda_0 := \inf\{\mu \mid E_\lambda^B \le E_\mu\} \le \lambda.$$

But then $F_\lambda \leq E_{\lambda_0} \leq E_\lambda$. Conversely, if $F_\lambda \leq E_\lambda$ for all $\lambda \in \mathbb{R}$, then it is obvious that

$$f_E(a) = \inf\{\mu \in \mathbb{R} \mid a \leq E_\mu\} \leq \{\mu \in \mathbb{R} \mid a \leq F_\mu\} = f_F(a)$$

for all $a \in \mathbb{L} \setminus \{0\}$. ∎

Note that the abstract characterization of observable functions implies that the relation $f_E \leq f_F$ is independent of the domain $\mathbb{L} \setminus \{0\}$, $\mathcal{Q}(\mathbb{L})$, or $\mathcal{D}(\mathbb{L})$ on which we consider observable functions. We define a partial order on $\mathcal{E}(\mathbb{L})$ by

**Definition 5.6.2.** The partial order on $\mathcal{E}(\mathbb{L})$, given by

$$E \leq_s F \quad :\Longleftrightarrow \quad f_E \leq f_F \quad \Longleftrightarrow \quad \forall \lambda \in \mathbb{R} : F_\lambda \leq E_\lambda,$$

is called the *spectral order* on $\mathcal{E}(\mathbb{L})$.

If $\mathbb{L}$ is the projection lattice of a von Neumann algebra $\mathcal{R}$, the spectral order has been defined directly for self-adjoint operators by Olson [19] (see also [3]) by their spectral families. This has the advantage that the definition of the lattice operations causes no problem:

**Definition 5.6.3.** A family $(E^\kappa)_{k \in \mathbb{K}}$ in $\mathcal{E}(\mathbb{L})$ is called bounded if $\bigcup_{k \in \mathbb{K}} sp(E^\kappa)$ is a bounded set. For a bounded family $\mathcal{E} = (E^\kappa)_{k \in \mathbb{K}}$ in $\mathcal{E}(\mathbb{L})$, we define

$$E_\lambda^\wedge := \bigwedge_{\mu > \lambda} \bigvee_{k \in \mathbb{K}} E_\mu^\kappa \quad \text{and} \quad E_\lambda^\vee := \bigwedge_{k \in \mathbb{K}} E_\lambda^\kappa.$$

In what follows, let $\mathbb{M}$ be a complete orthomodular sublattice of $\mathbb{L}$ containing 0 and 1.

**Proposition 5.6.3 [3, 19].** $E^\wedge := (E_\lambda^\wedge)_{\lambda \in \mathbb{R}}$ and $E^\vee := (E_\lambda^\vee)_{\lambda \in \mathbb{R}}$ are bounded spectral families in $\mathbb{L}$. $E^\wedge$ is the infimum of the bounded family $\mathcal{E}$ and $E^\vee$ is its supremum. Thus, $\mathcal{E}(\mathbb{L})$ is a boundedly complete lattice.

Now we come to our basic definition.

**Definition 5.6.4.** Let $E \in \mathcal{E}(\mathbb{L})$ and $f_E : \mathbb{L} \setminus \{0\} \to \mathbb{R}$ be its corresponding observable function. Then, $\varrho_\mathbb{M} f_E := f_E \mid_{\mathbb{M} \setminus \{0\}}$ is an observable function on $\mathbb{M} \setminus \{0\}$. The corresponding bounded spectral family in $\mathbb{M}$ is denoted by $\varrho_\mathbb{M} E$ and called the algebraic restriction of $E$ to $\mathbb{M}$.

This definition, although completely natural, is rather abstract. To describe the restricted spectral family explicitly, we need some more notions.

**Definition 5.6.5.** Let $a \in \mathbb{L}$. Then, $c_\mathbb{M}(a) := \bigvee\{b \in \mathbb{M} \mid b \leq a\}$ is called the $\mathbb{M}$-core of $a$ and $s_\mathbb{M}(a) := \bigwedge\{b \in \mathbb{M} \mid b \geq a\}$ is called the $\mathbb{M}$-support of $a$.

**Remark 5.6.2.** There is a simple relation between core and support of an element $a \in \mathbb{L}$:

$$c_\mathbb{M}(a^\perp) = s_\mathbb{M}(a)^\perp.$$

**Lemma 5.6.3.** Let $E = (E_\lambda)_{\lambda \in \mathbb{R}}$ be a spectral family in $\mathcal{R}$ and for $\lambda \in \mathbb{R}$, define

$$(c_\mathbb{M} E)_\lambda := c_\mathbb{M}(E_\lambda), \quad (s_\mathbb{M} E)_\lambda := \bigwedge_{\mu > \lambda} s_\mathbb{M}(E_\mu).$$

Then, $c_\mathbb{M} E := ((c_\mathbb{M} E)_\lambda)_{\lambda \in \mathbb{R}}$ and $s_\mathbb{M} E := ((s_\mathbb{M} E)_\lambda)_{\lambda \in \mathbb{R}}$ are spectral families in $\mathbb{M}$.

**PROOF.** If $\lambda < \mu$, then $c_\mathbb{M}(E_\lambda) \leq E_\lambda \leq E_\mu$ and, therefore, $c_\mathbb{M}(E_\lambda) \leq c_\mathbb{M}(E_\mu)$. Moreover, $\bigwedge_{\mu > \lambda} c_\mathbb{M}(E_\mu) \leq \bigwedge_{\mu > \lambda} E_\mu = E_\lambda$; hence,

$$\bigwedge_{\mu > \lambda} c_\mathbb{M}(E_\mu) \leq c_\mathbb{M}(E_\lambda) \leq \bigwedge_{\mu > \lambda} c_\mathbb{M}(E_\mu).$$

The other assertions are obvious. Note, however, that $\lambda \mapsto s_\mathbb{M}(E_\lambda)$ is not a spectral family in general! ∎

**Definition 5.6.6.** Let $\mathcal{J}$ be a dual ideal in $\mathbb{M}$. Then,

$$C_\mathbb{L}(\mathcal{J}) := \{b \in \mathbb{L} \mid \exists a \in \mathcal{J} : a \leq b\}$$

is called the *cone over* $\mathcal{J}$ in $\mathbb{L}$. It is the smallest dual ideal in $\mathbb{L}$ that contains $\mathcal{J}$.

**Proposition 5.6.4.** If $f : \mathbb{L} \setminus \{0\} \to \mathbb{R}$ is an observable function, then

$$\varrho_\mathbb{M} f(\mathcal{I}) = f(C_\mathbb{L}(\mathcal{I}))$$

for all dual ideals $\mathcal{I}$ in $\mathbb{M}$.

**PROOF.** Because $f$ is an increasing function on $\mathbb{L} \setminus \{0\}$, the assertion follows from the abstract characterization of observable functions:

$$f(C_\mathbb{L}(\mathcal{I})) = \inf\{f(b) \mid b \in C_\mathbb{L}(\mathcal{I})\} = \inf\{f(a) \mid a \in \mathcal{I}\} = \varrho_\mathbb{M} f(\mathcal{I}). \quad \blacksquare$$

Note that we could equally well choose this property as the definition of the restriction of $f : \mathcal{D}(\mathbb{L}) \to \mathbb{R}$ to an observable function on $\mathcal{D}(\mathbb{M})$.

**Corollary 5.6.1.** The spectral family $\varrho_\mathbb{M} E$ is given by

$$(\varrho_\mathbb{M} E)_\lambda = c_\mathbb{M}(E_\lambda).$$

**PROOF.** It is enough to show that the observable functions of $\varrho_\mathbb{M} E$ and $c_\mathbb{M} E$ coincide. Because $f_{\varrho_\mathbb{M} E}(\mathcal{I}) = f_E(C_\mathbb{L}(\mathcal{I})$ for every $\mathcal{I} \in \mathcal{D}(\mathbb{M})$, we obtain

$$f_{\varrho_\mathbb{M} E}(\mathcal{I}) = \inf\{\lambda \mid E_\lambda \in C_\mathbb{L}(\mathcal{I})\}$$
$$= \inf\{\lambda \mid \exists a \in \mathcal{I} : a \leq E_\lambda\}$$
$$= \inf\{\lambda \mid c_\mathbb{M}(E_\lambda) \in \mathcal{I}\}. \quad \blacksquare$$

The spectral family $s_\mathbb{M} E = (\bigwedge_{\mu > \lambda} s_\mathbb{M} E_\mu)_{\lambda \in \mathbb{R}}$ suggests another restriction of spectral families: $E \mapsto s_\mathbb{M} E$. This type of restriction can be derived from the concept of *antonymous functions*. These were introduced in [11] for self-adjoint operators $A \in \mathcal{L}(\mathcal{H})$, but their definition is the same for spectral families $E \in \mathcal{E}(\mathbb{L})$:

$$\forall \mathcal{J} \in \mathcal{D}(\mathbb{L}) : g_E(\mathcal{J}) := \sup\{\lambda \in \mathbb{R} \mid E_\lambda^\perp \in \mathcal{J}\}.$$

There is a simple relation between antonymous functions and observable functions:

**Remark 5.6.3.**
$$g_E = -f_{\neg E},$$
where $\neg E := (\bigwedge_{\mu > \lambda} E^\perp_{-\mu})_{\lambda \in \mathbb{R}}$ is the *complement* of $E \in \mathcal{E}(\mathbb{L})$.

PROOF. Using elementary properties of sup and inf, we obtain for all dual ideals $\mathcal{J} \in \mathcal{D}(\mathbb{L})$:

$$\begin{aligned} g_E(\mathcal{J}) &= \sup\{\lambda \in \mathbb{R} \mid E^\perp_\lambda \in \mathcal{J}\} \\ &= \sup\{-\lambda \in \mathbb{R} \mid E^\perp_{-\lambda} \in \mathcal{J}\} \\ &= -\inf\{\lambda \in \mathbb{R} \mid E^\perp_{-\lambda} \in \mathcal{J}\} \\ &= -\inf\{\lambda \in \mathbb{R} \mid \bigwedge_{\mu > \lambda} E^\perp_{-\mu} \in \mathcal{J}\} \\ &= -f_{\neg E}(\mathcal{J}). \end{aligned}$$ ∎

If $\mathcal{R}$ is a von Neumann algebra and if $E$ is the spectral family of $A \in \mathcal{R}_{sa}$, then $\neg E$ is the spectral family of $-A$. The foregoing relation between antonymous functions and observable functions then reads:

$$g_A = -f_{-A}.$$

Therefore, we prefer to call $g_A$, as well as $g_E$, the *mirrored observable function* of $A$ and $E$, respectively.

We denote by $\mathcal{O}(\mathbb{L})$ the set of all observable functions on $\mathcal{Q}(\mathbb{L})$ (or $\mathcal{D}(\mathbb{L})$ or $\mathbb{L} \setminus \{0\}$). Then, according to Remark 5.6.3, the set of all mirrored observable functions is $-\mathcal{O}(\mathbb{L})$, and it follows easily from the topological characterization of observable functions that $\mathcal{O}(\mathbb{L}) = -\mathcal{O}(\mathbb{L})$ if and only if $\mathbb{L}$ is a distributive lattice. In particular, also the mirrored observable functions on $\mathcal{Q}(\mathcal{R})$ coincide with the Gelfand transforms of self-adjoint elements of an abelian von Neumann algebra $\mathcal{R}$. But note that for a distributive complete orthomodular lattice $\mathbb{L}$, we have $g_E = f_E$ ($E \in \mathcal{E}(\mathbb{L})$) **as functions on $\mathcal{Q}(\mathbb{L})$, but not as functions on $\mathcal{D}(\mathbb{L})$**!

Mirrored observable functions have an abstract characterization that is quite analogous to that of observable functions:

**Theorem 5.6.1.** A function $g : \mathcal{D}(\mathbb{L}) \to \mathbb{R}$ is a mirrored observable function if and only if it satisfies the following two conditions:

1. $g(\mathcal{J}) = \sup_{a \in \mathcal{J}} g(H_a)$ for all dual ideals $\mathcal{J}$ in $\mathcal{D}(\mathbb{L})$.
2. $g(\bigcap_{k \in \mathbb{K}} \mathcal{J}_k) = \inf_{k \in \mathbb{K}} g(\mathcal{J}_k)$ for all families $(\mathcal{J}_k)_{k \in \mathbb{K}}$ in $\mathcal{D}(\mathbb{L})$.

This result follows immediately from the abstract characterization of observable functions using Remark 5.6.3. In particular, mirrored observable functions can be described by *completely decreasing* functions $s : \mathbb{L} \setminus \{0\} \to \mathbb{R}$—that is, by bounded functions that satisfy

$$s(\bigvee_{k \in \mathbb{K}} a_k) = \inf_{k \in \mathbb{K}} s(a_k)$$

for all families $(a_k)_{k \in \mathbb{K}}$ in $\mathbb{L} \setminus \{0\}$.

We can define a restriction of bounded spectral families, $\sigma_\mathbb{M} : \mathcal{E}(\mathbb{L}) \to \mathcal{E}(\mathbb{M})$ by restricting completely decreasing functions on $\mathbb{L} \setminus \{0\}$ to functions on $\mathbb{M} \setminus \{0\}$. It can be shown [9] that this new restriction of bounded spectral families in $\mathbb{L}$ coincides on the level of spectral families with their $E \mapsto s_\mathbb{M} E$ and that it is closely related to the restriction $\varrho_\mathbb{M}$:

$$\sigma_\mathbb{M} E = \neg(\varrho_\mathbb{M}(\neg E)).$$

Finally, we emphasize that $\varrho_\mathbb{M} E$ and $\sigma_\mathbb{M} E$ are the best approximations of $E$ by spectral families in $\mathbb{M}$ from above and from below, respectively, with respect to the spectral order:

**Proposition 5.6.5 [9].** Let $\mathbb{M}$ be a complete orthomodular sublattice of the orthomodular lattice $\mathbb{L}$. Then,

1. $\varrho_\mathbb{M} E = \bigwedge \{F \in \mathcal{E}(\mathbb{M}) \mid E \leq_{sp} F\}$ and
2. $\sigma_\mathbb{M} E = \bigvee \{F \in \mathcal{E}(\mathbb{M}) \mid E \geq_{sp} F\}$.

for all $E \in \mathcal{E}(\mathbb{L})$.

We consider now the particular case of an abelian von Neumann algebra $\mathcal{R}$ and a von Neumann subalgebra $\mathcal{M}$ of $\mathcal{R}$. In this case, the observable function $f_A$ of a self-adjoint element $A \in \mathcal{R}$ is the Gelfand transform of $A$.

**Proposition 5.6.6.** Let $\mathcal{R}$ be an abelian von Neumann algebra and $\mathcal{M}$ a von Neumann subalgebra of $\mathcal{R}$. Then, for all $A \in \mathcal{R}_{sa}$ and for all $\beta \in \mathcal{Q}(\mathcal{A})$,

$$\varrho_\mathcal{M} f_A(\beta) = \sup_{\gamma \in \zeta_\mathcal{M}^{-1}(\beta)} f_A(\gamma)$$

and

$$\sigma_\mathcal{M} f_A(\beta) = \inf_{\gamma \in \zeta_\mathcal{M}^{-1}(\beta)} f_A(\gamma).$$

PROOF. The *proof* rests on the fact that in a distributive lattice, any dual ideal is the intersection of the maximal dual ideals that contain it [9]:

$$\varrho_\mathcal{M} f_A(\beta) = f_A(C_\mathcal{R}(\beta))$$
$$= f_A(\bigcap \{\gamma \in \mathcal{Q}(\mathcal{R}) \mid \beta \subseteq \gamma\})$$
$$= \sup \{f_A(\gamma) \mid \beta \subseteq \gamma\}.$$

Similarly,

$$\sigma_\mathcal{M} f_A(\beta) = \sigma_\mathcal{M} g_A(\beta)$$
$$= g_A(C_\mathcal{R}(\beta))$$
$$= g_A(\bigcap \{\gamma \in \mathcal{Q}(\mathcal{R}) \mid \beta \subseteq \gamma\})$$
$$= \inf \{f_A(\gamma) \mid \beta \subseteq \gamma\},$$

where we have used that $f_A = g_A$ on quasipoints. ∎

### 5.6.2.2 The Algebraic Presheaf

Now let $\mathfrak{A}(\mathbb{L})$ be the set of all complete Boolean subalgebras of the complete orthomodular lattice $\mathbb{L}$. We consider $\mathfrak{A}(\mathbb{L})$ as a (small) category, whose morphisms are simply the inclusions. Let $\mathbb{D}, \mathbb{E} \in \mathfrak{A}(\mathbb{L})$ such that $\mathbb{D} \subseteq \mathbb{F}$. The restrictions $\varrho_{\mathbb{D}}^{\mathbb{E}} : \mathcal{E}(\mathbb{E}) \to \mathcal{E}(\mathbb{D})$ have the properties

$$\varrho_{\mathbb{D}}^{\mathbb{D}} = id_{\mathcal{E}(\mathbb{D})}, \quad \varrho_{\mathbb{D}}^{\mathbb{E}} \circ \varrho_{\mathbb{E}}^{\mathbb{F}} = \varrho_{\mathbb{D}}^{\mathbb{F}} \quad \text{for} \quad \mathbb{D} \subseteq \mathbb{E} \subseteq \mathbb{F}.$$

Therefore, the set $\{\mathcal{E}(\mathbb{D}) \mid \mathbb{E} \in \mathfrak{A}(\mathbb{L})\}$, together with the restriction maps $\varrho_{\mathbb{D}}^{\mathbb{E}} : \mathcal{E}(\mathbb{E}) \to \mathcal{E}(\mathbb{D})$ ($\mathbb{D} \subseteq \mathbb{E}$), forms a presheaf $\mathcal{O}_{\mathbb{L}}^+$; that is, a contravariant functor $\mathcal{O}_{\mathbb{L}}^+ : \mathfrak{A}(\mathbb{L}) \to Sets$, called the *upper observable presheaf*. Similarly, we obtain the *lower observable presheaf*, $\mathcal{O}_{\mathbb{L}}^-$, by choosing the maps $\sigma_{\mathbb{D}}^{\mathbb{E}} : \mathcal{E}(\mathbb{E}) \to \mathcal{E}(\mathbb{D})$ as restrictions. The relation

$$g_E = -f_{\neg E}$$

implies

$$\sigma_{\mathbb{D}}^{\mathbb{E}}(E) = \neg \varrho_{\mathbb{D}}^{\mathbb{E}}(\neg E)$$

for all $E \in \mathcal{E}(\mathbb{E})$; hence, the mapping $\neg : E \mapsto \neg E$ induces an isomorphism $\neg : \mathcal{O}_{\mathbb{L}}^+ \to \mathcal{O}_{\mathbb{L}}^-$ of presheaves over $\mathfrak{A}(\mathbb{L})$. Therefore, it suffices to consider only one of them. Because we are concerned mainly with observable functions, it is natural to consider $\mathcal{O}_{\mathbb{L}}^+$. We call this presheaf simply the *observable presheaf of* $\mathbb{L}$ and denote it by $\mathcal{O}_{\mathbb{L}}$. An important special case is given by $\mathbb{L} = \mathcal{P}(\mathcal{R})$ for a von Neumann algebra $\mathcal{R}$. Here, the objects of the presheaf $\mathcal{O}_{\mathcal{P}(\mathcal{R})}$ can be regarded as the self-adjoint elements of $\mathcal{R}$.

### 5.6.3 Contextual Observables and Contextual Physical Quantities

#### 5.6.3.1 Contextual Observables

**Definition 5.6.7.** Let $\mathbf{C}$ be a category and $\mathcal{S} : \mathbf{C} \to \mathbf{Set}$ a presheaf; that is, a contravariant functor from $\mathbf{C}$ to the category $\mathbf{Set}$ of sets. A global section of $\mathcal{S}$ assigns to every object $a$ of $\mathbf{C}$ an element $\sigma(a)$ of the set $\mathcal{S}(a)$ such that for every morphism $\varphi : b \to a$ of $\mathbf{C}$

$$\sigma(b) = \mathcal{S}(\varphi)(\sigma(a))$$

holds.

Not every presheaf admits global sections. An important example is the **spectral presheaf** of the von Neumann algebra $\mathcal{R}$. This is the presheaf $\mathcal{S} : \mathfrak{A}(\mathcal{R}) \to \mathbf{CO}$ from the category $\mathfrak{A}(\mathcal{R})$ of abelian von Neumann subalgebras of $\mathcal{R}$ to the category $\mathbf{CO}$ of compact Hausdorff spaces, which is defined by

1. $\mathcal{S}(\mathcal{A}) := \mathcal{Q}(\mathcal{A})$ for all $\mathcal{A} \in \mathfrak{A}(\mathcal{R})$,
2. $\mathcal{S}(\mathcal{A} \hookrightarrow \mathcal{B}) := \pi_{\mathcal{A}}^{\mathcal{B}}$, where the mapping $\pi_{\mathcal{A}}^{\mathcal{B}} : \mathcal{Q}(\mathcal{B}) \to \mathcal{Q}(\mathcal{A})$ is defined by $\beta \mapsto \beta \cap \mathcal{A}$.

We know from Theorem 5.2.1 that there is a canonical homeomorphism $\omega_{\mathcal{A}} : \mathcal{Q}(\mathcal{A}) \to \Omega(\mathcal{A})$. This homeomorphism intertwines the ordinary restriction

$$r_{\mathcal{A}}^{\mathcal{B}} : \Omega(\mathcal{B}) \to \Omega(\mathcal{A})$$
$$\tau \mapsto \tau|_{\mathcal{A}}$$

with $\pi_{\mathcal{A}}^{\mathcal{B}}$:

$$r_{\mathcal{A}}^{\mathcal{B}} \circ \omega_{\mathcal{B}} = \omega_{\mathcal{A}} \circ \pi_{\mathcal{A}}^{\mathcal{B}}.$$

This shows, according to a reformulation of the Kochen–Specker theorem by Hamilton, Isham, and Butterfield [10, 15] that the presheaf $\mathcal{S} : \mathfrak{A}(\mathcal{R}) \to \mathbf{CO}$ admits no global sections.

In the case of the observable presheaf $\mathcal{O}_\mathcal{R}$ of self-adjoint elements of the von Neumann algebra $\mathcal{R}$, there are plenty of global sections because each $A \in \mathcal{R}_{sa}$ induces one. Here, the natural question arises of whether all global sections of $\mathcal{O}_\mathcal{R}$ are induced by self-adjoint elements of $\mathcal{R}$. This is certainly not true if the Hilbert space $\mathcal{H}$ has dimension two. For, in this case, the constraints that define a global section are void; therefore, *any* function on the complex projective line defines a global section of $\mathcal{O}_{\mathcal{L}(\mathcal{H})}$. But Gleason's (or Kochen–Specker's) theorem teaches us that the dimension two is something peculiar. It can be shown, however, that the phenomenon—that there are global sections of $\mathcal{O}_\mathcal{R}$ that are not induced by self-adjoint elements of $\mathcal{R}$—is not restricted to dimension two (see [9]). Döring and Isham [12, 13] used these results (already announced in [4]) for a new access, based on topos theory, to quantum physics.

**Definition 5.6.8.** If $\mathcal{R}$ is a von Neumann algebra, a global section of the observable presheaf $\mathcal{O}_\mathcal{R}$ is called a *contextual observable*.

Of course, these considerations can be generalized to define *contextual spectral families in a complete orthomodular lattice* $\mathbb{L}$ as global sections of the presheaf $\mathcal{O}_\mathbb{L}$.

The analogue of the Gelfand transformation appears here as a presheaf-isomorphism of the observable presheaf onto the Gelfand presheaf:

**Definition 5.6.9.** The *Gelfand presheaf* is a presheaf $\mathcal{G}_\mathcal{R}$ over the category $\mathfrak{A}(\mathcal{R})$, which is defined on objects by

$$\mathcal{G}_\mathcal{R}(\mathcal{A}) := C(\mathcal{Q}(\mathcal{A}), \mathbb{R}),$$

and whose restrictions are defined by

$$\varrho_{\mathcal{A}}^{\mathcal{B}}(\varphi)(\beta) := \sup\{\varphi(\gamma) \mid \gamma \in \mathcal{Q}(\mathcal{B}), \beta \subseteq \gamma\}$$

for all $\mathcal{A}, \mathcal{B} \in \mathfrak{A}(\mathcal{R})$, $\mathcal{A} \subseteq \mathcal{B}$, and all $\varphi \in C(\mathcal{Q}(\mathcal{B}), \mathbb{R})$, $\beta \in \mathcal{Q}(\mathcal{A})$.

To be more precise, one should speak of upper Gelfand presheaf $\mathcal{G}_\mathcal{R}^+$ because there is an equally justified lower Gelfand presheaf, $\mathcal{G}_\mathcal{R}^-$, defined by the restrictions

$$\sigma_{\mathcal{A}}^{\mathcal{B}}\varphi(\beta) := \inf\{\varphi(\gamma) \mid \gamma \in \mathcal{Q}(\mathcal{B}), \beta \subseteq \gamma\}$$

for all $\varphi \in C(\mathcal{Q}(\mathcal{B}), \mathbb{R})$ and $\beta \in \mathcal{Q}(\mathcal{A})$. But, as with the observable presheaves, $\mathcal{G}_\mathcal{R}^-$ is isomorphic to $\mathcal{G}_\mathcal{R}^+$, so we will consider only $\mathcal{G}_\mathcal{R} = \mathcal{G}_\mathcal{R}^+$.

The Gelfand transformation induces an isomorphism of presheaves

$$f_* : \mathcal{O}_\mathcal{R} \to \mathcal{G}_\mathcal{R},$$

which is defined at each level $\mathcal{A} \in \mathfrak{A}(\mathcal{R})$ by the usual Gelfand transformation $A \mapsto f_A$. We call $f_*$ the *contextual Gelfand transformation for* $\mathcal{R}$.

One can complexify all this by writing an arbitrary element $A \in \mathcal{R}$ as $A = A_1 + iA_2$ with self-adjoint $A_1, A_2 \in \mathcal{R}$ and defining $f_A := f_{A_1} + f_{A_2}$ as the observable function of $A$. We can generalize the foregoing results to the non–self-adjoint case by applying the previous constructions to the real and imaginary parts of the operators. Of course, this method is limited to *normal* operators, but this requires only typing work, so we omit it.

### 5.6.3.2 The State Presheaf

As is well known, a state of a von Neumann algebra $\mathcal{R}$, acting on a Hilbert space $\mathcal{H}$, is a positive linear functional $\varphi : \mathcal{R} \to \mathbb{C}$ such that $\varphi(I) = 1$. Each state is continuous with respect to the norm-topology. It is called *regular* (or *normal*) if it is continuous with respect to the weak topology on $\mathcal{R}$. The set of states of $\mathcal{R}$, $\mathcal{S}(\mathcal{R})$, is a convex weak*-compact set.

**Definition 5.6.10.** The contravariant functor $\mathcal{S}_\mathcal{R} : \mathfrak{A}(\mathcal{R}) \to \text{Set}$, defined on objects by

$$\mathcal{S}_\mathcal{R}(\mathcal{A}) := \mathcal{S}(\mathcal{A})$$

and on morphisms by

$$\mathcal{S}_\mathcal{R}(\mathcal{A} \hookrightarrow \mathcal{B})(\varphi) := \varphi_{|\mathcal{A}} \quad \text{for all} \quad \varphi \in \mathcal{S}(\mathcal{B})$$

is called the *state presheaf* of the von Neumann algebra $\mathcal{R}$.

In contrast to the spectral presheaf, the state presheaf has a lot of global sections: every $\varphi \in \mathcal{S}(\mathcal{R})$ induces a global section by restrictions. Using a generalization of Gleason's theorem:

**Theorem 5.6.2 ([18] thm. 12.1).** Let $\mathcal{R}$ be a von Neumann algebra without direct summand of type $I_2$ and let $\mu : \mathcal{P}(\mathcal{R}) \to [0, 1]$ be a finitely additive probability measure on $\mathcal{P}(\mathcal{R})$. Then, $\mu$ can be extended to a unique state of $\mathcal{R}$.

We can prove that global sections of $\mathcal{S}_\mathcal{R}$ are induced by states of $\mathcal{R}$:

**Theorem 5.6.2 [9].** The states of a von Neumann algebra $\mathcal{R}$ without direct summand of type $I_2$ are in one-to-one correspondence to the set $\Gamma(\mathcal{S}_\mathcal{R})$ of global sections of the state presheaf $\mathcal{S}_\mathcal{R}$. This correspondence is given by the bijective map

$$\begin{aligned} \Gamma_\mathcal{R} : \mathcal{S}(\mathcal{R}) &\to \quad \Gamma(\mathcal{S}_\mathcal{R}) \\ \varphi &\mapsto (\varphi|_\mathcal{A})_{\mathcal{A} \in \mathfrak{A}(\mathcal{R})}. \end{aligned}$$

Clearly, the core of this theorem is the surjectivity of the mapping $\Gamma_\mathcal{R}$. We can show that this property implies that each probability measure on $\mathcal{P}(\mathcal{R})$ can be extended to a state of $\mathcal{R}$. Thus, Theorem 5.6.2 is indeed equivalent to Theorem 5.6.2.

**Proposition 5.6.7 [9].** The following properties of a von Neumann algebra $\mathcal{R}$ are equivalent:
1. Every finitely additive measure on $\mathcal{P}(\mathcal{R})$ extends to a state of $\mathcal{R}$.
2. Every global section of $\mathcal{S}_\mathcal{R}$ is induced by a state of $\mathcal{R}$.

### 5.6.3.3 Contextual Physical Quantities

Let $\mathcal{A} \in \mathfrak{A}(\mathcal{R})$ and let

$$\downarrow \mathcal{A} := \{\mathcal{B} \in \mathfrak{A}(\mathcal{R}) \mid \mathcal{B} \subseteq \mathcal{A}\}$$

be the sieve defined by $\mathcal{A}$ in the poset $\mathfrak{A}(\mathcal{R})$. We define a presheaf $\mathcal{F}^{\leq}_{\mathfrak{A}(\mathcal{R})}$ on objects by

$$\mathcal{F}^{\leq}(\mathcal{A}) := \{f : \downarrow \mathcal{A} \to \mathbb{R} \mid f \text{ is order-reversing}\}$$

and on morphisms $\mathcal{A} \hookrightarrow \mathcal{B}$ by

$$\varrho^{\mathcal{B}}_{\mathcal{A}} f := f_{|\downarrow \mathcal{A}}.$$

It is evident that $\mathcal{F}^{\leq}_{\mathfrak{A}(\mathcal{R})}$ is a presheaf over $\mathfrak{A}(\mathcal{R})$.[7]

If $\mathcal{A} \in \mathfrak{A}(\mathcal{R})$ and $\varphi_{\mathcal{A}}$ is a state of $\mathcal{A}$, we denote by $\mu_{\varphi_{\mathcal{A}}}$ as the positive Radon measure on $\mathcal{Q}(\mathcal{A})$, corresponding to $\varphi_{\mathcal{A}}$.

**Definition 5.6.11.** Let $\mathcal{S}_{\mathcal{R}}$ be the state presheaf of the von Neumann algebra $\mathcal{R}$ and let $(A_{\mathcal{A}})_{\mathcal{A} \in \mathfrak{A}(\mathcal{R})}$ be a global section of the (upper) observable presheaf $\mathcal{O}^{+}_{\mathcal{R}}$. Then, for all $\mathcal{A} \in \mathfrak{A}(\mathcal{R})$, the global section $A^{\varrho}$ induces a natural transformation

$$\underline{A^{\varrho}} : \mathcal{S}_{\mathcal{R}} \to \mathcal{F}^{\leq}_{\mathfrak{A}(\mathcal{R})},$$

defined by

$$\underline{A^{\varrho}}_{\mathcal{A}} : \mathcal{S}(\mathcal{A}) \to \mathcal{F}^{\leq}(\mathcal{A})$$
$$\varphi_{\mathcal{A}} \mapsto (\int_{\mathcal{Q}(\mathcal{B})} f_{A_{\mathcal{B}}} d\mu_{\varphi_{\mathcal{B}}})_{\mathcal{B} \hookrightarrow \mathcal{A}}.$$

$\underline{A^{\varrho}}$ is called the *contextual physical quantity* corresponding to $A^{\varrho}$.

Switching to global sections $A^{\sigma}$, based on mirrored observable functions, we get another notion of a contextual physical quantity that is intimately connected with the foregoing one:

$$\underline{A^{\sigma}} : \mathcal{S}_{\mathcal{R}} \to \mathcal{F}^{\geq}_{\mathfrak{A}(\mathcal{R})},$$

where $\mathcal{F}^{\geq}_{\mathfrak{A}(\mathcal{R})}$ denotes the presheaf of order-preserving real-valued functions on subsieves of $\mathfrak{A}(\mathcal{R})$. Moreover, following Döring and Isham, we can vary the quantity-value object in the same way as in [13].

The main advantage of using the state presheaf instead of the spectral presheaf in the definition of a contextual physical quantity is that superposition of states is no more problematic. There arises another problem with the use of the spectral presheaf as the state object: If $\mathcal{A} \in \mathfrak{A}(\mathcal{R})$ and $\beta \in \mathcal{Q}(\mathcal{A})$, then the character corresponding to $\beta$ is regular if and only if $\beta$ is an isolated point of $\mathcal{Q}(\mathcal{A})$. But there are examples of abelian von Neumann algebras whose Stone spectrum has no isolated points.

---

[7] This presheaf was introduced by Döring and Isham in [13] and denoted by $\underline{\mathbb{R}^{\leq}}$.

## 5.7 Concluding Remarks

We have seen that both quantum and classical observables have a dual description: either as spectral families in a certain lattice or as continuous functions on the Stone spectrum of that lattice. These two descriptions of observables are the two sides of the same coin. They are connected by the mapping $f_* : E \mapsto f_E$, which is (essentially) a canonical generalization of the Gelfand transformation. Thus, we have found *two* equivalent common background structures for classical and quantum observables. At first glance, there is a fundamental difference between quantum and classical observables: the lattices corresponding to classical observables are *distributive*, whereas the lattices corresponding to quantum observables are not. But this difference vanishes if we consider quantum observables as global sections of the observable presheaf. In that perspective for quantum observables, the nondistributive lattice $\mathcal{P}(\mathcal{R})$ is replaced by the presheaf $\mathcal{O}_{\mathcal{P}(\mathcal{R})}$ of complete Boolean algebras $\mathcal{P}(\mathcal{A})$ ($\mathcal{A} \in \mathfrak{A}(\mathcal{R})$). In that case, the generalization of the Gelfand transformation appears in the form of a presheaf-isomorphism $f_* : \mathcal{O}_{\mathcal{R}} \to \mathcal{G}_{\mathcal{R}}$. The approach to observables as spectral families in an appropriate lattice seems to be the more general one because it allows consideration of spectral families in an arbitrary $\sigma$-algebra as generalized measurable functions.

There are a lot of open problems in this field of research. We mention only some of them:

1. **Stone spectra of von Neumann algebras.** The Stone spectra of finite von Neumann algebras of type I have been characterized. The characterization of Stone spectra for all other types seems to be a much more difficult problem. We know that the Stone spectrum of an abelian von Neumann algebra is compact. We *conjecture* that also the converse is true: if $P \in \mathcal{P}(\mathcal{R})$, then $\mathcal{Q}_P(\mathcal{R})$ is compact if and only if $P$ is abelian.
2. **The trace problem.** This problem has probably some connections with the foregoing one. If $\mathcal{R}$ is a von Neumann algebra, $\mathcal{M}$ is a von Neumann subalgebra of $\mathcal{R}$, and $\mathfrak{B} \in \mathcal{Q}(\mathcal{R})$, what are the conditions that $\mathfrak{B} \cap \mathcal{M}$ is not only a dual ideal but also even a quasipoint of $\mathcal{P}(\mathcal{M})$? The following subproblem, which is the proper trace problem, shows that there must be some condition: Given $\mathfrak{B} \in \mathcal{Q}(\mathcal{R})$, is there a *maximal abelian von Neumann subalgebra* $\mathcal{M}$ of $\mathcal{R}$ such that $\mathfrak{B} \cap \mathcal{M}$ is a quasipoint of $\mathcal{P}(\mathcal{M})$? If $\mathcal{R}$ is not of type $I_2$, there must be a maximal abelian von Neumann subalgebra $\mathcal{M}$ of $\mathcal{R}$ such that $\mathfrak{B} \cap \mathcal{M} \notin \mathcal{Q}(\mathcal{M})$. For otherwise, $\mathfrak{B}$ would induce a global section of the spectral presheaf of $\mathcal{R}$, contradicting the Kochen–Specker theorem. If $\mathcal{R}$ is a finite von Neumann algebra of type $I_n$, the proper trace problem has a positive answer [8].
3. **Quasipoints as generalized pure states.** If $\mathcal{R}$ is an abelian von Neumann algebra and $A \in \mathcal{R}_{sa}$, the range $f_A(\mathcal{Q}(\mathcal{R}))$ is the spectrum of $A$ and $\mathcal{Q}(\mathcal{R})$ is the set of all pure states of $\mathcal{R}$. Therefore, if $\mathcal{R}$ is an arbitrary von Neumann algebra, it is tempting to consider $\mathfrak{B} \in \mathcal{Q}(\mathcal{R})$ as a sort of generalized pure state. This point of view is supported by the fact that also in the general case $f_A(\mathfrak{B}) \in sp(A)$, so $f_A(\mathfrak{B}) \geq 0$ for $A \geq 0$, and $f_I(\mathfrak{B}) = 1$. But, unfortunately, the mapping $A \mapsto f_A(\mathfrak{B})$ is not linear—except in the abelian case. The problem is to find a connection with the traditional notion of state. Again, there are connections with the proper trace problem.
4. **Stone spectrum of $\mathfrak{A}(\mathcal{R})$.** We have defined several presheaves over the category $\mathfrak{A}(\mathcal{R})$ of abelian von Neumann subalgebras of a von Neumann algebra $\mathcal{R}$. The sheafification

of these presheaves gives sheaves over the Stone spectrum $\mathcal{Q}(\mathfrak{A}(\mathcal{R}))$ of $\mathfrak{A}(\mathcal{R})$. Our knowledge of the structure of $\mathcal{Q}(\mathfrak{A}(\mathcal{R}))$ is rather sparse. A systematic study of this structure should be worthwhile.

5. **Classical observables.** The main open problem in the description of classical observables by spectral families is the characterization of spectral families that are associated with smooth functions on a differentiable manifold.

6. **Observables in general relativity.** We mentioned already that the set of all causally closed subsets of a given spacetime $M$ forms an orthomodular lattice $\mathcal{C}(M)$ [2]. We can define observables for $M$ as (bounded) spectral families in $\mathcal{C}(M)$. The study of spectral families, as well as contextual spectral families in $\mathcal{C}(M)$, could be a substantial application to general relativity.

## Acknowledgments

I dedicate this chapter to my wife Karin. Without her love and care, I would not have been able to finish this work. Many thanks go also to my friend and colleague Prof. Dr. Malte Sieveking for his critical comments.

## References

[1] G. Birkhoff. *Lattice Theory*, American Mathematical Society Colloquium Publications, vol. 25. Providence, RI: AMS, 1974.

[2] H. Casini. The logic of causally closed space-time subsets. arXiv:gr-qc/0205013v2 (2002).

[3] H. F. de Groote. On a canonical lattice structure on the effect algebra of a von Neumann algebra. arXiv:mathph/0410018v2 (2004).

[4] H. F. de Groote. Observables. arXiv:math-ph/0507019v1 (2005).

[5] H. F. de Groote. Observables I: Stone spectra. arXiv:math-ph/0509020 (2005).

[6] H. F. de Groote. Observables II: Quantum observables. arXiv:math-ph/0509075 (2005).

[7] H. F. de Groote. Observables III: Classical observables. arXiv:math-ph/0601011 (2006).

[8] H. F. de Groote. Stone spectra of finite von Neumann algebras of type $I_n$. arXiv:math-ph/0605020 (2006).

[9] H. F. de Groote. Observables IV: The presheaf perspective. arXiv:0708.0677 [math-ph] (2007).

[10] A. Döring. Kochen–Specker theorem for von Neumann algebras. *International Journal of Theoretical Physics*, **44** (2005), 139–60. arXiv:quant-ph/0408106.

[11] A. Döring. Observables as functions: Antonymous functions. arXiv:quant-ph/0510102 (2005).

[12] A. Döring and C. J. Isham. A topos foundation for theories of physics: II. Daseinisation and the liberation of quantum theory. *Journal of Mathematical Physics*, **49** (2008), 053516. arXiv:quant-ph/0703062.

[13] A. Döring and C. J. Isham. A topos foundation for theories of physics: III. The representation of physical quantities with arrows $\breve{\delta}(A) : \underline{\Sigma} \to \underline{\mathbb{R}^{\succeq}}$. *Journal of Mathematical Physics*, **49** (2008), 053517. arXiv:quant-ph/0703064.

[14] G. G. Emch. *Mathematical and Conceptual Foundations of 20th Century Physics*, North-Holland Mathematics Studies, vol. 100. Amsterdam: North Holland, 1984.

[15] J. Hamilton, C. J. Isham, and J. Butterfield. A topos perspective on the Kochen–Specker theorem: III. Von Neumann algebras as the base category. *International Journal of Theoretical Physics*, **39** (2000), 1413–36. arXiv:quant-ph/9911020.

[16] R. V. Kadison and J. R. Ringrose. *Fundamentals of the Theory of Operator Algebras, Vol. III: Special Topics, Elementary Theory—An Exercise Approach*. Boston: Birkhäuser, 1991.
[17] G. Kalmbach. *Orthomodular Lattices*, L. M. S. Monographs, vol. 18. London: Academic Press, 1983.
[18] S. Maeda. Probability measures on projections in von Neumann algebras. *Reviews in Mathematical Physics*, **1** (1990), 235–90.
[19] M. P. Olson. The self-adjoint operators of a von Neumann algebra form a conditionally complete lattice. *Proceedings of the AMS*, **28** (1971), 537–44.
[20] R. Sikorski. *Boolean Algebras*. Springer Verlag, 1964.
[21] M. H. Stone. The theory of representations for Boolean algebras. *Transactions of the AMS*, **40** (1936), 37–111.

# CHAPTER 6
# Bohrification

Chris Heunen, Nicolaas P. Landsman, and Bas Spitters

## 6.1 Introduction

More than a decade ago, Chris Isham proposed a topos-theoretic approach to quantum mechanics, initially in the context of the Consistent Histories approach [56], and subsequently (in collaboration with Jeremy Butterfield) in relationship with the Kochen–Specker Theorem [21–23] (see also [20] with John Hamilton). More recently, jointly with Andreas Döring, Isham expanded the topos approach so as to provide a new mathematical foundation for all of physics [38, 39]. One of the most interesting features of their approach is, in our opinion, the so-called Daseinisation map, which should play an important role in determining the empirical content of the formalism.

Over roughly the same period, in an independent development, Bernhard Banaschewski and Chris Mulvey published a series of papers on the extension of Gelfand duality (which in its usual form establishes a categorical duality between unital commutative C*-algebras and compact Hausdorff spaces; see, e.g., [57], [65]) to arbitrary toposes (with natural numbers object) [6–8]. One of the main features of this extension is that the Gelfand spectrum of a commutative C*-algebra is no longer defined as a space but rather as a locale (i.e., a lattice satisfying an infinite distributive law [57]; see also Section 6.2). Briefly, locales describe spaces through their topologies instead of through their points, and the notion of a locale continues to make sense even in the absence of points (whence the alternative name of *pointfree topology* for the theory of locales). It then becomes apparent that Gelfand duality in the category **Set** of sets and functions is exceptional (compared with the situation in arbitrary toposes), in that the localic Gelfand spectrum of a commutative C*-algebra is spatial (i.e., it is fully described by its points). In the context of constructive mathematics (which differs from topos theory in a number of ways, notably in the latter being impredicative), the work of Banaschewski and Mulvey was taken up by Thierry Coquand [29]. He provided a direct lattice-theoretic description of the localic Gelfand spectrum, which will form the basis of its explicit computation in Section 6.4 to follow. This, finally, led to a completely constructive version of Gelfand duality [30, 32].

The third development that fed the research reported here was the program of relating Niels Bohr's ideas on the foundations of quantum mechanics [13, 14] (and, more generally, the problem of explaining the appearance of the classical world [67]) to the formalism of algebraic quantum theory [66, 68]. Note that this formalism was initially developed in response to the mathematical difficulties posed by quantum field theory [48], but it subsequently turned out to be relevant to a large number of issues in quantum theory, including its axiomatization and its relationship with classical physics [26, 49, 64].

The present work merges these three tracks, which (to the best of our knowledge) so far have been pursued independently. It is based on an ab initio redevelopment of quantum physics in the setting of topos theory, published in a series of papers [24, 53–55] (see also [52]), of which the present chapter forms a streamlined and self-contained synthesis, written with the benefit of hindsight.

Our approach is based on a specific mathematical interpretation of Bohr's "doctrine of classical concepts" [79], which in its original form states, roughly speaking, that the empirical content of a quantum theory is entirely contained in its effects on classical physics. In other words, the quantum world can be seen only through classical glasses. In view of the obscure and wholly unmathematical way of Bohr's writings, it is not a priori clear what this means mathematically, but we interpret this doctrine as follows: all physically relevant information contained in a noncommutative (unital) C*-algebra $A$ (in its role of the algebra of observables of some quantum system) is contained in the family of its commutative unital C*-algebras.

The role of topos theory, then, is to describe this family as a single commutative unital C*-algebra, as follows. Let $\mathcal{C}(A)$ be the poset of all commutative unital C*-algebras of $A$, partially ordered by inclusion. This poset canonically defines the topos $[\mathcal{C}(A), \mathbf{Set}]$ of covariant functors from $\mathcal{C}(A)$ (seen as a category, with a unique arrow from $C$ to $D$ if $C \subseteq D$ and no arrow otherwise) to the category $\mathbf{Set}$ of sets and functions. Perhaps the simplest such functor is the tautological one, mapping $C \in \mathcal{C}(A)$ to $C \in \mathbf{Set}$ (with slight abuse of notation), and mapping an arrow $C \subseteq D$ to the inclusion $C \hookrightarrow D$. We denote this functor by $\underline{A}$ and call it the "Bohrification" of $A$. The point is that $\underline{A}$ is a (unital) commutative C*-algebra internal to the topos $[\mathcal{C}(A), \mathbf{Set}]$ under natural operations and, as such, it has a localic Gelfand spectrum $\underline{\Sigma(A)}$ by the Gelfand duality theorem of Banaschewski and Mulvey aforementioned.

The easiest way to study this locale is by means of its external description [59], which is a locale map $f : \Sigma_A \to \mathcal{C}(A)$ (where the poset $\mathcal{C}(A)$ is seen as a topological space in its Alexandrov topology). Denoting the frame or Heyting algebra associated to $\Sigma_A$ by $\mathcal{O}(\Sigma_A)$, we now identify the (formal) open subsets of $\Sigma_A$, defined as the elements of $\mathcal{O}(\Sigma_A)$, with the atomic propositions about the quantum system $A$. The logical structure of these propositions is then controlled by the Heyting algebra structure of $\mathcal{O}(\Sigma_A)$, so that we have found a quantum analogue of the logical structure of classical physics, the locale $\Sigma_A$ playing the role of a quantum phase space. As in the classical case, this object carries both spatial and logical aspects, corresponding to the locale $\Sigma_A$ and the Heyting algebra (or frame) $\mathcal{O}(\Sigma_A)$, respectively.

The key difference between the classical and the quantum case lies in the fact that $\mathcal{O}(\Sigma_A)$ is non-Boolean whenever $A$ is noncommutative. It has to be emphasized, though, that the lattice $\mathcal{O}(\Sigma_A)$ is always distributive; this makes our intuitionistic approach to quantum logic fundamentally different from the traditional one initiated by Birkhoff and von Neumann [11]. Indeed, if $A$ contains sufficiently many projections (as in the case in which $A$ is a von Neumann algebra or, more generally, a Rickart C*-algebra), then the orthomodular lattice Proj($A$) of projections in $A$ (which is the starting point for quantum logic in the context of algebraic quantum theory [73]) is nondistributive whenever $A$ is noncommutative. This feature of quantum logic leads to a number of problems with its interpretation as well as with its structure as a deductive theory, which are circumvented in our approach (see [54] for a detailed discussion of the conceptual points involved).

The plan of this chapter is as follows. Section 6.2 is a brief introduction to locales and toposes. In Section 6.3, we give a constructive definition of C*-algebras that can be interpreted in any topos and review the topos-valid Gelfand duality theory already mentioned. In Section 6.4, we construct the internal C*-algebra $\underline{A}$ and its localic Gelfand spectrum $\underline{\Sigma}(\underline{A})$, computing the external description $\Sigma_A$ of the latter explicitly. Section 6.5 gives a detailed mathematical comparison of the intuitionistic quantum logic $\mathcal{O}(\Sigma_A)$ with its traditional counterpart Proj($A$). Finally, in Section 6.6, we discuss how a state on $A$ gives rise to a probability integral on $\underline{A}_{\text{sa}}$ within the topos $[\mathcal{C}(A), \mathbf{Set}]$, give our analogue of the Daseinisation map of Döring and Isham, and formulate and compute the associated state-proposition pairing.

## 6.2 Locales and Toposes

This section introduces locales and toposes by summarizing well-known results. Both are generalizations of the concept of topological space, and both also carry logical structures. We start with complete Heyting algebras. These can be made into categories in several ways. We consider a logical, an order theoretical, and a spatial perspective.

**6.2.1 Definition.** A partially ordered set $X$ is called a *lattice* when it has binary joins (least upper bounds, suprema) and meets (greatest lower bounds, infima). It is called a *bounded lattice* when it, moreover, has a least element 0 and a greatest element 1. It is called a *complete lattice* when it has joins and meets of arbitrary subsets of $X$. A bounded lattice $X$ is called a *Heyting algebra* when, regarding $X$ as a category, $(\_) \wedge x$ has a right adjoint $x \Rightarrow (\_)$ for every $x \in X$. Explicitly, a Heyting algebra $X$ comes with a monotone function $\Rightarrow : X^{\text{op}} \times X \to X$ satisfying $x \leq (y \Rightarrow z)$ if and only if $x \wedge y \leq z$.

**6.2.2 Definition.** A *Boolean algebra* is a Heyting algebra in which $\neg\neg x = x$ for all $x$, where $\neg x$ is defined to be $(x \Rightarrow 0)$.

**6.2.3 Definition.** A *morphism of complete Heyting algebras* is a function that preserves the operations $\wedge, \bigvee$, and $\Rightarrow$, as well as the constants 0 and 1. We denote the category of complete Heyting algebras and their morphisms by **CHey**. This gives a logical perspective on complete Heyting algebras.

**6.2.4 Definition.** Heyting algebras are necessarily distributive; that is, $x \wedge (y \vee z) = (x \wedge y) \vee (x \wedge z)$ because $(\_) \wedge x$ has a right adjoint and hence preserves colimits. When a Heyting algebra is complete, arbitrary joins exist, whence the following infinitary distributive law holds:

$$\left(\bigvee_{i \in I} y_i\right) \wedge x = \bigvee_{i \in I} (y_i \wedge x). \tag{6.1}$$

Conversely, a complete lattice that satisfies this infinitary distributive law is a Heyting algebra by defining $y \Rightarrow z = \bigvee \{x \mid x \wedge y \leq z\}$. This gives an order-theoretical perspective on complete Heyting algebras. The category **Frm** of *frames* has complete Heyting algebras as objects; morphisms are functions that preserve finite meets and arbitrary joins. The categories **Frm** and **CHey** are not identical because a morphism of frames does not necessarily have to preserve the Heyting implication.

**6.2.5 Definition.** The category **Loc** of *locales* is the opposite of the category of frames. This gives a spatial perspective on complete Heyting algebras.

**6.2.6 Example.** To see why locales provide a spatial perspective, let $X$ be a topological space. Denote its topology, that is, the collection of open sets in $X$, by $\mathcal{O}(X)$. Ordered by inclusion, $\mathcal{O}(X)$ satisfies Equation (6.1) and is therefore a frame. If $f \colon X \to Y$ is a continuous function between topological spaces, then its inverse image $f^{-1} \colon \mathcal{O}(Y) \to \mathcal{O}(X)$ is a morphism of frames. We can also consider $\mathcal{O}(f) = f^{-1}$ as a morphism $\mathcal{O}(X) \to \mathcal{O}(Y)$ of locales, in the same direction as the original function $f$. Thus, $\mathcal{O}(\_)$ is a covariant functor from the category **Top** of topological spaces and continuous maps to the category **Loc** of locales.

**6.2.7 Convention.** To emphasize the spatial aspect of locales, we follow the convention that a locale is denoted by $X$ and the corresponding frame by $\mathcal{O}(X)$ (whether or not the frame comes from a topological space) [70, 86]. Also, we denote a morphism of locales by $f \colon X \to Y$, and the corresponding frame morphism by $f^{-1} \colon \mathcal{O}(Y) \to \mathcal{O}(X)$ (whether or not $f^{-1}$ is indeed the pullback of a function between topological spaces). A fortiori, we will write $C(X, Y)$ for $\mathbf{Loc}(X, Y) = \mathbf{Frm}(\mathcal{O}(Y), \mathcal{O}(X))$.

**6.2.8** A point $x$ of a topological space $X$ may be identified with a continuous function $1 \to X$, where 1 is a singleton set with its unique topology. Extending this to locales, a *point* of a locale $X$ is a locale map $1 \to X$ or, equivalently, a frame map $\mathcal{O}(X) \to \mathcal{O}(1)$. Here, $\mathcal{O}(1) = \{0, 1\} = \Omega$ is the subobject classifier of **Set**, as we will see in Example 6.2.18.

Likewise, an *open* of a locale $X$ is defined as a locale morphism $X \to S$, where $S$ is the locale defined by the *Sierpinski space* (i.e., $\{0, 1\}$ with $\{1\}$ as the only nontrivial open). The corresponding frame morphism $\mathcal{O}(S) \to \mathcal{O}(X)$ is determined by its value at 1 so that we may consider opens in $X$ as morphisms $1 \to \mathcal{O}(X)$ in **Set**. If $X$ is a genuine topological space and $\mathcal{O}(X)$ its collection of opens, then each such morphism $1 \to \mathcal{O}(X)$ corresponds to an open subset of $X$ in the usual sense.

The set $\mathrm{Pt}(X)$ of points of a locale $X$ may be topologized in a natural way, by declaring its opens to be the sets of the form $\mathrm{Pt}(U) = \{p \in \mathrm{Pt}(X) \mid p^{-1}(U) = 1\}$ for some open $U \in \mathcal{O}(X)$. This defines a functor $\mathrm{Pt} \colon \mathbf{Loc} \to \mathbf{Top}$

[57, Theorem II.1.4]. In fact, there is an adjunction

$$\mathbf{Top} \underset{\mathrm{Pt}}{\overset{\mathcal{O}(\_)}{\rightleftarrows}} \mathbf{Loc}.$$

It restricts to an equivalence between so-called *spatial* locales and *sober* topological spaces. Any Hausdorff topological space is sober [57, Lemma I.1.6].

**6.2.9 Example.** Let $(P, \leq)$ be a partially ordered set. This can be turned into a topological space by endowing it with the *Alexandrov topology*, in which open subsets are upper sets in $P$; principal upper sets form a basis for the topology. The associated locale $\mathrm{Alx}(P) = \mathcal{O}(P)$ thus consists of the upper sets $UP$ in $P$.

If we give a set $P$ the discrete order, then the Alexandrov topology on it is the discrete topology (in which every subset is open), and so $\mathcal{O}(P)$ is just the power set $\mathcal{P}(P)$.

As another example, we now study a way to construct frames (locales) by generators and relations. The generators form a meet-semilattice, and the relations are combined into one suitable so-called covering relation. This technique has been developed in the context of formal topology [77, 78].

**6.2.10 Definition.** Let $L$ be a meet-semilattice. A *covering relation* on $L$ is a relation $\triangleleft \subseteq L \times \mathcal{P}(L)$, written as $x \triangleleft U$ when $(x, U) \in \triangleleft$, satisfying:

(a) If $x \in U$, then $x \triangleleft U$.
(b) If $x \triangleleft U$ and $U \triangleleft V$ (i.e., $y \triangleleft V$ for all $y \in U$), then $x \triangleleft V$.
(c) If $x \triangleleft U$, then $x \wedge y \triangleleft U$.
(d) If $x \triangleleft U$ and $x \triangleleft V$, then $x \triangleleft U \wedge V$ (where $U \wedge V = \{x \wedge y \mid x \in U, y \in V\}$).

**6.2.11 Example.** If $X \in \mathbf{Top}$, then $\mathcal{O}(X)$ has a covering relation defined by $U \triangleleft \mathcal{U}$ iff $U \subseteq \bigcup \mathcal{U}$ (i.e., iff $\mathcal{U}$ covers $U$).

**6.2.12 Definition.** Let $DL$ be the partially ordered set of all lower sets in a meet-semilattice $L$, ordered by inclusion. A covering relation $\triangleleft$ on $L$ induces a closure operation $\overline{(\_)}\colon DL \to DL$; namely, $\overline{U} = \{x \in L \mid x \triangleleft U\}$. We define

$$\mathcal{F}(L, \triangleleft) = \{U \in DL \mid \overline{U} = U\} = \{U \in \mathcal{P}(L) \mid x \triangleleft U \Rightarrow x \in U\}. \tag{6.2}$$

Because $\overline{(\_)}$ is a closure operation and $DL$ is a frame [57, Section 1.2], so is $\mathcal{F}(L, \triangleleft)$.

**6.2.13 Proposition.** The frame $\mathcal{F}(L, \triangleleft)$ is the free frame on a meet-semilattice $L$ satisfying $x \leq \bigvee U$ whenever $x \triangleleft U$ for the covering relation $\triangleleft$. The canonical inclusion $i\colon L \to \mathcal{F}(L, \triangleleft)$, defined by $i(x) = \overline{(\downarrow x)}$, is the universal map satisfying $i(x) \leq \bigvee U$ whenever $x \triangleleft U$. That is, if $f\colon L \to F$ is a morphism of meet-semilattices into a frame $F$ satisfying $f(x) \leq \bigvee f(U)$ if $x \triangleleft U$, then $f$ factors uniquely through $i$:

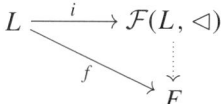

If $f$ generates $F$, in the sense that $V = \bigvee \{f(x) \mid x \in L, f(x) \leq V\}$ for each $V \in F$, there is an isomorphism of frames $F \cong \mathcal{F}(L, \lhd)$ where $x \lhd U$ iff $f(x) \leq \bigvee f(U)$.

**PROOF.** Given $f$, define $g \colon \mathcal{F}(L, \lhd) \to F$ by $g(U) = f(\bigvee U)$. For $x, y \in L$ satisfying $x \lhd \downarrow y$, one then has $f(x) \leq g(\bigvee \downarrow y) = f(y)$, whence $g \circ i(y) = \bigvee \{f(x) \mid x \lhd \downarrow y\} \leq f(y)$. Conversely, $y \lhd \downarrow y$ because $y \in \downarrow y$, so that $f(y) \leq \bigvee \{f(x) \mid x \lhd \downarrow y\} = g \circ i(y)$. Therefore, $g \circ i = f$. Moreover, $g$ is the unique such frame morphism. The second claim is proven in [4, Theorem 12]. □

**6.2.14 Definition.** Let $(L, \lhd)$ and $(M, \blacktriangleleft)$ be meet-semilattices with covering relations. A *continuous map* $f \colon (M, \blacktriangleleft) \to (L, \lhd)$ is a function $f^* \colon L \to \mathcal{P}(M)$ with:

(a) $f^*(L) = M$.
(b) $f^*(x) \wedge f^*(y) \blacktriangleleft f^*(x \wedge y)$.
(c) If $x \lhd U$, then $f^*(x) \blacktriangleleft f^*(U)$ (where $f^*(U) = \bigcup_{u \in U} f^*(U)$).

We identify two such functions if $f_1^*(x) \blacktriangleleft f_2^*(x)$ and $f_2^*(x) \blacktriangleleft f_1^*(x)$ for all $x \in L$.

**6.2.15 Proposition.** Each continuous map $f \colon (M, \blacktriangleleft) \to (L, \lhd)$ is equivalent to a frame morphism $\mathcal{F}(f) \colon \mathcal{F}(L, \lhd) \to \mathcal{F}(M, \blacktriangleleft)$ given by $\mathcal{F}(f)(U) = \overline{f^*(U)}$.

**6.2.16** In fact, the previous proposition extends to an equivalence $\mathcal{F}$ between the category of frames and that of formal topologies, which is a generalization of the previous triples $(L, \leq, \lhd)$, where $\leq$ is merely required to be a preorder. In this more general case, the axioms on the covering relation $\lhd$ take a slightly different form. For this, including the proof of the previous proposition, we refer to [4, 9, 72].

We now generalize the concept of locales by introducing toposes.

**6.2.17** A *subobject classifier* in a category **C** with a terminal object 1 is a monomorphism $\top \colon 1 \to \Omega$ such that for any mono $m \colon M \to X$, there is a unique $\chi_m \colon X \to \Omega$ such that the following diagram is a pullback:

$$\begin{array}{ccc} M & \longrightarrow & 1 \\ {\scriptstyle m} \downarrow & & \downarrow {\scriptstyle \top} \\ X & \xrightarrow{\chi_m} & \Omega. \end{array}$$

Sometimes, the object $\Omega$ alone is referred to as the subobject classifier [58, A1.6]. Hence, a subobject classifier $\Omega$ induces a natural isomorphism $\mathrm{Sub}(X) \cong \mathbf{C}(X, \Omega)$, where the former functor acts on morphisms by pullback, the latter acts by precomposition, and the correspondence is the specific pullback $[m] \mapsto \chi_m$ shown previously.

**6.2.18 Example.** The category **Set** has a subobject classifier $\Omega = \{0, 1\}$, with the morphism $\top \colon 1 \to \Omega$ determined by $\top(*) = 1$.

For any small category **C**, the functor category $[\mathbf{C}, \mathbf{Set}]$ has a subobject classifier, which we now describe. A *cosieve* $S$ on an object $X \in \mathbf{C}$ is a collection of morphisms with domain $X$ such that $f \in S$ implies $g \circ f \in S$ for any morphism $g$ that is composable with $f$. For $X \in \mathbf{C}$, elements of $\Omega(X)$ are the cosieves on $X$ [58, A1.6.6]. On a morphism $f \colon X \to Y$, the action $\Omega(f) \colon \Omega(X) \to \Omega(Y)$ is given by

$$\Omega(f)(S) = \{g \colon Y \to Z \mid Z \in \mathbf{C}, g \circ f \in S\}.$$

Moreover, one has $F \in \mathrm{Sub}(G)$ for functors $F, G \colon \mathbf{C} \rightrightarrows \mathbf{Set}$ if and only $F$ is a *subfunctor* of $G$, in that $F(X) \subseteq G(X)$ for all $X \in \mathbf{C}$.

In the especially easy case that $\mathbf{C}$ is a partially ordered set, seen as a category, a cosieve $S$ on $X$ is just an *upper set* above $X$, in the sense that $Y \in S$ and $Y \leq Z$ imply $Z \in S$ and $X \leq Y$.

**6.2.19 Definition.** A *topos* is a category that has finite limits, exponentials (i.e., right adjoints $(\_)^X$ to $(\_) \times X$), and a subobject classifier (see 6.2.17).

**6.2.20 Example.** The category $\mathbf{Set}$ of sets and functions is a topos: the exponential $Y^X$ is the set of functions $X \to Y$, and the set $\Omega = \{0, 1\}$ is a subobject classifier (see Example 6.2.18).

For any small category $\mathbf{C}$, the functor category $[\mathbf{C}, \mathbf{Set}]$ is a topos. Limits are computed pointwise [15, Theorem 2.15.2], exponentials are defined via the Yoneda embedding [70, Proposition I.6.1], and the cosieve functor $\Omega$ of Example 6.2.18 is a subobject classifier.

**6.2.21 Example.** Without further explanation, let us mention that a *sheaf* over a locale $X$ is a functor from $X^{\mathrm{op}}$ (where the locale $X$ is regarded as a category via its order structure) to $\mathbf{Set}$ that satisfies a certain continuity condition. The category $\mathrm{Sh}(X)$ of sheaves over a locale $X$ is a topos. Its subobject classifier is $\Omega(x) = \mathop{\downarrow} x$ [16, Example 5.2.3].

The categories $\mathrm{Sh}(X)$ and $\mathrm{Sh}(Y)$ are equivalent if and only if the locales $X$ and $Y$ are isomorphic. Thus, toposes are generalizations of locales and hence of topological spaces. Moreover, a morphism $X \to Y$ of locales induces morphisms $\mathrm{Sh}(X) \to \mathrm{Sh}(Y)$ of a specific form: a *geometric morphism* $\mathbf{S} \to \mathbf{T}$ between toposes is a pair of functors $f^* \colon \mathbf{T} \to \mathbf{S}$ and $f_* \colon \mathbf{S} \to \mathbf{T}$, of which $f^*$ preserves finite limits, with $f^* \dashv f_*$. We denote the category of toposes and geometric morphisms by **Topos**.

**6.2.22** If $X$ is the locale resulting from putting the Alexandrov topology on a poset $P$, then $[P, \mathbf{Set}] \cong \mathrm{Sh}(X)$. In this sense, Example 6.2.20 is a special case of Example 6.2.21. We call the category $[P, \mathbf{Set}]$ for a poset $P$ a *Kripke topos*.

One could say that sheaves are the prime example of a topos in that they exhibit its spatial character as a generalization of topology. However, this chapter is primarily concerned with functor toposes and therefore will not mention sheaves again. We now switch to the logical aspect inherent in toposes, by sketching their internal language and its semantics. For a precise description, we refer the reader to [58, Part D], [70, Chapter VI], or [16, Chapter 6].

**6.2.23** In a (cocomplete) topos $\mathbf{T}$, each subobject lattice $\mathrm{Sub}(X)$ is a (complete) Heyting algebra. Moreover, pullback $f^{-1} \colon \mathrm{Sub}(Y) \to \mathrm{Sub}(X)$ along $f \colon X \to Y$

is a morphism of (complete) Heyting algebras. Finally, there are always both left and right adjoints $\exists_f$ and $\forall_f$ to $f^{-1}$. This means that we can write down properties about objects and morphisms in **T** using familiar first-order logic. For example, the formula $\forall_{x \in M} \forall_{y \in M}.x \cdot y = y \cdot x$ makes sense for any object $M$ and morphism $\cdot : M \times M \to M$ in **T** and is interpreted as follows. First, the subformula $x \cdot y = y \cdot x$ is interpreted as the subobject $a : A \rightarrowtail M \times M$ given by the equalizer of $M \times M \xrightarrow{\cdot} M$ and $M \times M \xrightarrow{\gamma} M \times M \xrightarrow{\cdot} M$. Next, the subformula $\forall_{y \in M}.x \cdot y = y \cdot x$ is interpreted as the subobject $b = \forall_{\pi_1}(a) \in \mathrm{Sub}(M)$, where $\pi_1 : M \times M \to M$. Finally, the whole formula $\forall_{x \in M} \forall_{y \in M}.x \cdot y = y \cdot x$ is interpreted as the subobject $c = \forall_\pi(b) \in \mathrm{Sub}(1)$, where $\pi : M \to 1$. The subobject $c \in \mathrm{Sub}(1)$ is classified by a unique $\chi_c : 1 \to \Omega$. This, then, is the *truth value* of the formula. In general, a formula $\varphi$ is said to *hold* in the topos **T**, denoted by $\Vdash \varphi$, when its truth value factors through the subobject classifier $\top : 1 \to \Omega$.

If **T** = **Set**, the subobject $a$ is simply the set $\{(x, y) \in M \times M \mid x \cdot y = y \cdot x\}$; therefore, the truth value of the formula is $1 \in \Omega$ if for all $x, y \in M$ we have $x \cdot y = y \cdot x$, and $0 \in \Omega$ otherwise. But the preceding interpretation can be given in any topos **T**, even if there are few or no "elements" $1 \to M$. Thus, we can often reason about objects in a topos **T** as if they were sets. Indeed, the fact that a topos has exponentials and a subobject classifier means that we can use higher order logic to describe properties of its objects, by interpreting a power set $\mathcal{P}(X)$ as the exponential $\Omega^X$ and the inhabitation relation $\in$ as the subobject of $X \times \Omega^X$ that is classified by the transpose $X \times \Omega^X \to \Omega$ of $\mathrm{id} : \Omega^X \to \Omega^X$. All this can be made precise by defining the *internal* or *Mitchell-Bénabou language* of a topos, which prescribes in detail which logical formulae about the objects and morphisms of a topos are "grammatically correct" and which ones hold.

**6.2.24** The interpretation of the internal language takes an especially easy form in Kripke toposes. We now give this special case of the so-called *Kripke-Joyal semantics*. First, let us write $[\![t]\!]$ for the interpretation of a term $t$ as in 6.2.23. For example, in the notation of 6.2.23, $[\![x]\!]$ is the morphism $\mathrm{id} : M \to M$, and $[\![x \cdot y]\!]$ is the morphism $\cdot : M \times M \to M$. We now inductively define $p \Vdash \varphi(\vec{a})$ for $p \in P$, a formula $\varphi$ in the language of $[P, \mathbf{Set}]$ with free variables $x_i$ of type $X_i$, and $\vec{a} = (a_1, \ldots, a_n)$ with $a_i \in X_i(p)$:

- $p \Vdash (t = t')(\vec{a})$ if and only if $[\![t]\!]_p(\vec{a}) = [\![t']\!]_p(\vec{a})$;
- $p \Vdash R(t_1, \ldots, t_k)(\vec{a})$ if and only if $([\![t_1]\!]_p(\vec{a}), \ldots, [\![t_k]\!](\vec{a})) \in R(p)$, where $R$ is a relation on $X_1 \times \cdots \times X_n$ interpreted as a subobject of $X_1 \times \cdots \times X_n$;
- $p \Vdash (\varphi \wedge \psi)(\vec{a})$ if and only if $p \Vdash \varphi(\vec{a})$ and $p \Vdash \varphi(\vec{a})$;
- $p \Vdash (\varphi \vee \psi)(\vec{a})$ if and only if $p \Vdash \varphi(\vec{a})$ or $p \Vdash \varphi(\vec{a})$;
- $p \Vdash (\varphi \Rightarrow \psi)(\vec{a})$ if and only if $q \Vdash \varphi(\vec{a})$ implies $q \Vdash \psi(\vec{a})$ for all $q \geq p$;
- $p \Vdash \neg\varphi(\vec{a})$ if and only if $q \Vdash \varphi(\vec{a})$ for no $q \geq p$;
- $p \Vdash \exists_{x \in X}.\varphi(\vec{a})$ if and only if $p \Vdash \varphi(a, \vec{a})$ for some $a \in X(p)$;
- $p \Vdash \forall_{x \in X}.\varphi(\vec{a})$ if and only if $q \Vdash \varphi(a, \vec{a})$ for all $q \geq p$ and $a \in X(q)$.

It turns out that $\varphi$ holds in $[P, \mathbf{Set}]$; that is, $\Vdash \varphi$, precisely when $p \Vdash \varphi(\vec{a})$ for all $p \in P$ and all $\vec{a} \in X_1(p) \times \cdots \times X_n(p)$.

**6.2.25** The axioms of intuitionistic logic hold when interpreted in any topos, and there are toposes in whose internal language formulae that are not derivable from the axioms of intuitionistic logic do not hold. For example, the principle of excluded middle $\varphi \vee \neg \varphi$ does not hold in the topos Sh($\mathbb{R}$) [16, 6.7.2]. Thus, we can derive properties of objects of a topos as if they were sets, using the usual higher-order logic, as long as our reasoning is *constructive*, in the sense that we use neither the axiom of choice nor the principle of excluded middle.

The astute reader will have noticed that the account of this chapter up to now has been constructive in this sense (including the material around Proposition 6.2.13). In particular, we can speak of objects in a topos **T** that satisfy the defining properties of locales as *locales within that topos*. Explicitly, these are objects $L$ that come with morphisms $0, 1: 1 \rightrightarrows L$ and $\bigwedge, \bigvee: \Omega^L \rightrightarrows L$ for which the defining formulae of locales, such as (6.1), hold in **T** [16, Section 6.11]. The category of such objects is denoted by Loc(**T**), so that Loc(**Set**) $\cong$ **Loc**. For the rest of this chapter, we also take care to use constructive reasoning whenever we reason in the internal language of a topos.

**6.2.26** We have two ways of proving properties of objects and morphisms in toposes. First, we can take an *external* point of view. This occurs, for example, when we use the structure of objects in [$P$, **Set**] as **Set**-valued functors. Second, we can adopt the *internal* logic of the topos, as previously discussed. In this viewpoint, we regard the topos as a "universe of discourse." At least, intuitionistic reasoning is valid, but more logical laws might hold, depending on the topos one is studying. To end this section, we consider the internal and external points of view in several examples.

**6.2.27 Example.** Let **T** be a topos and $X$ an object in it. Externally, one simply looks at Sub($X$) as a set, equipped with the structure of a Heyting algebra *in the category* **Set**. Internally, Sub($X$) is described as the exponential $\Omega^X$, or $\mathcal{P}(X)$, which is a Heyting algebra object *in the topos* **T** [70, p. 201].

**6.2.28 Example.** For any poset $P$, the category Loc([$P$, **Set**]) is equivalent to the slice category **Loc**/Alx($P$) of locale morphisms $L \to$ Alx($P$) from some locale $L$ to the Alexandrov topology on $P$ (by 6.2.22 and [59]). Therefore, an internal locale object $\underline{L}$ in [$P$, **Set**] is described externally as a locale morphism $f: L \to$ Alx($P$), determined as follows. First, $\mathcal{O}(\underline{L})(P)$ is a frame in **Set**, and for $U$ in Alx($P$), the action $\mathcal{O}(\underline{L})(P) \to \mathcal{O}(\underline{L})(U)$ on morphisms is a frame morphism. Because $\mathcal{O}(\underline{L})$ is complete, there is a left adjoint $l_U^{-1}: \mathcal{O}(\underline{L})(U) \to \mathcal{O}(\underline{L})(P)$, which in turn defines a frame morphism $f^{-1}: \mathcal{O}(\text{Alx}(P)) \to \mathcal{O}(\underline{L})(P)$ by $f^{-1}(U) = l_U^{-1}(1)$. Taking $L = \mathcal{O}(\underline{L})(P)$ then yields the desired locale morphism.

**6.2.29 Example.** Let $L$ be a locale object in the Kripke topos over a poset $P$. Internally, a point of $L$ is a locale morphism $1 \to L$, which is the same thing as an internal frame morphism $\mathcal{O}(L) \to \Omega$. Externally, one looks at $\Omega$ as the frame Sub(1) in **Set**. Because Sub(1) $\cong \mathcal{O}(\text{Alx}(P))$ in [$P$, **Set**], one finds Loc([$P$, **Set**]) $\cong$ **Loc**/Alx($P$). By Example 6.2.28, $L$ has an external description as a locale morphism $f\colon K \to L$, so that points in $L$ are described externally by sections of $f$; that is, locale morphisms $g\colon L \to K$ satisfying $f \circ g = \mathrm{id}$.

**6.2.30** Locales already possess a logical aspect as well as a spatial one because the logical perspective on complete Heyting algebras translates to the spatial perspective on locales. Elements $1 \to \mathcal{O}(L)$ of the Heyting algebra $\mathcal{O}(L)$ are the opens of the associated locale $L$, to be thought of as propositions, whereas points of the locale correspond to models of the logical theory defined by these propositions [86].

More precisely, recall that a formula is *positive* when it is built from atomic propositions by the connectives $\wedge$ and $\vee$ only, where $\vee$ but not $\wedge$ is allowed to be indexed by an infinite set. This can be motivated observationally: to verify a proposition $\bigvee_{i \in I} p_i$, one only needs to find a single $p_i$, whereas to verify $\bigwedge_{i \in I} p_i$, the validity of each $p_i$ needs to be established [3], an impossible task in practice when $I$ is infinite. A *geometric formula* then is one of the form $\varphi \Rightarrow \psi$, where $\varphi$ and $\psi$ are positive formulae.

Thus, a frame $\mathcal{O}(L)$ defines a geometric propositional theory whose propositions correspond to opens in $L$, combined by logical connectives given by the lattice structure of $\mathcal{O}(L)$. Conversely, a propositional geometric theory $\mathfrak{T}$ has an associated *Lindenbaum algebra* $\mathcal{O}([\mathfrak{T}])$, defined as the poset of formulae of $\mathfrak{T}$ modulo provable equivalence, ordered by entailment. This poset turns out to be a frame, and the set-theoretical models of $\mathfrak{T}$ bijectively correspond to frame morphisms $\mathcal{O}([\mathfrak{T}]) \to \{0, 1\}$. Identifying $\{0, 1\}$ in **Set** with $\Omega = \mathcal{O}(1)$, one finds that a model of the theory $\mathfrak{T}$ is a point $1 \to [\mathfrak{T}]$ of the locale $[\mathfrak{T}]$. More generally, by Example 6.2.28, one may consider a model of $\mathfrak{T}$ in a frame $\mathcal{O}(L)$ to be a locale morphism $L \to [\mathfrak{T}]$.

**6.2.31 Example.** Consider models of a geometric theory $\mathfrak{T}$ in a topos **T**. Externally, these are given by locale morphisms $\mathrm{Loc}(\mathbf{T}) \to [\mathfrak{T}]$ [70, Theorem X.6.1 and Section IX.5]. One may also interpret $\mathfrak{T}$ in **T** and thus define a locale $[\mathfrak{T}]_\mathbf{T}$ internal to **T**. The points of this locale—that is, the locale morphisms $1 \to [\mathfrak{T}]_\mathbf{T}$ or frame morphisms $\mathcal{O}([\mathfrak{T}]_\mathbf{T}) \to \Omega$—describe the models of $\mathfrak{T}$ in **T** internally.

**6.2.32 Example.** Several important internal number systems in Kripke toposes are defined by geometric propositional theories $\mathfrak{T}$ and can be computed via Examples 6.2.21 and 6.2.22. Externally, the frame $\mathcal{O}([\mathfrak{T}])$ corresponding to the interpretation of $\mathfrak{T}$ in $[P, \mathbf{Set}]$ is given by the functor $\mathcal{O}([\mathfrak{T}])\colon p \mapsto \mathcal{O}(\uparrow p \times [\mathfrak{T}])$ [24, Appendix A].

**6.2.33 Example.** As an application of the previous example, we recall an explicit construction of the *Dedekind real numbers* (see [41] or [58, D4.7.4]). Define the propositional geometric theory $\mathfrak{T}_\mathbb{R}$ generated by formal symbols $(q, r) \in \mathbb{Q} \times \mathbb{Q}$ with $q < r$, ordered as $(q, r) \leq (q', r')$ iff $q' \leq q$ and $r \leq r'$, subject to the following relations:

$$(q_1, r_1) \wedge (q_2, r_2) = \begin{cases} (\max(q_1, q_2), \min(r_1, r_2)) & \text{if } \max(q_1, q_2) < \min(r_1, r_2) \\ 0 & \text{otherwise} \end{cases}$$

$$(q, r) = \bigvee \{(q', r') \mid q < q' < r' < r\}$$

$$1 = \bigvee \{(q, r) \mid q < r\}$$

$$(q, r) = (q, r_1) \vee (q_1, r) \qquad \text{if } q \leq q_1 \leq r_1 \leq r.$$

This theory may be interpreted in any topos **T** with a natural numbers object, defining an internal locale $\mathbb{R}_\mathbf{T}$. Points $p$ of $\mathbb{R}_\mathbf{T}$—that is, frame morphisms $p^{-1}\colon \mathcal{O}(\mathbb{R}_\mathbf{T}) \to \Omega$—correspond to Dedekind cuts $(L, U)$ by [70, p. 321]:

$$L = \{q \in \mathbb{Q} \mid p \models (q, \infty)\};$$
$$U = \{r \in \mathbb{Q} \mid p \models (-\infty, r)\},$$

where $(q, \infty)$ and $(-\infty, r)$ are defined in terms of the formal generators of the frame $\mathcal{O}(\mathbb{Q})$ by $(q, \infty) = \bigvee\{(q, r) \mid q < r\}$ and $(-\infty, r) = \bigvee\{(q, r) \mid q < r\}$. The notation $p \models (q, r)$ means that $m^{-1}(q, r)$ is the subobject classifier $\top \colon 1 \to \Omega$, where $(q, r)$ is seen as a morphism $1 \to \mathbb{Q} \times \mathbb{Q} \to \mathcal{O}(\mathbb{R}_\mathbf{T})$. Conversely, a Dedekind cut $(L, U)$ uniquely determines a point $p$ by $(q, r) \mapsto \top$ iff $(q, r) \cap U \neq \emptyset$ and $(q, r) \cap L \neq \emptyset$. The Dedekind real numbers are therefore defined in any topos **T** as the subobject of $\mathcal{P}(\mathbb{Q}_\mathbf{T}) \times \mathcal{P}(\mathbb{Q}_\mathbf{T})$ consisting of those $(L, U)$ that are points of $\mathbb{R}_\mathbf{T}$.

One may identify $\mathrm{Pt}(\mathbb{R}_\mathbf{Set})$ with the field $\mathbb{R}$ in the usual sense and $\mathcal{O}(\mathbb{R}_\mathbf{Set})$ with the usual Euclidean topology on $\mathbb{R}$.

In case $\mathbf{T} = [P, \mathbf{Set}]$ for a poset $P$, one finds that $\mathcal{O}(\mathbb{R}_\mathbf{T})$ is the functor $p \mapsto \mathcal{O}(\uparrow p \times \mathbb{R}_\mathbf{Set})$; cf. Example 6.2.32. The latter set may be identified with the set of monotone functions $\uparrow p \to \mathcal{O}(\mathbb{R}_\mathbf{Set})$. When $P$ has a least element, the functor $\mathrm{Pt}(\mathbb{R}_\mathbf{T})$ may be identified with the constant functor $p \mapsto \mathbb{R}_\mathbf{Set}$.

## 6.3 C*-Algebras

This section considers a generalization of the concept of topological space different from locales and toposes—namely, so-called C*-algebras [36, 60, 85]. These operator algebras also play a large role in quantum theory [48, 64, 81]. We first give a constructive definition of C*-algebras that can be interpreted in any topos (with a natural numbers object), after [6–8].

**6.3.1** In any topos (with a natural numbers object), the rationals $\mathbb{Q}$ can be interpreted [70, Section VI.8], as can the *Gaussian rationals* $\mathbb{C}_\mathbb{Q} = \{q + ri \mid q, r \in \mathbb{Q}\}$. For example, the interpretation of $\mathbb{C}_\mathbb{Q}$ in the Kripke topos over a poset $P$ is the constant functor that assigns the set $\mathbb{C}_\mathbb{Q}$ to each $p \in P$.

**6.3.2** A monoid in $\mathbf{Vect}_K$ for some $K \in \mathbf{Fld}$ is called a (unital) *K-algebra*—not to be confused with Eilenberg–Moore algebras of a monad. It is called *commutative* when the multiplication of its monoid structure is. A *\*-algebra* is an algebra $A$ over an involutive field, together with an antilinear involution $(\_)^*: A \to A$.

**6.3.3 Definition.** A *seminorm* on a *-algebra $A$ over $\mathbb{C}_\mathbb{Q}$ is a relation $N \subseteq A \times \mathbb{Q}^+$ satisfying

$$(0, p) \in N,$$
$$\exists_{q \in \mathbb{Q}^+}.(a, q) \in N,$$
$$(a, q) \in N \Rightarrow (a^*, q) \in N,$$
$$(a, r) \in N \iff \exists_{q < r}.(a, q) \in N,$$
$$(a, q) \in N \land (b, r) \in N \Rightarrow (a + q, p + r) \in N,$$
$$(a, q) \in N \land (b, r) \in N \Rightarrow (ab, qr) \in N,$$
$$(a, q) \in N \Rightarrow (za, qr) \in N \qquad (|z| < r),$$
$$(1, q) \in N \qquad (q > 1),$$

for all $a, b \in A$, $q, r \in \mathbb{Q}^+$, and $z \in \mathbb{C}_\mathbb{Q}$. If this relation furthermore satisfies

$$(a^*a, q^2) \in N \iff (a, q) \in N$$

for all $a \in A$ and $q \in \mathbb{Q}^+$, then $A$ is said to be a *pre-semi-C\*-algebra*.

A seminorm $N$ is called a *norm* if $a = 0$ whenever $(a, q) \in N$ for all $q \in \mathbb{Q}^+$. One can then formulate a suitable notion of completeness in this norm that does not rely on the axiom of choice—namely, by considering Cauchy sequences of sets instead of Cauchy sequences [8]. A *C\*-algebra* is a pre-semi-C\*-algebra $A$ whose seminorm is a norm in which $A$ is complete. Notice that a C\*-algebra by definition has a unit; what we defined as a C\*-algebra is sometimes called a unital C\*-algebra in the literature.

A morphism between C\*-algebras $A$ and $B$ is a linear function $f : A \to B$ satisfying $f(ab) = f(a)f(b)$, $f(a^*) = f(a)^*$, and $f(1) = 1$. C\*-algebras and their morphisms form a category **CStar**. We denote its full subcategory of commutative C\*-algebras by **cCStar**.

**6.3.4** Classically, a seminorm induces a norm, and vice versa, by $(a, q) \in N$ if and only if $\|a\| < q$.

**6.3.5** The geometric theory $\mathfrak{T}_\mathbb{R}$ of Example 6.2.33 can be extended to a geometric theory $\mathfrak{T}_\mathbb{C}$ describing the complexified locale $\mathbb{C} = \mathbb{R} + i\mathbb{R}$. There are also direct descriptions that avoid a defining role of $\mathbb{R}$ [8]. In **Set**, the frame $\mathcal{O}(\mathbb{C})$ defined by $\mathfrak{T}_\mathbb{C}$ is the usual topology on the usual complex field $\mathbb{C}$. As a consequence of its completeness, a C\*-algebra is automatically an algebra over $\mathbb{C}$ (and not just over $\mathbb{C}_\mathbb{Q}$, as is inherent in the definition).

**6.3.6 Example.** The continuous linear operators **Hilb**$(H, H)$ on a Hilbert space $H$ form a C\*-algebra. In fact, by the classical Gelfand–Naimark theorem, any C\*-algebra can be embedded into one of this form [45].

**6.3.7 Example.** A locale $X$ is *compact* if every subset $S \subseteq X$ with $\bigvee S = 1$ has a finite subset $F \subseteq S$ with $\bigvee F = 1$. It is *regular* if $y = \bigvee(\downarrow y)$ for all $y \in X$, where

$\downarrow y = \{x \in X \mid x \ll y\}$ and $x \ll y$ iff there is a $z \in X$ with $z \wedge x = 0$ and $z \vee y = 1$. If the axiom of dependent choice is available—as in Kripke toposes [42]—then regular locales are automatically completely regular. Assuming the full axiom of choice, the category **KRegLoc** of compact regular locales in **Set** is equivalent to the category **KHausTop** of compact Hausdorff topological spaces. In general, if $X$ is a completely regular compact locale, then $C(X, \mathbb{C})$ is a commutative C*-algebra. In fact, the following theorem shows that all commutative C*-algebras are of this form. This so-called Gelfand duality justifies regarding C*-algebras as "noncommutative" generalizations of topological spaces [28].

**6.3.8 Theorem** [6–8]. There is an equivalence

$$\mathbf{cCStar} \xrightleftharpoons[C(\_,\mathbb{C})]{\Sigma} \mathbf{KRegLoc}^{\mathrm{op}}.$$

The locale $\Sigma(A)$ is called the *Gelfand spectrum* of $A$. □

The previous theorem is proved in such a way that it applies in any topos. This means that we can give an explicit description of the Gelfand spectrum. The rest of this section is devoted to just that, following the reformulation that is fully constructive [29, 32].

**6.3.9** To motivate the following description, we mention that the classical proof [44, 45] defines $\Sigma(A)$ to be the set of *characters* of $A$ (i.e., nonzero multiplicative functionals $\rho \colon A \to \mathbb{C}$). This set becomes a compact Hausdorff topological space by the sub-base consisting of $\{\rho \in \Sigma(A) \mid |\rho(a) - \rho_0(a)| < \varepsilon\}$ for $a \in A$, $\rho_0 \in \Sigma$, and $\varepsilon > 0$. A much simpler choice of sub-base would be $\mathcal{D}_a = \{\rho \in \Sigma \mid \rho(a) > 0\}$ for $a \in A_{\mathrm{sa}} = \{a \in A \mid a^* = a\}$. Both the property that the $\rho$ are multiplicative and the fact that the $\mathcal{D}_a$ form a sub-base may then be expressed lattice-theoretically by letting $\mathcal{O}(\Sigma(A))$ be the frame freely generated by the formal symbols $\mathcal{D}_a$ for $a \in A_{\mathrm{sa}}$, subject to the relations

$$\mathcal{D}_1 = 1, \tag{6.3}$$

$$\mathcal{D}_a \wedge \mathcal{D}_{-a} = 0, \tag{6.4}$$

$$\mathcal{D}_{-b^2} = 0, \tag{6.5}$$

$$\mathcal{D}_{a+b} \leq \mathcal{D}_a \vee \mathcal{D}_b, \tag{6.6}$$

$$\mathcal{D}_{ab} = (\mathcal{D}_a \wedge \mathcal{D}_b) \vee (\mathcal{D}_{-a} \wedge \mathcal{D}_{-b}), \tag{6.7}$$

supplemented with the "regularity rule"

$$\mathcal{D}_a \leq \bigvee_{r \in \mathbb{Q}^+} \mathcal{D}_{a-r}. \tag{6.8}$$

**6.3.10** Classically, the *Gelfand transform* $A \xrightarrow{\cong} C(\Sigma(A), \mathbb{C})$ is given by $a \mapsto \hat{a}$ with $\hat{a}(\rho) = \rho(a)$, and restricting to $A_{\mathrm{sa}}$ yields an isomorphism $A_{\mathrm{sa}} \cong C(\Sigma(A), \mathbb{R})$. Hence, classically $\mathcal{D}_a = \{\rho \in \Sigma(A) \mid \hat{a}(\rho) > 0\}$. In a constructive setting, we must associate a locale morphism $\hat{a} \colon \Sigma(A) \to \mathbb{R}$ to each $a \in A_{\mathrm{sa}}$, which is, by definition, a frame morphism $\hat{a}^{-1} \colon \mathcal{O}(\mathbb{R}) \to \mathcal{O}(\Sigma(A))$. Aided by the intuition of 6.3.9, one finds that $\hat{a}^{-1}(-\infty, s) = \mathcal{D}_{s-a}$ and $\hat{a}^{-1}(r, \infty) = \mathcal{D}_{a-r}$

for basic opens. Hence, $\hat{a}^{-1}(r, s) = \mathcal{D}_{s-a} \wedge \mathcal{D}_{a-r}$ for rationals $r < s$. By Example 6.2.33, we have $A_{sa} \cong C(\Sigma(A)), \mathbb{R}) = \Gamma(\mathrm{Pt}(\mathbb{R})_{\mathrm{Sh}(\Sigma(A))})$, where $\Gamma$ is the global sections functor. Hence, $A_{sa}$ is isomorphic (through the Gelfand transform) to the global sections of the real numbers in the topos of sheaves on its spectrum (and $A$ itself "is" the complex numbers in the same sense).

**6.3.11** To describe the Gelfand spectrum more explicitly, we start with the distributive lattice $L_A$ freely generated by the formal symbols $\mathrm{D}_a$ for $a \in A_{sa}$, subject to the relations (6.3)–(6.7). Being an involutive ring, $A_{sa}$ has a positive cone $A^+ = \{a \in A_{sa} \mid a \geq 0\} = \{a^2 \mid a \in A_{sa}\}$. (For $A = \mathbf{Hilb}(H, H)$, one has $a \in A^+$ iff $\langle x \mid a(x) \rangle \geq 0$ for all $x \in H$.) The given definition of $A^+$ induces a partial order $\leq$ on $A^+$ by $a \leq b$ iff $0 \leq a - b$, with respect to which $A^+$ is a distributive lattice. Now we define a partial order $\preccurlyeq$ on $A^+$ by $a \preccurlyeq b$ iff $a \leq nb$ for some $n \in \mathbb{N}$. Define an equivalence relation on $A^+$ by $a \approx b$ iff $a \preccurlyeq b$ and $b \preccurlyeq a$. The lattice operations on $A^+$ respect $\approx$ and, hence, $A^+/\approx$ is a lattice. We have

$$L_A \cong A^+/\approx .$$

The image of the generator $\mathrm{D}_a$ in $L_A$ corresponds to the equivalence class $[a^+]$ in $A^+/\approx$, where $a = a^+ - a^-$ with $a^{\pm} \in A^+$ in the usual way. Theorem 6.4.12 shows that the lattice $L_A$ can be computed locally in certain Kripke toposes. In preparation, we now work toward Lemma 6.3.16 to follow.

**6.3.12** Extending the geometric *propositional* logic of 6.2.30, the positive formulae of a geometric *predicate* logic may furthermore involve finitely many free variables and the existential quantifier $\exists$, and its axioms take the form $\forall_{x \in X}.\varphi(x) \Rightarrow \psi(x)$ for positive formulae $\varphi, \psi$. Geometric formulae form an important class of logical formulae because they are precisely the ones whose truth value is preserved by inverse images of geometric morphisms between toposes. From their syntactic form alone, it follows that their external interpretation is determined locally in Kripke toposes, as the following lemma shows.

**6.3.13 Lemma.** See [58, Corollary D1.2.14]. Let $\mathfrak{T}$ be a geometric theory and denote the category of its models in a topos $\mathbf{T}$ by $\mathbf{Model}(\mathfrak{T}, \mathbf{T})$. For any category $\mathbf{C}$, there is a canonical isomorphism of categories $\mathbf{Model}(\mathfrak{T}, [\mathbf{C}, \mathbf{Set}]) \cong [\mathbf{C}, \mathbf{Model}(\mathfrak{T}, \mathbf{Set})]$. □

**6.3.14 Definition.** A *Riesz space* is a vector space $R$ over $\mathbb{R}$ that is simultaneously a distributive lattice, such that $f \leq g$ implies $f + h \leq g + h$ for all $h$, and $f \geq 0$ implies $rf \geq 0$ for all $r \in \mathbb{R}^+$ [69, Definition 11.1].

An *f-algebra* is a commutative $\mathbb{R}$-algebra $R$ whose underlying vector space is a Riesz space in which $f, g \geq 0$ implies $fg \geq 0$, and $f \wedge g = 0$ implies $hf \wedge g = 0$ for all $h \geq 0$. Moreover, the multiplicative unit 1 has to be *strong* in the sense that for each $f \in R$, one has $-n1 \leq f \leq n1$ for some $n \in \mathbb{N}$ [87, Definition 140.8].

**6.3.15 Example.** If $A$ is a commutative C*-algebra, then $A_{sa}$ becomes an f-algebra over $\mathbb{R}$ under the order defined in 6.3.11. Conversely, by the Stone–Yosida representation theorem, every f-algebra over $\mathbb{R}$ can be densely embedded in $C(X, \mathbb{R})$ for some compact locale $X$ [30]. Like commutative C*-algebras, f-algebras have a spectrum, for the definition of which we refer to [32].

**6.3.16 Lemma.** Let $A$ be a commutative C*-algebra.

(a) The Gelfand spectrum of $A$ coincides with the spectrum of the f-algebra $A_{sa}$.
(b) The theory of f-algebras is geometric.

**PROOF.** Part (a) is proven in [32]. For (b), notice that an f-algebra over $\mathbb{Q}$ is precisely a uniquely divisible lattice-ordered ring [29, p. 151] because unique divisibility turns a ring into a $\mathbb{Q}$-algebra. The definition of a lattice-ordered ring can be written using equations only. The theory of torsion-free rings—that is, if $n > 0$ and $nx = 0$, then $x = 0$—is also algebraic. The theory of divisible rings is obtained by adding infinitely many geometric axioms $\exists_y.ny = x$, one for each $n > 0$, to the algebraic theory of rings. Finally, a torsion-free divisible ring is the same as a uniquely divisible ring: if $ny = x$ and $nz = x$, then $n(y - z) = 0$, so that $y - z = 0$. We conclude that the theory of uniquely divisible lattice-ordered rings—that is, f-algebras—is geometric, establishing (b). □

**6.3.17 Proposition.** The lattice $L_A$ generating the spectrum of a commutative C*-algebra $A$ is preserved under inverse images of geometric morphisms.

**PROOF.** By the previous lemma $A_{sa}$ and, hence, $A^+$ are definable by a geometric theory. Because the relation $\approx$ of 6.3.11 is defined by an existential quantification, $L_A \cong A^+/\approx$ is preserved under inverse images of geometric morphisms. □

We now turn to the regularity condition (6.8), which is to be imposed on $L_A$. This condition turns out to be a special case of the relation $\ll$ (see Example 6.3.7).

**6.3.18 Lemma.** For all $D_a, D_b \in L_A$, the following are equivalent:

(a) There exists $D_c$ with $D_c \vee D_a = 1$ and $D_c \wedge D_b = 0$.
(b) There exists a rational $q > 0$ with $D_b \leq D_{a-q}$.

**PROOF.** Assuming (a), there exists a rational $q > 0$ with $D_{c-q} \vee D_{a-q} = 1$ by [29, Corollary 1.7]. Hence, $D_c \vee D_{a-q} = 1$, so $D_b = D_b \wedge (D_c \vee D_{a-q}) = D_b \wedge D_{a-q} \leq D_{a-q}$, establishing (b). For the converse, choose $D_c = D_{q-a}$. □

**6.3.19** In view of the preceding lemma, we henceforth write $D_b \ll D_a$ if there exists a rational $q > 0$ such that $D_b \leq D_{a-q}$, and note that the regularity condition (6.8) just states that the frame $\mathcal{O}(\Sigma(A))$ is regular [29].

We recall that an *ideal* of a lattice $L$ is a lower set $U \subseteq L$ that is closed under finite joins; the collection of all ideals in $L$ is denoted by $\mathrm{Idl}(L)$. An ideal $U$ of a distributive lattice $L$ is *regular* when $\downarrow x \subseteq U$ implies $x \in U$. Any ideal $U$ can be turned into a regular ideal $\overline{U}$ by means of the closure operator $\overline{(\_)}: DL \to DL$ defined by $\overline{U} = \{x \in L \mid \forall_{y \in L}.y \ll x \Rightarrow y \in U\}$ [25], with a canonical inclusion as in Proposition 6.2.13.

**6.3.20 Theorem.** The Gelfand spectrum $\mathcal{O}(\Sigma(A))$ of a commutative C*-algebra $A$ is isomorphic to the frame $\mathrm{RIdl}(L_A)$ of all regular ideals of $L_A$; that is,

$$\mathcal{O}(\Sigma(A)) \cong \{U \in \mathrm{Idl}(L_A) \mid (\forall_{D_b \in L_A}.D_b \ll D_a \Rightarrow D_b \in U) \Rightarrow D_a \in U\}.$$

In this realization, the canonical map $f: L_A \to \mathcal{O}(\Sigma(A))$ is given by

$$f(D_a) = \{D_c \in L_A \mid \forall_{D_b \in L_A}.D_b \ll D_c \Rightarrow D_b \leq D_a\}.$$

**PROOF.** For a commutative C*-algebra $A$, the lattice $L_A$ is *strongly normal* [29, Theorem 1.11] and, hence, *normal*. (A distributive lattice is *normal* if for all $b_1, b_2$ with $b_1 \vee b_2 = 1$ there are $c_1, c_2$ such that $c_1 \wedge c_2 = 0$ and $c_1 \vee b_1 = 1$ and $c_2 \vee b_2 = 1$.) By [25, Theorem 27], regular ideals in a normal distributive lattice form a compact regular frame. The result now follows from [29, Theorem 1.11]. □

**6.3.21 Corollary.** The Gelfand spectrum of a commutative C*-algebra $A$ is given by

$$\mathcal{O}(\Sigma(A)) \cong \{U \in \mathrm{Idl}(L_A) \mid \forall_{a \in A_{sa}} \forall_{q>0}.D_{a-q} \in U \Rightarrow D_a \in U\}.$$

**PROOF.** By combining Lemma 6.3.18 with Theorem 6.3.20. □

The following theorem is the key to explicitly determining the external description of the Gelfand spectrum $\mathcal{O}(\Sigma(A))$ of a C*-algebra $A$ in a topos.

**6.3.22 Theorem.** For a commutative C*-algebra $A$, define a covering relation $\triangleleft$ on $L_A$ by $x \triangleleft U$ iff $f(x) \leq \bigvee f(U)$, in the notation of Theorem 6.3.20.

1. One has $\mathcal{O}(\Sigma(A)) \cong \mathcal{F}(L_A, \triangleleft)$, under which $D_a \mapsto \downarrow D_a$.
2. Then, $D_a \triangleleft U$ iff for all rational $q > 0$ there is a (Kuratowski) finite $U_0 \subseteq U$ such that $D_{a-q} \leq \bigvee U_0$.

**PROOF.** Part (a) follows from Proposition 6.2.13. For (b), first assume $D_a \triangleleft U$ and let $q \in \mathbb{Q}$ satisfy $q > 0$. From (the proof of) Lemma 6.3.16, we have $D_a \vee D_{q-a} = 1$, whence $\bigvee f(U) \vee f(D_{q-a}) = 1$. Because $\mathcal{O}(\Sigma(A))$ is compact, there is a finite $U_0 \subseteq U$ for which $\bigvee f(U_0) \vee f(D_{q-a}) = 1$. Because $f(D_a) = 1$ if and only if $D_a = 1$ by Theorem 6.3.20, we have $D_b \vee D_{q-a} = 1$, where $D_b = \bigvee U_0$. By (6.4), we have $D_{a-q} \wedge D_{q-a} = 0$ and, hence,

$$D_{a-q} = D_{a-q} \wedge 1 = D_{a-q} \wedge (D_b \vee D_{q-a}) = D_{a-q} \wedge D_b \leq D_b = \bigvee U_0.$$

For the converse, notice that $f(D_a) \leq \bigvee \{f(D_{a-q}) \mid q \in \mathbb{Q}, q > 0\}$ by construction. So, from the assumption, we have $f(D_a) \leq \bigvee f(U)$ and, hence, $D_a \triangleleft U$. □

## 6.4 Bohrification

This section explains the technique of *Bohrification*. For a (generally) noncommutative C*-algebra $A$, Bohrification constructs a topos in which $A$ becomes commutative. More precisely, to any C*-algebra $A$, we associate a particular commutative C*-algebra $\underline{A}$ in the Kripke topos $[\mathcal{C}(A), \mathbf{Set}]$, where $\mathcal{C}(A)$ is the set of commutative C*-subalgebras of $A$. By Gelfand duality, the commutative C*-algebra $\underline{A}$ has a spectrum $\Sigma(\underline{A})$, which is a locale in $[\mathcal{C}(A), \mathbf{Set}]$.

**6.4.1** To introduce the idea, we outline the general method of Bohrification. We subsequently give concrete examples.

Let $\mathfrak{T}_1$ and $\mathfrak{T}_2$ be geometric theories whose variables range over only one type, apart from constructible types such as $\mathbb{N}$ and $\mathbb{Q}$. Suppose that $\mathfrak{T}_1$ is a subtheory of $\mathfrak{T}_2$. There is a functor $\mathcal{C}\colon \mathbf{Model}(\mathfrak{T}_1, \mathbf{Set}) \to \mathbf{Poset}$, defined on objects as $\mathcal{C}(A) = \{C \subseteq A \mid C \in \mathbf{Model}(\mathfrak{T}_2, \mathbf{Set})\}$, ordered by inclusion. On a morphism $f\colon A \to B$ of $\mathbf{Model}(\mathfrak{T}_1, \mathbf{Set})$, the functor $\mathcal{C}$ acts as $\mathcal{C}(f)\colon \mathcal{C}(A) \to \mathcal{C}(B)$ by the direct image $C \mapsto f(C)$. Hence, there is a functor $\mathcal{T}\colon \mathbf{Model}(\mathfrak{T}_1, \mathbf{Set}) \to \mathbf{Topos}$, defined on objects by $\mathcal{T}(A) = [\mathcal{C}(A), \mathbf{Set}]$ and determined on morphisms by $\mathcal{T}(f)^* = (\_) \circ \mathcal{C}(f)$. Define the canonical object $\underline{A} \in \mathcal{T}(A)$ by $\underline{A}(C) = C$, acting on a morphism $D \subseteq C$ of $\mathcal{C}(A)$ as the inclusion $\underline{A}(D) \hookrightarrow \underline{A}(C)$. Then, $\underline{A}$ is a model of $\mathfrak{T}_2$ in the Kripke topos $\mathcal{T}(A)$ by Lemma 6.3.13.

**6.4.2 Example.** Let $\mathfrak{T}_1$ be the theory of groups and $\mathfrak{T}_2$ be the theory of Abelian groups. Both are geometric theories, and $\mathfrak{T}_1$ is a subtheory of $\mathfrak{T}_2$. Then, $\mathcal{C}(G)$ is the collection of Abelian subgroups $C$ of $G$, ordered by inclusion, and the functor $\underline{G}\colon C \mapsto C$ is an Abelian group in $\mathcal{T}(G) = [\mathcal{C}(G), \mathbf{Set}]$.

This resembles the so-called microcosm principle, according to which structure of an internal entity depends on similar structure of the ambient category [5, 51].

We now turn to the setting of our interest: (commutative) C*-algebras. The theory of C*-algebras is not geometric, so it does not follow from the arguments of 6.4.1 that $\underline{A}$ will be a commutative C*-algebra in $\mathcal{T}(A)$. Theorem 6.4.8 shows that the latter is nevertheless true.

**6.4.3 Proposition.** There is a functor $\mathcal{C}\colon \mathbf{CStar} \to \mathbf{Poset}$, defined on objects as

$$\mathcal{C}(A) = \{C \in \mathbf{cCStar} \mid C \text{ is a C*-subalgebra of } A\},$$

ordered by inclusion. Its action $\mathcal{C}(f)\colon \mathcal{C}(A) \to \mathcal{C}(B)$ on a morphism $f\colon A \to B$ of $\mathbf{CStar}$ is the direct image $C \mapsto f(C)$. Hence, there is a functor $\mathcal{T}\colon \mathbf{CStar} \to \mathbf{Topos}$, defined by $\mathcal{T}(A) = [\mathcal{C}(A), \mathbf{Set}]$ on objects and $\mathcal{T}(f)^* = (\_) \circ \mathcal{C}(f)$ on morphisms.

**PROOF.** It suffices to show that $\mathcal{T}(f)^*$ is part of a geometric morphism, which follows from [70, Theorem VII.2.2]. □

**6.4.4 Example.** The following example determines $\mathcal{C}(A)$ for $A = \mathbf{Hilb}(\mathbb{C}^2, \mathbb{C}^2)$, the C*-algebra of complex 2 by 2 matrices. Any C*-algebra has a single one-dimensional commutative C*-subalgebra—namely, $\mathbb{C}$, the scalar multiples of the unit. Furthermore, any two-dimensional C*-subalgebra is generated by a pair of orthogonal one-dimensional projections. The one-dimensional projections in $A$ are of the form

$$p(x, y, z) = \frac{1}{2} \begin{pmatrix} 1+x & y+iz \\ y-iz & 1-x \end{pmatrix}, \tag{6.9}$$

where $(x, y, z) \in \mathbb{R}^3$ satisfies $x^2 + y^2 + z^2 = 1$. Thus, the one-dimensional projections in $A$ are precisely parametrized by $S^2$. Because $1 - p(x, y, z) = p(-x, -y, -z)$, and pairs $(p, 1-p)$ and $(1-p, p)$ define the same C*-subalgebra, the two-dimensional elements of $\mathcal{C}(A)$ are parametrized by $S^2/\sim$, where $(x, y, z) \sim (-x, -y, -z)$. This space, in turn, is homeomorphic with the real projective plane $\mathbb{RP}^2$—that is, the set of lines in $\mathbb{R}^3$ passing through the origin.[1] Parametrizing $C(A) \cong \{\mathbb{C}\} + \mathbb{RP}^2$, a point $[x, y, z] \in S^2/\sim$ then corresponds to the C*-algebra $C_{[x,y,z]}$ generated by the projections $\{p(x, y, z), p(-x, -y, -z)\}$. The order of $C(A)$ is flat: $C < D$ iff $C = \mathbb{C}$.

**6.4.5 Example.** We now generalize the previous example to $A = \mathbf{Hilb}(\mathbb{C}^n, \mathbb{C}^n)$ for any $n \in \mathbb{N}$. In general, one has $C(A) = \coprod_{k=1}^{n} C(k, n)$, where $C(k, n)$ denotes the collection of all $k$-dimensional commutative unital C*-subalgebras of $A$. To parametrize $C(k, n)$, we first show that each of its elements $C$ is a unitary rotation $C = UDU^*$, where $U \in SU(n)$ and $D$ is some subalgebra contained in the algebra of all diagonal matrices. This follows from the case $k = n$ because each element of $C(k, n)$ with $k < n$ is contained in some maximal commutative subalgebra. For $k = n$, note that $C \in C(n, n)$ is generated by $n$ mutually orthogonal projections $p_1, \ldots, p_n$ of rank 1. Each $p_i$ has a single unit eigenvector $u_i$ with eigenvalue 1; its other eigenvalues are 0. Put these $u_i$ as columns in a matrix, called $U$. Then, $U^* p_i U$ is diagonal for all $i$, for if $(e_i)$ is the standard basis of $\mathbb{C}^n$, then $Ue_i = u_i$ for all $i$ and, hence, $U^* p_i U e_i = U^* p_i u_i = U^* u_i = e_i$, whereas for $i \neq j$, one finds $U^* p_i U e_j = 0$. Hence, the matrix $U^* p_i U$ has a 1 at location $ii$ and a zero everywhere else. All other elements $a \in C$ are functions of the $p_i$, so that $U^* a U$ is equally well diagonal. Hence, $C = U D_n U^*$, with $D_n$ the algebra of all diagonal matrices. Thus,

$$C(n, n) = \{U D_n U^* \mid U \in SU(n)\},$$

with $D_n = \{\mathrm{diag}(a_1, \ldots, a_n) \mid a_i \in \mathbb{C}\}$, and $C(k, n)$ for $k < n$ is obtained by partitioning $\{1, \ldots, n\}$ into $k$ nonempty parts and demanding $a_i = a_j$ for $i, j$ in the same part. However, because of the conjugation with arbitrary $U \in SU(n)$, two such partitions induce the same subalgebra precisely when they permute parts of equal size. Such permutations may be handled using Young tableaux [43]. The size of a part is of more interest than the part itself, so we define

$$Y(k, n) = \{(i_1, \ldots, i_k) \mid 0 < i_1 < i_2 < \cdots < i_k = n, \quad i_{j+1} - i_j \leq i_j - i_{j-1}\}$$

(where $i_0 = 0$) as the set of partitions inducing different subalgebras. Hence,

$$C(k, n) \cong \big\{(p_1, \ldots, p_k) : p_j \in \mathrm{Proj}(A), \quad (i_1, \ldots, i_k) \in Y(k, n)$$
$$\mid \dim(\mathrm{Im}(p_j)) = i_j - i_{j-1}, \quad p_j \wedge p_{j'} = 0 \text{ for } j \neq j'\big\}.$$

Now, because $d$-dimensional orthogonal projections in $\mathbb{C}^n$ bijectively correspond to the $d$-dimensional (closed) subspaces of $\mathbb{C}^n$ they project onto, we can write

$$C(k, n) \cong \big\{(V_1, \ldots, V_k) : (i_1, \ldots, i_k) \in Y(k, n), V_j \in \mathrm{Gr}(i_j - i_{j-1}, n)$$
$$\mid V_j \cap V_{j'} = 0 \text{ for } j \neq j'\big\},$$

where $\mathrm{Gr}(d, n) = U(n)/(U(d) \times U(n - d))$ is the well-known Grassmannian; that is, the set of all $d$-dimensional subspaces of $\mathbb{C}^n$ [47]. In terms of the partial

---

[1] This space has an interesting topology that is quite different from the Alexandrov topology on $C(A)$ but that we nevertheless ignore.

flag manifold

$$G(i_1, \ldots, i_k; n) = \prod_{j=1}^{k} \text{Gr}(i_j - i_{j-1}, n - i_{j-1}),$$

for $(i_1, \ldots, i_k) \in Y(k, n)$ (see [43]), we finally obtain

$$\mathcal{C}(k, n) \cong \{V \in G(i; n) : i \in Y(k, n)\}/\sim,$$

where $i \sim i'$ if one arises from the other by permutations of equal-sized parts.

This indeed generalizes the previous example $n = 2$. First, for any $n$, the set $\mathcal{C}(1, n)$ has a single element because there is only one Young tableau for $k = 1$. Second, we have $Y(2, 2) = \{(1, 2)\}$, so that

$$\mathcal{C}(2, 2) \cong (\text{Gr}(1, 2) \times \text{Gr}(1, 1))/S(2) \cong \text{Gr}(1, 2)/S(2) \cong \mathbb{CP}^1/S(2) \cong \mathbb{RP}^2.$$

**6.4.6 Definition.** Let $A$ be a C*-algebra. Define the functor $\underline{A} \colon \mathcal{C}(A) \to \mathbf{Set}$ by acting on objects as $\underline{A}(C) = C$, and acting on morphisms $C \subseteq D$ of $\mathcal{C}(A)$ as the inclusion $\underline{A}(C) \hookrightarrow \underline{A}(D)$. We call $\underline{A}$, or the process of obtaining it, the *Bohrification* of $A$.

**6.4.7 Convention.** We will underline entities internal to $\mathcal{T}(A)$ to distinguish between the internal and external points of view.

The particular object $\underline{A}$ turns out to be a commutative C*-algebra in the topos $\mathcal{T}(A)$, even though the theory of C*-algebras is not geometric.

**6.4.8 Theorem.** Operations inherited from $A$ make $\underline{A}$ a commutative C*-algebra in $\mathcal{T}(A)$. More precisely, $\underline{A}$ is a vector space over the complex field $\text{Pt}(\underline{\mathbb{C}}) \colon C \mapsto \mathbb{C}$ by

$$0 \colon \underline{1} \to \underline{A}, \qquad + \colon \underline{A} \times \underline{A} \to \underline{A}, \qquad \cdot \colon \text{Pt}(\underline{\mathbb{C}}) \times \underline{A} \to \underline{A},$$
$$0_C(*) = 0, \qquad a +_C b = a + b, \qquad z \cdot_C a = z \cdot a,$$

and an involutive algebra through

$$\cdot \colon \underline{A} \times \underline{A} \to \underline{A}, \qquad (\_)^* \colon \underline{A} \to \underline{A}$$
$$a \cdot_C b = a \cdot b, \qquad (a^*)_C = a^*.$$

The norm relation is the subobject $N \in \text{Sub}(\underline{A} \times \underline{\mathbb{Q}^+})$ given by

$$N_C = \{(a, q) \in C \times \mathbb{Q}^+ \mid \|a\| < q\}.$$

**PROOF.** Recall (Definition 6.3.3) that a pre-semi-C*-algebra is a C*-algebra that is not necessarily Cauchy complete and whose seminorm is not necessarily a norm. Because the theory of pre-semi-C*-algebras is geometric, Lemma 6.3.13 shows that $\underline{A}$ is a commutative pre-semi-C*-algebra in $\mathcal{T}(A)$, as in 6.4.1. Let us prove that $\underline{A}$ is, in fact, a pre-C*-algebra; that is, that the seminorm is a norm. It suffices to show that $C \Vdash \forall_{a \in \underline{A}_{\text{sa}}} \forall_{q \in \underline{Q^+}}.(a, q) \in N \Rightarrow a = 0$ for all $C \in \mathcal{C}(A)$.

By 6.2.24, this means

for all $C' \supseteq C$ and $a \in C'$, if $C' \Vdash \forall_{q \in \mathbb{Q}^+}.(a,q) \in N$, then $C' \Vdash a = 0$;

that is, for all $C' \supseteq C$ and $a \in C'$, if $C'' \Vdash (a,q) \in N$ for all $C'' \supseteq C'$ and $q \in \mathbb{Q}^+$, then $C' \Vdash a = 0$;

that is, for all $C' \supseteq C$ and $a \in C'$, if $\|a\| = 0$, then $a = 0$.

But this holds because every $C'$ is a C*-algebra.

Finally, we prove that $\underline{A}$ is, in fact, a C*-algebra. Because the axiom of dependent choice holds in $\mathcal{T}(A)$ [42], it suffices to prove that every *regular* Cauchy sequence converges, where a sequence $(x_n)$ is regular Cauchy when $\|x_n - x_m\| \leq 2^{-n} + 2^{-m}$ for all $n, m \in \mathbb{N}$. Thus, we need to prove

$$C \Vdash \forall_{n,m \in \underline{\mathbb{N}}}.\|x_n - x_m\| \leq 2^{-n} + 2^{-m} \Rightarrow \exists_{x \in \underline{A}}.\forall_{n \in \underline{\mathbb{N}}}.\|x - x_n\| \leq 2^{-n};$$

that is, for all $C' \supseteq C$, if $C' \Vdash (\forall_{n,m \in \underline{\mathbb{N}}}.\|x_n - x_m\| \leq 2^{-n} + 2^{-m})$, then $C' \Vdash \exists_{x \in \underline{A}}.\forall_{n \in \underline{\mathbb{N}}}.\|x - x_n\| \leq 2^{-n}$;

that is, for all $C' \supseteq C$, if $C' \Vdash$ "$(x)_n$ is regular," then $C' \Vdash$ "$(x)_n$ converges."

Once again, this holds because every $C'$ is a C*-algebra. □

**6.4.9** Applying 6.3.8 to the commutative C*-algebra $\underline{A}$ in the topos $\mathcal{T}(A)$, we obtain a locale $\underline{\Sigma}(\underline{A})$ in that topos. As argued in the Introduction, $\underline{\Sigma}(\underline{A})$ is the "state space" carrying the logic of the physical system whose observable algebra is $A$.

An important property of $\underline{\Sigma}(\underline{A})$ is that it is typically highly nonspatial, as the following theorem proves. This theorem is a localic extension of a topos-theoretic reformulation of the Kochen–Specker theorem [63] attributable to Jeremy Butterfield and Chris Isham [20–23].

**6.4.10 Theorem.** Let $H$ be a Hilbert space with $\dim(H) > 2$, and $A =$ **Hilb**$(H, H)$. The locale $\underline{\Sigma}(\underline{A})$ has no points.

**PROOF.** A point $\rho\colon 1 \to \underline{\Sigma}(\underline{A})$ of the locale $\underline{\Sigma}(\underline{A})$ (see 6.2.8) may be combined with $a \in \underline{A}$sa, with Gelfand transform $\hat{a}\colon \underline{\Sigma}(\underline{A}) \to \overline{\mathbb{R}}$ as in 6.3.10, so as to produce a point $\hat{a} \circ \rho\colon 1 \to \overline{\mathbb{R}}$ of the locale $\overline{\mathbb{R}}$. This yields a map $\underline{V}_\rho\colon \underline{A}$sa $\to$ Pt$(\overline{\mathbb{R}})$, which turns out to be a multiplicative functional [6, 8, 29]. Being a morphism in $\mathcal{T}(A)$, the map $\underline{V}_\rho$ is a natural transformation, with components $\underline{V}_\rho(C)\colon \underline{A}$sa$(C) \to$ Pt$(\overline{\mathbb{R}})(C)$; by Definition 6.4.6 and Example 6.2.33, this is just $\underline{V}_\rho(C)\colon C_{sa} \to \mathbb{R}$. Hence, one has a multiplicative functional $\underline{V}_\rho(C)$ for each $C \in \mathcal{C}(A)$ in the usual sense, with the naturality, or "noncontextuality," property that if $C \subseteq D$, then the restriction of $\underline{V}_\rho(D)$ to $C_{sa}$ is $\underline{V}_\rho(C)$. But that is precisely the kind of function on **Hilb**$(H, H)$ of which the Kochen–Specker theorem proves the nonexistence [63]. □

**6.4.11** The previous theorem holds for more general C*-algebras than **Hilb**$(H, H)$ (for large enough Hilbert spaces $H$); see [37] for results on von Neumann algebras. A C*-algebra $A$ is called *simple* when its closed two-sided ideals are trivial and *infinite* when there is an $a \in A$ with $a^*a = 1$ but $aa^* \neq 1$ [33].

A simple infinite C*-algebra does not admit a dispersion-free quasi-state [50]; whence the previous theorem holds for such C*-algebras as well.

The rest of this section is devoted to describing the structure of the Gelfand spectrum $\underline{\Sigma}(\underline{A})$ of the Bohrification $\underline{A}$ of $A$ from the external point of view.

**6.4.12 Theorem.** For a C*-algebra $A$ and each $C \in \mathcal{C}(A)$, one has $\underline{L_A}(C) = L_C$. Moreover $\underline{L_A}(C \subseteq D)\colon L_C \to L_D$ is a frame morphism that maps each generator $D_c$ for $c \in C_{sa}$ to the same generator for the spectrum of $D$.

**PROOF.** This follows from Lemma 6.3.13 and Proposition 6.3.17. □

**6.4.13** The next corollary interprets $D_a \triangleleft U$ in our situation, showing that also the covering relation $\triangleleft$ can be computed locally. To do so, we introduce the notation $\underline{L_A}_{|\uparrow C}$ for the restriction of the functor $\underline{L_A}\colon \mathcal{C}(A) \to \mathbf{Set}$ to $\uparrow C \subseteq \mathcal{C}(A)$. Then, $\underline{\Omega^{L_A}}(C) \cong \mathrm{Sub}(\underline{L_A}_{|\uparrow C})$ by [70, Section II.8]. Hence, by Kripke–Joyal semantics, cf. 6.2.24, the formal variables $D_a$ and $U$ in $C \Vdash D_a \triangleleft U$ for $C \in \mathcal{C}(A)$ are to be instantiated with actual elements $D_c \in L_C = \underline{L_A}(C)$ and a subfunctor $\underline{U}\colon \uparrow C \to \mathbf{Set}$ of $\underline{L_A}_{|\uparrow C}$. Because $\triangleleft$ is a subfunctor of $\underline{L_A} \times \underline{\mathcal{P}}(\underline{L_A})$, we can speak of $\triangleleft_C$ for $C \in \mathcal{C}(A)$ as the relation $\underline{L_A}(C) \times \underline{\mathcal{P}}(\underline{L_A})$ induced by evaluation at $C$.

**6.4.14 Corollary.** The covering relation $\triangleleft$ of Theorem 6.3.22 is computed locally. That is, for $C \in \mathcal{C}(A)$, $D_c \in L_C$, and $\underline{U} \in \mathrm{Sub}(\underline{L_A}_{|\uparrow C})$, the following are equivalent:

(a) $C \Vdash D_a \triangleleft U(D_c, \underline{U})$.

(b) $D_c \triangleleft_C \underline{U}(C)$.

(c) For every rational $q > 0$, there is a finite $U_0 \subseteq \underline{U}(C)$ with $D_{c-q} \leq \bigvee U_0$.

**PROOF.** The equivalence of (b) and (c) follows from Theorem 6.3.22. We prove the equivalence of (a) and (c). Assume, without loss of generality, that $\bigvee U_0 \in U$, so that $U_0$ may be replaced by $D_b = \bigvee U_0$. Hence, the formula $D_a \triangleleft U$ in (a) means

$$\forall_{q>0} \exists_{D_b \in L_A}.(D_b \in U \wedge D_{a-q} \leq D_b).$$

We interpret this formula step by step, as in 6.2.24. First, $C \Vdash (D_a \in U)(D_c, \underline{U})$ iff for all $D \supseteq C$ one has $D_c \in \underline{U}(D)$. Because $\underline{U}(C) \subseteq \underline{U}(D)$, this is the case iff $D_c \in \underline{U}(C)$. Also, one has $C \Vdash (D_b \leq D_a)(D_{c'}, D_c)$ iff $D_{c'} \leq D_c$ in $L_C$. Hence, $C \Vdash (\exists_{D_b \in L_A}.D_b \in U \wedge D_{a-q} \leq D_b)(D_c, \underline{U})$ iff there is $D_{c'} \in \underline{U}(C)$ with $D_{c-q} \leq D_{c'}$. Finally, $C \Vdash (\forall_{q>0} \exists_{D_b \in L_A}.D_b \in U \wedge D_{a-q} \leq D_b)(D_c, \underline{U})$ iff for all $D \supseteq C$ and all rational $q > 0$ there is $D_d \in \underline{U}(D)$ such that $D_{c-q} \leq D_d$, where $D_c \in L_C \subseteq L_D$ by Theorem 6.4.12 and $\underline{U} \in \mathrm{Sub}(\underline{L_A}_{|\uparrow C}) \subseteq \mathrm{Sub}(\underline{L_A}_{|\uparrow D})$ by restriction. This holds at all $D \supseteq C$ iff it holds at $C$, because $\underline{U}(C) \subseteq \underline{U}(D)$, whence one can take $D_d = D_{c'}$. □

**6.4.15** The following theorem explicitly determines the Gelfand spectrum $\underline{\Sigma}(\underline{A})$ from the external point of view. It turns out that the functor $\underline{\Sigma}(\underline{A})$ is completely determined by its value $\underline{\Sigma}(\underline{A})(\mathbb{C})$ at the least element $\mathbb{C}$ of $\mathcal{C}(A)$. Therefore, we abbreviate $\underline{\Sigma}(\underline{A})(\mathbb{C})$ by $\Sigma_A$ and call it the *Bohrified state space* of $A$.

**6.4.16 Theorem.** For a C*-algebra $A$:

(a) At $C \in \mathcal{C}(A)$, the set $\mathcal{O}(\underline{\Sigma}(A))(C)$ consists of the subfunctors $\underline{U} \in \mathrm{Sub}(\underline{L}_{A|\uparrow C})$ satisfying $D_d \triangleleft_D \underline{U}(D) \Rightarrow D_d \in \underline{U}(D)$ for all $D \supseteq C$ and $D_d \in L_D$.

(b) In particular, the set $\mathcal{O}(\underline{\Sigma}(A))(\mathbb{C})$ consists of the subfunctors $\underline{U} \in \mathrm{Sub}(\underline{L}_A)$ satisfying $D_c \triangleleft_C \underline{U}(C) \Rightarrow D_c \in \underline{U}(C)$ for all $C \in \mathcal{C}(A)$ and $D_c \in L_C$.

(c) The action $\mathcal{O}(\underline{\Sigma}(A)) \to \mathcal{O}(\underline{\Sigma}(A))$ of $\mathcal{O}(\underline{\Sigma}(A))$ on a morphism $C \subseteq D$ of $\mathcal{C}(A)$ is given by truncating $\underline{U} \colon \uparrow C \to \mathbf{Set}$ to $\uparrow D$.

(d) The external description of $\mathcal{O}(\underline{\Sigma}(A))$ is the frame morphism

$$f^{-1} \colon \mathcal{O}(\mathrm{Alx}(\mathcal{C}(A))) \to \mathcal{O}(\underline{\Sigma}(A))(\mathbb{C}),$$

given on basic opens $\uparrow D \in \mathcal{O}(\mathrm{Alx}(\mathcal{C}(A)))$ by

$$f^{-1}(\uparrow D)(E) = \begin{cases} L_E & \text{if } E \supseteq D, \\ \emptyset & \text{otherwise.} \end{cases}$$

**PROOF.** By Theorem 6.3.22(a) and (6.2), $\mathcal{O}(\underline{\Sigma}(A))$ is the subobject of $\underline{\Omega}^{\underline{L}_A}$ defined by the formula $\forall_{D_a \in L_A}. D_a \triangleleft U \Rightarrow D_a \in U$. As in 6.4.13, elements $\underline{U} \in \mathcal{O}(\underline{\Sigma}(A))(C)$ may be identified with subfunctors of $\underline{L}_{A|\uparrow C}$. Hence, by Corollary 6.4.14, we have $\underline{U} \in \mathcal{O}(\underline{\Sigma}(A))$ if and only if

$$\forall_{D \supseteq C} \forall_{D_d \in L_D} \forall_{E \supseteq D}. D_d \triangleleft_E \underline{U}(E) \Rightarrow D_d \in \underline{U}(E),$$

where $D_d$ is regarded as an element of $L_E$. This is equivalent to the apparently weaker condition

$$\forall_{D \supseteq C} \forall_{D_d \in L_D}. D_d \triangleleft_D \underline{U}(D) \Rightarrow D_d \in \underline{U}(D),$$

because the latter applied at $D = E$ actually implies the former condition as $D_d \in L_D$ also lies in $L_E$. This proves (a), (b), and (c). Part (d) follows from Example 6.2.28. $\square$

## 6.5 Projections

This section compares the quantum state spaces $\mathcal{O}(\underline{\Sigma}(A))$ with quantum logic in the sense of [11]. In the setting of operator algebras, this more traditional quantum logic is concerned with projections; we denote the set of projections of a C*-algebra $A$ by

$$\mathrm{Proj}(A) = \{p \in A \mid p^* = p = p \circ p\}.$$

A generic C*-algebra may not have enough projections: for example, if $A$ is a commutative C*-algebra whose Gelfand spectrum $\Sigma(A)$ is connected, then $A$ has no projections except for 0 and 1. Hence, we need to specialize to C*-algebras that have enough projections. The best-known such class consists of von Neumann algebras but, in fact, the most general class of C*-algebras that are generated by their projections *and* can easily be Bohrified turns out to consist of so-called Rickart C*-algebras. To motivate this choice, we start by recalling several types of C*-algebras and known results about their spectra.

**6.5.1 Definition.** Let $A$ be a C*-algebra. Define $R(S) = \{a \in A \mid \forall_{s \in S}.sa = 0\}$ to be the *right annihilator* of some subset $S \subseteq A$. Then, $A$ is said to be:

(a) a *von Neumann algebra* if it is the dual of some Banach space [76];
(b) an *AW*-algebra* if for each nonempty $S \subseteq A$ there is a $p \in \text{Proj}(A)$ satisfying $R(S) = pA$ [62];
(c) a *Rickart C*-algebra* if for each $x \in A$ there is a $p \in \text{Proj}(A)$ satisfying $R(\{x\}) = pA$ [74];
(d) a *spectral C*-algebra* if for each $a \in A^+$ and each $r, s \in (0, \infty)$ with $r < s$, there is a $p \in \text{Proj}(A)$ satisfying $ap \geq rp$ and $a(1-p) \leq s(1-p)$ [83].

In all cases, the projection $p$ turns out to be unique. Each class contains the previous one(s).

**6.5.2** To prepare for what follows, we recall the *Stone representation theorem* [57]. This theorem states that any Boolean algebra $B$ (in the topos **Set**) is isomorphic to the lattice $\mathcal{B}(X)$ of clopen subsets of a *Stone space* $X$—that is, a compact Hausdorff space that is *totally* disconnected, in that its only connected subsets are singletons. Equivalently, a Stone space is compact, $T_0$, and has a basis of clopen sets. The space $X$ is uniquely determined by $B$ up to homeomorphism and hence may be written $\hat{\Sigma}(B)$; one model for it is given by the set of all maximal filters in $B$, topologized by declaring that for each $b \in B$, the set of all maximal filters containing $b$ is a basic open for $\hat{\Sigma}(B)$. Another description is based on the isomorphism

$$\mathcal{O}(\hat{\Sigma}(B)) \cong \text{Idl}(B) \tag{6.10}$$

of locales, where the left-hand side is the topology of $\hat{\Sigma}(B)$, and the right-hand side is the ideal completion of $B$ (seen as a distributive lattice). This leads to an equivalent model of $\hat{\Sigma}(B)$—namely, as $\text{Pt}(\text{Idl}(B))$ with its canonical topology (cf. 6.2.8); see Corollaries II.4.4 and II.3.3 and Proposition II.3.2 in [57]. Compare this with Theorem 6.3.20, which states that the Gelfand spectrum $\Sigma(A)$ of a unital commutative C*-algebra $A$ may be given as

$$\mathcal{O}(\Sigma(A)) \cong \text{RIdl}(L_A). \tag{6.11}$$

The analogy between Equations (6.10) and (6.11) is more than an optical one. A Stone space $X$ gives rise to a Boolean algebra $\mathcal{B}(X)$ as well as to a commutative C*-algebra $C(X, \mathbb{C})$. Conversely, if $A$ is a commutative C*-algebra, then $\text{Proj}(A)$ is isomorphic with the Boolean lattice $\mathcal{B}(\Sigma(A))$ of clopens in $\Sigma(A)$. If we regard $\Sigma(A)$ as consisting of characters as in 6.3.9, then this isomorphism is given by

$$\text{Proj}(A) \xrightarrow{\cong} \mathcal{B}(\Sigma(A))$$
$$p \mapsto \{\sigma \in \Sigma(A) \mid \sigma(p) \neq 0\},$$

where $\hat{p}$ is the Gelfand transform of $p$ as in 6.3.10.

To start with a familiar case, a von Neumann algebra $A$ is commutative if and only if $\text{Proj}(A)$ is a Boolean algebra ([73], Proposition 4.16). In that case, the Gelfand spectrum $\Sigma(A)$ of $A$ may be identified with the Stone spectrum of Proj($A$); passing to the respective topologies, in view of (6.10), we therefore have

$$\mathcal{O}(\Sigma(A)) \cong \mathrm{Idl}(\mathrm{Proj}(A)). \tag{6.12}$$

In fact, this holds more generally in two different ways: first, it is true for the larger class of Rickart C*-algebras; second, the proof is constructive and, hence, the result holds in arbitrary toposes; see Theorem 6.5.15 to follow. As to the first point, in **Set** (where the locales in question are spatial), we may conclude from this theorem that for a commutative Rickart C*-algebra $A$, one has a homeomorphism

$$\Sigma(A) \cong \hat{\Sigma}(\mathrm{Proj}(A)), \tag{6.13}$$

a result that so far had only been known for von Neumann algebras.

**6.5.3** One reason for dissatisfaction with von Neumann algebras is that the preceding correspondence between Boolean algebras and commutative von Neumann algebras is not bijective. Indeed, if $A$ is a commutative von Neumann algebra, then Proj($A$) is complete, so that $\Sigma(A)$ is not merely Stone but *Stonean*—that is, compact, Hausdorff, and *extremely* disconnected—in that the closure of every open set is open. (The Stone spectrum of a Boolean algebra $L$ is Stonean if and only if $L$ is complete.) But commutative von Neumann algebras do not correspond bijectively to complete Boolean algebras either because the Gelfand spectrum of a commutative von Neumann algebra is not merely Stone but has the stronger property of being *hyperstonean* in that it admits sufficiently many positive normal measures ([85], Definition 6.14). Indeed, a commutative C*-algebra $A$ is a von Neumann algebra if and only if its Gelfand spectrum (and, hence, the Stone spectrum of its projection lattice) is hyperstonean.

**6.5.4 Theorem.** A commutative C*-algebra $A$ is:
1. a von Neumann algebra if and only if $\Sigma(A)$ is hyperstonean ([85], Section III.1);
2. an AW*-algebra if and only if $\Sigma(A)$ is Stonean, if and only if $\Sigma(A)$ is Stone and $\mathcal{B}(\Sigma(A))$ is complete ([10], Theorem 1.7.1);
3. a Rickart C*-algebra if and only if $\Sigma(A)$ is Stone and $\mathcal{B}(\Sigma(A))$ is countably complete ([10], Theorem 1.8.1);
4. a spectral C*-algebra if and only if $\Sigma(A)$ is Stone ([83], Section 9.7). □

**6.5.5** Although spectral C*-algebras are the most general class in Definition 6.5.1, their projections may not form a lattice in the noncommutative case. A major advantage of Rickart C*-algebras is that their projections do, as in the following proposition. Rickart C*-algebras are also of interest for classification programs, as follows. The class of so-called *real rank zero* C*-algebras has been classified using K-theory. This is a functor $K$ from **CStar** to graded Abelian groups. In fact, it is currently believed that real rank zero C*-algebras are the widest class of C*-algebras for which $A \cong B$ if and only if $K(A) \cong K(B)$ ([75], Section 3). Rickart C*-algebras are always real rank zero ([12], Theorem 6.1.2).

**6.5.6 Proposition.** Let $A$ be a Rickart C*-algebra.
(a) If it is ordered by $p \leq q \Leftrightarrow pA \subseteq qA$, then Proj($A$) is a countably complete lattice ([10], Proposition 6.3.7, and Lemma 6.8.3).

(b) If $A$ is commutative, then it is the (norm-)closed linear span of Proj($A$) ([10], Proposition 1.8.1.(3)).

(c) If $A$ is commutative, then it is monotone countably complete; that is, each increasing bounded sequence in $A_{\text{sa}}$ has a supremum in $A$ ([83], Proposition 9.2.6.1). □

**6.5.7** Definition 6.5.1(a) requires the so-called ultraweak or $\sigma$-weak topology, which is hard to internalize to a topos. There are constructive definitions of von Neumann algebras [35, 82], but they rely on the strong operator topology, which is hard to internalize too. Furthermore, the latter rely on the axiom of dependent choice. Although this holds in Kripke toposes, we prefer to consider Rickart C*-algebras. All one loses in this generalization is that the projection lattice is only countably complete instead of complete—this is not a source of tremendous worry because countable completeness of Proj($A$) implies completeness if $A$ has a faithful representation on a separable Hilbert space. Moreover, Rickart C*-algebras can easily be Bohrified, as Theorem 6.5.10 shows.

**6.5.8 Proposition.** For a commutative C*-algebra $A$, the following are equivalent:

1. $A$ is Rickart.
2. For each $a \in A$, there is a (unique) $[a = 0] \in \text{Proj}(A)$ such that $a[a = 0] = 0$, and $b = b[a = 0]$ when $ab = 0$.
3. For each $a \in A_{\text{sa}}$, there is a (unique) $[a > 0] \in \text{Proj}(A)$ such that $[a > 0]a = a^+$ and $[a > 0][-a > 0] = 0$.

PROOF. For the equivalence of (a) and (b), we refer to [10, Proposition 1.3.3]. Assuming (b) and defining $[a > 0] = 1 - [a^+ = 0]$, we have

$$[a > 0]a = (1 - [a^+ = 0])(a^+ - a^-)$$
$$= a^+ - a^- - a^+[a^+ = 0] + a^-[a^+ = 0]$$
$$= a^+, \qquad \text{(because } a^-a^+ = 0, \text{ so that } a^-[a^+ = 0] = a^-\text{)}$$

and, similarly, $a^-[a > 0] = a^- - a^-[a^+ = 0] = 0$, whence

$$[a > 0][-a > 0] = [a > 0](1 - [(-a)^+ = 0])$$
$$= [a > 0] - [a > 0][a^- = 0] = 0, \qquad \text{(because } [a^-[a > 0] = 0\text{)}$$

establishing (c). For the converse, notice that it suffices to handle the case $a \in A^+$: decomposing general $a \in A$ into four positives, we obtain $[a = 0]$ by multiplying the four associated projections. Assuming (c) and $a \in A^+$, define $[a = 0] = 1 - [a > 0]$. Then, $a[a = 0] = (1 - [a > 0]) = a^+ - a[a > 0] = 0$. If $ab = 0$ for $b \in A$, then

$$D_{b[a>0]} = D_{b \wedge [a>0]} = D_b \wedge D_{[a>0]} = D_b \wedge D_a = D_{ba} = D_0,$$

so that $b[a < 0] \preccurlyeq 0$ by 6.3.11. That is, $b[a < 0] \leq n \cdot 0 = 0$ for some $n \in \mathbb{N}$. □

**6.5.9** Parallel to Proposition 6.4.3, we define $\mathcal{C}_R(A)$ to be the collection of all commutative Rickart C*-subalgebras $C$ of $A$, and $\mathcal{T}_R(A) = [\mathcal{C}_R(A), \textbf{Set}]$. The

Bohrification $\underline{A}$ of a Rickart C*-algebra $A$ is then defined by $\underline{A}(C) = C$, just as in Definition 6.4.6.

**6.5.10 Theorem.** Let $A$ be a Rickart C*-algebra. Then, $\underline{A}$ is a commutative Rickart C*-algebra in $\mathcal{T}_R(A)$.

**PROOF.** By Theorem 6.4.8, we already know that $\underline{A}$ is a commutative C*-algebra in $\mathcal{T}_R(A)$. Proposition 6.5.8 captures the property of a commutative C*-algebra being Rickart in a geometric formula. Hence, by Lemma 6.3.13, $\underline{A}$ is Rickart because every $C \in \mathcal{C}_R(A)$ is. □

We now work toward an explicit formula for the external description of the Gelfand spectrum of the Bohrification of a Rickart C*-algebra.

**6.5.11 Lemma.** Let $A$ be a commutative Rickart C*-algebra and $a, b \in A$ self-adjoint. If $ab \geq a$, then $a \preccurlyeq b$, i.e. $D_a \leq D_b$.

**PROOF.** If $a \leq ab$, then certainly $a \preccurlyeq ab$. Hence, $D_a \leq D_{ab} = D_a \wedge D_b$. In other words, $D_a \leq D_b$, whence $a \preccurlyeq b$. □

**6.5.12 Definition.** Recall that a function $f$ between posets satisfying $f(x) \geq f(y)$ when $x \leq y$ is called *antitone*. A *pseudocomplement* on a distributive lattice $L$ is an antitone function $\neg \colon L \to L$ satisfying $x \wedge y = 0$ iff $x \leq \neg y$. Compare 6.2.2.

**6.5.13 Proposition.** For a commutative Rickart C*-algebra $A$, the lattice $L_A$ has a pseudocomplement, determined by $\neg D_a = D_{[a=0]}$ for $a \in A^+$.

**PROOF.** Without loss of generality, let $b \leq 1$. Then,

$$\begin{aligned}
D_a \wedge D_b = 0 &\iff D_{ab} = D_0 \\
&\iff ab = 0 \\
&\iff b[a=0] = b \qquad (\Rightarrow \text{ by Proposition 6.5.8}) \\
&\iff b \preccurlyeq [a=0] \qquad (\Leftarrow \text{ because } b \leq 1, \Rightarrow \text{ by Lemma 6.5.11}) \\
&\iff D_b \leq D_{[a=0]} = \neg D_a.
\end{aligned}$$

To see that $\neg$ is antitone, suppose that $D_a \leq D_b$. Then, $a \preccurlyeq b$, so $a \leq nb$ for some $n \in \mathbb{N}$. Hence, $[b=0]a \leq [b=0]bn = 0$, so that $\neg D_b \wedge D_a = D_{[b=0]a} = 0$ and therefore $\neg D_b \leq \neg D_a$. □

**6.5.14 Lemma.** If $A$ is a commutative Rickart C*-algebra, then the lattice $L_A$ satisfies $D_a \leq \bigvee_{r \in \mathbb{Q}^+} D_{[a-r>0]}$ for all $a \in A^+$.

**PROOF.** Because $[a > 0]a = a^+ \geq a$, Lemma 6.5.11 gives $a \preccurlyeq [a > 0]$ and therefore $D_a \leq D_{[a>0]}$. Also, for $r \in \mathbb{Q}^+$ and $a \in A^+$, one has $1 \leq \frac{2}{r}((r-a) \vee a)$, whence

$$[a - r > 0] \leq \frac{2}{r}((r-a) \vee a)[a-r > 0] = \frac{2}{r}(a[a-r > 0]).$$

Lemma 6.5.11 then yields $D_{[a-r>0]} \leq D_{\frac{2}{r}a} = D_a$. In total, we have $D_{[a-r>0]} \leq D_a \leq D_{[a>0]}$ for all $r \in \mathbb{Q}^+$, from which the statement follows. □

The following simplifies Theorem 6.3.20 by restricting to Rickart C*-algebras.

**6.5.15 Theorem.** The Gelfand spectrum $\mathcal{O}(\Sigma(A))$ of a commutative Rickart C*-algebra $A$ is isomorphic to the frame $\mathrm{Idl}(\mathrm{Proj}(A))$ of ideals of $\mathrm{Proj}(A)$. Hence, the regularity condition may be dropped if one uses $\mathrm{Proj}(A)$ instead of $L_A$. Moreover, $\mathcal{O}(\Sigma(A))$ is generated by the sublattice $P_A = \{D_a \in L_A \mid a \in A^+, \neg\neg D_a = D_a\}$ of "clopens" of $L_A$, which is Boolean by construction.

**PROOF.** Because $\neg D_p = D_{1-p}$ for $p \in \mathrm{Proj}(A)$, we have $\neg\neg D_p = D_p$. Conversely, $\neg\neg D_a = D_{[a>0]}$, so that each element of $P_A$ is of the form $D_a = D_p$ for some $p \in \mathrm{Proj}(A)$. So $P_A = \{D_p \mid p \in \mathrm{Proj}(A)\} \cong \mathrm{Proj}(A)$ because each projection $p \in \mathrm{Proj}(A)$ may be selected as the unique representative of its equivalence class $D_p$ in $L_A$. By Lemma 6.5.14, we may use $\mathrm{Proj}(A)$ instead of $L_A$ as the generating lattice for $\mathcal{O}(\Sigma(A))$. So $\mathcal{O}(\Sigma(A))$ is the collection of regular ideals of $\mathrm{Proj}(A)$ by Theorem 6.3.20. But, because $\mathrm{Proj}(A) \cong P_A$ is Boolean, all its ideals are regular, as $D_p \ll D_p$ for each $p \in \mathrm{Proj}(A)$ [57]. This establishes the statement $\mathcal{O}(\Sigma(A)) \cong \mathrm{Idl}(\mathrm{Proj}(A))$. □

We can now give a concise external description of the Gelfand spectrum of the Bohrification of a Rickart C*-algebra $A$, simplifying Theorem 6.4.16.

**6.5.16 Theorem.** The Bohrified state space $\Sigma_A$ of a Rickart C*-algebra $A$ is given by

$$\mathcal{O}(\Sigma_A) \cong \{F : \mathcal{C}(A) \to \mathbf{Set} \mid F(C) \in \mathcal{O}(\Sigma(C)) \text{ and}$$
$$\Sigma(C \subseteq D)(F(C)) \subseteq F(D) \text{ if } C \subseteq D\}.$$

It has a basis given by

$$\mathcal{B}(\Sigma_A) = \{G : \mathcal{C}(A) \to \mathrm{Proj}(A) \mid G(C) \in \mathrm{Proj}(C) \text{ and } G(C) \leq G(D) \text{ if } C \subseteq D\}.$$

More precisely, there is an injection $f : \mathcal{B}(\Sigma_A) \to \mathcal{O}(\Sigma_A)$ given by $f(G)(C) = \mathrm{supp}(\widehat{G(C)})$, using the Gelfand transform of 6.3.10 in **Set**. Each $F \in \mathcal{O}(\Sigma_A)$ can be expressed as $F = \bigvee\{f(G) \mid G \in \mathcal{B}(\Sigma_A), f(G) \leq F\}$.

**PROOF.** By (the proof of) Theorem 6.5.15, one can use $\mathrm{Proj}(C)$ instead of $\underline{L_A}(C)$ as a generating lattice for $\mathcal{O}(\Sigma(A))$. Translating Theorem 6.4.16(b) in these terms yields that $\mathcal{O}(\Sigma_A)$ consists of subfunctors $\underline{U}$ of $\underline{L_A}$ for which $\underline{U}(C) \in \mathrm{Idl}(\mathrm{Proj}(C))$ at each $C \in \mathcal{C}(A)$. Notice that Theorem 6.4.16 holds in $\mathcal{T}_R(A)$ as well as in $\mathcal{T}(A)$ (by interpreting Theorem 6.3.22 in the former instead of in the latter topos). Thus, we obtain a frame isomorphism $\mathrm{Idl}(\mathrm{Proj}(C)) \cong \mathcal{O}(\Sigma(C))$ and the description in the statement. □

**6.5.17 Corollary.** If $A$ is finite-dimensional, then

$$\mathcal{O}(\Sigma_A) \cong \{G : \mathcal{C}(A) \to \mathrm{Proj}(A) \mid G(C) \in \mathrm{Proj}(C) \text{ and } G(C) \leq G(D) \text{ if } C \subseteq D\}.$$

This is a complete Heyting algebra under pointwise order with respect to the usual ordering of projections. As shown in [24], the lattice $\mathcal{O}(\Sigma_A)$ is not Boolean whenever $A$ is noncommutative, so that the intrinsic logical structure carried by $\Sigma_A$ is intuitionistic. This fact may be related conceptually to the fact that the passage from the initial noncommutative C*-algebra $A$ to its Bohrification $\underline{A}$ involves some loss of information. Furthermore, compared with the standard formalism of von Neumann, in which single projections are interpreted as (atomic) propositions, it now appears that in our "Bohrified" description, each atomic proposition $G \in \mathcal{O}(\Sigma_A)$ consists of a famiy of projections, one (namely, $G(C)$) for each classical context $C \in \mathcal{C}(A)$.

We now examine the connection with quantum logic in the usual sense in some more detail. To do so, we assume that $A$ is a Rickart C*-algebra, in which case it follows from Example 6.3.6 that Proj($A$) is a countably complete orthomodular lattice. This includes the situation in which $A$ is a von Neumann algebra, in which case Proj($A$) is a complete orthomodular lattice [73]. For the sake of completeness, we recall:

**6.5.18 Definition.** A (complete) lattice $X$ is called *orthomodular* when it is equipped with a function $\perp \colon X \to X$ that satisfies:

1. $x^{\perp\perp} = x$.
2. $y^\perp \leq x^\perp$ when $x \leq y$.
3. $x \wedge x^\perp = 0$ and $x \vee x^\perp = 1$.
4. $x \vee (x^\perp \wedge y) = y$ when $x \leq y$.

The first three requirements are sometimes called (1) "double negation," (2) "contraposition," (3) "noncontradiction" and "excluded middle;" but, as argued in the Introduction, one should refrain from names suggesting a logical interpretation. If these are satisfied, the lattice is called *orthocomplemented*. The requirement (4), called the orthomodular law, is a weakening of distributivity.

Hence, a Boolean algebra is a lattice that is at the same time a Heyting algebra and an orthomodular lattice with the same operations; that is, $x = x^{\perp\perp}$ for all $x$, where $x^\perp$ is defined to be $(x \Rightarrow 0)$. It is usual to denote the latter by $\neg x$ instead of $x^\perp$ in case the algebra is Boolean.

Using the description of the previous theorem, we are now in a position to compare our Bohrified state space $\mathcal{O}(\Sigma_A)$ with the traditional "quantum logic" Proj($A$). To do so, we recall an alternative characterization of orthomodular lattices.

**6.5.19 Definition.** A (complete) *partial Boolean algebra* is a family $(B_i)_{i \in I}$ of (complete) Boolean algebras whose operations coincide on overlaps:

- Each $B_i$ has the same least element 0.
- $x \Rightarrow_i y$ if and only if $x \Rightarrow_j y$, when $x, y \in B_i \cap B_j$.
- If $x \Rightarrow_i y$ and $y \Rightarrow_j z$, then there is a $k \in I$ with $x \Rightarrow_k z$.
- $\neg_i x = \neg_j x$ when $x \in B_i \cap B_j$.
- $x \vee_i y = x \vee_j y$ when $x, y \in B_i \cap B_j$.
- If $y \Rightarrow_i \neg_i x$ for some $x, y \in B_i$, and $x \Rightarrow_j z$ and $y \Rightarrow_k z$, then $x, y, z \in B_l$ for some $l \in I$.

**6.5.20** The requirements of a partial Boolean algebra imply that the amalgamation $\mathcal{A}(B) = \bigcup_{i \in I} B_i$ carries a well-defined structure $\vee, \wedge, 0, 1, \perp$, under which it becomes an orthomodular lattice. For example, $x^\perp = \neg_i x$ for $x \in B_i \subseteq \mathcal{A}(B)$. Conversely, any orthomodular lattice $X$ is a partial Boolean algebra, in which $I$ is the collection of all orthogonal subsets of $\mathcal{A}(B)$ and $B_i$ is the sublattice of $\mathcal{A}(B)$ generated by $I$. Here, a subset $E \subseteq \mathcal{A}(B)$ is called orthogonal when pairs

$(x, y)$ of different elements of $E$ are orthogonal (i.e., $x \leq y^\perp$). The generated sublattices $B_i$ are therefore automatically Boolean. If we order $I$ by inclusion, then $B_i \subseteq B_j$ when $i \leq j$. Thus, there is an isomorphism between the categories of orthomodular lattices and partial Boolean algebras [34, 40, 61, 63].

**6.5.21** A similar phenomenon occurs in the Heyting algebra $\mathcal{B}(\Sigma_A)$ of Theorem 6.5.16, when this is complete, which is the case for AW*-algebras and, in particular, for von Neumann algebras (provided, of course, that we require $\mathcal{C}(A)$ to consist of commutative subalgebras in the same class). Indeed, we can think of $\mathcal{B}(\Sigma_A)$ as an amalgamation of Boolean algebras: just as every $B_i$ in Definition 6.5.19 is a Boolean algebra, every Proj($C$) in Theorem 6.5.16 is a Boolean algebra. Hence, the fact that the set $I$ in Definition 6.5.19 is replaced by the partially ordered set $\mathcal{C}(A)$ and the requirement in Theorem 6.5.16 that $G$ be monotone are responsible for making the partial Boolean algebra $\mathcal{O}(\Sigma_A)$ into a Heyting algebra. Indeed, this construction works more generally, as the following theorem shows. (Compare also [46] and [88], which write an orthomodular lattice as a sheaf of Boolean and distributive ones, respectively.)

**6.5.22 Theorem.** Let $(I, \leq)$ be a partially ordered set and $B_i$ be an $I$-indexed family of complete Boolean algebras such that $B_i \subseteq B_j$ if $i \leq j$. Then,

$$\mathcal{B}(B) = \{f : I \to \bigcup_{i \in I} B_i \mid \forall_{i \in I}. f(i) \in B_i \text{ and } f \text{ monotone}\}$$

is a complete Heyting algebra, with Heyting implication

$$(g \Rightarrow h)(i) = \bigvee \{x \in B_i \mid \forall_{j \geq i}. x \leq g(j) \Rightarrow h(j)\}.$$

**PROOF.** Defining operations pointwise makes $Y$ into a frame. For example, $f \wedge g$, defined by $(f \wedge g)(i) = f(i) \wedge_i g(i)$, is again a well-defined monotone function whose value at $i$ lies in $B_i$. Hence, as in Definition 6.2.4, $\mathcal{B}(B)$ is a complete Heyting algebra by $(g \Rightarrow h) = \bigvee \{f \in Y \mid f \wedge g \leq h\}$. We now rewrite this Heyting implication:

$$(g \Rightarrow h)(i) = \left(\bigvee \{f \in \mathcal{B}(B) \mid f \wedge g \leq h\}\right)(i)$$
$$= \bigvee \{f(i) \mid f \in \mathcal{B}(B), f \wedge g \leq h\}$$
$$= \bigvee \{f(i) \mid f \in \mathcal{B}(B), \forall_{j \in I}. f(j) \wedge g(j) \leq h(j)\}$$
$$= \bigvee \{f(i) \mid f \in \mathcal{B}(B), \forall_{j \in I}. f(j) \leq g(j) \Rightarrow h(j)\}$$
$$\stackrel{*}{=} \bigvee \{x \in B_i \mid \forall_{j \geq i}. x \leq g(j) \Rightarrow h(j)\}.$$

To finish the proof, we establish the marked equation. First, suppose that $f \in \mathcal{B}(B)$ satisfies $f(j) \leq g(j) \Rightarrow h(j)$ for all $j \in I$. Take $x = f(i) \in B_i$. Then, for all $j \geq i$, we have $x = f(i) \leq f(j) \leq g(j) \Rightarrow h(j)$. Hence, the left-hand side of the marked equation is less than or equal to the right-hand side. Conversely, suppose that $x \in B_i$ satisfies $x \leq g(j) \Rightarrow h(j)$ for all $j \geq i$. Define $f : I \to \bigcup_{i \in I} B_i$ by $f(j) = x$ if $j \geq i$ and $f(j) = 0$ otherwise. Then, $f$ is monotone and $f(i) \in B_i$ for all $i \in I$, whence $f \in Y$. Moreover, $f(j) \leq g(j) \Rightarrow h(j)$ for all $j \in I$. Because $f(i) \leq x$, the right-hand side is less than or equal to the left-hand side. □

**6.5.23 Proposition.** Let $(I, \leq)$ be a partially ordered set. Let $(B_i)_{i \in I}$ be complete partial Boolean algebra and suppose that $B_i \subseteq B_j$ for $i \leq j$. Then, there is an injection $D \colon \mathcal{A}(B) \to \mathcal{B}(B)$. This injection reflects the order: if $D(x) \leq D(y)$ in $Y$, then $x \leq y$ in $X$.

**PROOF.** Define $D(x)(i) = x$ if $x \in B_i$ and $D(x)(i) = 0$ if $x \notin B_i$. Suppose that $D(x) = D(y)$. Then, for all $i \in I$, we have $x \in B_i$ iff $y \in B_i$. Because $x \in \mathcal{A}(B) = \bigcup_{i \in I} B_i$, there is some $i \in I$ with $x \in B_i$. For this particular $i$, we have $x = D(x)(i) = D(y)(i) = y$. Hence, $D$ is injective. If $D(x) \leq D(y)$ for $x, y \in \mathcal{A}(B)$, pick $i \in I$ such that $x \in B_i$. Unless $x = 0$, we have $x = D(x)(i) \leq D(y)(i) = y$. □

**6.5.24** In the situation of the previous proposition, the Heyting algebra $\mathcal{B}(B)$ comes with its Heyting implication, whereas the orthomodular lattice $\mathcal{A}(B)$ has a so-called *Sasaki hook* $\Rightarrow_S$, satisfying the adjunction $x \leq y \Rightarrow_S z$ iff $x \wedge y \leq z$ only for $y$ and $z$ that are compatible. This is the case if and only if $y$ and $z$ generate a Boolean subalgebra; that is, if and only if $y, z \in B_i$ for some $i \in I$. In that case, the Sasaki hook $\Rightarrow_S$ coincides with the implication $\Rightarrow$ of $B_i$. Hence,

$$(D(x) \Rightarrow D(y))(i) = \bigvee \{z \in B_i \mid \forall_{j \geq i}.z \leq D(x)(j) \Rightarrow D(y)(j)\}$$
$$= \bigvee \{z \in B_i \mid z \leq x \Rightarrow y\}$$
$$= (x \Rightarrow_S y).$$

In particular, we find that $\Rightarrow$ and $\Rightarrow_S$ coincide on $B_i \times B_i$ for $i \in I$; furthermore, this is precisely the case in which the Sasaki hook satisfies the defining adjunction for (Heyting) implications.

However, the canonical injection $D$ need not turn Sasaki hooks into implications in general. One finds that

$$D(x \Rightarrow_S y)(i) = \begin{bmatrix} x^\perp \vee (x \wedge y) & \text{if } x \Rightarrow_S y \in B_i \\ 0 & \text{otherwise} \end{bmatrix},$$

$$(D(x) \Rightarrow D(b))(i) = \bigvee \left\{ z \in B_i \mid \forall_{j \geq i}.z \leq \begin{bmatrix} 1 & \text{if } x \notin B_j \\ x^\perp & \text{if } x \in B_j, y \notin B_j \\ x^\perp \vee y & \text{if } x, y \in B_j \end{bmatrix} \right\}.$$

So, if $x \notin B_j$ for any $j \geq i$, we have $D(x \Rightarrow_S y)(i) = 0 \neq 1 = (D(x) \Rightarrow D(y))(i)$.

**6.5.25** To end this section, we consider the so-called Bruns–Lakser completion [17, 27, 84]. The Bruns–Lakser completion of a complete lattice is a complete Heyting algebra that contains the original lattice join-densely. It is the universal in that this inclusion preserves existing distributive joins. Explicitly, the Bruns–Lakser completion of a lattice $L$ is the collection $\mathrm{DIdl}(L)$ of its *distributive ideals*. Here, an ideal $M$ is called *distributive* when $(\bigvee M)$ exists and $(\bigvee M) \wedge x = \bigvee_{m \in M}(m \wedge x)$ for all $x \in L$. Now, consider the orthomodular lattice $X$ with the following Hasse diagram:

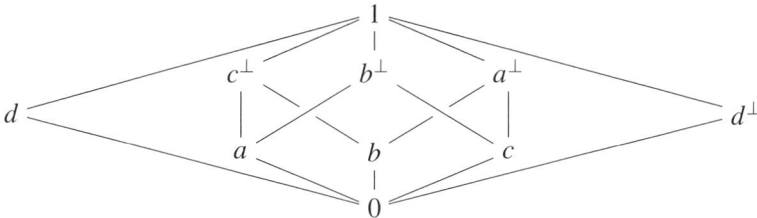

This contains precisely five Boolean algebras; namely, $B_0 = \{0, 1\}$ and $B_i = \{0, 1, i, i^\perp\}$ for $i \in \{a, b, c, d\}$. Hence, $X = \mathcal{A}(B)$ when we take $I = \{0, a, b, c, d\}$ ordered by $i < j$ iff $i = 0$. The monotony requirement in $\mathcal{B}(B)$ becomes $\forall_{i \in \{a,b,c,d\}}. f(0) \leq f(i)$. If $f(0) = 0 \in B_0$, this requirement is vacuous. But, if $f(0) = 1 \in B_0$, the other values of $f$ are already fixed. Thus, one finds that $\mathcal{B}(B) \cong (B_1 \times B_2 \times B_3 \times B_4) + 1$ has 257 elements.

Conversely, the distributive ideals of $X$ are given by

$$\mathrm{DI}(X) = \left\{ \left(\bigcup_{x \in A} \downarrow x\right) \cup \left(\bigcup_{y \in B} \downarrow y\right) \;\middle|\; A \subseteq \{a, b, c, d, d^\perp\}, B \subseteq \{a^\perp, b^\perp, c^\perp\} \right\}$$
$$- \{\emptyset\} + \{X\}.$$

In the terminology of [84],

$$\mathcal{J}_{\mathrm{dis}}(x) = \{S \subseteq \downarrow x \mid x \in S\};$$

that is, the covering relation is the trivial one and $\mathrm{DI}(X)$ is the Alexandrov topology (as a frame/locale). We are unaware of instances of the Bruns–Lakser completion of orthomodular lattices that occur naturally in quantum physics but lead to Heyting algebras different from ideal completions. The set $\mathrm{DI}(X)$ has 72 elements.

The canonical injection $D$ of Proposition 6.5.23 need not preserve the order; hence, it does not satisfy the universal requirement of which the Bruns–Lakser completion is the solution. Therefore, it is unproblemetic to conclude that the construction in Theorem 6.5.22 differs from the Bruns–Lakser completion.

## 6.6 States and Observables

This final section considers some relationships between the external C*-algebra $A$ and its Bohrification $\underline{A}$. For example, we discuss how a state on $A$ in the operator algebraic sense gives rise to a probability integral on $\underline{A}_{\mathrm{sa}}$ within $\mathcal{T}(A)$. The latter corresponds to a suitably adapted version of a probability measure on $\mathcal{O}(\Sigma(\underline{A}))$, justifying the name "Bohrified" state space. We also consider how so-called *Daseinisation* translates an external proposition about an observable $a \in A_{\mathrm{sa}}$ into a subobject of the Bohrified state space. The internalized state and observable are then combined to give a truth value.

**6.6.1 Definition.** A linear functional $\rho: A \to \mathbb{C}$ on a C*-algebra $A$ is called *positive* when $\rho(A^+) \subseteq \mathbb{R}^+$. It is a *state* when it is positive and satisfies $\rho(1) = 1$.

A state $\rho$ is *pure* when $\rho = t\rho' + (1-t)\rho''$ for some $t \in (0, 1)$ and some states $\rho', \rho''$ implies $\rho' = \rho''$. Otherwise, it is called *mixed*. A state is called *faithful* when $\rho(a) = 0$ implies $a = 0$ for all $a \in A^+$.

States are automatically Hermitian, in the sense that $\rho(a^*)$ is the complex conjugate of $\rho(a)$ or, equivalently, $\rho(a) \in \mathbb{R}$ for $a \in A_{\text{sa}}$.

**6.6.2 Example.** If $A = \mathbf{Hilb}(X, X)$ for some Hilbert space $X$, each unit vector $x \in X$ defines a pure state on $A$ by $\rho_x(a) = \langle x \mid a(x) \rangle$. Normal mixed states $\rho$ arise from countable sequences $(r_i)$ of numbers satisfying $0 \leq r_i \leq 1$ and $\sum_i r_i = 1$, coupled with a family $(x_i)$ of $x_i \in X$, through $\rho(a) = \sum_i r_i \rho_{x_i}(a)$. This state is faithful when $(x_i)$ comprises an orthonormal basis of $X$ and each $r_i > 0$.

Taking Bohr's doctrine of classical concepts seriously means accepting that two operators can only be added in a meaningful way when they commute, leading to the following notion [1, 18, 19, 71].

**6.6.3 Definition.** A *quasilinear* functional on a C*-algebra $A$ is a map $\rho: A \to \mathbb{C}$ that is linear on all commutative subalgebras and satisfies $\rho(a + ib) = \rho(a) + i\rho(b)$ for all (possibly noncommuting) $a, b \in A_{\text{sa}}$. It is called *positive* when $\rho(A^+) \subseteq A^+$, and it is called a *quasistate* when furthermore $\rho(1) = 1$.

This kind of quasilinearity determines when some property $P$ of $A$ descends to a corresponding property $\underline{P}$ for the Bohrification $\underline{A}$, as the following lemma shows. To be precise, for $P \subseteq A$, define $\underline{P} \in \text{Sub}(\underline{A})$ by $\underline{P}(C) = P \cap C$. A property $P \subseteq A$ is called *quasilinear* when $a, b \in P \cap A_{\text{sa}}$ implies $ra + isb \in P$ for all $r, s \in \mathbb{R}$.

**6.6.4 Lemma.** Let $A$ be a C*-algebra, and let $P \subseteq A$ be a quasilinear property. Then, $P = A$ if and only if $\underline{P} = \underline{A}$.

**PROOF.** One implication is trivial; for the other, suppose that $\underline{P} = \underline{A}$. For $a \in A$, denote by $C^*(a)$ the C*-subalgebra generated by $a$ (and 1). When $a$ is self-adjoint, $C^*(a)$ is commutative. So $A_{\text{sa}} \subseteq P$, whence by quasilinearity of $P$ and the unique decomposition of elements in a real and imaginary part, we have $A \subseteq P$. □

**6.6.5 Definition.** An *integral* on a Riesz space $R$ is a linear functional $I: R \to \mathbb{R}$ that is positive (i.e., if $f \geq 0$ then also $I(f) \geq 0$). If $R$ has a strong unit 1 (see Definition 6.3.14), then an integral $I$ satisfying $I(1) = 1$ is called a *probability integral*. An integral $I$ is *faithful* when $I(f) = 0$ and $f \geq 0$ imply $f = 0$.

**6.6.6 Example.** Except in the degenerate case $I(1) = 0$, any integral can obviously be normalized to a probability integral. The prime example of an integral is the Riemann or Lebesgue integral on the ordered vector space $C([0, 1], \mathbb{C})$. More generally, any positive linear functional on a commutative C*-algebra provides an example, states yielding probability integrals.

**6.6.7 Definition.** Let $R$ be a Riesz space. We now define the locale $\mathcal{I}(R)$ of probability integrals on $R$. First, let $\text{Int}(R)$ be the distributive lattice freely generated by symbols $P_f$ for $f \in R$, subject to the relations

$$P_1 = 1,$$
$$P_f \wedge P_{-f} = 0,$$
$$P_{f+g} \leq P_f \vee P_g,$$
$$P_f = 0 \qquad \text{(for } f \leq 0\text{)}.$$

This lattice generates a frame $\mathcal{O}(\mathcal{I}(R))$ by furthermore imposing the regularity condition

$$P_f = \bigvee \{P_{f-q} \mid q \in \mathbb{Q}, q > 0\}.$$

**6.6.8** Classically, points $p$ of $\mathcal{I}(R)$ correspond to probability integrals $I$ on $R$, by mapping $I$ to the point $p_I$ given by $p_I(P_f) = 1$ iff $I(f) > 0$. Conversely, a point $p$ defines an integral $I_p = (\{q \in \mathbb{Q} \mid p \models P_{f-q}\}, \{r \in \mathbb{Q} \mid p \models P_{r-f}\})$, which is a Dedekind cut by the relations imposed on $P_{(\_)}$, as in Example 6.2.33. Therefore, intuitively, $P_f = \{\rho \colon R \to \mathbb{R} \mid \rho(f) > 0, \rho \text{ positive linear}\}$.

Classically, for a locally compact Hausdorff space $X$, the Riesz-Markov theorem provides a duality between integrals on a Riesz space $\{f \in C(X, \mathbb{R}) \mid \text{supp}(f) \text{ compact}\}$ and regular measures on the Borel subsets of $X$. Constructively, one uses so-called valuations, which are only defined on open subsets of $X$, instead of measures. Theorem 6.6.13 gives a constructively valid version of the Riesz–Markov theorem. In preparation, we consider a suitable constructive version of measures.

**6.6.9** Classically, points of the locale $\mathbb{R}$ of Example 6.2.33 are Dedekind cuts $(L, U)$ (and $\mathcal{O}(\mathbb{R})$ is the usual Euclidean topology). We now introduce two variations on the locale $\mathbb{R}$. First, consider the locale $\mathbb{R}_l$ that is generated by formal symbols $q \in \mathbb{Q}$ subject to the following relations:

$$q \wedge r = \min(q, r), \qquad q = \bigvee \{r \mid r > q\}, \qquad 1 = \bigvee \{q \mid q \in \mathbb{Q}\}.$$

Classically, its points are *lower reals*, and locale morphisms to $\mathbb{R}_l$ correspond to lower-semicontinuous real-valued functions. Restricting generators to $0 \leq q \leq 1$ yields a locale denoted $[0, 1]_l$.

**6.6.10** Second, let $\mathbb{IR}$ be the locale defined by the very same generators $(q, r)$ and relations as in Example 6.2.33, except that we omit the fourth relation $(q, r) = (q, r_1) \vee (q_1, r)$ for $q \leq q_1 \leq r_1 \leq r$. The effect is that, classically, points of $\mathbb{IR}$ again correspond to pairs $(L, U)$ as in Example 6.2.33, except that the lower real $L$ and the upper real $U$ need not combine into a Dedekind cut, as the "kissing" requirement is no longer in effect. Classically, a point $(L, U)$ of $\mathbb{IR}$ corresponds to a compact interval $[\sup(L), \inf(U)]$ (including the singletons $[x, x] = \{x\}$). Ordered by reverse inclusion, the topology they carry is the *Scott topology* [2], whose closed sets are lower sets that are closed under directed joins. Hence, each open interval $(q, r)$ in $\mathbb{R}$ (with $q = -\infty$ and $r = \infty$ allowed) corresponds to a Scott open $\{[a, b] \mid q < a \leq b < r\}$ in $\mathbb{IR}$, and these form the basis of the Scott topology. Therefore, $\mathbb{IR}$ is also called the *interval domain* [72, 80]. One can think of it as approximations of real numbers by rational intervals, interpreting each individual interval as finitary information about the real number under scrutiny. The ordering by reverse inclusion is then explained as a smaller interval means that more information is available about the real number.

In a Kripke topos $[P, \mathbf{Set}]$ over a poset $P$ with a least element, one has $\mathcal{O}(\mathbb{IR})(p) = \mathcal{O}((\uparrow p) \times \mathbb{IR})$, which may be identified with the set of monotone functions from $\uparrow p$ to $\mathcal{O}(\mathbb{IR})$. This follows by carefully adapting the proof of [70, Theorem VI.8.2].

**6.6.11 Definition.** A *continuous probability valuation* on a locale $X$ is a monotone function $\mu \colon \mathcal{O}(X) \to \mathcal{O}([0,1]_l)$ such that $\mu(1) = 1$ as well as $\mu(U) + \mu(V) = \mu(U \wedge V) + \mu(U \vee V)$ and $\mu(\bigvee_i U_i) = \bigvee_i \mu(U_i)$ for a directed family $(U_i)$. Like integrals, continuous probability valuations organize themselves in a locale $\mathcal{V}(X)$.

**6.6.12 Example.** If $X$ is a compact Hausdorff space, a continuous probability valuation on $\mathcal{O}(X)$ is the same thing as a regular probability measure on $X$.

**6.6.13 Theorem.** [31] Let $R$ be an f-algebra and $\Sigma$ its spectrum. Then, the locales $\mathcal{I}(R)$ and $\mathcal{V}(\Sigma)$ are isomorphic. A continuous probability valuation $\mu$ gives a probability integral by

$$I_\mu(f) = (\sup_{(s_i)} \sum s_i \mu(D_{f-s_i} \wedge D_{s_{i+1}-f}), \inf_{(s_i)} \sum s_{i+1}(1 - \mu(D_{s_i - f}) - \mu(D_{f-s_{i+1}}))),$$

where $(s_i)$ is a partition of $[a, b]$ such that $a \leq f \leq b$. Conversely, a probability integral $I$ gives a continuous probability valuation

$$\mu_I(D_a) = \sup\{I(na^+ \wedge 1) \mid n \in \mathbb{N}\}. \qquad \square$$

**6.6.14 Corollary.** For a C*-algebra $A$, the locale $\mathcal{I}(\underline{A})$ in $\mathcal{T}(A)$ of probability integrals on $\underline{A}_{\mathrm{sa}}$ is isomorphic to the locale $\mathcal{V}(\underline{\Sigma}(A))$ in $\mathcal{T}(A)$ of continuous probability valuations on $\underline{\Sigma}(A)$.

PROOF. Interpret Theorem 6.6.13—whose proof is constructive—in $\mathcal{T}(A)$.
$\square$

**6.6.15 Theorem.** There is a bijective correspondence between (faithful) quasi-states on a C*-algebra $A$ and (faithful) probability integrals on $\underline{A}_{\mathrm{sa}}$.

PROOF. Every quasistate $\rho$ gives a natural transformation $I_\rho \colon \underline{A}_{\mathrm{sa}} \to \underline{\mathbb{R}}$ whose component $(I_\rho)_C \colon C_{\mathrm{sa}} \to \mathbb{R}$ is the restriction $\rho_{|C_{\mathrm{sa}}}$ of $\rho$ to $C_{\mathrm{sa}} \subseteq A_{\mathrm{sa}}$. Conversely, let $I \colon \underline{A}_{\mathrm{sa}} \to \underline{\mathbb{R}}$ be an integral. Define $\rho \colon A_{\mathrm{sa}} \to \mathbb{R}$ by $\rho(a) = I_{C^*(a)}(a)$, where $C^*(a)$ is the sub-C*-algebra generated by $a$. For commuting $a, b \in A_{\mathrm{sa}}$, then

$$\begin{aligned}
\rho(a+b) &= I_{C^*(a+b)}(a+b) \\
&= I_{C^*(a,b)}(a+b) \\
&= I_{C^*(a,b)}(a) + I_{C^*(a,b)}(b) \\
&= I_{C^*(a)}(a) + I_{C^*(b)}(b) \\
&= \rho(a) + \rho(b),
\end{aligned}$$

because $I$ is a natural transformation, $C^*(a) \cup C^*(b) \subseteq C^*(a,b) \supseteq C^*(a+b)$, and $I$ is locally linear. Moreover, $\rho$ is positive because $I$ is locally positive, by Lemma 6.6.4. Hence, we have defined $\rho$ on $A_{sa}$ and may extend it to $A$ by complex linearity. It is clear that the two maps $I \mapsto \rho$ and $\rho \mapsto I$ are each other's inverse. □

**6.6.16** Let $\rho$ be a (quasi-)state on a C*-algebra $A$. Then, $\mu_\rho$ is a continuous probability valuation on $\mathcal{O}(\underline{\Sigma}(\underline{A}))$. Hence, $\mu_\rho(\_) = 1$ is a term of the internal language of $\mathcal{T}(A)$ with one free variable of type $\mathcal{O}(\underline{\Sigma}(\underline{A}))$. Its interpretation $[\![\mu_\rho(\_) = 1]\!]$ defines a subobject of $\mathcal{O}(\underline{\Sigma}(\underline{A}))$ or, equivalently, a morphism $[\rho]\colon \mathcal{O}(\underline{\Sigma}(\underline{A})) \to \underline{\Omega}$.

For Rickart C*-algebras, we can make Theorem 6.6.15 a bit more precise.

**6.6.17 Definition.**

(a) A *probability measure* on a countably complete orthomodular lattice $X$ is a function $\mu\colon X \to [0,1]_l$ that on any countably complete Boolean sublattice of $X$ restricts to a probability measure (in the traditional sense).

(b) A *probability valuation* on an orthomodular lattice $X$ is a function $\mu\colon X \to [0,1]_l$ such that $\mu(0) = 0$, $\mu(1) = 1$, $\mu(x) + \mu(y) = \mu(x \wedge y) + \mu(x \vee y)$, and if $x \leq y$, then $\mu(x) \leq \mu(y)$.

**6.6.18 Lemma.** Let $\mu$ be a probability valuation on a Boolean algebra $X$. Then, $\mu(x)$ is a Dedekind cut for any $x \in X$.

**PROOF.** Because $X$ is Boolean, we have $\mu(\neg x) = 1 - \mu(x)$. Let $q, r \in \mathbb{Q}$, and suppose that $q < r$. We have to prove that $q < \mu(x)$ or $\mu(x) \leq r$. Because the inequalities concern rationals, it suffices to prove that $q < \mu(x)$ or $1 - r < 1 - \mu(x) = \mu(\neg x)$. This follows from $1 - (r - q) < 1 = \mu(1) = \mu(x \vee \neg x)$ and $q - r < 0 = \mu(0) = \mu(x \wedge \neg x)$. □

The following theorem relates Definition 6.6.11 and Definition 6.6.17. Definition 6.6.11 is applied to the Gelfand spectrum $\underline{\Sigma}(\underline{A})$ of the Bohrification of a Rickart C*-algebra $A$. Part (a) of Definition 6.6.17 is applied to Proj($A$) in **Set** for a Rickart C*-algebra $A$, and part (b) is applied to the lattice $\underline{P}_A$ of Theorem 6.5.15 in $\mathcal{T}(A)$.

**6.6.19 Theorem.** For a Rickart C*-algebra $A$, there is a bijective correspondence between:

(a) quasi-states on $A$
(b) probability measures on Proj($A$)
(c) probability valuations on $\underline{P}_A$
(d) continuous probability valuations on $\underline{\Sigma}(\underline{A})$

**PROOF.** The correspondence between (a) and (d) is Theorem 6.6.15. The correspondence between (c) and (d) follows from Theorem 6.5.15 and the observation that valuations on a compact regular frame are determined by their behavior on a generating lattice [31, Section 3.3]; indeed, if a frame $\mathcal{O}(X)$ is generated by $L$, then a probability measure $\mu$ on $L$ yields a continuous probability valuation $\nu$ on $\mathcal{O}(X)$ by $\nu(U) = \sup\{\mu(u) \mid u \in U\}$, where $U \subseteq L$ is regarded as an element of $\mathcal{O}(X)$. Finally, we turn to the correspondence between (b) and (c). Because $\mathbb{R}$ in $\mathcal{T}(A)$ is the constant functor $C \mapsto \mathbb{R}$ (as opposed to $\mathbb{R}_l$), according to the previous lemma a probability valuation $\mu \colon \underline{\mathrm{Idl}}(\mathrm{Proj}(\underline{A})) \to \overline{[0,1]}_l$ is defined by its components $\mu_C \colon \mathrm{Proj}(C) \to [0,1]$. By naturality, for $p \in \mathrm{Proj}(C)$, the real number $\mu_C(p)$ is independent of $C$, from which the correspondence between (b) and (c) follows immediately. □

**6.6.20** We now turn to internalizing an elementary proposition $a \in (q,r)$ concerning an observable $a \in A_{\mathrm{sa}}$ and rationals $q, r \in \mathbb{Q}$ with $q < r$. If $A$ were commutative, then $a$ would have a Gelfand transform $\hat{a} \colon \Sigma(A) \to \mathbb{R}$, and we could just internalize $\hat{a}^{-1}(q,r) \subseteq \Sigma(A)$ directly. For noncommutative $A$, there can be contexts $C \in \mathcal{C}(A)$ that do not contain $a$; therefore, the best we can do is approximate. Our strategy is to replace the reals $\mathbb{R}$ by the interval domain $\mathbb{IR}$ of 6.6.10. We will construct a locale morphism $\underline{\delta}(a) \colon \underline{\Sigma}(\underline{A}) \to \underline{\mathbb{IR}}$, called the *Daseinisation* of $a \in A_{\mathrm{sa}}$—this terminology stems from [38], but the morphism is quite different from the implementation in that article. The elementary proposition $a \in (q,r)$ is then internalized as the composite morphism

$$\underline{1} \xrightarrow{(q,r)} \mathcal{O}(\underline{\mathbb{IR}}) \xrightarrow{\underline{\delta}(a)^{-1}} \mathcal{O}(\underline{\Sigma}(\underline{A})),$$

where $(q,r)$ maps into the monotone function with constant value $\mathord{\downarrow}(q,r)$. (As in 6.6.10, $(q,r)$ is seen as an element of the generating semilattice, whereas $\mathord{\downarrow}(q,r)$ is its image in the frame $\mathcal{O}(\mathbb{IR})$ under the canonical inclusion of Proposition 6.2.13.)

**6.6.21** The interval domain $\mathcal{O}(\mathbb{IR})$ of 6.6.10 can be constructed as $\mathcal{F}(\mathbb{Q} \times_< \mathbb{Q}, \triangleleft)$, as in Definition 6.2.12 [72]. The pertinent meet-semilattice $\mathbb{Q} \times_< \mathbb{Q}$ consists of pairs $(q,r) \in \mathbb{Q} \times \mathbb{Q}$ with $q < r$, ordered by inclusion (i.e., $(q,r) \leq (q',r')$ iff $q' \leq q$ and $r \leq r'$), with a least element $0$ added. The covering relation $\triangleleft$ is defined by $0 \triangleleft U$ for all $U$, and $(q,r) \triangleleft U$ iff for all rational $q', r'$ with $q < q' < r' < r$ there exists $(q'',r'') \in U$ with $(q',r') \leq (q'',r'')$. In particular, we may regard $\mathcal{O}(\mathbb{IR})$ as a subobject of $\mathbb{Q} \times_< \mathbb{Q}$. As in 6.4.13:

$$\mathcal{O}(\underline{\mathbb{IR}})(\mathbb{C}) \cong \{ F \in \mathrm{Sub}(\underline{\mathbb{Q} \times_< \mathbb{Q}}) \mid \forall_{C \in \mathcal{C}(A)}. F(C) \in \mathcal{O}(\mathbb{IR}) \}.$$

**6.6.22 Lemma.** For a C*-algebra $A$ and a fixed element $a \in A_{\mathrm{sa}}$, the components $\underline{d}(a)_C \colon \mathbb{Q} \times_< \mathbb{Q} \to \mathrm{Sub}(\underline{L_{A|\uparrow C}})$ given by

$$\underline{d}(a)^*_C(q,r)(D) = \{ \mathrm{D}_{f-q} \wedge \mathrm{D}_{r-g} \mid f, g \in D_{\mathrm{sa}}, f \leq a \leq g \}$$
$$\underline{d}(a)^*_C(0)(D) = \{ \mathrm{D}_0 \}$$

form a morphism $\underline{d}(a)^* \colon \underline{\mathbb{Q} \times_< \mathbb{Q}} \to \underline{\Omega^{L_A}}$ in $\mathcal{T}(A)$ via 6.4.13. This morphism is a continuous map $(\underline{L_A}, \underline{\triangleleft}) \to (\underline{\mathbb{Q} \times_< \mathbb{Q}}, \blacktriangleleft)$ in the sense of Definition 6.2.14.

Notice that because $\underline{\mathbb{Q} \times_< \mathbb{Q}}(C) = \mathbb{Q} \times_< \mathbb{Q}$ for any $C \in \mathcal{C}(A)$, the natural transformation $\underline{d}(a)$ is completely determined by its component at $\mathbb{C} \in \mathcal{C}(A)$.

**PROOF.** We verify that the map defined in the statement satisfies the conditions of Definition 6.2.14.

(a) We have to show that $\Vdash \forall_{D_a \in \underline{L_A}} \exists_{(q,r) \in \underline{\mathbb{Q} \times_< \mathbb{Q}}}.D_a \in \underline{d}(a)^*(q, r)$. By interpreting via 6.2.24, we therefore have to prove: for all $C \in \mathcal{C}(A)$ and $D_c \in L_C$, there are $(q, r) \in \mathbb{Q} \times_< \mathbb{Q}$ and $f, g \in C_{\text{sa}}$ such that $f \leq a \leq g$ and $D_c = D_{f-q} \wedge D_{r-g}$. Equivalently, we have to find $(q, r) \in \mathbb{Q} \times_< \mathbb{Q}$ and $f, g \in C_{\text{sa}}$ such that $f + q \leq a \leq r + g$ and $D_c = D_f \wedge D_{-g}$. Choosing $f = c$, $g = -c$, $q = -\|c\| - \|a\|$ and $r = \|c\| + \|a\|$ does the job, because $D_c = D_c \wedge D_c$ and

$$f + q = c - \|c\| - \|a\| \leq -\|a\| \leq a \leq \|a\| \leq \|c\| + \|a\| - c = r + g.$$

(b) We have to show that

$$\Vdash \forall_{(q,r),(q',r') \in \underline{\mathbb{Q} \times_< \mathbb{Q}}} \forall_{u,v \in \underline{L_A}}.u \in \underline{d}(a)^*(q, r) \wedge v \in \underline{d}(a)^*(q', r')$$
$$\Rightarrow u \wedge v \vartriangleleft \underline{d}(a)^*((q, r) \wedge (q', r')).$$

Going through the motions of 6.2.24, that means we have to prove: for all $(q, r), (q', r') \in \mathbb{Q} \times_< \mathbb{Q}$, $C \subseteq D \in \mathcal{C}(A)$, and $f, f', g, g' \in C_{\text{sa}}$, if $(q'', r'') = (q, r) \wedge (q', r') \neq 0$, $f \leq a \leq g$ and $f' \leq a \leq g'$, then

$$D_{f-q} \wedge D_{r-g} \wedge D_{f'-q'} \wedge D_{r'-g'}$$
$$\vartriangleleft \{D_{f''-q''} \wedge D_{r''-g''} \mid f'', g'' \in D_{\text{sa}}, f'' \leq a \leq g''\}.$$

We distinguish the possible cases of $(q'', r'')$ (which distinction is constructively valid because it concerns rationals). For example, if $(q'', r'') = (q, r')$, then $q \leq q' \leq r \leq r'$. So $D_{f-q} \wedge D_{r'-g'} = D_{f''-q''} \wedge D_{r''-g''}$ for $f'' = f$, $g'' = g'$, $q'' = q$, and $r'' = r'$, whence the statement holds by (a) and (c) of Definition 6.2.10. The other cases are analogous.

(c) We have to show that

$$\Vdash \forall_{(q,r) \in \underline{\mathbb{Q} \times_< \mathbb{Q}}} \forall_{U \in \underline{\mathcal{P}(\mathbb{Q} \times_< \mathbb{Q})}}.(q, r) \vartriangleleft U \Rightarrow \underline{d}(a)^*(q, r) \vartriangleleft \bigcup_{(q',r') \in U} \underline{d}(a)^*(q', r').$$

By 6.2.24, we therefore have to prove: for all $(q, r) \in \mathbb{Q} \times_< \mathbb{Q}$, $U \subseteq U' \subseteq \mathbb{Q} \times_< \mathbb{Q}$, $D \in \mathcal{C}(A)$ and $f, g \in D_{\text{sa}}$, if $(q, r) \vartriangleleft U$ and $f \leq a \leq g$, then

$$D_{f-q} \wedge D_{r-g} \vartriangleleft \{D_{f'-q'} \wedge D_{r'-g'} \mid (q', r') \in U', f', g' \in D_{\text{sa}}, f' \leq a \leq g'\}.$$

To establish this, it suffices to show $D_{f-q} \wedge D_{r-g} \vartriangleleft \{D_{f-q'} \wedge D_{r-g} \mid (q', r') \in U\}$ when $(q, r) \vartriangleleft U$. Let $s \in \mathbb{Q}$ satisfy $0 < s$. Then, one has $(q, r - s) < (q, r)$. Because $(q, r) \vartriangleleft U$, 6.6.21 yields a $(q'', r'') \in U$ such that $(q, r - s) \leq (q'', r'')$, and so $r - s \leq r''$. Taking $U_0 = \{(q'', r'')\}$, one has $r - g - s \leq r'' - g$ and therefore $D_{r-g-s} \leq D_{r''-g} = \bigvee U_0$. So, by Corollary 6.4.14, we have $D_{r-g} \vartriangleleft \{D_{r'-g} \mid (q', r') \in U\}$. Similarly, one finds $D_{f-q} \vartriangleleft \{D_{f-q'} \mid (q', r') \in U\}$. Finally, $D_{f-q} \wedge D_{r-g} \vartriangleleft \{D_{f-q'} \wedge D_{r'-g} \mid (q', r') \in U\}$ by Definition 6.2.10(d). □

**6.6.23 Definition.** Let $A$ be a C*-algebra. The Daseinisation of $a \in A_{\text{sa}}$ is the locale morphism $\underline{\delta}(a) \colon \underline{\Sigma(A)} \to \underline{\mathbb{IR}}$, whose associated frame morphism $\underline{\delta}(a)^{-1}$ is given by $\mathcal{F}(\underline{d}(a)^*)$, where $\mathcal{F}$ is the functor of Proposition 6.2.15, and $\underline{d}(a)$ comes from Lemma 6.6.22.

**6.6.24 Example.** The locale $\underline{\Sigma}(\underline{A})$ is described externally by its value at $\mathbb{C} \in \mathcal{C}(A)$, see Theorem 6.4.16. The component at $\mathbb{C}$ of the Daseinisation $\underline{\delta}(a)$ is given by

$$\underline{\delta}(a)_\mathbb{C}^{-1}(q,r)(C) = \{D_{f-q} \wedge D_{r-g} \mid f,g \in C_{\text{sa}}, f \leq a \leq g\}.$$

Now, suppose that $A$ is commutative. Then, classically, $D_a = \{\rho \in \Sigma(A) \mid \rho(a) > 0\}$ as in 6.3.9. Hence, $D_{f-r} = \{\rho \in \Sigma(A) \mid \rho(f) > r\}$, so that

$$\underline{\delta}(a)_\mathbb{C}^{-1}(q,r)(C) = \bigcup_{\substack{f,g \in C_{\text{sa}} \\ f \leq a \leq g}} \{\rho \in \Sigma(A) \mid \rho(f) > q \text{ and } \rho(g) < r\}$$
$$= \{\rho \in \Sigma(A) \mid \exists_{f \leq a}.q < \rho(f) < r \text{ and } \exists_{g \geq a}.q < \rho(g) < r\}$$
$$= \{\rho \in \Sigma(A) \mid q < \rho(a) < r\}$$
$$= \hat{a}^{-1}(q,r).$$

**6.6.25 Proposition.** *The map* $\underline{\delta}: A_{\text{sa}} \to C(\Sigma(A), \mathbb{IR})$ *is injective. Moreover, $a \leq b$ if and only if $\underline{\delta}(a) \leq \underline{\delta}(b)$.*

**PROOF.** Suppose that $\underline{\delta}(a) = \underline{\delta}(b)$. Then, for all $C \in \mathcal{C}(A)$, the sets $L_a(C) = \{f \in C_{\text{sa}} \mid f \leq a\}$ and $U_a(C) = \{g \in C_{\text{sa}} \mid a \leq g\}$ must coincide with $L_b(C)$ and $U_b(C)$, respectively. Imposing these equalities at $C = C^*(a)$ and at $C = C^*(b)$ yields $a = b$. The order in $A_{\text{sa}}$ is clearly preserved by $\underline{\delta}$, whereas the converse implication can be shown by the same method as the first claim of the proposition. □

**6.6.26** Given a state $\rho$ of a C*-algebra $A$, an observable $a \in A_{\text{sa}}$, and an interval $(q,r)$ with rational endpoints $q,r \in \mathbb{Q}$, we can now compose the morphisms of 6.6.16, 6.6.20, and Definition 6.6.23 to obtain a truth value

$$1 \xrightarrow{(q,r)} \mathcal{O}(\mathbb{IR}) \xrightarrow{\underline{\delta}(a)^{-1}} \mathcal{O}(\underline{\Sigma}(\underline{A})) \xrightarrow{[\rho]} \underline{\Omega}.$$

Unfolding definitions, we find that at $\mathbb{C} \in \mathcal{C}(A)$, this truth value is given by

$$([\rho] \circ \underline{\delta}(a)^{-1} \circ (q,r))_\mathbb{C}(*)$$
$$= [\![\mu_\rho(\underline{d}(a)^*(q,r)) = 1]\!](C)$$
$$= \{C \in \mathcal{C}(A) \mid C \Vdash \mu_\rho(\underline{d}(a)^*(q,r)) = 1\}$$
$$= \{C \in \mathcal{C}(A) \mid C \Vdash \mu_\rho(\bigvee_{\substack{f,g \in C_{\text{sa}} \\ f \leq a \leq g}} D_{f-q} \wedge D_{r-g}) = 1\}$$
$$= \{C \in \mathcal{C}(A) \mid C \Vdash \mu_\rho(\bigvee_{f \leq a} D_{f-q}) = 1, \ C \Vdash \mu_\rho(\bigvee_{g \geq a} D_{r-g}) = 1\}.$$

By Theorem 6.6.13 and 6.2.24, $C \Vdash \mu_\rho(\bigvee_{f \leq a} D_{f-q}) = 1$ if and only if for all $n \in \mathbb{N}$ there are $m \in \mathbb{N}$ and $f \in C_{\text{sa}}$ such that $f \leq a$ and $\rho(m(f-q)^+ \wedge 1) > 1 - \frac{1}{n}$.

Hence, the preceding truth value is given by

$$\{C \in \mathcal{C}(A) \mid \forall_{n \in \mathbb{N}} \exists_{m \in \mathbb{N}} \exists_{f, g \in C_{\mathrm{sa}}} . f \le a \le g, \, \rho(m(f-q)^+ \wedge 1) > 1 - \frac{1}{n},$$
$$\rho(m(r-g)^+ \wedge 1) > 1 - \frac{1}{n}\}.$$

**6.6.27** If $A$ is a von Neumann algebra, the pairing formula of 6.6.26 simplifies further. Using the external description of the Bohrified state space in Theorem 6.5.16, one finds that the following are equivalent for a general open $F \in \mathcal{O}(\Sigma_A)$ and a state $\rho \colon A \to \mathbb{C}$:

$$C \Vdash \mu_\rho(F) = 1,$$
$$C \Vdash \forall_{q \in \mathbb{Q}, q < 1} . \mu_\rho(F) > q,$$
for all $D \supseteq C$ and rational $q < 1$: $D \Vdash \mu_\rho(F) > q,$
for all $D \supseteq C$ and rational $q < 1$: $D \Vdash \exists_{u \in F} . \mu_\rho(u) > q,$
for all $D \supseteq C$ and $q < 1$, there is $p \in F(D)$ with $D \Vdash \mu_\rho(p) > q,$
for all $q < 1$, there is $p \in F(C)$ with $\rho(p) > q,$
$$\sup_{p \in F(C)} \rho(p) = 1.$$

By Proposition 6.5.8 and Theorem 6.5.15, one may choose basic opens $\mathcal{D}_{f-q}$ of the spectrum $\underline{\Sigma}(A))$ corresponding to projections $[f - q > 0]$ of $\mathrm{Proj}(\underline{A})$. Let us now return to the case $F(C) = \{\mathcal{D}_{f-q} \mid f \in C_{\mathrm{sa}}, f \le a\}$. By Theorem 6.5.16, $F(C)$ is generated by projections; by Theorem 6.5.4, we can take their supremum, so that $(\bigvee F)(C) = \bigvee \{[f - q > 0] \mid f \in C_{\mathrm{sa}}, f \le a\}$. Hence, the preceding forcing condition $C \Vdash \mu_\rho(\bigvee_{f \le a} \mathcal{D}_{f-q}) = 1$ is equivalent to $\rho(\bigvee\{[f - q > 0] \mid f \in C_{\mathrm{sa}}, f \le a\}) = 1$. Thus, the pairing formula of 6.6.26 results in the truth value

$$\{C \in \mathcal{C}(A) \mid \rho(\bigvee\{[f - q > 0] \mid f \in C_{\mathrm{sa}}, f \le a\}) = 1,$$
$$\rho(\bigvee\{[r - g > 0] \mid g \in C_{\mathrm{sa}}, a \le g\}) = 1\},$$

listing the "possible worlds" $C$ in which the proposition $a \in (q, r)$ holds in state $\rho$ in the classical sense.

# References

[1] J. F. Aarnes. Quasi-states on C*-algebras. *Transactions of the American Mathematical Society*, **149** (1970), 601–25.

[2] S. Abramsky and A. Jung. Domain theory. In *Handbook of Logic in Computer Science*, vol. 3, eds. S. Abramsky, Dov M. Gabbay, and T. S. E. Maibaum. Oxford University Press, 1994, pp. 1–168.

[3] S. Abramsky and S. Vickers. Quantales, observational logic and process semantics. *Mathematical Structures in Computer Science*, **3** (1993), 161–227.

[4] P. Aczel. Aspects of general topology in constructive set theory. *Annals of Pure and Applied Logic*, **137** (2006), 3–29.

[5] J. C. Baez and J. Dolan. Higher-dimensional algebra III: $n$-categories and the algebra of opetopes. *Advances in Mathematics*, **135** (1998), 145–206.

[6] B. Banaschewski and C. J. Mulvey. The spectral theory of commutative C*-algebras: The constructive spectrum. *Quaestiones Mathematicae*, **23** (2000), 425–64.

[7] B. Banaschewski and C. J. Mulvey. The spectral theory of commutative C*-algebras: The constructive Gelfand–Mazur theorem. *Quaestiones Mathematicae*, **23** (2000), 465–88.

[8] B. Banaschewski and C. J. Mulvey. A globalisation of the Gelfand duality theorem. *Annals of Pure and Applied Logic*, **137** (2006), 62–103.

[9] G. Battilotti and G. Sambin. Pretopologies and uniform presentation of sup-lattices, quantales and frames. *Annals of Pure and Applied Logic*, **137** (2006), 30–61.

[10] S. K. Berberian. *Baer \*-rings*. Berlin, Heidelberg, New York: Springer-Verlag, 1972.

[11] G. Birkhoff and J. von Neumann. The logic of quantum mechanics. *Annals of Mathematics*, **37** (1936), 823–43.

[12] B. Blackadar. Projections in C*-algebras. In *C\*-Algebras: 1943–1993: A Fifty-Year Celebration*, Comtemporary Mathematics, vol. 167, ed. R. S. Doran. Providence, RI: American Mathematical Society, 1993, pp. 131–49.

[13] N. Bohr. *Collected Works*, Vol. 6: *Foundations of Quantum Physics I (1926–1932)*, ed. J. Kalckar. Amsterdam: North-Holland Physics Publishing (Elsevier), 1985.

[14] N. Bohr. *Collected Works*, Vol. 7: *Foundations of Quantum Physics II (1933–1958)*, ed. J. Kalckar. Amsterdam: North-Holland Physics Publishing/Elsevier, 1996.

[15] F. Borceux. *Handbook of Categorical Algebra 1: Basic Category Theory*, Encyclopedia of Mathematics and Its Applications 50. Cambridge: Cambridge University Press, 1994.

[16] F. Borceux. *Handbook of Categorical Algebra 3: Categories of Sheaves*, Encyclopedia of Mathematics and Its Applications 52. Cambridge: Cambridge University Press, 1994.

[17] G. Bruns and H. Lakser. Injective hulls of semilattices. *Canadian Mathematical Bulletin*, **13** (1970), 115–18.

[18] L. J. Bunce and J. D. Maitland Wright. The Mackey–Gleason problem for vector measures on projections in von Neumann algebras. *Journal of the London Mathematical Society 2*, **49** (1994), 133–49.

[19] L. J. Bunce and J. D. Maitland Wright. The quasi-linearity problem for C*-algebras. *Pacific Journal of Mathematics*, **172** (1996), 41–7.

[20] J. Butterfield, J. Hamilton, and C. J. Isham. A topos perspective on the Kochen–Specker theorem: III. Von Neumann algebras as the base category. *International Journal of Theoretical Physics*, **39** (2000), 1413–36.

[21] J. Butterfield and C. J. Isham. A topos perspective on the Kochen–Specker theorem: I. quantum states as generalized valuations. *International Journal of Theoretical Physics*, **37** (1998), 2669–733.

[22] J. Butterfield and C. J. Isham. A topos perspective on the Kochen–Specker theorem: II. Conceptual aspects and classical analogues. *International Journal of Theoretical Physics*, **38** (1999), 827–59.

[23] J. Butterfield and C. J. Isham. A topos perspective on the Kochen–Specker theorem: IV. Interval valuations. *International Journal of Theoretical Physics*, **41** (2002), 613–39.

[24] M. Caspers, C. Heunen, N. P. Landsman, and B. Spitters. Intuitionistic quantum logic of an $n$-level system. *Foundations of Physics*, **39** (2009), 731–59.

[25] J. Cederquist and T. Coquand. Entailment relations and distributive lattices. In *Logic Colloquium '98: Proceedings of the Annual European Summer Meeting of the Association for Symbolic Logic (9–15 August 1998, Prague, Czech Republic)*, Lecture Notes in Logic, vol. 13, ed. S. Buss, P. Hhajek, and P. Pudlak. Wellesley, MA: Association for Symbolic Logic/A.K. Peters, 2000, pp. 127–39.

[26] R. Clifton, J. Bub, and H. Halvorson. Characterizing quantum theory in terms of information-theoretic constraints. *Foundation Physics* (special issue dedicated to David Mermin, Part II), **33** (2003), 1561–91.

[27] B. Coecke. Quantum logic in intuitionistic perspective. *Studia Logica*, **70** (2002), 411–40.

[28] A. Connes. *Noncommutative Geometry*. San Diego, CA: Academic Press, 1994.

[29] T. Coquand. About Stone's notion of spectrum. *Journal of Pure and Applied Algebra*, **197** (2005), 141–58.

[30] T. Coquand and B. Spitters. Formal topology and constructive mathematics: The Gelfand and Stone–Yosida representation theorems. *Journal of Universal Computer Science*, **11** (2005), 1932–44.

[31] T. Coquand and B. Spitters. Integrals and valuations. *Journal of Logic and Analysis*, **1** (2009), 1–22.

[32] T. Coquand and B. Spitters. Constructive Gelfand duality for C*-algebras. To appear in *Mathematical Proceedings of the Cambridge Philosophical Society*, **147** (2), (2009), 323–37.

[33] J. Cuntz. The structure of multiplication and addition in simple C*-algebras. *Mathematica Scandinavica*, **40** (1977), 215–33.

[34] M. L. Dalla Chiara, R. Giuntini, and R. Greechie. *Reasoning in Quantum Theory: Sharp and Unsharp Quantum Logics*. Dordrecht: Kluwer: Academic Publishers, 2004.

[35] L. (Vîţă) Dediu and D. Bridges. Embedding a linear subset of $B(H)$ in the dual of its pre-dual. In *Reuniting the Antipodes—Constructive and Nonstandard Views of the Continuum*, eds. P. Schuster, U. Berger, and H. Osswald. Dordrecht: Kluwer, 2001, pp. 55–61.

[36] J. Dixmier. *C*-algebras*. Amsterdam: North-Holland Physics Publishing, 1977.

[37] A. Döring. Kochen–Specker theorem for Von Neumann algebras. *International Journal of Theoretical Physics*, **44** (2005), 139–60.

[38] A. Döring and C. J. Isham. A topos foundation for theories of physics. I–IV. *Journal of Mathematical Physics*, **49** (2008), 053515–053518.

[39] A. Döring and C. J. Isham. 'What is a thing?': Topos theory in the foundations of physics. in *New Structures in Physics*, Lecture Notes in Physics, vol. 813, ed. B. Coecke. Springer, 2011.

[40] P. D. Finch. On the structure of quantum logic. *Journal of Symbolic Logic*, **34** (1969), 275–82.

[41] M. P. Fourman and R. J. Grayson. Formal spaces. In *The L. E. J. Brouwer Centenary Symposium: Proceedings of the Conference Held in Noordwijkerhout* (8–13 June 1981, Noordwijkerhout), Studies in Logic and the Foundations of Mathematics, no. 110, eds. A. S. Troelstra and D. Dalen. Amsterdam: North-Holland Physics Publishing, 1982, pp. 107–22.

[42] M. P. Fourman and A. Ščedrov. The "world's simplest axiom of choice" fails. *Manuscripta Mathematica*, **38** (1982), 325–32.

[43] W. Fulton. *Young Tableaux*. Cambridge: Cambridge University Press, 1997.

[44] I. M. Gelfand. Normierte ringe. *Matematicheskii Sbornik*, **9** (1941), 3–24.

[45] I. M. Gelfand and M. A. Naimark. On the imbedding of normed rings into the ring of operators on a Hilbert space. *Matematicheskii Sbornik*, **12** (1943), 3–20.

[46] W. H. Graves and S. A. Selesnick. An extension of the Stone representation for orthomodular lattices. *Colloquium Mathematicum*, **27** (1973), 21–30.

[47] P. Griffiths and J. Harris. *Principles of Algebraic Geometry*. New York: Wiley, 1994.

[48] R. Haag. *Local Quantum Physics: Fields, Particles, Algebras*, 2nd ed., Texts and Monographs in Physics. Berlin, Heidelberg: Springer, 1996.

[49] H. Halvorson. On information-theoretic characterizations of physical theories. *Studies in History and Philosophy of Modern Physics*, **35** (2004), 277–93.

[50] J. Hamhalter. Traces, dispersions of states and hidden variables. *Foundations of Physics Letters*, **17** (2004), 581–97.

[51] I. Hasuo, C. Heunen, B. Jacobs, and A. Sokolova. Coalgebraic components in a many-sorted microcosm. *Proceedings of the 3rd International Conference on Algebra and Coalgebra in Computer Science* (7–10 September 2009, Udine, Italy), eds. A. Kurz, M. Lenisa, and A. Tarlecki. Berlin, Heidelberg: Springer-Verlag, 2009, pp. 64–80.

[52] C. Heunen. Categorical quantum models and logics. Ph.D. thesis, Radboud University Nijmegen, 2009.

[53] C. Heunen, N. P. Landsman, and B. Spitters. The principle of general tovariance. In *International Fall Workshop on Geometry and Physics XVI*, AIP Conference Proceedings, vol. 1023, eds. R. L. Fernandes and R. Picken. Melville, NY: American Institute of Physics, 2008, pp. 93–102.

[54] C. Heunen, N. P. Landsman, and B. Spitters. Bohrification of operator algebras and quantum logic (2011). Accepted for publication by *Synthese*.

[55] C. Heunen, N. P. Landsman, and B. Spitters. A topos for algebraic quantum theory. To appear in *Communications in Mathematical Physics*.

[56] C. J. Isham. Topos theory and consistent histories: The internal logic of the set of all consistent sets. *International Journal of Theoretical Physics*, **36** (1997), 785–814.

[57] P. T. Johnstone. *Stone Spaces*, Cambridge Studies in Advanced Mathematics 3. Cambridge: Cambridge University Press, 1982.

[58] P. T. Johnstone. *Sketches of an Elephant: A Topos Theory Compendium*. Oxford: Oxford University Press, 2002.

[59] A. Joyal and M. Tierney. An extension of the Galois theory of Grothendieck. *Memoirs of the American Mathematical Society*, no. 309, **51** (1984).

[60] R. V. Kadison and J. R. Ringrose. *Fundamentals of the Theory of Operator Algebras*. New York: Academic Press, 1983.

[61] G. Kalmbach. *Orthomodular Lattices*. Academic Press, 1983.

[62] I. Kaplansky. *Rings of Operators*. New York, Amsterdam: W. A. Benjamin, 1968.

[63] S. Kochen and E. Specker. The problem of hidden variables in quantum mechanics. *Journal of Mathematics and Mechanics*, **17** (1967), 59–87.

[64] N. P. Landsman. *Mathematical Topics between Classical and Quantum Mechanics*. New York: Springer-Verlag, 1998.

[65] N. P. Landsman. $c^*$-algebras and k-theory (2005). Available at http://www.science.uva.nl/~npl/CK.pdf.

[66] N. P. Landsman. When champions meet: Rethinking the Bohr–Einstein debate. *Studies in History and Philosophy of Modern Physics*, **37** (2006), 212–42.

[67] N. P. Landsman. Between classical and quantum. In *Handbook of Philosophy of Science*, vol. 2: Philosophy of Physics, eds. J. Earman and J. Butterfield. New York: Elsevier, 2007, pp. 417–553.

[68] N. P. Landsman. Macroscopic observables and the Born rule. I. Long run frequencies. *Reviews in Mathematical Physics*, **20** (2008), 1173–90.

[69] W. A. J. Luxemburg and A. C. Zaanen. *Riesz Spaces. I.* Amsterdam: North-Holland Physics Publishing, 1971.

[70] S. Mac Lane and I. Moerdijk. *Sheaves in Geometry and Logic: A First Introduction to Topos Theory*. New York: Springer-Verlag, 1992.

[71] G. W. Mackey. *Mathematical Foundations of Quantum Mechanics*. New York: W. A. Benjamin, 1963.

[72] S. Negri. Continuous domains as formal spaces. *Mathematical Structures in Computer Science*, **12** (2002), 19–52.

[73] M. Rédei. *Quantum Logic in Algebraic Approach*. Dordrecht, Boston: Kluwer Academic Publishers, 1998.

[74] C. E. Rickart. *General Theory of Banach Algebras*. Princeton, Toronto, London, New York: D. van Nostrand, 1960.

[75] M. Rørdam. Structure and classification of C*-algebras. In *Proceedings of the International Congress of Mathematicians*, vol. 2, Zurich: EMS Publishing House, 2006, pp. 1581–98.
[76] S. Sakai. *C*-algebras and W*-algebras*. Berlin, Heidelberg: Springer-Verlag, 1971.
[77] G. Sambin. Intuitionistic formal spaces—a first communication. In *Mathematical Logic and its Applications*, ed. D. Skordev. New York: Plenum Press, 1987, pp. 187–204.
[78] G. Sambin. Some points in formal topology. *Theoretical Computer Science*, **305** (2003), 347–408.
[79] E. Scheibe. *The Logical Analysis of Quantum Mechanics*. Oxford, New York: Pergamon, 1973.
[80] D. Scott. Lattice theory, data types and semantics. In *NYU Symposium on Formal Semantics*, ed. R. Rustin. Prentice-Hall, 1972, pp. 65–106.
[81] I. E. Segal. Postulates for general quantum mechanics. *Annals of Mathematics*, **48** (1947), 930–48.
[82] B. Spitters. Constructive results on operator algebras. *Journal of Universal Computer Science*, **11** (2005), 2096–113.
[83] S. Stratila and L. Zsido. *Operator Algebras*. Bucharest: Theta Foundation, 2009.
[84] I. Stubbe. The canonical topology on a meet-semilattice. *International Journal of Theoretical Physics*, **44** (2005), 2283–93.
[85] M. Takesaki. *Theory of Operator Algebra I*, Encyclopaedia of Mathematical Sciences. Berlin, Heidelberg, New York: Springer, 1979.
[86] S. Vickers. Locales and toposes as spaces. In *Handbook of Spatial Logics*, eds. M. Aiello, I. Pratt-Hartmann, and J. van Bentheim. Dordrecht: Springer, 2007, pp. 429–96.
[87] A. C. Zaanen. *Riesz Spaces. II*. Amsterdam: North-Holland Physics Publishing, 1983.
[88] E. Zafiris. Boolean coverings of quantum observable structure: A setting for an abstract differential geometric mechanism. *Journal of Geometry and Physics*, **50** (2004), 9–114.

# PART TWO
# Beyond the Hilbert Space Formalism: Operator Algebras

CHAPTER 7

# Yet More Ado about Nothing: The Remarkable Relativistic Vacuum State

Stephen J. Summers

## 7.1 Introduction

For millennia, the concept of nothingness, in many forms and guises, has occupied reflective minds, who have adopted an extraordinary range of stances toward the notion—from holding that it is the Godhead itself to rejecting it vehemently as a foul blasphemy. Even among more scientifically inclined thinkers, there has been a similar range of views [49]. We have no intention here to sketch this vast richness of thought about nothingness. Instead, we shall more modestly attempt to explain what mathematical physics has to say about nothingness in its modern scientific guise: the *relativistic vacuum state*.

What is the vacuum in modern science? Roughly speaking, it is that which is left over after all that can possibly be removed has been removed, where "possibly" refers to neither "technically possible" nor to "logically possible" but rather to "physically possible"—that which is possible in light of (the current understanding of) the laws of physics. The vacuum is therefore an idealization that is only approximately realized in the laboratory and in nature. But it is a most useful idealization and a surprisingly rich concept.

We discuss the vacuum solely in the context of the relativistic quantum theory of systems in four spacetime dimensional Minkowski space, although we briefly indicate how similar states for quantum systems in other spacetimes can be defined and studied. In a relativistic theory of systems in Minkowski space, the vacuum should appear to be the same at every position and in every direction for all inertial observers. In other words, it should be invariant under the Poincaré group, the group of isometries of Minkowski space. And because one can remove no further mass/energy from the vacuum, it should be the lowest possible (global) energy state. In a relativistic theory, when one removes all mass/energy, the total energy of the resultant state is 0.

These *desiderata* of a vacuum are intuitively appealing, but it remains to give mathematical content to these intuitions. Once this is done, it will be seen that this state with "nothing in it" manifests remarkable properties, most of which have been discovered only in the past twenty years and many of which are *not* intuitively appealing at first exposure. On the contrary, some properties of the vacuum state have proven to be decidedly controversial.

To formulate in a mathematically rigorous manner the notion of a vacuum state and to understand its properties, it is necessary to choose a mathematical framework that is sufficiently general to subsume large classes of models, is powerful enough to facilitate the proof of nontrivial assertions of physical interest, and yet is conceptually simple enough to have a direct, if idealized, interpretation in terms of operationally meaningful physical quantities. Such a framework is provided by algebraic quantum theory [3, 12, 13, 40, 52], also called *local quantum physics*, which is based on operator algebra theory, itself initially developed by J. von Neumann for the express purpose of providing quantum theory with a rigorous and flexible foundation [110, 111]. This framework is briefly described in the next section, in which a rigorous definition of a vacuum state in Minkowski space is given.

In Section 7.3, the earliest recognized consequences of such a definition are discussed, including such initially nonintuitive results as the Reeh–Schlieder Theorem. Rigorous results indicating that the vacuum is a highly entangled state are presented in Section 7.4. Indeed, by many measures, it is a maximally entangled state. Although some of these results have been proven quite recently, readers who are familiar with the heuristic picture of the relativistic vacuum as a seething broth of virtual particle–antiparticle pairs causing wide-ranging vacuum correlations may not be entirely surprised by their content. But there are concepts available in algebraic quantum field theory (AQFT) that have no known counterpart in heuristic quantum field theory, such as the mathematical objects that arise in the modular theory of M. Tomita and M. Takesaki [103], which is applicable in the setting of AQFT. As explained in Sections 7.5 and 7.6, the modular objects associated with the vacuum state encode a truly astonishing amount of physical information and also serve to provide an intrinsic characterization of the vacuum state, which admits a generalization to quantum fields on arbitrary spacetimes. In addition, it is shown in Section 7.7 how these objects may be used to derive the spacetime itself, thereby providing—at least, in principle—a means to derive from the observables and their preparation (the state) a spacetime in which the former can be interpreted as being localized and evolving without any a priori input on the nature or even existence of a spacetime. We make some concluding remarks in Section 7.8.

## 7.2 The Mathematical Framework

The operationally fundamental objects in a laboratory are the preparation apparata—devices that prepare in a repeatable manner the individual quantum systems that are to be examined—and the measuring apparata—devices that are applied to the prepared systems and that measure the "value" of some observable property of the system. The physical notion of a "state" can be viewed as a certain equivalence class of such preparation devices, and the physical notion of an "observable" (or "effect") can be viewed as a certain equivalence class of such measuring (or registration) devices [3, 71]. In principle, therefore, these quantities are operationally determined.

In AQFT, such observables are represented by self-adjoint elements of certain algebras of operators, either $W^*$- or $C^*$-algebras.[1] In this chapter, we restrict our attention primarily to concretely represented $W^*$-algebras, which are commonly called *von Neumann algebras* in honor of the person who initiated their study [111]. The reader unfamiliar with these notions may simply think of algebras $\mathcal{M}$ of bounded operators[2] on some (separable) Hilbert space $\mathcal{H}$ (or see [59, 60, 104–106] for a thorough background). We denote by $\mathcal{B}(\mathcal{H})$ the algebra of all bounded operators on $\mathcal{H}$. Physical states are represented by mathematical *states* $\phi$; that is, linear, continuous maps $\phi : \mathcal{M} \to \mathbb{C}$ from the algebra of observables to the complex number system that take the value 1 on the identity map $I$ on $\mathcal{H}$ and are positive in the sense that $\phi(A^*A) \geq 0$ for all $A \in \mathcal{M}$. An important subclass of states consists of *normal states*; these are states such that $\phi(A) = \text{Tr}(\rho A)$, $A \in \mathcal{M}$, for some *density matrix* $\rho$ acting on $\mathcal{H}$—that is, a bounded operator on $\mathcal{H}$ satisfying the conditions $0 \leq \rho = \rho^*$ and $\text{Tr}(\rho) = 1$. A special case of such normal states is constituted by the *vector states*: if $\Phi \in \mathcal{H}$ is a unit vector and $P_\Phi \in \mathcal{B}(\mathcal{H})$ is the orthogonal projection onto the one dimensional subspace of $\mathcal{H}$ spanned by $\Phi$, the corresponding vector state is given by

$$\phi(A) = \langle \Phi, A\Phi \rangle = \text{Tr}(P_\Phi A), \ A \in \mathcal{M}.$$

Generally speaking, theoretical physicists tacitly restrict their attention to normal states.[3]

In AQFT, the spacetime localization of the observables is taken into account. Let $\mathbb{R}^4$ represent four dimensional Minkowski space and $\mathcal{O}$ denote an open subset of $\mathbb{R}^4$. Because any measurement is carried out in a finite spatial region and in a finite time, for every observable $A$ there exist bounded regions $\mathcal{O}$ containing this "localization" of $A$.[4] We say that the observable $A$ is localized in any such region $\mathcal{O}$ and denote by $\mathcal{R}(\mathcal{O})$ the von Neumann algebra generated by all observables localized in $\mathcal{O}$. Clearly, it follows that if $\mathcal{O}_1 \subset \mathcal{O}_2$, then $\mathcal{R}(\mathcal{O}_1) \subset \mathcal{R}(\mathcal{O}_2)$. This yields a net $\mathcal{O} \mapsto \mathcal{R}(\mathcal{O})$ of observable algebras associated with the experiment(s) in question. In turn, this net determines the smallest von Neumann algebra $\mathcal{R}$ on $\mathcal{H}$ containing all $\mathcal{R}(\mathcal{O})$. The preparation procedures in the experiment(s) then determine states $\phi$ on $\mathcal{R}$, the global observable algebra.

Given a state $\phi$ on $\mathcal{R}$, one can construct [40, 59, 104] a Hilbert space $\mathcal{H}_\phi$, a distinguished unit vector $\Omega_\phi \in \mathcal{H}_\phi$, and a $(C^*$-$)$homomorphism $\pi_\phi \colon \mathcal{R} \to \mathcal{B}(\mathcal{H}_\phi)$, so that $\pi_\phi(\mathcal{R})$ is a $(C^*$-$)$algebra acting on the Hilbert space $\mathcal{H}_\phi$, the set of vectors $\pi_\phi(\mathcal{R})\Omega_\phi = \{\pi_\phi(A)\Omega_\phi \mid A \in \mathcal{R}\}$ is dense in $\mathcal{H}_\phi$, and

$$\phi(A) = \langle \Omega_\phi, \pi_\phi(A)\Omega_\phi \rangle, \ A \in \mathcal{R}.$$

---

[1] Other sorts of algebras have also been seriously considered for various reasons; see, for example, [40, 86, 87].

[2] Technicalities concerning topology will be systematically suppressed in this chapter. We therefore do not discuss the difference between $C^*$- and $W^*$-algebras.

[3] It is in the context of von Neumann algebras and normal states that classical probability theory has a natural generalization to noncommutative probability theory; see, for example, [79].

[4] It is clear from operational considerations that one could not expect to determine a minimal localization region for a given observable experimentally. In [64], the possibility of determining such a minimal localization region in the idealized context of AQFT is discussed at length. However, the existence of such a region is not necessary for any results in AQFT known to the author.

The triple $(\mathcal{H}_\phi, \Omega_\phi, \pi_\phi)$ is uniquely determined up to unitary equivalence by these properties, and $\pi_\phi$ is called the GNS representation of $\mathcal{R}$ determined by $\phi$. Only if $\phi$ is a normal state is $\pi_\phi(\mathcal{R})$ a von Neumann algebra and can $(\mathcal{H}_\phi, \Omega_\phi, \pi_\phi)$ be identified with (a subrepresentation of) $(\mathcal{H}, \Omega, \mathcal{R})$ such that $\Omega_\phi \in \mathcal{H}$.[5] Hence, a state determines a concrete, although idealized, representation of the experimental setting in a Hilbert space.

In this setting, relativistic covariance is expressed through the presence of a representation $\mathcal{P}_+^\uparrow \ni \lambda \mapsto \alpha_\lambda$ of the identity component $\mathcal{P}_+^\uparrow$ of the Poincaré group by automorphisms $\alpha_\lambda : \mathcal{R} \to \mathcal{R}$ of $\mathcal{R}$ such that

$$\alpha_\lambda(\mathcal{R}(\mathcal{O})) = \mathcal{R}(\lambda\mathcal{O}),$$

for all $\mathcal{O}$ and $\lambda$, where $\alpha_\lambda(\mathcal{R}(\mathcal{O})) = \{\alpha_\lambda(A) \mid A \in \mathcal{R}(\mathcal{O})\}$ and $\lambda\mathcal{O} = \{\lambda(x) \mid x \in \mathcal{O}\}$. One says that a state $\phi$ is *Poincaré invariant* if $\phi(\alpha_\lambda(A)) = \phi(A)$ for all $A \in \mathcal{R}$ and $\lambda \in \mathcal{P}_+^\uparrow$. In this case, there then exists a unitary representation $\mathcal{P}_+^\uparrow \ni \lambda \mapsto U_\phi(\lambda)$ acting on $\mathcal{H}_\phi$, leaving $\Omega_\phi$ invariant, and implementing the action of the Poincaré group:

$$U_\phi(\lambda)\pi_\phi(A)U_\phi(\lambda)^{-1} = \pi_\phi(\alpha_\lambda(A)),$$

for all $A$ and $\lambda$. If the joint spectrum of the self-adjoint generators of the translation subgroup $U_\phi(\mathbb{R}^4)$ is contained in the forward lightcone, then $U_\phi(\mathcal{P}_+^\uparrow)$ is said to satisfy the (relativistic) *spectrum condition*. This condition is a relativistically invariant way of requiring that the total energy in the theory be nonnegative with respect to every inertial frame of reference and that the quantum system is stable in the sense that it cannot decay to energies below that of the vacuum state.

We can now present a standard rigorous definition of a vacuum state, which incorporates all of the intuitive *desiderata* already discussed.

**Definition 7.2.1.** A vacuum state is a Poincaré invariant state $\phi$ on $\mathcal{R}$ such that $U_\phi(\mathcal{P}_+^\uparrow)$ satisfies the spectrum condition.[6] The corresponding GNS representation $(\mathcal{H}_\phi, \Omega_\phi, \pi_\phi)$ is called a vacuum representation of the net of observable algebras.

Note that after choosing an inertial frame of reference, the self-adjoint generator $H$ of the time translation subgroup $U_\phi(t)$, $t \in \mathbb{R}$, carries the interpretation of the total energy operator and that, by definition, $H\Omega_\phi = 0$ if $\phi$ is a vacuum state. Moreover, the total momentum operator $\vec{P}$ and the total mass operator $M \equiv \sqrt{H^2 - \vec{P}^2} \geq 0$ also annihilate the vacuum ($M\Omega_\phi = 0 = \vec{P}\Omega_\phi$).

Such vacuum states and, hence, such vacuum representations actually exist. In the case of four-dimensional Minkowski space, vacuum representations for quantum field models with trivial $S$-matrix have been rigorously constructed by various means (cf., e.g., [2, 8, 17, 48, 114]) and, more recently, the same has been accomplished

---

[5] If the state $\phi$ is not normal, then the state vectors in $\mathcal{H}_\phi$ are, in a certain mathematically rigorous sense, *orthogonal* to those in $\mathcal{H}$. The vacuum state of a fully interacting model, as opposed to an interacting theory with various kinds of cutoffs introduced precisely so that it may be realized in Fock space, is not normal with respect to Fock space, which is the representation space for the corresponding free theory—see, for example, [8,47]. For further perspective on this issue, see [93].

[6] Some authors just require of a vacuum state that it be invariant under the translation group and satisfy the spectrum condition. For the purposes of this chapter, it is convenient to adopt the more restrictive of the two standard definitions.

for quantum field models with nontrivial scattering matrices [31, 32, 50]. For two, resp. three, dimensional Minkowski space, fully interacting quantum field models in vacuum representations have been constructed (cf., e.g., [8, 47, 48, 69]). Moreover, general conditions are known under that to a quantum field model without a vacuum state can be (under certain conditions uniquely) associated a vacuum representation that is physically equivalent and locally unitarily equivalent to it [22, 33, 38]. Hence, the mathematical existence of a vacuum state is often assured even in models that are not initially provided with one.

It will be useful in the following to describe two special classes of spacetime regions in Minkowski space. A double cone is a (nonempty) intersection of an open forward lightcone with an open backward lightcone. Such regions are bounded, and the set $\mathcal{D}$ of all double cones is left invariant by the natural action of $\mathcal{P}_+^\uparrow$ on it. An important class of unbounded regions is specified as follows. After choosing an inertial frame of reference, one defines the right wedge to be the set $W_R = \{x = (t, x_1, x_2, x_3) \in \mathbb{R}^4 \mid x_1 > |t|\}$ and the set of wedges to be $\mathcal{W} = \{\lambda W_R \mid \lambda \in \mathcal{P}_+^\uparrow\}$. The set of wedges is independent of the choice of reference frame; only which wedge is designated the right wedge is frame dependent.

## 7.3 Immediate Consequences

We now turn to some immediate consequences of the definition of a vacuum state. One of the most controversial was also one of the first to be noted. To avoid a too heavily laden notation, and because in this and the next section our starting point is a vacuum representation, we drop the subscript $\phi$ and the symbol $\pi_\phi$ (i.e., we identify $\mathcal{R}(\mathcal{O})$ and $\pi_\phi(\mathcal{R}(\mathcal{O}))$). A vacuum representation is said to satisfy *weak additivity* if for each nonempty $\mathcal{O}$ the smallest von Neumann algebra containing

$$\{U(x)\mathcal{R}(\mathcal{O})U(x)^{-1} \mid x \in \mathbb{R}^4\}$$

coincides with $\mathcal{R}$. This is a weak technical assumption satisfied in most models; for example, it holds in any theory in which there is a Wightman field locally associated with the observable algebras (see, e.g., [6, 7, 37]).

Let $\mathcal{O}$ be an open subset of $\mathbb{R}^4$ and let $\mathcal{O}'$ denote the interior of its causal complement, the set of all points in $\mathbb{R}^4$ that are spacelike separated from all points in $\mathcal{O}$. A net $\mathcal{O} \mapsto \mathcal{R}(\mathcal{O})$ is said to be *local* (or to satisfy *locality*) if whenever $\mathcal{O}_1 \subset \mathcal{O}_2'$ one has $\mathcal{R}(\mathcal{O}_1) \subset \mathcal{R}(\mathcal{O}_2)'$, where $\mathcal{R}(\mathcal{O})'$, the commutant of $\mathcal{R}(\mathcal{O})$, represents the set of all bounded operators on $\mathcal{H}$ that commute with all elements of $\mathcal{R}(\mathcal{O})$. Ordinarily, this property of locality is viewed as a manifestation of Einstein causality, which posits that signals and causal influences cannot propagate faster than the speed of light and, therefore, spacelike separated quantum systems must be independent in some sense. As is the case with so many received notions, there is much more here than initially meets the eye; but this is not the place to address this matter (cf. [34, 91, 94] for certain aspects of this point). We emphasize that locality is not a standing assumption in this chapter. If a net is not explicitly assumed to be local, then the property is not necessary for the respective result. In fact, locality will be *derived* in the settings discussed in Sections 7.6 and 7.7.[7]

For vacuum representations of local nets in which weak additivity is satisfied, the Reeh–Schlieder Theorem holds (cf. [3, 8, 52, 58, 89]).

**Theorem 7.3.1.** Consider a vacuum representation of a local net fulfilling the condition of weak additivity. For every nonempty region $\mathcal{O}$ such that $\mathcal{O}' \neq \emptyset$, the vector $\Omega$ is cyclic and separating for $\mathcal{R}(\mathcal{O})$; that is the set of vectors $\mathcal{R}(\mathcal{O})\Omega$ is dense in $\mathcal{H}$, resp. $A \in \mathcal{R}(\mathcal{O})$ and $A\Omega = 0$ entail $A = 0$.[8]

There are two distinct aspects to this theorem. First, the fact that the vacuum is separating for local observables means exactly that no nonzero local observable can annihilate $\Omega$. Hence, any event represented by a nonzero projection $P \in \mathcal{R}(\mathcal{O})$ must have nonzero expectation in the vacuum state: $\langle \Omega, P\Omega \rangle > 0$. In the vacuum, any local event can occur! Moreover, $0 < C = C^* \in \mathcal{R}(\mathcal{O})$ entails the existence of an element $0 \neq A \in \mathcal{R}(\mathcal{O})$ such that $C = A^*A$; thus, $\langle \Omega, C\Omega \rangle = \|A\Omega\|^2$, which also yields $\langle \Omega, C\Omega \rangle > 0$ in this more general case. Therefore, the stress–energy density tensor $T(x)$ smeared with any test function with compact support cannot be a positive operator in a vacuum representation [41] (in fact, it is unbounded from below), in contrast to the situation in classical physics, because its vacuum expectation is zero. Furthermore, in light of the fact that the vacuum state contains no real particles ($M\Omega = 0$), it follows that there can be no localized particle counters. Indeed, if $C \in \mathcal{R}$ is a particle counter for a particle described in the model, then $C = C^* > 0$ and $\langle \Omega, C\Omega \rangle = 0$. Therefore, $C$ cannot be an element of any algebra $\mathcal{R}(\mathcal{O})$ with $\mathcal{O}'$ nonempty. Hence, the notion of particle in relativistic quantum field theory cannot be quite as simple as classical mechanics would have it. It has even been argued that the notion is nonsensical in relativistic quantum field theory, but this is not the place for further discussion of this point either. (See, however, [18, 26, 45, 52, 55, 83].)

Second, there is the cyclicity of the vacuum for all local algebras: every vector state in the vacuum representation can be arbitrarily well approximated using vectors of the form $A\Omega$, $A \in \mathcal{R}(\mathcal{O})$, no matter how small in extent $\mathcal{O}$ may be. Hence, the class of all states resulting from the action of arbitrary operations on the vacuum is effectively indistinguishable from the class of states resulting from operations performed in arbitrarily small spacetime regions upon the vacuum. Prima facie, such a state would seem to be different from the vacuum only in a region that one can make as small as one desires. In our view, a reasonable physical picture of this situation is indicated in this way: an experimenter in any given region $\mathcal{O}$ can, in principle, perform measurements designed to exploit nonlocal vacuum fluctuations (see the next section) in such a manner that any prescribed state can be reproduced with any given accuracy. These consequences of cyclicity also unleashed some controversy, some of which is well discussed in [53] (see also [80]). We do not elaborate on these matters here, except to point out the fact that the existing proposals to avoid Reeh–Schlieder by changing the notion of localization (1) are necessarily restricted to free quantum field models, and (2) introduce at least as many problems as they "solve" (e.g., [53]).

---

[7] For a very different derivation of *locality*, see [30].
[8] Note that even if the net of observable algebras is not local, $\Omega$ is still cyclic for $\mathcal{R}(\mathcal{O})$.

We emphasize that these (for some readers disturbing) properties are by no means unique to the vacuum—the Reeh–Schlieder Theorem is valid for *any* vector in the vacuum representation that is analytic for the energy [9]; in particular, it holds for any vector with finite energy content. So its conclusions and various consequences are true of *all* physically realizable vector states in the vacuum representation because any preparation can only implement a finite exchange of energy!

## 7.4 Vacuum Correlations

We turn to what is rigorously known about the nature of vacuum correlations, preparing first some definitions to be used in this section. Given a pair $(\mathcal{M}, \mathcal{N})$ of algebras representing the observable algebras of two subsystems of a given quantum system, a state $\phi$ is said to be a *product state* across $(\mathcal{M}, \mathcal{N})$ if $\phi(MN) = \phi(M)\phi(N)$ for all $M \in \mathcal{M}, N \in \mathcal{N}$. In such states, the observables of the two subsystems are not correlated and the subsystems manifest a certain kind of independence—see, for example, [91]. A normal state $\phi$ on $\mathcal{M} \bigvee \mathcal{N}$ is *separable*[9] if it is in the norm closure of the convex hull of the normal product states across $(\mathcal{M}, \mathcal{N})$; that is, it is a mixture of normal product states. Otherwise, $\phi$ is said to be *entangled* (across $(\mathcal{M}, \mathcal{N})$).[10] From the point of view of what is now called *quantum information theory*, the primary difference between classical and quantum theory is the existence of entangled states in quantum theory. In fact, only if both subsystems are quantum (i.e., both algebras are noncommutative so there exist entangled states on the composite system) [77]. Although not understood at that time in this manner, some of the founders of quantum theory realized as early as 1935 [39, 82] that such entangled states were the source of the "paradoxical" behavior of quantum theory (as viewed from the vantage point of classical physics). Today, entangled states are regarded as a resource to be employed to carry out tasks that cannot be done classically, that is, only with separable states (cf. [57, 62, 112]).

Another direct consequence of the Reeh–Schlieder Theorem is that for all nonempty spacelike separated $\mathcal{O}_1, \mathcal{O}_2$ with nonempty causal complements, no matter how far spacelike separated they may be, there exist many projections $P_i \in \mathcal{R}(\mathcal{O}_i)$ that are positively correlated in the vacuum state; that is, such that $\phi(P_1 P_2) > \phi(P_1)\phi(P_2)$.

**Theorem 7.4.1.** Consider a vacuum representation of a local net fulfilling the condition of weak additivity, and let $\mathcal{O}_1, \mathcal{O}_2$ be any nonempty spacelike separated regions with nonempty causal complements. Let $\phi$ be any state induced by a vector analytic for the energy (e.g., the vacuum state). Then, for any projection $P_1 \in \mathcal{R}(\mathcal{O}_1)$ with $0 \neq P_1 \neq I$ there exists a projection $P_2 \in \mathcal{R}(\mathcal{O}_2)$ such that $\phi(P_1 P_2) > \phi(P_1)\phi(P_2)$.

---

[9] Also termed *decomposable*, classically correlated, or unentangled by various authors.

[10] This terminology is becoming standard in quantum information theory, but there are still physicists who tacitly restrict their attention to vector states on mutually commuting algebras of observables that are isomorphic to full matrix algebras (i.e., they consider only *pure* states, which are entangled if and only if they are not product states).

This is an immediate consequence of Theorem 7.3.1 and the following lemma, the proof of which is implicit in the proof of Theorem 5 in [80]. For the convenience of the reader, we make this explicit here.

**Lemma 7.4.2.** Let $\mathcal{M}$ and $\mathcal{N}$ be von Neumann algebras on $\mathcal{H}$ with $\Omega \in \mathcal{H}$ a unit vector cyclic for $\mathcal{N}$ and separating for $\mathcal{M}$, and let $\omega$ be the corresponding state induced on $\mathcal{B}(\mathcal{H})$. Then, for any projection $P \in \mathcal{M}$ with $0 \neq P \neq I$, there exists a projection $Q \in \mathcal{N}$ such that $\omega(PQ) > \omega(P)\omega(Q)$.

**PROOF.** Let $P \in \mathcal{M}$ be a projection with $0 \neq P \neq I$. It suffices to establish the existence of a projection $Q \in \mathcal{N}$ such that $\omega(PQ) \neq \omega(P)\omega(Q)$ because, if necessary, $Q$ can be replaced by $I - Q \in \mathcal{N}$ to yield the assertion. So assume for the sake of contradiction that $\omega(PQ) = \omega(P)\omega(Q)$, for all such $Q$. Then, with $\widehat{P} = P - \omega(P) \cdot I \in \mathcal{M}$, one has $\omega(\widehat{P}Q) = 0$, for all projections $Q \in \mathcal{N}$. By the spectral theorem, this entails $\omega(\widehat{P}N) = 0$, for all $N \in \mathcal{N}$; that is,

$$\langle \widehat{P}\Omega, N\Omega \rangle = 0, \, N \in \mathcal{N}.$$

Because $\Omega$ is cyclic for $\mathcal{N}$, this yields $\widehat{P}\Omega = 0$, so that $\widehat{P} = 0$; that is, $P = \omega(P) \cdot I$. Also, because $P = P^2$, this entails $\|P\Omega\|^2 = \langle P\Omega, P\Omega \rangle = \omega(P) \in \{0, 1\}$; that is, either $P\Omega = 0$ or $P\Omega = \Omega$ and, because $\Omega$ is separating for $\mathcal{M}$, this implies either $P = 0$ or $P = I$ holds, a contradiction in either case. ∎

The fact that vacuum fluctuations enable such generic "superluminal correlations" has also generated controversy because they seem to challenge received notions of causality. This is another complex matter that we cannot go into here, but at least some forms of causality have been proven in AQFT (for recent discussions, see, e.g., [34, 78]) and therefore are completely compatible with such correlations.

Of course, Theorem 7.4.1 entails that the vacuum is not a product state across $(\mathcal{R}(\mathcal{O}_1), \mathcal{R}(\mathcal{O}_2))$ but not yet that it is entangled across $(\mathcal{R}(\mathcal{O}_1), \mathcal{R}(\mathcal{O}_2))$. Much finer analyses of the nature and degree of the entanglement of the vacuum state have been carried out in the literature, and we explain some of these. A quantitative measure of entanglement is provided by using *Bell correlations*. The following definition was made in [96].

**Definition 7.4.3.** Let $\mathcal{M}, \mathcal{N} \subset \mathcal{B}(\mathcal{H})$ be von Neumann algebras such that $\mathcal{M} \subset \mathcal{N}'$. The maximal Bell correlation of the pair $(\mathcal{M}, \mathcal{N})$ in the state $\phi$ is

$$\beta(\phi, \mathcal{M}, \mathcal{N}) \equiv \sup \frac{1}{2} \phi(M_1(N_1 + N_2) + M_2(N_1 - N_2)),$$

where the supremum is taken over all self-adjoint $M_i \in \mathcal{M}$, $N_j \in \mathcal{N}$ with norm less than or equal to 1.

As explained in, for example, [97], the CHSH version of Bell's inequalities can be formulated in AQFT as

$$\beta(\phi, \mathcal{M}, \mathcal{N}) \leq 1. \tag{7.1}$$

If $\phi$ is separable across $(\mathcal{M}, \mathcal{N})$, then $\beta(\phi, \mathcal{M}, \mathcal{N}) = 1$ [97]. Hence, states that violate Bell's inequalities are necessarily entangled, although the converse is not true (cf. [112]

for a discussion and references). Whenever at least one of the systems is classical, the bound Equation (7.1) is satisfied in *every* state:

**Proposition 7.4.4 [97].** Let $\mathcal{M}, \mathcal{N} \subset \mathcal{B}(\mathcal{H})$ be mutually commuting von Neumann algebras. If either $\mathcal{M}$ or $\mathcal{N}$ is abelian, then $\beta(\phi, \mathcal{M}, \mathcal{N}) = 1$ for all states $\phi$ on $\mathcal{B}(\mathcal{H})$.

If, conversely, both algebras are nonabelian, then there always exists a state in which the inequality Equation (7.1) is (maximally) violated, as long as the Schlieder property holds; that is, $MN = 0$ for $M \in \mathcal{M}$ and $N \in \mathcal{N}$ entail either $M = 0$ or $N = 0$ [68]. Because it is known [35, 97] that $1 \leq \beta(\phi, \mathcal{M}, \mathcal{N}) \leq \sqrt{2}$, for all states $\phi$ on $\mathcal{B}(\mathcal{H})$, one says that if $\beta(\phi, \mathcal{M}, \mathcal{N}) = \sqrt{2}$, then the pair $(\mathcal{M}, \mathcal{N})$ maximally violates Bell's inequalities in the state $\phi$.

In [98], it is shown under quite general physical assumptions that in a vacuum representation of a local net one has $\beta(\phi, \mathcal{R}(W), \mathcal{R}(W')) = \sqrt{2}$, for every wedge $W$ and *every* normal state $\phi$. In particular, Bell's inequalities are *maximally* violated in the vacuum state. In addition, under somewhat more restrictive but still general assumptions satisfied by include free quantum field theories and other physically relevant models, it is shown in [98] that $\beta(\phi, \mathcal{R}(\mathcal{O}_1), \mathcal{R}(\mathcal{O}_2)) = \sqrt{2}$, for any two spacelike separated double cones whose closures intersect (i.e., tangent double cones) and *all* normal states $\phi$. Hence, such pairs of observable algebras also maximally violate Bell's inequalities in the vacuum.

Commonly, physicists say that theories violating Bell's inequalities are "nonlocal"; yet, here are fully local models maximally violating Bell's inequalities. This linguistic confusion is probably so profoundly established by usage that it cannot be repaired, but the reader should be aware of the distinct meanings of these two uses of "local." The former refers to nonlocalities in certain correlations (in certain states), whereas the latter refers to the commensurability of observables localized in spacelike separated spacetime regions. So, the former is a property of states, whereas the latter is a property of observable algebras. The results discussed herein establish the generic compatibility of the former sort of "nonlocality" with the latter kind of "locality." The wary reader should always ascertain which sense of "local" is being employed by a given author.

In the now quite extensive quantum information theory literature, there are various attempts to quantify the degree of entanglement of a given state (cf., e.g., [57, 62]), but these agree that maximal violation of inequality Equation (7.1) entails maximal entanglement. Thus, the vacuum state is maximally entangled and thereby describes a maximally nonclassical situation.

The localization regions for the observable algebras that have been proven to manifest maximal violation of Bell's inequality in the vacuum (indeed, in every state) are spacelike separated but tangent. If the double cones have nonzero spacelike separation, any violation of Bell's inequality in the vacuum cannot be maximal:

**Proposition 7.4.5 [96, 97, 99].** Let $\mathcal{O} \mapsto \mathcal{R}(\mathcal{O})$ be a local net in an irreducible vacuum representation with a lowest mass $m > 0$. Then, for any pair $(\mathcal{O}_1, \mathcal{O}_2)$ of spacelike separated regions, one has

$$\beta(\phi, \mathcal{R}(\mathcal{O}_1), \mathcal{R}(\mathcal{O}_2)) \leq \sqrt{2} - \frac{\sqrt{2}}{7 + 4\sqrt{2}}(1 - e^{-md(\mathcal{O}_1, \mathcal{O}_2)})$$

(optimal for smaller $d(\mathcal{O}_1, \mathcal{O}_2)$) and

$$\beta(\phi, \mathcal{R}(\mathcal{O}_1), \mathcal{R}(\mathcal{O}_2)) \leq 1 + 2e^{-md(\mathcal{O}_1, \mathcal{O}_2)}$$

(optimal for larger $d(\mathcal{O}_1, \mathcal{O}_2)$), where $\phi$ is a vacuum state and $d(\mathcal{O}_1, \mathcal{O}_2)$ is the maximal timelike distance $\mathcal{O}_1$ can be translated before it is no longer spacelike separated from $\mathcal{O}_2$.

Hence, if $d(\mathcal{O}_1, \mathcal{O}_2)$ is much larger than a few Compton wavelengths of the lightest particle in the theory, then any violation of Bell's inequality in the vacuum would be too small to be observed. As explained in [97], if there are massless particles in the theory, then the best decay in the vacuum Bell correlation one can expect is proportional to $d(\mathcal{O}_1, \mathcal{O}_2)^{-2}$. Although the decay in the massless case is much weaker, experimental apparata have nonzero lower bounds on the particle energies they can effectively measure. Such nonzero sensitivity limits would serve as an effective lowest mass, leading to an exponential decay once again [97]. Nonetheless, attempts have been made to obtain lower bounds on the Bell correlation $\beta(\phi, \mathcal{R}(\mathcal{O}_1), \mathcal{R}(\mathcal{O}_2))$ as a function of $d(\mathcal{O}_1, \mathcal{O}_2)$. Because the published results have only treated some very special models and very special observables, we shall refrain from discussing these here (but cf. [81] and references given there).

Nonetheless, using properties of $\beta(\phi, \mathcal{M}, \mathcal{N})$ established by the author and R. F. Werner [99], H. Halvorson and R. Clifton have proven the following result, which entails that in a vacuum representation in which weak additivity and locality hold, the vacuum state (and any state induced by a vector analytic for the energy) is entangled across $(\mathcal{R}(\mathcal{O}_1), \mathcal{R}(\mathcal{O}_2))$ for arbitrary nonempty spacelike separated regions $\mathcal{O}_1, \mathcal{O}_2$.

**Theorem 7.4.6 [54].** Let $\mathcal{M}$ and $\mathcal{N}$ be nonabelian von Neumann algebras acting on $\mathcal{H}$ such that $\mathcal{M} \subset \mathcal{N}'$. If $\Omega \in \mathcal{H}$ is cyclic for $\mathcal{M}$ and $\omega$ is the state on $\mathcal{B}(\mathcal{H})$ induced by $\Omega$, then $\omega$ is entangled across $(\mathcal{M}, \mathcal{N})$.

The proof does not provide a lower bound on $\beta(\phi, \mathcal{M}, \mathcal{N})$. For further discussion and references concerning the violation of Bell's inequalities in algebraic quantum theory, see [54, 80, 92, 99].

Although model independent lower bounds on $\beta(\phi, \mathcal{R}(\mathcal{O}_1), \mathcal{R}(\mathcal{O}_2))$ are not yet available, R. Verch and Werner [109] have obtained model independent results on the nature of the entanglement of the vacuum state across nontangent pairs $(\mathcal{R}(\mathcal{O}_1), \mathcal{R}(\mathcal{O}_2))$ in terms of some further notions currently employed in quantum information theory, which go beyond Theorem 7.4.6. They proposed the following definition [109].

**Definition 7.4.7.** Let $\mathcal{M}$ and $\mathcal{N}$ be von Neumann algebras acting on a Hilbert space $\mathcal{H}$. A state $\phi$ on $\mathcal{B}(\mathcal{H})$ has the positive partial transpose (ppt) property if for any choice of finitely many $M_1, \ldots, M_k \in \mathcal{M}$ and $N_1, \ldots, N_k \in \mathcal{N}$, one has

$$\sum_{\alpha, \beta} \phi(M_\beta M_\alpha^* N_\alpha^* N_\beta) \geq 0.$$

They show that this generalizes the notion of states with ppt familiar from quantum information theory [74], a notion restricted to finite dimensional Hilbert spaces before [109]. They also show that if a state is ppt, then it satisfies Bell's inequalities, and they prove that any separable state is ppt. Indeed, in general, the class of ppt states properly contains the class of separable states.

Another notion from quantum information theory is that of distillability (of entanglement). Roughly speaking, this refers to being able to operate on a given state in certain (local) ways to increase its entanglement across two subsystems. Separable states are not distillable; they are not entangled, and operating on them in the allowable manner will not result in an entangled state. We refer the reader to [109] for a discussion of the general case and restrict ourselves here to a discussion of the following special case.

**Definition 7.4.8 [109].** Let $\mathcal{M}$ and $\mathcal{N}$ be von Neumann algebras acting on a Hilbert space $\mathcal{H}$. A state $\phi$ on $\mathcal{B}(\mathcal{H})$ is 1-distillable if there exist completely positive maps $T : \mathcal{B}(\mathbb{C}^2) \to \mathcal{M}$ and $S : \mathcal{B}(\mathbb{C}^2) \to \mathcal{N}$ such that the functional $\omega(X \otimes Y) \equiv \phi(T(X)S(Y))$, $X \otimes Y \in \mathcal{B}(\mathbb{C}^2) \otimes \mathcal{B}(\mathbb{C}^2)$ is not ppt.

Verch and Werner show that 1-distillable states are distillable and not ppt. They also prove the following theorem.

**Proposition 7.4.9 [109].** Let $\mathcal{O} \mapsto \mathcal{R}(\mathcal{O})$ be a local net in a vacuum representation satisfying weak additivity. Then, if $\mathcal{O}_1$ and $\mathcal{O}_2$ are strictly spacelike separated double cones, the vacuum state is 1-distillable across the pair $(\mathcal{R}(\mathcal{O}_1), \mathcal{R}(\mathcal{O}_2))$.

Hence, the vacuum is distillable and not ppt across $(\mathcal{R}(\mathcal{O}_1), \mathcal{R}(\mathcal{O}_2))$ no matter how large $d(\mathcal{O}_1, \mathcal{O}_2)$ is. We remark that once again, this theorem is valid also for states induced by vectors in the vacuum representation that are analytic for the energy [109]. For a discussion of some further aspects of the entanglement of the vacuum in AQFT, we refer the reader to [36].

## 7.5 Geometric Modular Action

We emphasize that nearly all of the remarkable properties of the vacuum state discussed to this point are shared by all vector states that are analytic for the energy. In the remainder of this chapter, we deal with properties unique to the vacuum.

A crucial breakthrough in the theory of operator algebras was the Tomita–Takesaki theory [103] (see also [60, 105]), which is proving itself to be equally powerful and productive for the purposes of mathematical quantum theory. One of the settings subsumed by this theory is a von Neumann algebra $\mathcal{M}$ with a cyclic and separating vector $\Omega \in \mathcal{H}$. The data $(\mathcal{M}, \Omega)$ then uniquely determine an antiunitary involution[11] $J \in \mathcal{B}(\mathcal{H})$ and a strongly continuous group of unitaries $\Delta^{it}$, $t \in \mathbb{R}$,[12] such that $J\Omega = \Omega = \Delta^{it}\Omega$, $JMJ = \mathcal{M}'$, and $\Delta^{it}\mathcal{M}\Delta^{-it} = \mathcal{M}$, for all $t \in \mathbb{R}$, along with further significant properties. From the Reeh–Schlieder Theorem (Theorem 7.3.1), this theory is applicable to the pair $(\mathcal{R}(\mathcal{O}), \Omega)$, under the indicated conditions. As explained already, the algebras and states are operationally determined (in principle), the corresponding modular objects $J_\mathcal{O}$, $\Delta_\mathcal{O}^{it}$ are as well.

---

[11] Commonly called the modular conjugation or modular involution associated with $(\mathcal{M}, \Omega)$.
[12] $\Delta$ is a certain, typically unbounded, positive operator called the modular operator associated with $(\mathcal{M}, \Omega)$.

In pathbreaking work [6,7], J. J. Bisognano and E. H. Wichmann showed that for a net of von Neumann algebras $\mathcal{O} \mapsto \mathcal{R}(\mathcal{O})$ locally associated with a finite-component quantum field satisfying the Wightman axioms [8,58,89] (and therefore in a vacuum representation), the modular objects $J_W$, $\Delta_W^{it}$ determined by the wedge algebras $\mathcal{R}(W)$, $W \in \mathcal{W}$, and the vacuum vector $\Omega$ have a geometric interpretation[13]:

$$\Delta_W^{it} = U(\lambda_W(2\pi t)), \tag{7.2}$$

for all $t \in \mathbb{R}$ and $W \in \mathcal{W}$, where $\{\lambda_W(2\pi t) \mid t \in \mathbb{R}\} \subset \mathcal{P}_+^{\uparrow}$ is the one-parameter subgroup of boosts leaving $W$ invariant. Explicitly for $W = W_R$,

$$\lambda_{W_R}(t) = \begin{pmatrix} \cosh t & \sinh t & 0 & 0 \\ \sinh t & \cosh t & 0 & 0 \\ 0 & 0 & 1 & 0 \\ 0 & 0 & 0 & 1 \end{pmatrix}.$$

The relation Equation (7.2) has come to be referred to as *modular covariance*. Moreover, for scalar Boson fields,[14] one has

$$J_{W_R} = \Theta U_\pi, \tag{7.3}$$

where $\Theta$ is the PCT-operator associated to the Wightman field and $U_\pi$ implements the rotation through the angle $\pi$ about the 1-axis, with similar results for general wedge $W \in \mathcal{W}$. Hence, one has

$$J_{W_R} \mathcal{R}(\mathcal{O}) J_{W_R} = \mathcal{R}(\theta_R \mathcal{O}), \tag{7.4}$$

for all $\mathcal{O}$, where $\theta_R \in \mathcal{P}_+$ is the reflection through the edge $\{(0, 0, x_2, x_3) \mid x_2, x_3 \in \mathbb{R}\}$ of the wedge $W_R$. This implies in turn that for all $W \in \mathcal{W}$, one has

$$J_W \{\mathcal{R}(\widetilde{W}) \mid \widetilde{W} \in \mathcal{W}\} J_W = \{\mathcal{R}(\widetilde{W}) \mid \widetilde{W} \in \mathcal{W}\}. \tag{7.5}$$

Thus, the adjoint action of the modular involutions $J_W$, $W \in \mathcal{W}$, leaves the set $\{\mathcal{R}(W) \mid W \in \mathcal{W}\}$ of observable algebras associated with wedges invariant; that is, wedge algebras are transformed to wedge algebras by this adjoint action.

Although in the special case of the massless free scalar field [56] (and, more generally, for conformally invariant quantum field theories [15]) also the modular objects corresponding to $(\mathcal{R}(\mathcal{O}), \Omega)$ for $\mathcal{O} \in \mathcal{D}$ have geometric meaning, some explicit computations in the free massive field have indicated that this is not true in general. Moreover, as we shall see in the next section, *only* the vacuum vector $\Omega$ yields modular objects having any geometric content. This fact yields an intrinsic characterization of the vacuum state.

But before we explore this noteworthy state of affairs, let us examine some of the more striking consequences of the previous relations. For simplicity, we shall restrict these remarks to the case of nets of algebras locally associated with a scalar Bose field. Because every element $\widetilde{\lambda} \in \mathcal{L}_+ \setminus \mathcal{L}_+^{\uparrow}$ of the complement of the identity component $\mathcal{L}_+^{\uparrow}$ of the Lorentz group in the proper Lorentz group $\mathcal{L}_+$ can be factored uniquely into a product $\widetilde{\lambda} = \theta_R \lambda$, with $\lambda \in \mathcal{L}_+^{\uparrow}$, it follows that by defining $U(\widetilde{\lambda}) = J_{W_R} U(\lambda)$, one

---

[13] See also [37] for later advances in this particular setting.
[14] See [7] for arbitrary finite-component Wightman fields.

obtains an (anti-)unitary representation of the proper Poincaré group $\mathcal{P}_+$, which acts covariantly on the original net of observables. Moreover, denoting by $\mathcal{J}$ the group generated by $\{J_W \mid W \in \mathcal{W}\}$ and $\mathcal{J}_+$ as the subgroup of $\mathcal{J}$ consisting of products of even numbers of the generating involutions $\{J_W \mid W \in \mathcal{W}\}$, one has

$$\mathcal{J} = U(\mathcal{P}_+) \text{ and } \mathcal{J}_+ = U(\mathcal{P}_+^\uparrow). \tag{7.6}$$

Hence, the modular involutions $\{J_W \mid W \in \mathcal{W}\}$ encode the isometries of the underlying spacetime as well as a representation of the isometry group that acts covariantly on the observables. So, in particular, $U(\mathbb{R}^4) \subset \mathcal{J}_+$. Recalling that the subgroup of translations $U(\mathbb{R}^4)$ determines the dynamics of the quantum field, one sees that the modular involutions also encode the dynamics of the model! The dynamics need not be posited but instead can be derived from the observables and preparations of the quantum system, at least in principle, using the modular involutions.

If the quantum field model is such that a scattering theory can be defined for it and satisfies asymptotic completeness [3,8,58], then the original fields and the asymptotic fields act on the same Hilbert space and have the same vacuum. Letting $\mathcal{R}^{(0)}(W)$, $W \in \mathcal{W}$, denote the observable algebras associated with the free asymptotic field and $J_W^{(0)}$ represent the modular involution corresponding to $(\mathcal{R}^{(0)}(W), \Omega)$, one has, as was pointed out by Schroer [84],

$$S = J_{W_R} J_{W_R}^{(0)},$$

where $S$ is the *scattering matrix* for the original field model. Hence, the modular involutions associated with the wedge algebras and the vacuum state also encode all information about the results of scattering processes in the given model![15]

In addition, because of the connection between Tomita–Takesaki modular theory and KMS–states [13], modular covariance entails that when the vacuum state is restricted to $\mathcal{R}(W)$ for any wedge $W$, then with respect to the automorphism group on $\mathcal{R}(W)$ generated by the boosts $U(\lambda_W(t))$, it is an equilibrium state at temperature $1/2\pi$ (in suitable units). Hence, any uniformly accelerated observers find when testing the vacuum that it has a nonzero temperature [88]. This striking fact is called the Unruh effect [108]. Moreover, because KMS–states are passive [75], the vacuum satisfies the second law of thermodynamics with respect to boosts—an additional stability property.

Modular covariance and/or the geometric action of the modular conjugations in Equation (7.4) have also been derived under other sets of assumptions (in addition to those discussed in the next section) that do not refer to the Wightman axioms; that is, purely algebraic settings in which no appeal to Wightman fields is made [16,65,72,107] (see [10] for a review). Thus, these properties and their many consequences hold quite generally. It is also of interest that some of these settings provide algebraic versions of the PCT Theorem and the Spin–Statistics connection [16,51,63,67,72], but we shall not enter on this topic here. We now turn to those conditions that provide an intrinsic characterization of the vacuum state.

---

[15] Note that the same is *not* true about the modular unitaries because both the original field and the asymptotic field are covariant under the same representation of $\mathcal{P}_+^\uparrow$.

## 7.6 Intrinsic Characterization of the Vacuum State

Although the definition of a vacuum state given in Definition 7.2.1 is standard, it is not quite satisfactory because it is not (operationally) *intrinsic*. It has been seen in Section 7.2 that the elements of quantum theory that are closest to its operational foundations are states and observables. However, in the definition of the vacuum state, one finds such notions as the spectrum condition and automorphic (and unitary) representations of the Poincaré group, all of which are not expressed solely in terms of these states and observables. This may not disturb some readers, so let us step back and locate the notion of Minkowski space vacuum state in a larger context.

One of the primary roles of the vacuum state in quantum field theory has been to serve as a physically distinguished reference state with respect to which other physical states can be defined and referred. Let us recall as an example of this that perturbation theory is performed with respect to the vacuum state; that is, computations performed for general states of interest in quantum field theory are carried out by suitably perturbing the vacuum. This role has proven to be so central that when theorists tried to formulate quantum field theory in spacetimes other than Minkowski space,[16] they tried to find analogous states in these new settings, thereby running into some serious conceptual and mathematical problems. This is not the place to explain the range and scope of these difficulties, but one noteworthy problem is indicated by the question: What could replace the large isometry group (the Poincaré group) of Minkowski space in the definition of "vacuum state," in light of the fact that the isometry group of a generic spacetime is trivial? A further point is that in the definition of "vacuum state," the spectrum condition serves as a stability condition; what could replace it even in such highly symmetric spacetimes as de Sitter space, where the isometry group, although large, does not contain any translations?

After much effort, a number of interesting selection criteria have been isolated and studied; see, for example, [11, 14, 19, 21, 28, 29, 43, 61, 66, 70, 76, 90]. Of these, all but one either select an entire folium of states—that is, a representation, instead of a state—or are explicitly limited to a particular subclass of spacetimes (or both). Here, we discuss the selection criterion provided by the Condition of Geometric Modular Action (CGMA), which in the special case of Minkowski space selects the vacuum state (as opposed to selecting the entire vacuum representation) but which can be formulated for general spacetimes.

As we now no longer have a vacuum state/representation given, we return to the notation of Section 7.2 and the initial data of a net $\mathcal{O} \mapsto \mathcal{R}(\mathcal{O})$ of observable algebras and a state $\phi$ on $\mathcal{R}$. The question we are now examining is: Under which conditions, stated solely in terms of mathematical quantities completely determined by these initial data, is $\phi$ a vacuum state? Surprisingly, the core of the answer to this question is the relation in Equation (7.5). It will be convenient to introduce the notation $\mathcal{R}_\phi(\mathcal{O}) \equiv \pi_\phi(\mathcal{R}(\mathcal{O}))'' = (\pi_\phi(\mathcal{R}(\mathcal{O}))')'$. We consider a special case of the condition first discussed in [27] and subsequently further generalized in [19].

---

[16] After all, the spacetime in which we find ourselves is not Minkowski space.

**Definition 7.6.1.** A state $\phi$ on a net $\mathcal{O} \mapsto \mathcal{R}(\mathcal{O})$ satisfies the CGMA if the vector $\Omega_\phi$ is cyclic and separating for $\mathcal{R}_\phi(W)$, $W \in \mathcal{W}$, and if the modular conjugation $J_W$ corresponding to $(\mathcal{R}_\phi(W), \Omega_\phi)$ satisfies

$$J_W \{\mathcal{R}_\phi(\widetilde{W}) \mid \widetilde{W} \in \mathcal{W}\} J_W \subset \{\mathcal{R}_\phi(\widetilde{W}) \mid \widetilde{W} \in \mathcal{W}\} \tag{7.7}$$

for all $W \in \mathcal{W}$.

Note that there is no prima facie reason why Equation (7.5) should imply Equation (7.4). Indeed, why should the action in Equation (7.5) even be implemented by point transformations on $\mathbb{R}^4$, much less by Poincaré transformations? And, because all Poincaré transformations map wedges to wedges, why should Equation (7.4) be the only solution, even if one did find oneself in the latter, fortunate situation?

The following theorem was proven in [19,28]. The interested reader may consult [28] for the definition of the weak technical property referred to in hypothesis (c) of the following theorem—a property that involves only the net $W \mapsto \mathcal{R}_\phi(W)$ itself.[17]

**Theorem 7.6.2 [19,28].** Let $\phi$ be a state on a net $\mathcal{O} \mapsto \mathcal{R}(\mathcal{O})$ that satisfies the following constraints:

(a) The map $\mathcal{W} \ni W \mapsto \mathcal{R}_\phi(W) \in \{\mathcal{R}_\phi(W) \mid W \in \mathcal{W}\}$ is an order-preserving bijection.
(b) If $W_1 \cap W_2 \neq \emptyset$, then $\Omega_\phi$ is cyclic and separating for $\mathcal{R}_\phi(W_1) \cap \mathcal{R}_\phi(W_2)$. Conversely, if $\Omega_\phi$ is cyclic and separating for $\mathcal{R}_\phi(W_1) \cap \mathcal{R}_\phi(W_2)$, then $\overline{W_1} \cap \overline{W_2} \neq \emptyset$, where the bar denotes closure.
(c) The net $W \mapsto \mathcal{R}_\phi(W)$ is locally generated.
(d) The adjoint action of the modular conjugations $J_W$, $W \in \mathcal{W}$, acts transitively upon the set $\{\mathcal{R}_\phi(W) \mid W \in \mathcal{W}\}$; that is, there exists a wedge $W_0 \in \mathcal{W}$ such that

$$\{J_W \mathcal{R}_\phi(W_0) J_W \mid W \in \mathcal{W}\} = \{\mathcal{R}_\phi(W) \mid W \in \mathcal{W}\}.$$

Then, there exists a continuous (anti-)unitary representation $U$ of $\mathcal{P}_+$ that leaves $\Omega_\phi$ invariant and acts covariantly upon the net:

$$U(\lambda)\mathcal{R}_\phi(\mathcal{O})U(\lambda)^{-1} = \mathcal{R}_\phi(\lambda\mathcal{O}),$$

for all $\mathcal{O}$ and $\lambda \in \mathcal{P}_+$. Moreover, $\mathcal{J} = U(\mathcal{P}_+)$, $\mathcal{J}_+ = U(\mathcal{P}_+^\uparrow)$ and

$$J_{W_R} \mathcal{R}(\mathcal{O}) J_{W_R} = \mathcal{R}(\theta_R \mathcal{O}),$$

for all $\mathcal{O}$. Furthermore, the wedge duality condition holds:

$$\mathcal{R}_\phi(W') = \mathcal{R}_\phi(W)',$$

for all $W \in \mathcal{W}$, which entails that the net $W \mapsto \mathcal{R}_\phi(W)$ is local.

Hence, from the state and net are *derived* the isometry group of the spacetime; a unitary representation of the isometry group formed from the modular involutions, leaving the state invariant and acting covariantly on the net; the specific geometric

---

[17] In fact, hypothesis (c) may be dispensed with if the Modular Stability Condition (see later) is satisfied [28].

action of the modular involutions found in a special case by Bisognano and Wichmann; the locality of the net; and even the dynamics and so forth of the theory (see Section 7.5).

The conceptually crucial observation is that all conditions in the hypothesis of this theorem are expressed solely in terms of the initial net and state, or algebraic quantities completely determined by them. Condition (a) entails that the adjoint action of the modular involutions $J_W$ upon the net induces an inclusion preserving bijection on the set $\mathcal{W}$. Condition (b) assures that this bijection can be implemented by point transformations (indeed, Poincaré transformations) [19], and (c) implies that the representation $U(\mathcal{P}_+)$ is continuous [28].[18] Condition (d) strengthens the CGMA. Without this strengthening, the adjoint action of the $J_W$ can still be shown to be implemented by Poincaré transformations [19], but the group $\mathcal{J}$ can then be isomorphic to a proper subgroup of $\mathcal{P}_+$ [44].

Although such a state $\phi$ is clearly a physically distinguished state, the spectrum condition and modular covariance need not be fulfilled [19]. As an *intrinsic* stability condition, the Modular Stability Condition has been proposed.

**Definition 7.6.3 [19].** For any $W \in \mathcal{W}$, the elements $\Delta_W^{it}, t \in \mathbb{R}$, of the modular group corresponding to $(\mathcal{R}_\phi(W), \Omega_\phi)$ are contained in $\mathcal{J}$.

Note that in this condition, no reference is made to the spacetime, its isometry group, or any representation of the isometry group. This condition can be posed for models on any spacetime [19]. Together with the CGMA, this modular stability condition then yields both the spectrum condition and modular covariance in Equation (7.2).

**Theorem 7.6.4 [19, 28].** If, in addition to the hypothesis of Theorem 7.6.2, the Modular Stability Condition is satisfied, then after choosing suitable coordinates on $\mathbb{R}^4$, the spectrum condition is satisfied by $U(\mathcal{P}_+)$ and modular covariance holds. The associated representation $(\mathcal{H}_\phi, \pi_\phi, \Omega_\phi)$ is therefore a vacuum representation.

Of course, this is not, strictly speaking, a characterization of arbitrary vacuum states; this theorem provides an intrinsic characterization of those vacuum states that manifest further desirable properties, properties that are also manifested in the models in the special circumstances considered by Bisognano and Wichmann. But because these latter circumstances are precisely those expected to arise in standard quantum field theory, the vacuum states characterized in Theorems 7.6.2 and 7.6.4 are probably the vacuum states of most direct physical interest.

## 7.7 Deriving Spacetime from States and Observables

Although the hypothesis of Theorem 7.6.4 makes no explicit or implicit reference to an underlying spacetime, Theorem 7.6.2 does so implicitly through use of the set of wedges $\mathcal{W}$.[19] However, the results of the preceding section did suggest the possibility

---

[18] Note that the continuity of the representation of the translation group follows without condition (c) [20].

[19] In fact, only a four-dimensional real manifold with a coordinatizon is required to formulate and prove the theorems in Section 7.6, but it is nonetheless clear that the introduction of wedges as defined tacitly appeals to Minkowski space.

that without any a priori reference to a spacetime, the spacetime itself as well as an assignment of localization regions for the observable algebras—along with all of the previously mentioned results—could be derived from the modular conjugations associated with a collection of algebras and a suitable state, as long as the set of modular conjugations verifies certain purely algebraic relations. And, if the Modular Stability Condition is also satisfied, the state would then be a vacuum state, and the CGMA and modular covariance would be satisfied. In fact, this program has been carried out for a few spacetimes in [100–102, 113]. To minimize technical complications that would distract attention away from the essential conceptual point to be made, we discuss this approach only in the example of three-dimensional Minkowski space.

To eliminate any reference to a spacetime and to strengthen the purely operational nature of the initial data, we consider a collection $\{\mathcal{A}_i\}_{i \in I}$ of unital $C^*$-algebras indexed by "laboratories" $i \in I$. $\mathcal{A}_i$ is interpreted as the algebra generated by all observables measurable in the laboratory $i$.[20] It makes sense to speak of one laboratory as being contained in another, so the set $I$ of laboratories is provided with a natural partial order $\leq$. It is then immediate that if $i \leq j$, then $\mathcal{A}_i \subset \mathcal{A}_j$. Hence, the map $I \ni i \mapsto \mathcal{A}_i \in \{\mathcal{A}_i\}_{i \in I}$ is order preserving. We shall assume that this map is a bijection because otherwise there would be some redundancy in the description of the system. If $(I, \leq)$ is a directed set, then $\{\mathcal{A}_i\}_{i \in I}$ is a net and the inductive limit $\mathcal{A}$ of $\{\mathcal{A}_i\}_{i \in I}$ exists and may be used as a reference algebra. But even if $\{\mathcal{A}_i\}_{i \in I}$ is not a net, it is possible [46] naturally to embed $\mathcal{A}_i$, $i \in I$, into a $C^*$-algebra $\mathcal{A}$ so that the inclusion relations are preserved. It is not necessary to distinguish between these cases in the results, and we refer to states $\phi$ on $\mathcal{A}$ as being states on the net $\{\mathcal{A}_i\}_{i \in I}$.

Given such a state $\phi$, we proceed to the corresponding GNS representation and define $\mathcal{R}_i = \pi_\phi(\mathcal{A}_i)''$, $i \in I$. We assume that the implementing vector $\Omega_\phi$ is cyclic and separating for all $\mathcal{R}_i$, $i \in I$, and denote by $J_i$, $\Delta_i$, the corresponding modular objects. Again, let $\mathcal{J}$ denote the group generated by the involutions $J_i$, $i \in I$. Note that $J\Omega_\phi = \Omega_\phi$, for all $J \in \mathcal{J}$. In this abstract context, the CGMA is the requirement that the adjoint action of each $J_i$ upon the elements of $\{\mathcal{R}_i\}_{i \in I}$ leaves the set $\{\mathcal{R}_i\}_{i \in I}$ invariant [19]. Among other matters, the CGMA here entails that the set $\{J_i\}_{i \in I}$ is an invariant generating set for the group $\mathcal{J}$,[21] and such a structure is the starting point for the investigations of the branch of geometry known as absolute geometry (see e.g., [1, 4, 5]). From such a group and a suitable set of axioms to be satisfied by the generators of that group, absolute geometers derive various "metric" spaces such as Minkowski spaces and Euclidean spaces on which the abstract group $\mathcal{J}$ now acts as the isometry group of the metric space. Different sets of axioms on the group yield different metric spaces. This affords us with the possibility of deriving a spacetime from the group $\mathcal{J}$, so that the operational data $(\phi, \{\mathcal{A}_i\}_{i \in I})$ would determine the spacetime in which the quantum systems could naturally be considered to be evolving. We emphasize that different groups $\mathcal{J}$ would verify different sets of algebraic relations and would thus lead to different spacetimes.

---

[20] The index set can be naturally refined by further encoding the time (with respect to some reference clock in the laboratory) during which the measurement is carried out without changing the validity of the following assertions.

[21] In other words, the smallest group containing $\{J_i\}_{i \in I}$ is $\mathcal{J}$ and $J\{J_i\}_{i \in I}J^{-1} \subset \{J_i\}_{i \in I}$ for all $J \in \mathcal{J}$.

For the convenience of the reader, we summarize our standing assumptions, which refer solely to objects that are completely determined by the data $(\phi, \{\mathcal{A}_i\}_{i \in I})$.

**Standing Assumptions.** For the net $\{\mathcal{A}_i\}_{i \in I}$ of nonabelian $C^*$-algebras and the state $\phi$ on $\mathcal{A}$, we assume

(i) $i \mapsto \mathcal{R}_i$ is an order-preserving bijection.
(ii) $\Omega_\phi$ is cyclic and separating for each algebra $\mathcal{R}_i$, $i \in I$.
(iii) the adjoint action of each $J_i$ leaves the set $\{\mathcal{R}_i\}_{i \in I}$ invariant.

Already these assumptions restrict significantly the class of admissible groups $\mathcal{J}$ [19]. In general, it may be necessary to pass to a suitable subcollection of $\{\mathcal{R}_i\}_{i \in I}$ for the Standing Assumptions to be satisfied [19] (if, indeed, they are satisfied at all); see [100] for a brief discussion of this point.

We must introduce some notation to formulate concisely the algebraic requirements on $\mathcal{J}$ that lead to the construction of three-dimensional Minkowski space. We use lower-case Latin letters to denote arbitrary modular involutions $J_i$, $i \in I$, upper-case Latin letters to denote involutions in $\mathcal{J}$ of the form $ab$, and lower-case Greek letters for arbitrary elements of $\mathcal{J}$. By $\xi \mid \eta$, we mean "$\xi\eta$ is an involution," and $\alpha, \beta \mid \xi, \eta$ is shorthand for "$\alpha \mid \xi, \beta \mid \xi, \alpha \mid \eta$, and $\beta \mid \eta$."

**Theorem 7.7.1 [100].** Assume in the previous setting that the following relations hold in $\mathcal{J}$:

1. For every $P, Q$ there exists a $g$ with $P, Q \mid g$.
2. If $P, Q \mid g, h$, then $P = Q$ or $g = h$.
3. If $a, b, c \mid P$, then $abc \in \{J_i : i \in I\}$.
4. If $a, b, c \mid g$, then $abc \in \{J_i : i \in I\}$.
5. There exist $g, h, j$ such that $g \mid h$ but $j \mid g, h, gh$ are all false.
6. For each $P$ and $g$ with $P \mid g$ false, there exist exactly two distinct elements $h_1, h_2$ such that $h_1, h_2 \mid P$ is true and $g, h_i \mid R, c$ are false for all $R, c, i = 1, 2$.

Then there exists a model (based on $\mathcal{J}$) of three-dimensional Minkowski space in which each $J_i$, $i \in I$ is identified as a spacelike line (and every spacelike line is such an element) and on which each $J_i$, $i \in I$ acts adjointly as the reflection about the spacelike line to which it corresponds. $\mathcal{J}$ is isomorphic to $\mathcal{P}_+$[22] and forms in a canonical manner a strongly continuous (anti-)unitary representation $U$ of $\mathcal{P}_+$. Moreover, there exists a bijection $\chi : I \to \mathcal{W}$[23] such that after defining $\mathcal{R}(\chi(i)) = \mathcal{R}_i$, the resultant net $\{\mathcal{R}(\chi(i))\}$ of wedge algebras on Minkowski space is covariant under the action of the representation $U(\mathcal{P}_+)$. Furthermore, one has $\mathcal{R}(\chi(i))' = \mathcal{R}(\chi(i)')$ for all $i \in I$. Thus, if the map $\chi : I \to \mathcal{W}$ is order preserving, then the net $\{\mathcal{R}(\chi(i))\}$ is local.

If, further, $\Delta_j^{it} \in \mathcal{J}$ for all $j \in I, t \in \mathbb{R}$, then modular covariance is satisfied and the state $\omega$ is a vacuum state on the net $\{\mathcal{R}(\chi(i))\}$.

We emphasize that assumptions 1–6 are purely algebraic in nature and involve only the group $\mathcal{J}$, which is completely determined by the initial data $(\phi, \{\mathcal{A}_i\}_{i \in I})$. Although

---

[22] The proper Poincaré group for three-dimensional Minkowski space.
[23] The set of wedges in three-dimensional Minkowski space.

we do not propose the verification of such conditions as a practical procedure to determine spacetime, it is, in our view, a noteworthy conceptual point that such a derivation is possible in principle. It is also noteworthy that the derived structure is so rigid and provides such a complete basis for physical interpretation. Indeed, from the observables and state can be derived a spacetime, an identification of the localizations of the observables in that spacetime, and a continuous unitary representation of the isometry group of the spacetime such that the resultant, reinterpreted net is covariant under the action of the isometry group and the reinterpreted state is a vacuum state. It is perhaps worth mentioning that the modular symmetry group $\mathcal{J}$ of a theory on four-dimensional Minkowski space as discussed in Section 7.6 *does not* verify assumptions 1–6. Moreover, models on three-dimensional Minkowski space satisfying the CGMA do verify assumptions 1–6.

In [101, 102, 113], sets of algebraic conditions on $\mathcal{J}$ have also been found so that the space derived is three-dimensional de Sitter space, respectively, four-dimensional Minkowski space. We anticipate that similar results can be proven for other highly symmetric spacetimes such as anti–de Sitter space and the Einstein universe, but not for general spacetimes.

## 7.8 Concluding Remarks

It is a striking fact that in the aforementioned senses, the modular involutions associated with the vacuum state (and *only* the vacuum state) encode the following physically significant matters:

- the spacetime in which the quantum systems may be viewed as evolving
- the isometry group of the spacetime
- a strongly continuous unitary representation of this isometry group that acts covariantly on the net of observable algebras and leaves the state invariant
- the locality (i.e., the Einstein causality, of the quantum systems)
- the dynamics of the quantum systems
- the scattering behavior of the quantum systems
- the spin–statistics connection in the quantum systems
- the stability of the quantum systems
- the thermodynamic behavior of the quantum systems

It has also become clear through examples—quantum field theories on de Sitter space [11, 19, 44], anti–de Sitter space [21, 29], a class of positively curved Robertson–Walker spacetimes [24, 25], as well as others [90, 95]—that the encoding of crucial physical information by modular objects and the subsequent utility of this approach are not limited to Minkowski space theories.

It is necessary to distinguish between the (in some sense) maximal results of Section 7.7 and those of Section 7.6. The former cannot be expected to be reproducible in most spacetimes because the isometry groups are not large enough to determine the spacetime, and the arguments in Section 7.7 rely tacitly on the possibility of interpreting the modular group $\mathcal{J}$ as (a suitably large subgroup of) the isometry group of some spacetime. However, most of the results of Section 7.6 and, hence, most of the previous list can be expected to be attainable in more general spacetimes, without regard to the size of the isometry group of the spacetime. As has been verified in a class of models in a family of Robertson–Walker spacetimes [24], the CGMA and the encoding of crucial physical information by modular involutions associated with certain observable algebras and select states can hold even when the modular symmetry group $\mathcal{J}$ is strictly larger than the isometry group of the spacetime (in fact, in these examples, a significant portion of $\mathcal{J}$ is not associated with any kind of pointlike transformations on the spacetime). In other words, it is quite possible that the fact that the modular symmetry group gives no more than (a subgroup of) the isometry group of the spacetime in the presence of the CGMA for theories on Minkowski or de Sitter space is an accident due to the fact that these spacetimes are maximally symmetric. Moreover, it is possible that using the CGMA and Modular Stability Condition to select states of physical interest yields a modular symmetry group $\mathcal{J}$ containing—along with the standard symmetries expected from classical theory—new and purely quantum symmetries encoding unexpected physical information (further evidence for this speculation that goes beyond [24] can be adduced in [42]).

Finally, we mention that modular objects associated with privileged algebras of observables and states (usually the vacuum) are also proving to be useful in the construction of quantum field models in two-, three-, and four-dimensional Minkowski space, which cannot be constructed by previously known techniques of constructive quantum field theory [17, 23, 31, 32, 50, 69, 73, 85]. But, such matters go well beyond the scope of this chapter.

# References

[1] J. Ahrens. Begründung der absoluten Geometrie des Raumes aus dem Spiegelungsbegriff. *Mathematische Zeitschrift*, **71** (1959), 154–85.

[2] H. Araki. Von Neumann algebras of local observables for free scalar field. *Journal of Mathematical Physics*, **5** (1964), 1–13.

[3] H. Araki. *Mathematical Theory of Quantum Fields*. Oxford: Oxford University Press, 1999.

[4] F. Bachmann. *Aufbau der Geometrie aus dem Spiegelungsbegriff*, 2nd ed. Berlin, New York: Springer-Verlag, 1973.

[5] F. Bachmann, A. Baur, W. Pejas, and H. Wolff. Absolute geometry. In *Fundamentals of Mathematics*, vol. 2, eds. H. Behnke, F. Bachmann, K. Fladt, and H. Kunle. Cambridge, MA: MIT Press, 1986, pp. 129–73.

[6] J. J. Bisognano and E. H. Wichmann. On the duality condition for a hermitian scalar field. *Journal of Mathematical Physics*, **16** (1975), 985–1007.

[7] J. J. Bisognano and E. H. Wichmann. On the duality condition for quantum fields. *Journal of Mathematical Physics*, **17** (1976), 303–21.

[8] N. N. Bogolubov, A. A. Logunov, and I. T. Todorov. *Introduction to Axiomatic Quantum Field Theory*. Reading, MA: W. A. Benjamin, 1975. (Translation of Russian original, published in 1969.)

[9] H.-J. Borchers. On the converse of the Reeh–Schlieder theorem. *Communications in Mathematical Physics*, **10** (1968), 269–73.

[10] H.-J. Borchers. On revolutionizing quantum field theory with Tomita's modular theory. *Journal of Mathematical Physics*, **41** (2000), 3604–73.

[11] H.-J. Borchers and D. Buchholz. Global properties of vacuum states in de Sitter space. *Annales de l'Institut Henri Poincaré*, **70** (1999), 23–40.

[12] O. Bratteli and D. W. Robinson. *Operator Algebras and Quantum Statistical Mechanics I*. Berlin, Heidelberg, New York: Springer Verlag, 1979.

[13] O. Bratteli and D. W. Robinson. *Operator Algebras and Quantum Statistical Mechanics II*. Berlin, Heidelberg, New York: Springer Verlag, 1981.

[14] R. Brunetti, K. Fredenhagen, and M. Köhler. The microlocal spectrum condition and Wick polynomials of free fields on curved spacetimes. *Communications in Mathematical Physics*, **180** (1996), 633–52.

[15] R. Brunetti, D. Guido, and R. Longo. Modular structure and duality in conformal quantum field theory. *Communications in Mathematical Physics*, **156** (1993), 201–19.

[16] R. Brunetti, D. Guido, and R. Longo. Group cohomology, modular theory and space-time symmetries. *Reviews in Mathematical Physics*, **7** (1995), 57–71.

[17] R. Brunetti, D. Guido, and R. Longo. Modular localization and Wigner particles. *Reviews in Mathematical Physics*, **14** (2002), 759–85.

[18] D. Buchholz. On the manifestations of particles. In *Mathematical Physics Towards the 21st Century*, eds. R. Sen and A. Gersten. Beer-Sheva: Ben Gurion University Press, 1994.

[19] D. Buchholz, O. Dreyer, M. Florig, and S. J. Summers. Geometric modular action and spacetime symmetry groups. *Reviews in Mathematical Physics*, **12** (2000), 475–560.

[20] D. Buchholz, M. Florig, and S. J. Summers. An algebraic characterization of vacuum states in Minkowski space, II: Continuity aspects. *Letters in Mathematical Physics*, **49** (1999), 337–50.

[21] D. Buchholz, M. Florig, and S. J. Summers. The second law of thermodynamics, TCP and Einstein causality in anti–de Sitter space-time. *Classical and Quantum Gravity*, **17** (2000), L31–L37.

[22] D. Buchholz and K. Fredenhagen. Locality and the structure of particle states. *Communications in Mathematical Physics*, **84** (1982), 1–54.

[23] D. Buchholz and G. Lechner. Modular nuclearity and localization. *Annales Henri Poincaré*, **5** (2004), 1065–80.

[24] D. Buchholz, J. Mund, and S. J. Summers. Transplantation of local nets and geometric modular action on Robertson–Walker space-times. In *Mathematical Physics in Mathematics and Physics (Siena)*, ed. R. Longo. *Fields Institute Communications*, **30** (2001), 65–81.

[25] D. Buchholz, J. Mund, and S. J. Summers. Covariant and quasi-covariant quantum dynamics in Robertson–Walker space-times. *Classical Quantum and Gravity*, **19** (2002), 6417–34.

[26] D. Buchholz, M. Porrmann, and U. Stein. Dirac versus Wigner: Towards a universal particle concept in local quantum field theory. *Physical Letters*, **B, 267** (1991), 377–81.

[27] D. Buchholz and S. J. Summers. An algebraic characterization of vacuum states in Minkowski space. *Communications in Mathematical Physics*, **155** (1993), 449–58.

[28] D. Buchholz and S. J. Summers. An algebraic characterization of vacuum states in Minkowski space, III: Reflection maps. *Communications in Mathematical Physics*, **246** (2004), 625–41.

[29] D. Buchholz and S. J. Summers. Stable quantum systems in anti–de Sitter space: Causality, independence and spectral properties. *Journal of Mathematical Physics*, **45** (2004), 4810–31.

[30] D. Buchholz and S. J. Summers. Quantum statistics and locality. *Physics Letters*, **A 337** (2005), 17–21.

[31] D. Buchholz and S. J. Summers. String– and brane–localized causal fields in a strongly nonlocal model. *Journal of Physics A*, **40** (2007), 2147–63.

[32] D. Buchholz and S. J. Summers. Warped convolutions: A novel tool in the construction of quantum field theories. In *Quantum Field Theory and Beyond*, eds. E. Seiler and K. Sibold. Singapore: World Scientific, 2008, pp. 107–21.

[33] D. Buchholz and R. Wanzenberg. The realm of the vacuum. *Communications in Mathematical Physics*, **143** (1992), 577–89.

[34] J. Butterfield. Stochastic Einstein locality revisited. *British Journal of Philosophy of Science*, **58** (2007), 805–67.

[35] B. S. Cirel'son. Quantum generalization of Bell's inequality. *Letters in Mathematical Physics*, **4** (1980), 93–100.

[36] R. Clifton and H. Halvorson. Entanglement and open systems in algebraic quantum field theory. *Studies in History and Philosophy of Modern Physics*, **32** (2001), 1–31.

[37] W. Driessler, S. J. Summers, and E. H. Wichmann. On the connection between quantum fields and von Neumann algebras of local operators. *Communications in Mathematical Physics*, **105** (1986), 49–84.

[38] W. Dybalski. A sharpened nuclearity condition and the uniqueness of the vacuum in QFT. *Communications in Mathematical Physics*, **283** (2008), 523–42.

[39] A. Einstein, B. Podolsky, and N. Rosen. Can quantum mechanical description of physical reality be considered complete? *Physical Review*, **47** (1935), 777–80.

[40] G. G. Emch. *Algebraic Methods in Statistical Mechanics and Quantum Field Theory*. New York: John Wiley & Sons, 1972.

[41] H. Epstein, V. Glaser, and A. Jaffe. Nonpositivity of the energy density in quantized field theories. *Nuovo Cimento*, **36** (1965), 1016–22.

[42] L. Fassarella and B. Schroer. The modular origin of chiral diffeomorphisms and their fuzzy analogs in higher-dimensional quantum field theories. *Physics Letters*, **B, 538** (2002), 415–25.

[43] C. J. Fewster and R. Verch. Stability of quantum systems at three scales: Passivity, quantum weak energy inequalities and the microlocal spectrum condition. *Communications in Mathematical Physics*, **240** (2003), 329–75.

[44] M. Florig. *Geometric Modular Action*. Ph.D. Dissertation, University of Florida, 1999.

[45] D. Fraser. The fate of "particles" in quantum field theories with interactions. *Studies in History and Philosophy of Modern Physics*, **39** (2008), 841–59.

[46] K. Fredenhagen. Global observables in local quantum physics. In *Quantum and Non-Commutative Analysis*. Amsterdam: Kluwer Academic Publishers, 1993.

[47] J. Glimm and A. Jaffe. Boson quantum field theory models. In *Mathematics of Contemporary Physics*, ed. R. F. Streater. London: Academic Press, 1972.

[48] J. Glimm and A. Jaffe. *Quantum Physics*. Berlin, Heidelberg, New York: Springer Verlag, 1981.

[49] E. Grant. *Much Ado About Nothing: Theories of Space and Vacuum from the Middle Ages to the Scientific Revolution*. Cambridge: Cambridge University Press, 1981.

[50] H. Grosse and G. Lechner. Wedge-local quantum fields and noncommutative Minkowski space. *Journal of High Energy Physics*, **11** (2007), 012.

[51] D. Guido and R. Longo. An algebraic spin and statistics theorem, I. *Communications in Mathematical Physics*, **172** (1995), 517–33.

[52] R. Haag. *Local Quantum Physics*. Berlin, Heidelberg, New York: Springer Verlag, 1992.

[53] H. Halvorson. Reeh-Schlieder defeats Newton-Wigner: On alternative localization schemes in relativistic quantum field theory. *Philosophy of Science*, **68** (2001), 111–33.

[54] H. Halvorson and R. Clifton. Generic Bell correlation between arbitrary local algebras in quantum field theory. *Journal of Mathematical Physics*, **41** (2000), 1711–17.

[55] H. Halvorson and R. Clifton. No place for particles in relativistic quantum theory? *Philosophy of Science*, **69** (2002), 1–28.

[56] P. D. Hislop and R. Longo. Modular structure of the local algebras associated with the free massless scalar field theory. *Communications in Mathematical Physics*, **84** (1982), 71–85.

[57] R. Horodecki, P. Horodecki, M. Horodecki, and K. Horodecki. Quantum entanglement. *Reviews of Modern Physics*, **81** (2009), 865–942. arXiv:quant-ph/0702225.

[58] R. Jost. *General Theory of Quantized Fields*. Providence, RI: American Mathematical Society, 1965.

[59] R. V. Kadison and J. R. Ringrose. *Fundamentals of the Theory of Operator Algebras*, vol. I. Orlando, FL: Academic Press, 1983.

[60] R. V. Kadison and J. R. Ringrose. *Fundamentals of the Theory of Operator Algebras*, vol. II. Orlando, FL: Academic Press, 1986.

[61] B. S. Kay and R. M. Wald. Theorems on the uniqueness and thermal properties of stationary, nonsingular, quasi-free states on spacetimes with a bifurcate Killing horizon. *Physics Reports*, **207** (1991), 49–136.

[62] M. Keyl. Fundamentals of quantum information theory. *Physics Reports*, **369** (2002), 431–548.

[63] B. Kuckert. A new approach to spin & statistics. *Letters in Mathematical Physics*, **35** (1995), 319–31.

[64] B. Kuckert. Localization regions of local observables. *Communications in Mathematical Physics*, **215** (2000), 197–216; erratum, **228** (2002), 589–90.

[65] B. Kuckert. Two uniqueness results on the Unruh effect and on PCT-symmetry. *Communications in Mathematical Physics*, **221** (2001), 77–100.

[66] B. Kuckert. Covariant thermodynamics of quantum systems: Passivity, semipassivity and the Unruh effect. *Annals of Physics*, **295** (2002), 216–29.

[67] B. Kuckert and R. Lorenzen. Spin, statistics and reflections, II, Lorentz invariance. *Communications in Mathematical Physics*, **269** (2007), 809–31.

[68] L. J. Landau. On the violation of Bell's inequality in quantum theory. *Physics Letters*, **A, 120** (1987), 54–6.

[69] G. Lechner. Construction of quantum field theories with factorizing S-matrices. *Communications in Mathematical Physics*, **277** (2008), 821–60.

[70] C. Lüders and J. E. Roberts. Local quasiequivalence and adiabatic vacuum states. *Communications in Mathematical Physics*, **134** (1990), 29–63.

[71] G. Ludwig. *Foundations of Quantum Mechanics*, vol. I. Berlin, Heidelberg, New York: Springer-Verlag, 1983.

[72] J. Mund. The Bisognano–Wichmann theorem for massive theories. *Annales Henri Poincaré*, **2** (2001), 907–26.

[73] J. Mund, B. Schroer, and J. Yngvason. String-localized quantum fields and modular localization. *Communications in Mathematical Physics*, **268** (2006), 621–72.

[74] A. Peres. Separability criterion for density matrices. *Physical Review Letters*, **77** (1996), 1413–15.

[75] W. Pusz and S. L. Woronowicz. Passive states and KMS states for general quantum systems. *Communications in Mathematical Physics*, **58** (1978), 273–90.

[76] M. J. Radzikowski. Micro-local approach to the Hadamard condition in quantum field theory on curved space-time. *Communications in Mathematical Physics*, **179** (1996), 529–53.

[77] G. A. Raggio. A remark on Bell's inequality and decomposable normal states. *Letters in Mathematical Physics*, **15** (1988), 27–9.

[78] M. Rédei and S. J. Summers. Local primitive causality and the common cause principle in quantum field theory. *Foundations of Physics*, **32** (2002), 335–55.

[79] M. Rédei and S. J. Summers. Quantum probability theory. *Studies in History and Philosophy of Modern Physics*, **38** (2007), 390–417.

[80] M. Redhead. More ado about nothing. *Foundations of Physics*, **25** (1995), 123–37.

[81] B. Reznik, A. Retzker, and J. Silman. Violating Bell's inequality in vacuum. *Physical Review A*, **71** (2005), 042104.

[82] E. Schrödinger. Die gegenwärtige Situation in der Quantenmechanik. *Naturwissenschaften*, **23** (1935), 807–12, 812–28, 844–9.

[83] B. Schroer. Infrateilchen in der Quantenfeldtheorie. *Fortschritte der Physik*, **11** (1963), 1–31.

[84] B. Schroer. Wigner representation theory of the Poincare group, localization, statistics and the S-matrix. *Nuclear Physics B*, **499** (1997), 519–46.

[85] B. Schroer. Modular localization and the bootstrap–formfactor program. *Nuclear Physics B*, **499** (1997), 547–68.

[86] J. Schwinger. The algebra of microscopic measurements. *Proceedings of the National Academic Science of USA*, **45** (1959), 1542–53.

[87] I. E. Segal. Postulates for general quantum mechanics. *Annals of Mathematics*, **48** (1947), 930–48.

[88] G. L. Sewell. Quantum fields on manifolds: PCT and gravitationally induced thermal states. *Annals of Physics*, **141** (1982), 201–24.

[89] R. F. Streater and A. S. Wightman. *PCT, Spin and Statistics, and All That*. Reading, MA: Benjamin/Cummings, 1964.

[90] R. Strich. Passive states for essential observers. *Journal of Mathematical Physics*, **49** (2008), 022301.

[91] S. J. Summers. On the independence of local algebras in quantum field theory. *Reviews in Mathematical Physics*, **2** (1990), 201–47.

[92] S. J. Summers. Bell's inequalities and algebraic structure. In *Operator Algebras and Quantum Field Theory*, Lecture Notes in Mathematics, vol. 1441, eds. S. Doplicher, R. Longo, J. E. Roberts, and L. Zsido. Cambridge, MA: International Press, 1997, pp. 633–46. (Distributed by the American Mathematical Society, Providence, RI.)

[93] S. J. Summers. On the Stone–von Neumann uniqueness theorem and its ramifications. In *John von Neumann and the Foundations of Quantum Physics*, eds. M. Rédei and M. Stoelzner. Dordrecht: Kluwer Academic Publishers, 2001, pp. 135–52.

[94] S. J. Summers. Subsystems and independence in relativistic microscopic physics. *Studies in History and Philosophy of Modern Physics*, **40** (2009), 133–41.

[95] S. J. Summers and R. Verch. Modular inclusion, the Hawking temperature and quantum field theory in curved space-time. *Letters in Mathematical Physics*, **37** (1996), 145–58.

[96] S. J. Summers and R. Werner. The vacuum violates Bell's inequalities. *Physics Letters*, **A, 110** (1985), 257–9.

[97] S. J. Summers and R. Werner. Bell's inequalities and quantum field theory, I. General setting. *Journal of Mathematical Physics*, **28** (1987), 2440–7.

[98] S. J. Summers and R. Werner. Maximal violation of Bell's inequalities for algebras of observables in tangent spacetime regions. *Annales de l'Institut Henri Poincaré*, **49** (1988), 215–43.

[99] S. J. Summers and R. F. Werner. On Bell's inequalities and algebraic invariants. *Letters in Mathematical Physics*, **33** (1995), 321–34.

[100] S. J. Summers and R. K. White. On deriving space-time from quantum observables and states. *Communications in Mathematical Physics*, **237** (2003), 203–20.

[101] S. J. Summers and R. K. White. On deriving space-time from quantum observables and states, II: Three-dimensional de Sitter space. Manuscript in preparation.

[102] S. J. Summers and R. K. White. On deriving space-time from quantum observables and states, III: Four-dimensional Minkowski space. Manuscript in preparation.

[103] M. Takesaki. *Tomita's Theory of Modular Hilbert Algebras and Its Applications*. Berlin, Heidelberg, New York: Springer-Verlag, 1970.

[104] M. Takesaki. *Theory of Operator Algebras I*. Berlin, Heidelberg, New York: Springer-Verlag, 1979.

[105] M. Takesaki. *Theory of Operator Algebras II*. Berlin, Heidelberg, New York: Springer-Verlag, 2003.

[106] M. Takesaki. *Theory of Operator Algebras III*. Berlin, Heidelberg, New York: Springer-Verlag, 2003.

[107] S. Trebels. *Über die geometrische Wirkung modularer Automorphismen*. Ph.D. Dissertation, University of Göttingen, 1997.

[108] W. G. Unruh. Notes on black-hole evaporation. *Physical Review D*, **14** (1976), 870–92.

[109] R. Verch and R. F. Werner. Distillability and positivity of partial transposes in general quantum field systems. *Reviews in Mathematical Physics*, **17** (2005), 545–76.

[110] J. von Neumann. *Mathematische Grundlagen der Quantenmechanik*. Berlin: Springer-Verlag, 1932; English translation: *Mathematical Foundations of Quantum Mechanics*. Princeton, NJ: Princeton University Press, 1955.

[111] J. von Neumann. *Collected Works*, vol. I, ed. A. H. Taub. New York, Oxford: Pergamon Press, 1962.

[112] R. F. Werner and M. M. Wolf. Bell inequalities and entanglement. *Quantation Information and Computing*, **1** (2001), 1–25.

[113] R. White. *An Algebraic Characterization of Minkowski Space*. Ph.D. Dissertation, University of Florida, 2001.

[114] A. S. Wightman and L. Gårding. Fields as operator-valued distributions in relativistic quantum field theory. *Arkiv för Fysik*, **28** (1964), 129–84.

CHAPTER 8

# Einstein Meets von Neumann: Locality and Operational Independence in Algebraic Quantum Field Theory

Miklós Rédei

## 8.1 Main Claim

I argue in this chapter that Einstein and von Neumann meet in algebraic relativistic quantum field theory in the following metaphorical sense: algebraic quantum field theory was created in the late 1950s/early 1960s and was based on the theory of "rings of operators," which von Neumann established in 1935–1940 (partly in collaboration with J. Murray). In the years 1936–1949, Einstein criticized standard, nonrelativistic quantum mechanics, arguing that it does not satisfy certain criteria that he regarded as necessary for any theory to be compatible with a field theoretical paradigm. I claim that algebraic quantum field theory (AQFT) does satisfy those criteria and hence that AQFT can be viewed as a theory in which the mathematical machinery created by von Neumann made it possible to express in a mathematically explicit manner the physical intuition about field theory formulated by Einstein.

The argument in favor of this claim has two components:

1. **Historical**: An interpretation of Einstein's (semi)formal wordings of his critique of nonrelativistic quantum mechanics.

   This interpretation results in mathematically explicit operational independence definitions, which, I claim, express independence properties of systems that are localized in causally disjoint spacetime regions. Einstein regarded these as necessary for a theory to comply with field theoretical principles.

2. **Systematic**: The presentation of several propositions formulated in terms of AQFT that state that the operational independence conditions in question do in fact typically hold in AQFT.

This chapter is structured as follows:
After presenting some historical comments in Section 8.2, I quote from Einstein's famous 1948 *Dialectica* paper in Section 8.3 and isolate from the text three

requirements Einstein thought a physical theory must satisfy: spatiotemporality, independence, and local operations. After recalling some basic notions of algebraic quantum mechanics in Section 8.4, the main axioms of AQFT are recalled in Section 8.5, and it is argued here that spatiotemporality is the very principle on which the key notion of AQFT (the definition of local net of observables) is based.

Section 8.6 reviews several of the most important definitions in the hierarchy of independence concepts formulated in algebraic quantum mechanics and concludes with two propositions stating that the independence properties typically hold in AQFT. Section 8.7 interprets the *local operations* requirement by identifying it with what is called "operational separability," the definition of which is formulated in this section in terms of operations (understood as completely positive unit preserving linear maps). Linking operational separability to the independence condition called *operational independence*, Section 8.7 concludes that the local-operations condition also is typically satisfied in AQFT. Section 8.8 raises the problem of relation of operational independence and operational separability in general (i.e., irrespective of the quantum field theoretical context) and, distinguishing a stronger and a weaker version of operational separability, argues that the weaker version is strictly weaker than operational separability.

I am aware that the historical aspect of the main claim is somewhat controversial because it rests on a particular interpretation of Einstein's wording of his criticism of standard, nonrelativistic quantum mechanics. Section 8.9 indicates other possible interpretations of Einstein's criticism and will qualify the main claim. The systematic part of the main claim seems to me meaningful, however, irrespective of its historical accuracy: The operational separability notions are intuitively physically motivated and mathematically well-defined concepts; investigating them and their relation to other independence concepts raises nontrivial questions, showing the richness and beauty of AQFT.

## 8.2 Preliminary Historical Comments

Einstein and von Neumann played very different but crucial roles in establishing nonrelativistic quantum mechanics: Einstein's Nobel Prize–winning explanation of the photoelectric effect in 1905 was a decisive step in lending credibility to the quantum hypothesis, whereas von Neumann's three papers [43–45] and his subsequent book [46] clarified and summarized the mathematical foundations of quantum theory. In addition to their contribution to (mathematical) physics proper, both Einstein and von Neumann were deeply interested in interpretational–philosophical problems related to quantum mechanics: Einstein's dissatisfaction with quantum mechanics and his criticism of this theory as a complete description of physical reality are well known and have been the subject of intensive scrutiny by historians and philosophers of physics. Von Neumann's foundational work also was deeply philosophical: his famous no-go theorem on hidden variables presented in [46] and his idea of a nonclassical (quantum) logic associated with quantum mechanics and the problems he had seen in this connection are classical examples of his philosophical attitude and treatment of interpretational problems.[1]

Given their interest in philosophical–foundational issues related to quantum mechanics and the fact that both were members of the Institute for Advanced Study in Princeton from 1933 on, one would expect that the two had exchanged ideas and had discussions about the interpretation and foundations of quantum mechanics, especially around 1935–1936. This was the time when Einstein was working with Podolsky and Rosen on the EPR article [15] and von Neumann was working on the theory of "rings of operators" (called today *von Neumann algebras*) and on quantum logic [3]—both motivated by Hilbert space quantum mechanics. Curiously, however, Einstein and von Neumann did not seem to have had discussions about the foundations of quantum mechanics. To be more precise, the only record I am aware of that shows specifically that Einstein and von Neumann did possibly talk about interpretational problems of quantum mechanics is von Neumann's letter to Schrödinger dated April 11, 1936, in which von Neumann writes:

> Einstein has kindly shown me your letter as well as a copy of the Pr. Cambr. Phil. Soc. manuscript. I feel rather more over-quoted than under-quoted and I feel that my merits in the subject are over-emphasized. [22, p. 211] (von Neumann to Schrödinger, April 11, 1936)

Von Neumann refers here to the second of Schrödinger's two papers on the problem of probabilistic correlations between spatially separated quantum systems [30, 31]. Einstein and Schrödinger corresponded about the EPR article in the summer of 1935 (see Jammer's article [18]), and Schrödinger's two articles were motivated by his correspondence with Einstein. Schrödinger was bothered by the EPR-type correlations between spatially distant systems predicted by nonrelativistic quantum mechanics, and he seems to have thought that quantum field theory will also be problematic for this reason:

> Though in the mean time some progress seemed to have been made in the way of coping with this condition (quantum electrodynamics), there now appears to be a strong probability (as P. A. M. Dirac[2] has recently pointed out on a special occasion) that this progress is futile. [31, p. 451]

But von Neumann did not share Schrödinger's concern:

> I think that the difficulties you hint at are "pseudo-problems." The "action at distance" in the case under consideration says only that even if there is no dynamical interaction between two systems (e.g., because they are far removed from each other), the systems can display statistical correlations. This is not at all specific for quantum mechanics, it happens classically as well. [22, p. 212] (von Neumann to Schrödinger, April 11, 1936)

To illustrate his point, von Neumann gives a simple example of a nonproblematic correlation between spatially distant systems—the example shows that von Neumann

---

[1] The literature on both Einstein's and von Neumann's work on quantum mechanics is enormous, so I make no attempt here to list even the most important works.
[2] Schrödinger's footnote: P. A. M. Dirac, *Nature*. **137** (1936), 298.

regarded distant correlations nonproblematic as long as one can give an explanation of them in terms of common causes (see [23] for this interpretation of von Neumann's position). Whether the EPR correlations can be given an interpretation in terms of common causes is a subtle matter to which I briefly return in Section 8.9; now I wish to point out that von Neumann reacted explicitly to Schrödinger's skeptical remark about the prospects of relativistic quantum field theory as well:

> And of course quantum electrodynamics proves that quantum mechanics and the special theory of relativity are compatible "philosophically"—quantum electrodynamic fails only because of the concrete form of Maxwell's equations in the vicinity of a charge. [22, p. 213] (von Neumann to Schrödinger, April 11, 1936)

So it seems that von Neumann thought that the real problem with relativistic quantum field theory was the presence of singularities caused by the assumption of point-like charges and the related infinitely sharp localization of fields (both originating in classical field theory), rather than some irreconcilable "philosophical incompatibility" between quantum theory and principles of causality. However, von Neumann did not attempt to make the compatibility explicit by formulating postulates that a quantum field theory should satisfy in order to be acceptable "philosophically." This was done by Einstein in his critique of standard quantum mechanics as a complete theory.

## 8.3 Einstein's Contrasting Standard Nonrelativistic Quantum Mechanics with Field Theory in 1948

Einstein's dissatisfaction with quantum mechanics and his attempts to show that quantum mechanics is an incomplete theory are among the most analyzed aspects of the history and philosophy of quantum mechanics. Einstein gave his argument against the completeness of quantum mechanics (at least) four times after the publication of the EPR article [15] in 1935. The first formulation is contained in his 1935 letter to Schrödinger, which he wrote just a few weeks after the EPR paper had appeared (see [17] and [18] for the details of the Einstein–Schrödinger correspondence). This was followed by formulations in 1936 [12], 1948 [13], and 1949 [14]. All the formulations are informal; and, although the core idea remains the same, they are slightly different. One can see a gradual shift toward what I claim is a formulation of several criteria that Einstein thought must be satisfied by a physical theory if it is to be compatible with a field theoretical paradigm. These criteria are most explicitly present in his 1948 *Dialectica* paper [13], in which he writes:

> If one asks what is characteristic of the realm of physical ideas independently of the quantum theory, then above all the following attracts our attention: the concepts of physics refer to a real external world, i.e. ideas are posited of things that claim a "real existence" independent of the perceiving subject (bodies, fields, etc.), and these ideas are, on the other hand, brought into as secure a relationship as possible with sense impressions. Moreover, it is characteristic of these physical things that they are conceived of as being arranged in a spacetime continuum. Further, it appears to be essential for this arrangement of the things introduced in physics that, at a specific time, these things claim an existence independent

of one another, insofar as these things "lie in different parts of space." Without such an assumption of mutually independent existence (the "being-thus") of spatially distant things, an assumption which originates in everyday thought, physical thought in the sense familiar to us would not be possible. Nor does one see how physical laws could be formulated and tested without such a clean separation. Field theory has carried out this principle to the extreme, in that it localizes within infinitely small (four dimensional) space-elements the elementary things existing independently of one another that it takes as basic as well as the elementary laws it postulates for them.

For the relative independence of spatially distant things ($A$ and $B$), this idea is characteristic: an external influence on $A$ has no *immediate* effect on $B$; this is known as the "principle of local action," which is applied consistently only in field theory. The complete suspension of this basic principle would make impossible the idea of the existence of (quasi-)closed systems and, thereby, the establishment of empirically testable laws in the sense familiar to us.

...

Matters are different, however, if one seeks to hold on principle II—the autonomous existence of the real states of affairs present in two separated parts of space $R_1$ and $R_2$—simultaneously with the principles of quantum mechanics. In our example the complete measurement on $S_1$ of course implies a physical interference which only effects the portion of space $R_1$. But such an interference cannot immediately influence the physically real in the distant portion of space $R_2$. From that it would follow that every measurement regarding $S_2$ which we are able to make on the basis of a complete measurement on $S_1$ must also hold for the system $S_2$ if, after all, no measurement whatsoever ensued on $S_1$. That would mean that for $S_2$ all statements that can be derived from the postulation of $\psi_2$ or $\psi'_2$, etc. must hold simultaneously. This is naturally impossible, if $\psi_2, \psi'_2$, are supposed to signify mutually distinct real states of affairs of $S_2, \ldots$ [13] (Translation taken from [17].)

In the preceding passage, Einstein formulates (informally) three requirements for a physical theory to be compatible with a field theoretical paradigm:

1. **Spatiotemporality**: "physical things ... are conceived of as being arranged in a space-time continuum. ..."
2. **Independence**: "essential for this arrangement of the things introduced in physics is that, at a specific time, these things claim an existence independent of one another, insofar as these things 'lie in different parts of space'."
3. **Local operation**: "an external influence on $A$ has no *immediate* effect on $B$; this is known as the 'principle of local action'"; "... measurement on $S_1$ of course implies a physical interference which only effects the portion of space $R_1$. But such an interference cannot immediately influence the physically real in the distant portion of space $R_2$."

These three requirements are not independent: *independence* presupposes *spatiotemporality* conceptually: only if physical things are assumed to be arranged in a spacetime continuum can one ask whether the things *so arranged* have the feature independence; and it also is more or less clear that the local-operations requirement is an independence condition—independence of system $S_2$ of (measurement) operations carried out on system $S_1$.

It is true that standard (nonrelativistic) Hilbert space quantum mechanics is *not* field theoretical in the sense that observables in nonrelativistic quantum theory are *not* "conceived of as being arranged in a spacetime continuum": the observable quantities in quantum theory do *not* carry labels that would indicate their spatiotemporal location in a four-dimensional spacetime continuum and, accordingly, quantum measurements and operations are also *not* conceived of in quantum mechanics as possessing spatiotemporal tags explicitly. Neither is Hilbert space quantum mechanics covariant with respect to a relativistic symmetry group. Thus, quantum mechanics does *not* meet requirements of relativistic locality interpreted in the sense of a field theoretical paradigm, but AQFT does, or so I argue in the rest of this chapter.

## 8.4 Some Notions of Algebraic Quantum Mechanics

In what follows, $\mathcal{A}$ denotes a unital $C^*$-algebra; $\mathcal{A}_1, \mathcal{A}_2$ are assumed to be $C^*$-subalgebras of $\mathcal{A}$ (with common unit). $\mathcal{N}$ denotes a von Neumann algebra; algebras $\mathcal{N}_1, \mathcal{N}_2$ are assumed to be von Neumann subalgebras of $\mathcal{N}$ (with common unit). $\mathcal{A}_1 \vee \mathcal{A}_2$ (respectively, $\mathcal{N}_1 \vee \mathcal{N}_2$) denotes the $C^*$-algebra (respectively, von Neumann algebra) generated by $\mathcal{A}_1$ and $\mathcal{A}_2$ (respectively, by $\mathcal{N}_1$ and $\mathcal{N}_2$). A $W^*$-algebra $\mathcal{N}$ is hyperfinite (or *approximately finite dimensional*) if there exists a series of finite-dimensional full matrix algebras $M_n$ ($n = 1, 2, \ldots$) such that $\cup_n M_n$ is weakly dense in $\mathcal{N}$. $\mathcal{B}(\mathcal{H})$ is the $C^*$-algebra (and von Neumann algebra) formed by the set of *all* bounded operators on Hilbert space $\mathcal{H}$. $\mathcal{B}(\mathcal{H})$ is hyperfinite if $\mathcal{H}$ is a separable Hilbert space. For von Neumann algebra, $\mathcal{N} \subseteq \mathcal{B}(\mathcal{H})$, $\mathcal{N}'$ stands for the commutant of $\mathcal{N}$ in $\mathcal{B}(\mathcal{H})$. $S(\mathcal{A})$ is the state space of $C^*$-algebra $\mathcal{A}$. The self-adjoint elements in a $C^*$-algebra are interpreted as representatives of physical observables; the elements of the state space $S(\mathcal{A})$ represent physical states. (For the operator algebraic notions, see [4, 5, 19].)

In what follows, $T$ will denote a completely positive (CP) map on a $C^*$-algebra $\mathcal{A}$; such a $T$ will also be assumed to preserve the identity: $T(I) = I$ (where $I$ is the unit of $\mathcal{A}$). A (unit preserving) CP map is called a (nonselective) *operation*. An operation $T$ on a von Neumann algebra $\mathcal{N}$ is called a *normal* operation if it is $\sigma$ weakly continuous.

The dual $T^*$ of an operation defined by

$$S(\mathcal{A}) \ni \phi \mapsto T^*\phi \doteq \phi \circ T \in S(\mathcal{A})$$

maps the state space $S(\mathcal{A})$ into itself. If $T$ is a normal operation on the von Neumann algebra $\mathcal{N}$, then $T^*$ takes normal states into normal states.

Operations are the mathematical representatives of physical operations: physical processes that take place as a result of physical interactions with the system. For a detailed description and physical interpretation of the notion of operation, see [20].

**Remark 1.** In sharp contrast to states, operations defined on a subalgebra of an arbitrary $C^*$-algebra are *not*, in general, extendable to an operation on the larger algebra [2]. A $C^*$-algebra $\mathcal{B}$ is said to be *injective* if for any $C^*$-algebras $\mathcal{A}_1 \subset \mathcal{A}$ every CP unit-preserving linear map $T_1 : \mathcal{A}_1 \to \mathcal{B}$ has an extension to a CP unit-preserving linear map $T : \mathcal{A} \to \mathcal{B}$. A von Neumann algebra is injective (by definition) if it is injective as a $C^*$-algebra. It was shown in [2] that $\mathcal{B}(\mathcal{H})$ is injective. Hyperfiniteness of a von Neumann algebra entails injectivity in general, and a von Neumann algebra acting on a separable Hilbert space is injective if and only if it is hyperfinite [9], [10, Theorem 6]—this is why injectivity of the double-cone algebras in AQFT (Proposition 4) is important.

The following is a classic result characterizing operations:

**Proposition 1 (Stinespring's representation theorem).** $T: \mathcal{A} \to \mathcal{B}(\mathcal{H})$ is a CP linear map from $C^*$-algebra $\mathcal{A}$ into $\mathcal{B}(\mathcal{H})$ if and only if it has the form

$$T(X) = V^* \pi(X) V \qquad X \in \mathcal{A}$$

where $\pi: \mathcal{A} \to \mathcal{B}(\mathcal{K})$ is a representation of $\mathcal{A}$ on Hilbert space $\mathcal{K}$ and $V: \mathcal{H} \to \mathcal{K}$ is a bounded linear map. If $\mathcal{A}$ is a von Neumann algebra and $T$ is normal, then $\pi$ is a normal representation.

The following is a corollary of Stinespring's theorem:

**Proposition 2 (Kraus's representation theorem).** $T: \mathcal{B}(\mathcal{H}) \to \mathcal{B}(\mathcal{H})$ is a normal operation if and only if there exists bounded operators $W_i$ on $\mathcal{H}$ such that

$$T(X) = \sum_i W_i^* X W_i \qquad \sum_i W_i^* W_i = I \qquad (8.1)$$

The infinite sums are taken to converge in the $\sigma$-weak topology. $W_i$ are sometimes called "Kraus operators."

It is important in Stinespring's theorem (and, hence, also in Kraus's theorem) that $T$ takes its value in the set of all bounded operators $\mathcal{B}(\mathcal{H})$ on a Hilbert space. Stinespring's theorem does not hold for an arbitrary von Neumann algebra in place of $\mathcal{B}(\mathcal{H})$ because if it did, then this would entail that operations defined on subalgebras are always extendable from the subalgebra to the superalgebra, which however is not the case (cf. Remark 1). To put it differently: Kraus's representation theorem does not hold for an arbitrary von Neumann algebra; hence, general operations on a von Neumann algebra are *not* of the form Equation (8.1).

A special case of operations are *measurements*: If one measures a (possibly unbounded) observable $Q$ defined on the Hilbert space $\mathcal{H}$ with purely discrete spectrum $\lambda_i$ and corresponding spectral projections $P_i$, then the "projection postulate" is described by the operation $T_{proj}$ defined by

$$\mathcal{N} \ni X \mapsto T_{proj}(X) = \sum_i P_i X P_i. \qquad (8.2)$$

$T_{proj}$ is a normal operation from $\mathcal{B}(\mathcal{H})$ into the commutative von Neumann algebra $\{P_i\}'$ consisting of operators that commute with each $P_i$:

$$\{P_i\}' = \{X \in \mathcal{B}(\mathcal{H}) : X P_i = P_i X \text{ for all } i\}.$$

It is known that the projection postulate has limited applicability: not every interaction with (operation on) a quantum system can be described by a $T_{proj}$ of the form Equation (8.2). (The Kraus operators need not be projections; an operation might not even have a Kraus representation at all.) For instance, if the observable to be measured does not have a discrete spectrum, then one cannot directly generalize formula (8.2) to obtain a CP map [11]. But one can generalize some of the characteristic features of $T_{proj}$:

A positive, linear unit preserving map $T$ from $C^*$-algebra $\mathcal{A}$ onto a $C^*$-subalgebra $\mathcal{A}_0$ of $\mathcal{A}$ whose restriction to $\mathcal{A}_0$ is equal to the identity map is called a *conditional expectation* from $\mathcal{A}$ onto $\mathcal{A}_0$. Such a map will be denoted by $T_c$. Conditional expectations are completely positive.

If for a state $\varphi$ on $\mathcal{A}$ the $T_c$ conditional expectation also preserves $\varphi$—that is, $\varphi(T_c(X)) = \varphi(X)$ for all $X \in \mathcal{A}$—then $T_c$ is called a *$\varphi$-preserving conditional expectation* from $\mathcal{A}$ onto $\mathcal{A}_0$, and it will be denoted by $T_c^\varphi$.

It is known that for a given $\mathcal{A}$, $\mathcal{A}_0$, and $\varphi$, a $\varphi$-preserving conditional expectation from $\mathcal{A}$ onto $\mathcal{A}_0$ does not necessarily exist. But if $\varphi$ is a faithful normal state on von Neumann algebra $\mathcal{N}$, then there always exists a $\varphi$-preserving CP map $T^\varphi$ from $\mathcal{N}$ into any subalgebra $\mathcal{N}_0$: the so-called Accardi–Cecchini $\varphi$-conditional expectation [1]. (Note that $T^\varphi$ is *not* a conditional expectation; that is, it is not a projection to $\mathcal{N}_0$. So "$\varphi$-preserving conditional expectation" and "$\varphi$-conditional expectation" are different concepts, although the terms are deceptively close.)

## 8.5 Algebraic Quantum Field Theory

The basic idea of algebraic quantum field theory is precisely what Einstein requires in *spatiotemporality*: "physical things... are conceived of as being arranged in a space-time continuum...": In AQFT, observables—interpreted as self-adjoint parts of $C^*$-algebras—are assumed to be localized in regions $V$ of the Minkowski spacetime $M$. The basic object in the mathematical model of a quantum field is thus the association of a $C^*$-algebra $\mathcal{A}(V)$ to (open, bounded) regions $V$ of $M$, and *all* the physical content of the theory is assumed to be contained in the assignment $V \mapsto \mathcal{A}(V)$. In particular, relativistic covariance of the theory also is expressed in terms of the *net of algebras* $(\{A(V)\}, V \subset M)$, the net of algebras of local observables.

The net $(\{A(V)\}, V \subset M)$ is specified by imposing on it physically motivated postulates. The following are these postulates:

1. **Isotony**: $\mathcal{A}(V_1)$ is a $C^*$-subalgebra (with common unit) of $\mathcal{A}(V_2)$ if $V_1 \subseteq V_2$.
2. **Local commutativity** (also called *Einstein causality* or *microcausality*):
   $\mathcal{A}(V_1)$ commutes with $\mathcal{A}(V_2)$ if $V_1$ and $V_2$ are spacelike separated.

Let
$$\mathcal{A}_0 \equiv \cup_V \mathcal{A}(V);$$

then, $\mathcal{A}_0$ is a normed $*$-algebra, completion of which (in norm) is a $C^*$-algebra, called the *quasilocal algebra* determined by the net $(\{A(V)\}, V \subset M)$.

3. **Relativistic covariance**: There exists a continuous representation $\alpha$ of the identity-connected component of the Poincaré group $\mathcal{P}$ by automorphisms $\alpha(g)$ on $\mathcal{A}$ such that
$$\alpha(g)\mathcal{A}(V) = \mathcal{A}(gV) \qquad (8.3)$$

for every $g \in \mathcal{P}$ and for every $V$.

4. **Vacuum**: It also is postulated that there exists at least one physical representation of the algebra $\mathcal{A}$; that is, it is required that there exist a Poincaré invariant state $\phi_0$ (vacuum) such that the spectrum condition (5. below) is fulfilled in the corresponding cyclic (GNS) representation $(\mathcal{H}_{\phi_0}, \Omega_{\phi_0}, \pi_{\phi_0})$. In this representation of the quasilocal algebra, the representation $\alpha$ is given as a unitary representation, and there exist the generators $P_i$, $(i = 0, 1, 2, 3)$ of the translation subgroup of the Poincaré group $\mathcal{P}$. The spectrum condition is formulated in terms of these generators as the following.
5. **Spectrum condition**:

$$P_0 \geq 0, \qquad P_0^2 - P_1^2 - P_2^2 - P_3^2 \geq 0 \tag{8.4}$$

Given a state $\phi$, one can consider the net in the representation $\pi_\phi$ determined by $\phi$. If the particular state $\phi$ is not important, then the local von Neumann algebras $\pi_\phi(\mathcal{A}(V))''$ will be denoted by $\mathcal{N}(V)$.

It is a remarkable feature of these axioms that (under some additional assumptions) they are very rich in consequences: they entail a number of nontrivial features of the net. We mention two sorts of consequences that will be referred to in what follows: one is the Reeh–Schlieder theorem.

**Proposition 3 (Reeh–Schlieder theorem).** The vacuum state $\phi_0$ (more generally, any state of bounded energy) is faithful on local algebras $\mathcal{A}(V)$ pertaining to open bounded spacetime regions $V$.

The other type of result concerns the type and structure of certain local algebras. To state the proposition in this direction, recall that a double-cone region $D(x, y)$ of the Minkowski spacetime determined by points $x, y \in M$ such that $y$ is in the forward light cone of $x$ is, by definition, the interior of the intersection of the forward light cone of $x$ with the backward light cone of $y$. A general double cone is denoted by $D$.

**Proposition 4 [6], [16, p. 225].** The local von Neumann algebras $\mathcal{N}(D)$ associated with double cones $D$ are hyperfinite.

## 8.6 Independence

Einstein's *independence* requirement demands that if $V_1$ and $V_2$ are spacelike separated regions, then the systems represented by algebras $\mathcal{A}(V_1)$ and $\mathcal{A}(V_2)$ must be "independent." The *local commutativity* postulate is such an independence requirement; however, as it turns out, there are many other stronger, nonequivalent (but also nonindependent) concepts of independence that one can formulate for two $C^*$- and $W^*$-algebras. In the following, we list a few from the rich hierarchy; the list is far from being complete (for an extensive review of the independence concepts, see [33]).

**Definition 1.** A pair $(\mathcal{A}_1, \mathcal{A}_2)$ of $C^*$-subalgebras of $C^*$-algebra $\mathcal{A}$ is called $C^*$-independent if for any state $\phi_1$ on $\mathcal{A}_1$ and for any state $\phi_2$ on $\mathcal{A}_2$ there exists a state $\phi$ on $\mathcal{A}$ such that

$$\phi(X) = \phi_1(X) \quad \text{for any } X \in \mathcal{A}_1,$$
$$\phi(Y) = \phi_2(Y) \quad \text{for any } Y \in \mathcal{A}_2.$$

$C^*$-independence expresses that any two partial states (states on $\mathcal{A}_1$ and $\mathcal{A}_2$) are co-possible as states of the larger system $\mathcal{A}$. The next independence condition is a strengthening of $C^*$-independence:

**Definition 2.** A pair $(\mathcal{A}_1, \mathcal{A}_2)$ of $C^*$-subalgebras of a $C^*$-algebra $\mathcal{A}$ is called $C^*$-*independent in the product sense* if the map $\eta$ defined by

$$\eta: \mathcal{A}_1 \vee \mathcal{A}_2 \to \mathcal{A}_1 \otimes \mathcal{A}_2 \tag{8.5}$$

$$\eta(XY) \doteq X \otimes Y \tag{8.6}$$

extends to a $C^*$-isomorphism of $\mathcal{A}_1 \vee \mathcal{A}_2$ with $\mathcal{A}_1 \otimes \mathcal{A}_2$, where $\mathcal{A}_1 \otimes \mathcal{A}_2$ denotes the tensor product of $\mathcal{A}_1$ and $\mathcal{A}_2$ with the minimal $C^*$-norm.

If the $C^*$-algebras happen to be von Neumann algebras, then states can have stronger continuity properties; this makes it possible to formulate independence notions that fit the von Neumann algebra category more naturally and are in analogy of the previous independence concepts.

**Definition 3.** Two von Neumann subalgebras $\mathcal{N}_1, \mathcal{N}_2$ of the von Neumann algebra $\mathcal{N}$ are called $W^*$-*independent* if for any *normal* state $\phi_1$ on $\mathcal{N}_1$ and for any *normal* state $\phi_2$ on $\mathcal{N}_2$ there exists a *normal* state $\phi$ on $\mathcal{N}$ such that

$$\phi(X) = \phi_1(X) \quad \text{for any } X \in \mathcal{N}_1.$$
$$\phi(Y) = \phi_2(Y) \quad \text{for any } Y \in \mathcal{N}_2.$$

The following is the analogue of Definition 2 in the von Neumann algebra setting.

**Definition 4.** Two von Neumann subalgebras $\mathcal{N}_1, \mathcal{N}_2$ of the von Neumann algebra $\mathcal{N}$ are called $W^*$-*independent* in the product sense if for any normal state $\phi_1$ on $\mathcal{N}_1$ and for any normal state $\phi_2$ on $\mathcal{N}_2$ there exists a normal product state $\phi$ on $\mathcal{N}$ (i.e., a normal state $\phi$ on $\mathcal{M}$) such that

$$\phi(XY) = \phi_1(X)\phi_2(Y) \quad \text{for any} \quad X \in \mathcal{N}_1, Y \in \mathcal{N}_2.$$

The following notion is an even stronger independence property.

**Definition 5.** Two von Neumann subalgebras $\mathcal{N}_1, \mathcal{N}_2$ of the von Neumann algebra $\mathcal{N}$ are called $W^*$-*independent* in the spatial product sense if there exist faithful normal representations $(\pi_1, \mathcal{H}_1)$ of $\mathcal{N}_1$ and $(\pi_2, \mathcal{H}_2)$ of $\mathcal{N}_2$ such that the map

$$\mathcal{N} \ni XY \mapsto \pi_1(X) \otimes \pi_2(Y) \qquad X \in \mathcal{N}_1 \quad Y \in \mathcal{N}_2$$

extends to a spatial isomorphism of $\mathcal{N}_1 \vee \mathcal{N}_2$ with $\pi(\mathcal{N}_1) \otimes \pi(\mathcal{N}_2)$.

If $\mathcal{N}_1, \mathcal{N}_2$ are commuting von Neumann subalgebras of von Neumann algebra $\mathcal{N}$, then they are $W^*$-independent in the spatial product sense if and only if they satisfy the following so-called split property:

**Definition 6.** $\mathcal{N}_1, \mathcal{N}_2$ have the split property if there exists a type I factor $\mathcal{N}$ such that

$$\mathcal{N}_1 \subset \mathcal{N} \subset \mathcal{N}_2'.$$

The next four definitions formulate an idea of operational independence; these definitions were proposed recently in [28], and they are closely related to the local-operation requirement, as we will see.

**Definition 7.** A pair $(\mathcal{A}_1, \mathcal{A}_2)$ of $C^*$-subalgebras of $C^*$-algebra $\mathcal{A}$ is operationally $C^*$-independent in $\mathcal{A}$ if any two operations on $\mathcal{A}_1$ and $\mathcal{A}_2$, respectively, have a joint extension to an operation on $\mathcal{A}$; that is, if for any two CP unit preserving maps

$$T_1 : \mathcal{A}_1 \to \mathcal{A}_1$$
$$T_2 : \mathcal{A}_2 \to \mathcal{A}_2$$

there exists a CP unit preserving map

$$T : \mathcal{A} \to \mathcal{A}$$

such that

$$T(X) = T_1(X) \quad \text{for all } X \in \mathcal{A}_1,$$
$$T(Y) = T_2(Y) \quad \text{for all } Y \in \mathcal{A}_2.$$

The following is a natural strengthening of operational independence:

**Definition 8.** A pair $(\mathcal{A}_1, \mathcal{A}_2)$ of $C^*$-subalgebras of $C^*$-algebra $\mathcal{A}$ is operationally $C^*$-independent in $\mathcal{A}$ in the product sense if any two operations on $\mathcal{A}_1$ and $\mathcal{A}_2$, respectively, have a joint extension to an operation on $\mathcal{A}$ that factorize across the algebras $(\mathcal{A}_1, \mathcal{A}_2)$ in the sense that

$$T(XY) = T(X)T(Y) \quad X \in \mathcal{A}_1, Y \in \mathcal{A}_2. \tag{8.7}$$

The corresponding von Neumann algebra versions of these two operational independence notions are as follows.

**Definition 9.** A pair $(\mathcal{N}_1, \mathcal{N}_2)$ of von Neumann subalgebras of a von Neumann algebra $\mathcal{N}$ is operationally $W^*$-independent in $\mathcal{N}$ if any two *normal* operations on $\mathcal{N}_1$ and $\mathcal{N}_2$, respectively, have a joint extension to a *normal* operation on $\mathcal{N}$.

**Definition 10.** If normal extensions exist that factorize across the pair $(\mathcal{N}_1, \mathcal{N}_2)$ in the manner of Eq. (8.7), then the pair $(\mathcal{N}_1, \mathcal{N}_2)$ is called operationally $W^*$-independent in $\mathcal{N}$ in the product sense.

*Operational $C^*$-independence* expresses that any operation (e.g., procedure, state preparation) on system $S_1$ is co-possible with any such operation on system $S_2$—if these systems are represented by $C^*$-algebras (similarly for $W^*$-algebras). For a more detailed motivation of operational independence, see [28].

The characterization of the interdependence relations of the preceding independence notions is a highly nontrivial task, and there are still a number of open problems related to this issue. We refer to [33] for a review of the main results known by 1990 on this problem. What is important from the perspective of the current discussion is that local algebras in AQFT pertaining to spacelike separated spacetime regions *typically* do satisfy the independence notions. We just recall, in the form of two propositions, the status of these independence properties in AQFT.

**Proposition 5.** If $\mathcal{N}(D_1), \mathcal{N}(D_2)$ are two von Neumann algebras associated with *strictly* spacelike separated double-cone regions $D_1$ and $D_2$ in a local net of von Neumann algebras satisfying the standard axioms, then $\mathcal{N}(D_1), \mathcal{N}(D_2)$ are *typically* independent in the sense of *all* of the Definitions 1–10, where operational $C^*$-independence of $\mathcal{N}(D_1), \mathcal{N}(D_2)$ (and operational $C^*$-independence in the product sense) is understood to hold in the von Neumann algebra $\mathcal{N}(D)$, with $D$ being a double cone containing the double cones $D_1$ and $D_2$, and operational $W^*$-independence (and operational $W^*$-independence in the product sense) is understood to hold in $\mathcal{N}(D_1)\overline{\otimes}\mathcal{N}(D_2)$.

**Remark 2.** Note that hyperfiniteness of the double-cone algebras (Proposition 4) is crucial in the preceding claim that operational $C^*$-independence of $\mathcal{N}(D_1), \mathcal{N}(D_2)$ holds in $\mathcal{N}(D)$, and it is *not* known whether $\mathcal{N}(D_1), \mathcal{N}(D_2)$ are also operationally $W^*$-independent in $\mathcal{N}(D)$—although they are operationally $W^*$-independent in $\mathcal{N}(D_1)\overline{\otimes}\mathcal{N}(D_2)$. The reason why this latter fact does not entail the former is that although operations on $\mathcal{N}(D_1), \mathcal{N}(D_2)$ are extendable to $\mathcal{N}(D)$ by injectivity of $\mathcal{N}(D)$ ensured by hyperfiniteness of $\mathcal{N}(D)$ (Remark 1), it is unclear whether extendability obtains under the additional requirement of the operations being normal.

"Typically" in Proposition 5 means "in all physically nonpathological models," and possibly with a spacelike distance of the double cones that is above a certain threshold. In those physically nonpathological models, the algebras associated with strictly spacelike separated double cones are split (Definition 6), which entails all of the other independence properties. (See [35] for a detailed nontechnical discussion of the split property and for further discussion of what "[non-]pathological" means in this context.)

**Proposition 6.** If $\mathcal{N}(D_1), \mathcal{N}(D_2)$ are two von Neumann algebras associated with spacelike separated *tangent* double-cone regions $D_1$ and $D_2$ in a local net of von Neumann algebras satisfying the standard axioms, then $\mathcal{N}(D_1), \mathcal{N}(D_2)$ are independent in the sense of Definitions 1 and 3 but *not* independent in the sense of Definitions 4, 5, 8, and 10. It is *not* known whether operational $C^*$- or $W^*$-independence is violated by these von Neumann algebras.

## 8.7 Local Operations

In what follows, $V_1$, $V_2$, and $V$ are assumed to be open-bounded spacetime regions, with $V_1$ and $V_2$ spacelike separated and $V_1, V_2 \subseteq V$. Let $T$ be an operation on $\mathcal{A}(V)$

and $\phi$ be a state on $\mathcal{A}(V)$. Then,

$$(\mathcal{A}(V), \mathcal{A}(V_1), \mathcal{A}(V_2), \phi, T) \tag{8.8}$$

is called a *local system*.

Given such a local system, let $\phi_1$ and $\phi_2$ be the restrictions of $\phi$ to $\mathcal{A}(V_1)$ and $\mathcal{A}(V_2)$, respectively. Suppose $T_1$ is an operation on $\mathcal{A}(V_1)$. Carrying out this operation changes the state $\phi_1$ into $T_1^*\phi_1$. By the requirement of isotony, $\mathcal{A}(V_1)$ is a subalgebra of $\mathcal{A}(V)$, so the operation $T_1$ is an operation that is carried out on the elements of $\mathcal{A}(V)$ that are localized in region $V_1 \subset V$. Assume that $T$ is an operation on $\mathcal{A}(V)$ that is an extension of $T_1$ from $\mathcal{A}(V_1)$ to $\mathcal{A}(V)$.[3] Then, $T$ changes the state $\phi$ into $T^*\phi$ and, because $T$ is an extension of $T_1$, the restriction of $T^*\phi$ to $\mathcal{A}(V_1)$ coincides with $T_1^*\phi$. $V_1$ and $V_2$ are spacelike separated, hence, causally independent regions; therefore, one would like to have the extension $T$ of $T_1$ be such that the change $\phi \mapsto T^*\phi$ caused by the operation $T$ in state $\phi$ is restricted to $\mathcal{A}(V_1)$; that is to say, causally well-behaving systems are the ones for which $T^*\phi(X) = \phi_2(X)$ for every $X \in \mathcal{A}(V_2)$. The next definition of *operational separatedness* formulates this idea precisely.

**Definition 11.** The local system $(\mathcal{A}(V), \mathcal{A}(V_1), \mathcal{A}(V_2), \phi, T)$ with an operation $T$ on $\mathcal{A}(V)$ is defined to be *operationally separated* if both of the following two conditions are satisfied:

1. If $T$ is an extension of an operation on $\mathcal{A}(V_1)$, then the operation conditioned state $T^*\phi = \phi \circ T$ coincides with $\phi$ on $\mathcal{A}(V_2)$—that is,

$$\phi(T(A)) = \phi(A) \quad \text{for all} \quad A \in \mathcal{A}(V_2). \tag{8.9}$$

2. If $T$ is an extension of an operation on $\mathcal{A}(V_2)$, then the operation conditioned state $T^*\phi = \phi \circ T$ coincides with $\phi$ on $\mathcal{A}(V_1)$—that is,

$$\phi(T(A)) = \phi(A) \quad \text{for all} \quad A \in \mathcal{A}(V_1). \tag{8.10}$$

Given a local system $(\mathcal{A}(V), \mathcal{A}(V_1), \mathcal{A}(V_2), \phi, T)$ with the operation $T$ defined by Kraus operators (Equation (8.1)), the local commutativity requirement of AQFT entails that if the operation $T$ is defined by Kraus operators $W_i$ in $\mathcal{A}(V_1)$, then $T$ is the identity map on $\mathcal{A}(V_2)$ (and if $T$ is defined by Kraus operators $W_i$ in $\mathcal{A}(V_2)$, then it is the identity on $\mathcal{A}(V_1)$). This is well known and is sometimes referred to as the *no-signaling theorem* it is typically cited as the motivation for the local commutativity (Einstein causality) axiom in AQFT. We state it explicitly as follows.

**Proposition 7 (no-signaling theorem).** The local system

$$(\mathcal{A}(V), \mathcal{A}(V_1), \mathcal{A}(V_2), \phi, T)$$

with $T$ given by Kraus operators in $\mathcal{A}(V_1)$ or $\mathcal{A}(V_2)$ is operationally separated for *every* state $\phi$.

---

[3] Because operations defined on $C^*$-subalgebras of $C^*$-algebras are not necessarily extendable from the subalgebra to the larger algebra (see Remark 1), it is not obvious that an operation on $\mathcal{A}(V_i)$ ($i = 1, 2$) can be extended to $\mathcal{A}(V)$; consequently, the *assumption* here (and later) that $T$ represents $T_1$ is not a redundant one.

Not every interaction with (operation on) a quantum system can be described by a $T$ given by Kraus operators, however (see Section 8.4); hence, it is not obvious that every local system in AQFT is operationally separated. In fact, one can show that operational separatedness fails in AQFT.

**Proposition 8 [29].** There exists a local system $(\mathcal{A}(V), \mathcal{A}(V_1), \mathcal{A}(V_2), \phi, T)$ that is not operationally separated.

Proposition 8 shows that a no-signaling theorem does not hold for spacelike separated local algebras for arbitrary operations. Furthermore, the failure is not exceptional or atypical because nonoperationally separated local systems are not rare: as the argument in [29] shows, for every locally faithful state $\phi$, there exists an operationally nonseparated local system, and such states are very typical by the Reeh–Schlieder theorem (Proposition 3). Thus, it would seem that the local-operations requirement is violated in AQFT—contrary to the claim in the introductory section. But this conclusion would be too quick. One can argue [24, 25, 29] that the mere existence of operationally not separated local systems should not be interpreted as a genuine incompatibility of AQFT with the local-operations requirement because one cannot expect a theory such as AQFT to exclude causally non–well-behaving local systems necessarily. But it is reasonable to demand that AQFT allow a locally equivalent and causally acceptable description of an operationally not separated local system. In other words, one can say that it may happen that the possible causal bad behavior of the local system $(\mathcal{A}(V), \mathcal{A}(V_1), \mathcal{A}(V_2), \phi, T)$ is due to the nonrelativistically conforming choice of the operation $T$ on $\mathcal{A}(V)$ representing an operation $T_1$ carried out in $\mathcal{A}(V_1)$, say, and there may exist another operation $T'$ on $\mathcal{A}(V)$ that has the same effect on $\mathcal{A}(V_1)$ as that of $T$—that is, $T'(X) = T(X)$ for all $X \in \mathcal{A}(V_1)$—and such that the system $(\mathcal{A}(V), \mathcal{A}(V_1), \mathcal{A}(V_2), \phi, T')$ is causally well behaving. This idea of reducibility of operational nonseparatedness is formulated explicitly in the form of the following weakening of Definition 11 (see [29]).

**Definition 12.** The local system $(\mathcal{A}(V), \mathcal{A}(V_1), \mathcal{A}(V_2), \phi, T)$ is called *operationally $C^*$-separable* if it is operationally separated in the sense of Definition 11, or if it is not operationally separated and $T$ is an extension of an operation in either $\mathcal{A}(V_1)$ or in $\mathcal{A}(V_2)$, then the following is true:

1. If $T$ is an extension of an operation in $\mathcal{A}(V_1)$, then there exists an operation $T' \colon \mathcal{A}(V) \to \mathcal{A}(V)$ such that $T'(X) = T(X)$ for all $X \in \mathcal{A}(V_1)$ and such that the system
   
   $$(\mathcal{A}(V), \mathcal{A}(V_1), \mathcal{A}(V_2), \phi, T')$$
   
   is operationally separated.

2. If $T$ is an extension of an operation in $\mathcal{A}(V_2)$, then there exists an operation $T' \colon \mathcal{A}(V) \to \mathcal{A}(V)$ such that $T'(X) = T(X)$ for all $X \in \mathcal{A}(V_2)$ and such that the system
   
   $$(\mathcal{A}(V), \mathcal{A}(V_1), \mathcal{A}(V_2), \phi, T')$$
   
   is operationally separated.

Operational $W^*$-separability is defined analogously, by requiring the operations and the state $\phi$ to be normal.

Einstein's requirement of local operations can now be interpreted as the requirement that local systems in algebraic quantum field theory should be operationally $C^*$- and $W^*$-separable in the sense of Definition 12. To answer the question of whether local systems in AQFT are operationally separable, one can relate operational separability to operational independence:

**Proposition 9 (Redei and Valente [29]).**

- If the pair $(\mathcal{A}(V_1), \mathcal{A}(V_2))$ is operationally $C^*$-independent in $\mathcal{A}(V)$, then for every $\phi$ and every $T$, the system $(\mathcal{A}(V), \mathcal{A}(V_1), \mathcal{A}(V_2), \phi, T)$ is operationally $C^*$-separable.
- If the pair $(\mathcal{N}(V_1), \mathcal{N}(V_2))$ is operationally $W^*$-independent in $\mathcal{N}(V)$, then for every normal state $\phi$ and every normal $T$, the system $(\mathcal{N}(V), \mathcal{N}(V_1), \mathcal{N}(V_2), \phi, T)$ is operationally $W^*$-separable.

Because operational $C^*$-independence in $\mathcal{A}(V)$ does hold for local algebras $\mathcal{A}(D_1)$ and $\mathcal{A}(D_2)$ associated with strictly spacelike separated double-cone regions $D_1, D_2$, and double-cone $D \supset D_1, D_2$ (Proposition 5), one concludes that AQFT *typically* satisfies the local-operations requirement—at least in the $C^*$-sense. As it was remarked (Remark 2), it is not clear at this point whether operational $W^*$-independence of the pair $(\mathcal{N}(D_1), \mathcal{N}(D_2))$ *in* $\mathcal{N}(D)$ also holds; so one cannot yet conclude in full generality that the local-operations condition also holds in AQFT typically, but it should be clear from these results that AQFT is a theory that displays exactly the features that Einstein thought were necessary for a physical theory to be compatible with a field theoretical paradigm.

## 8.8 Are Operational Separability and Operational Independence Equivalent?

Operational $C^*$- and $W^*$-independence was defined in Section 8.6 for a general pair of operator algebras, independently of AQFT. Although in the previous section the notion of operational separability was defined for local systems in AQFT, it is clear that operational separability also can be defined for a general pair $(\mathcal{A}_1, \mathcal{A}_2)$ of $C^*$- or $W^*$-algebras: Keeping in mind that the state space of a $C^*$-algebra (and the normal state space of a $W^*$-algebra) is separating and that states (also: normal states) defined on subalgebras are always extendable, it is clear that the following definition is the reformulation of Definition 12 in terms of general algebras:

**Definition 13.** The pair $(\mathcal{A}_1, \mathcal{A}_2)$ of $C^*$-subalgebras of $C^*$-algebra $\mathcal{A}$ is operationally $C^*$-separable in $\mathcal{A}$ if every operation $T_1$ that has an extension to $\mathcal{A}$ also has an extension $T'$, which is the identity map on $\mathcal{A}_2$, and every $T_2$ that has an extension to $\mathcal{A}$ also has an extension $T'$, which is the identity map on $\mathcal{A}_1$. (Operational $W^*$-separability is defined similarly by assuming the algebras to be $W^*$-algebras and requiring the operations to be normal.)

The argument in [29] leading to Proposition 9 applies to the general case, so one has:

**Proposition 10.** Let $\mathcal{A}_1, \mathcal{A}_2$ be $C^*$-subalgebras of $C^*$-algebra $\mathcal{A}$ and $\mathcal{N}_1, \mathcal{N}_2$ be $W^*$-subalgebras of $W^*$-algebra $\mathcal{N}$. Then, operational $C^*$-independence of $\mathcal{A}_1, \mathcal{A}_2$ in $\mathcal{A}$ entails operational $C^*$-separability of $\mathcal{A}_1, \mathcal{A}_2$ in $\mathcal{A}$, and operational $W^*$-independence of $\mathcal{N}_1, \mathcal{N}_2$ in $\mathcal{N}$ entails operational $W^*$-separability of $\mathcal{N}_1, \mathcal{N}_2$ in $\mathcal{N}$.

Whereas no counterexample is known, I conjecture that the converse of Proposition 10 does not hold: Operational $C^*$- and $W^*$-separability seem strictly weaker than operational $C^*$- and $W^*$-independence. To see why, consider the following strengthening of the definition of operational separability:

**Definition 14.** The pair $(\mathcal{A}_1, \mathcal{A}_2)$ of $C^*$-subalgebras of $C^*$-algebra $\mathcal{A}$ is *strongly* operationally $C^*$-separable in $\mathcal{A}$ if every operation $T_1$ has an extension $T'$ that is the identity map on $\mathcal{A}_2$ and every $T_2$ has an extension $T'$ that is the identity map on $\mathcal{A}_1$. (Strong operational $W^*$-separability is defined similarly by assuming the algebras to be $W^*$-algebras and requiring the operations to be normal.)

The difference between Definitions 14 and 13 is that strong operational separability requires that all operations on the subalgebras have extensions to the superalgebra, whereas Definition 13 does not require this. Because operations are not extendable in general (Remark 1), requiring the existence of extensions in Definition 14 is a highly nontrivial demand; therefore, strong operational separability seems much stronger than operational separability. In fact, it is easy to see that *strong* operational $C^*$-separability is *equivalent* to operational $C^*$-independence and that strong operational $W^*$-separability also is equivalent to operational $W^*$-independence (see [25] for details).

Because there are natural subclasses of operations, such as "measurements" and conditional expectations, one can further distinguish specific instances of the concepts of operational independence and operational separability by narrowing the admissible operations to these subclasses. One obtains in this way the notions of operational $C^*$- and $W^*$-independence in the sense of measurements and conditional expectations and operational $C^*$- and $W^*$-separability in the sense of measurements and conditional expectations (see [25] for details). The result is a hierarchy of operational independence and operational separability concepts with nontrivial logical interdependencies about which virtually nothing is known at this point.

## 8.9 Closing Comments

The main message of this chapter is that AQFT satisfies the criteria Einstein formulated in his critique of standard quantum mechanics as necessary for a physical theory to be compatible with a field theoretical paradigm. Here, I wish to qualify this claim.

The first qualification is emphasizing a trivial point: the conditions Einstein required are just necessary but not sufficient. No claim is made here that satisfying the three requirements (i.e., spatiotemporality, independence, and local operations) is *sufficient* to accept a theory as field-theory compatible. For instance, the time evolution of the system also must respect the causal structure of the spacetime, but constraints on the dynamics are independent of the three other conditions discussed here.

The second qualification is more substantial: we have seen that both the independence and the local operations requirements can be given different formulations and that the answer to the question of whether or not an independence condition holds depends sensitively on the particular specification of the independence notion. Certain independence conditions (e.g., $W^*$-independence in the product sense, the split property and operational $W^*$-independence in the product sense—see Proposition 6) do *not* hold for certain causally disjoint local systems, and it is not entirely clear how to interpret this fact.

One also can take the position that the "independence" Einstein refers to in the *Dialectica* paper is to be taken as lack of probabilistic correlation (especially lack of EPR correlation—that is, absence of entanglement) between casually disjoint local systems. This is the interpretation Howard [17] and Clifton and Halvorson [8] subscribe to. Under this interpretation, one must conclude (as Clifton and Halvorson [8] do) that, ironically, AQFT fares even *worse* than nonrelativistic quantum mechanics because entanglement is even more endemic in AQFT than in nonrelativistic quantum mechanics. (Bell's inequality is violated even more dramatically in AQFT than in nonrelativistic quantum mechanics; see [7, 32, 34, 36–40].) But, one can argue, as von Neumann did in his reply to Schrödinger's entanglement papers, that correlations between spacelike separated local systems do not in and by themselves make AQFT unacceptable from a field theoretical point of view—as long as it is possible, in principle at least, to give a causal explanation (in terms of local common causes) of the spacelike correlations (see [23] for an analysis of von Neumann's position from this perspective). Whether such local common-cause explanations of the spacelike correlations predicted by AQFT are possible within the framework of AQFT—that is, whether AQFT complies with Reichenbach's Common Cause Principle—is still an open question; only unsatisfactory results are known on the problem: one can show that there exist common causes for correlations between projections lying in von Neumann algebras associated with spacelike separated spacetime regions, but the common causes are known to be localized in the *union* rather than in the *intersection* of the backward light cones of the regions that contain the correlated projections (see [21], [26], and [27] for these results).

Yet another position one can take is that Proposition 8 *does* entail that the local operations requirement is violated in AQFT; hence, AQFT is not really compatible with the field theoretical paradigm after all; that is to say, one can regard the definition of operational $C^*$- and $W^*$-separability (Definition 12) as too weak. It would be interesting to explore how these definitions can be strengthened and to see whether physically well-motivated stronger definitions of operational separability are satisfied in AQFT.

# References

[1] L. Accardi and C. Cecchini. Conditional expectations in von Neumann algebras and a theorem of Takesaki. *Journal of Functional Analysis*, **45** (1982), 245–73.

[2] W. Arveson. Subalgebras of $C^*$-algebras. *Acta Mathematica*, **123** (1969), 141–224.

[3] G. D. Birkhoff and J. von Neumann. The logic of quantum mechanics. *Annals of Mathematics*, **37** (1936), 823–43. Reprinted in [42, No. 7].

[4] B. Blackadar. *Operator Algebras: Theory of $C^*$-Algebras and von Neumann Algebras*, Encyclopaedia of Mathematical Sciences, 1st edition. Berlin, Heidelberg: Springer-Verlag, 2005.

[5] O. Bratteli and D. W. Robinson. *Operator Algebras and Quantum Statistical Mechanics: $C^*$-Algebras and von Neumann Algebras*, vol. I. Berlin, Heidelberg: Springer-Verlag, 1979.

[6] D. Buchholz, C. D'Antoni, and K. Fredenhagen. The universal structure of local algebras. *Commununications in Mathematical Physics*, **111** (1987), 123–35.

[7] R. Clifton and H. Halvorson. Generic Bell correlation between arbitrary local algebras in quantum field theory. *Journal of Mathematical Physics*, **41** (2000), 1711–17.

[8] R. Clifton and H. Halvorson. Entanglement and open systems in algebraic quantum field theory. *Studies in History and Philosophy of Modern Physics*, **32** (2001), 1–31.

[9] A. Connes. Classification of injective factors. Cases $II_1$, $II_\infty$, $III_\lambda$, $\lambda \neq 1$. *The Annals of Mathematics*, **104** (1976), 73–115.

[10] A. Connes. *Noncommutative Geometry*. San Diego, CA: Academic Press, 1994.

[11] E. B. Davies. *Quantum Theory of Open Systems*. London: Academic Press, 1976.

[12] A. Einstein. Physik und Realität. *Journal of the Franklin Institute*, **221** (1936), 313–47.

[13] A. Einsten. Quanten-Mechanik und Wirklichkeit. *Dialectica*, **2** (1948), 320–4.

[14] A. Einstein. Autobiographisches. In *Albert Einstein: Philosopher-Scientist*, The Library of Living Philosophers, ed. P. A. Schilpp. Evanston, IL: Open Court, 1949, pp. 1–96.

[15] A. Einstein, B. Podolsky, and N. Rosen. Can quantum mechanical description of physical reality be considered complete? *Physical Review*, **47** (1935), 777–80. Reprinted in [47, pp. 138–41].

[16] R. Haag. *Local Quantum Physics*. Berlin, Heidelberg: Springer-Verlag, 1992.

[17] D. Howard. Einstein on locality and separability. *Studies in History and Philosophy of Science*, **16** (1985), 171–201.

[18] M. Jammer. The EPR problem in its historical development. In *Symposium on the Foundations of Modern Physics. 50 Years of the Einstein-Podolsky-Rosen Gedankenexperiment*, eds. P. Lahti and P. Mittelstaedt. Singapore: World Scientific, 1985, pp. 129–49.

[19] R. V. Kadison and J. R. Ringrose. *Fundamentals of the Theory of Operator Algebras*, vols. I. and II. Orlando, FL: Academic Press, 1986.

[20] K. Kraus. *States, Effects and Operations*, Lecture Notes in Physics, vol. 190. New York: Springer, 1983.

[21] M. Rédei. Reichenbach's Common Cause Principle and quantum field theory. *Foundations of Physics*, **27** (1997), 1309–21.

[22] M. Rédei (ed.). *John von Neumann: Selected Letters*, History of Mathematics, vol. 27, Providence, RI: American Mathematical Society and London Mathematical Society, 2005.

[23] M. Rédei. Von Neumann on quantum correlations. In *Physical Theory and Its Interpretation: Essays in Honor of Jeffrey Bub*, eds. W. Demopoulos and I. Pitowsky. Dordrecht: Springer, 2006, pp. 241–52.

[24] M. Rédei. Einstein's dissatisfaction with non-relativistic quantum mechanics and relativistic quantum field theory. *Philosophy of Science* (Supplement) **77** (2010), 1042–57.

[25] M. Rédei. Operational separability and operational independence in algebraic quantum mechanics. *Foundations of Physics* **40** (2011), 1439–49.

[26] M. Rédei and S. J. Summers. Local primitive causality and the Common Cause Principle in quantum field theory. *Foundations of Physics*, **32** (2002), 335–55.

[27] M. Rédei and S. J. Summers. Remarks on causality in relativistic quantum field theory. *International Journal of Theoretical Physics*, **46** (2007), 2053–62.

[28] M. Rédei and S. J. Summers. When are quantum systems operationally independent? *International Journal of Theoretical Physics*, **49** (2010), 3250–61.

[29] M. Rédei, G. Valente: How local are local operations in local quantum field theory?, *Studies in the History and Philosophy of Modern Physics* **41** (2010), 346–53.

[30] E. Schrödinger. Discussion of probability relations between separated systems. *Proceedings of the Cambridge Philosophical Society*, **31** (1935), 555–63.

[31] E. Schrödinger. Probability relations between separated systems. *Proceedings of the Cambridge Philosophical Society*, **32** (1936), 446–52.

[32] S. J. Summers. Bell's inequalities and quantum field theory. In *Quantum Probability and Applications V*, Lecture Notes in Mathematics, vol. 1441, Berlin: Springer, 1990, pp. 393–413.

[33] S. J. Summers. On the independence of local algebras in quantum field theory. *Reviews in Mathematical Physics*, **2** (1990), 201–47.

[34] S. J. Summers. Bell's inequalities and algebraic structure. In *Operator Algebras and Quantum Field Theory*, Lecture Notes in Mathematics, vol. 1441, eds. S. Doplicher, R. Longo, J. E. Roberts, and L. Zsido. Cambridge, MA: International Press, 1997, pp. 633–46. (Distributed by the American Mathematical Society, Providence, RI.)

[35] S. J. Summers. Subsystems and independence in relativistic microphysics. *Studies in History and Philosophy of Modern Physics*, **40** (2009), 133–41.

[36] S. J. Summers and R. Werner. The vacuum violates Bell's inequalities. *Physics Letters A*, **110** (1985), 257–79.

[37] S. J. Summers and R. Werner. Bell's inequalities and quantum field theory, I. General setting. *Journal of Mathematical Physics*, **28** (1987), 2440–7.

[38] S. J. Summers and R. Werner. Bell's inequalities and quantum field theory, II. Bell's inequalities are maximally violated in the vacuum. *Journal of Mathematical Physics*, **28** (1987), 2448–56.

[39] S. J. Summers and R. Werner. Maximal violation of Bell's inequalities is generic in quantum field theory. *Commununications in Mathematical Physics*, **110** (1987), 247–59.

[40] S. J. Summers and R. Werner. Maximal violation of Bell's inequalities for algebras of observables in tangent spacetime regions. *Annales de l'Institut Henri Poincaré—Physique théorique*, **49** (1988), 215–43.

[41] A. H. Taub (ed.). *John von Neumann: Collected Works*, vol. I, Logic, Theory of Sets and Quantum Mechanics. New York and Oxford: Pergamon Press, 1961.

[42] A. H. Taub (ed.). *John von Neumann: Collected Works*, vol. IV, Continuous Geometry and Other Topics. New York and Oxford: Pergamon Press, 1961.

[43] J. von Neumann. Mathematische Begründung der Quantenmechanik. *Göttinger Nachrichten*, 1927, pp. 1–57. Reprinted in [41, No. 9].

[44] J. von Neumann. Wahrscheinlichkeitstheoretischer Aufbau der Quantenmechanik. *Göttinger Nachrichten*, 1927, pp. 245–72. Reprinted in [41, No. 10].

[45] J. von Neumann. Thermodynamik quantenmechanischer Gesamtheiten. *Göttinger Nachrichten*, 1927, pp. 273–91. Reprinted in [41, No. 11].

[46] J. von Neumann. *Mathematische Grundlagen der Quantenmechanik*. Berlin: Springer-Verlag, 1932. English translation: *Mathematical Foundations of Quantum Mechanics*. Princeton, NJ: Princeton University Press, 1955.

[47] J. A. Wheeler and W. H. Zurek (ed.). *Quantum Theory and Measurement*. Princeton, NJ: Princeton University Press, 1983.

# PART THREE
# Behind the Hilbert Space Formalism

# CHAPTER 9

# Quantum Theory and Beyond: Is Entanglement Special?

Borivoje Dakić and Časlav Brukner

## 9.1 Introduction

The historical development of scientific progress teaches us that every theory that was established and broadly accepted at a certain time was later inevitably replaced by a deeper and more fundamental theory of which the old one remains a special case. One celebrated example is Newtonian (classical) mechanics, which was superseded by quantum mechanics at the beginning of the last century. It is natural to ask whether in a similar manner there could be logically consistent theories that are more generic than quantum theory itself. It could then turn out that quantum mechanics is an effective description of such a theory, only valid within our current restricted domain of experience.

At present, quantum theory has been tested against very specific alternative theories that, both mathematically and in their concepts, are distinctly different. Instances of such alternative theories are non-contextual hidden-variable theories [1], local hidden-variable theories [2], crypto-nonlocal hidden-variable theories [3,4], or some nonlinear variants of the Schrödinger equation [5–8]. Currently, many groups are working on improving experimental conditions to be able to test alternative theories based on various collapse models [9–14]. The common trait of all these proposals is to suppress one or the other counterintuitive feature of quantum mechanics and thus keep some of the basic notions of a classical worldview intact. Specifically, hidden-variable models would allow us to preassign definite values to outcomes of all measurements; collapse models are mechanisms for restraining superpositions between macroscopically distinct states, and nonlinear extensions of the Schrödinger equation may admit more localized solutions for wave-packet dynamics, thereby resembling localized classical particles.

In the last years, the new field of quantum information has initialized interest in generalized probabilistic theories that share certain features—such as the no-cloning and the no-broadcasting theorems [15, 16] or the trade-off between state disturbance and measurement [17]—generally thought of as specifically quantum, yet being shown to be present in all except classical theory. These generalized probabilistic theories can allow for stronger than quantum correlations in the sense that they can violate Bell's inequalities stronger than the quantum Cirelson bound (as it is the case for the celebrated "nonlocal boxes" of Popescu and Rohrlich [18]), although they all respect the "non-signaling" constraint according to which correlations cannot be used to send information faster than the speed of light.

Because most features that have been highlighted as "typically quantum" are actually quite generic for all non-classical probabilistic theories, one could conclude that additional principles must be adopted to single out quantum theory uniquely. Alternatively, these probabilistic theories indeed can be constructed in a logically consistent way and might even be realized in nature in a domain that is still beyond our observations. The vast majority of attempts to find physical principles behind quantum theory either fail to single out the theory uniquely or are based on highly abstract mathematical assumptions without an immediate physical meaning (e.g., [19]).

On the way to reconstructions of quantum theory from foundational physical principles rather than purely mathematical axioms, one finds interesting examples coming from an instrumentalist approach [20–22], where the focus is primarily on primitive laboratory operations such as preparations, transformations, and measurements. Whereas these reconstructions are based on a short set of simple axioms, they still partially use mathematical language in their formulation.

Evidently, added value of reconstructions for better understanding quantum theory originates from its power of explanation from which the structure of the theory comes. Candidates for foundational principles were proposed giving a basis for an understanding of quantum theory as a general theory of information supplemented by several information-theoretic constraints [23–28]. In a wider context, these approaches belong to attempts to find an explanation for quantum theory by putting primacy on the concept of information or on the concept of probability, which again can be seen as a way of quantifying information [29–41]. Other principles were proposed for separation of quantum correlations from general non-signaling correlations, such as that communication complexity is not trivial [42, 43]; that communication of $m$ classical bits causes information gain of, at most, $m$ bits ("information causality") [44]; or that any theory should recover classical physics in the macroscopic limit [45].

In his seminal article, Hardy [20] derives quantum theory from "five reasonable axioms" within the instrumentalist framework. He sets up a link between two natural numbers, $d$ and $N$, characteristics of any theory, where $d$ is the number of degrees of freedom of the system and is defined as the minimum number of real parameters needed to determine the state completely. The dimension $N$ is defined as the maximum number of states that can be reliably distinguished from one another in a single-shot experiment. A closely related notion is the information-carrying capacity of the system, which is the maximal number of bits encoded in the system and is equal to $\log N$ bits for a system of dimension $N$.

Examples of theories with an explicit functional dependence $d(N)$ are classical probability theory with the linear dependence $d = N - 1$ and quantum theory with the quadratic dependence for which it is necessary to use $d = N^2 - 1$ real parameters to

# INTRODUCTION

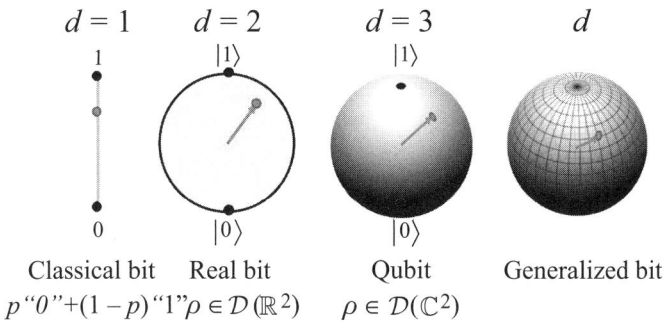

**Figure 9.1.** State spaces of a two-dimensional system in the generalized probabilistic theories analyzed here, where $d$ is the minimal number of real parameters necessary to determine the (generally mixed) state completely. From left to right: A classical bit with one parameter (the weight $p$ in the mixture of two-bit values); a real bit with two real parameters (state $\rho \in \mathcal{D}(\mathbb{R}^2)$ is represented by $2 \times 2$ real density matrix); a qubit (quantum bit) with three real parameters (state $\rho \in \mathcal{D}(\mathbb{C}^2)$ is represented by $2 \times 2$ complex density matrix); and a generalized bit for which $d$ real parameters are needed to specify the state. Note that when one moves continuously from one pure state (represented by a point on the surface of a sphere) to another, only in the classical probabilistic theory one must go through the set of mixed states [20]. Can probability theories that are more generic than quantum theory be extended in a logically consistent way to higher-dimensional and composite systems? Can entanglement exist in these theories? Where should we look in nature for potential empirical evidence of the theories?

completely characterize the quantum state.[1] Higher-order theories with more general dependencies $d(N)$ might exist as illustrated in Fig 9.1. Hardy's reconstruction resorts to a "simplicity axiom" that discards a large class of higher-order theories by requiring that for each given $N$, $d(N)$ takes the minimum value consistent with the other axioms. However, without making such an ad hoc assumption, the higher-order theories might be possible to be constructed in agreement with the rest of the axioms. In fact, an explicit quartic theory for which $d = N^4 - 1$ [46] and theories for generalized bit ($N = 2$) for which $d = 2^r - 1$ and $r \in N$ [47] were recently developed, although all of them are restricted to the description of individual systems only.

It is clear from the previous discussion that the question on the basis of which physical principles quantum theory can be separated from the multitude of possible generalized probability theories is still open. A particularly interesting unsolved problem is whether the higher-order theories [20,46,47] can be extended to describe non-trivial—that is, *entangled*—states of composite systems. Any progress in theoretical understanding of these issues would be very desirable, in particular because experimental research efforts in this direction have been sporadic. Although most experiments indirectly verify also the number of the degrees of freedom of quantum systems,[2] only a few dedicated attempts have been made at such a direct experimental verification. Quaternionic

---

[1] Hardy considers unnormalized states and for that reason takes $K = d + 1$ (in his notation) as the number of degrees of freedom.

[2] As noted by Zyczkowski [46], it is thinkable that within the time scales of standard experimental conditions "hyperdecoherence" may occur, which causes a system described in the framework of the higher-order theory to specific properties and behavior according to predictions of standard (complex) quantum theory.

quantum mechanics (for which $d = 2N^2 - N - 1$) was tested in a suboptimal setting [50] in a single neutron experiment in 1984 [48,49]. More recently, the generalized measure theory of Sorkin [51], in which higher-order interferences are predicted, was tested in a three-slit experiment with photons [52]. Both experiments put an upper bound on the extent of the observational effects that the two alternative theories may produce.

## 9.2 Basic Ideas and the Axioms

Here, we reconstruct quantum theory from three reasonable axioms. Following the general structure of any reconstruction, we first give a set of physical principles, then formulate their mathematical representation, and finally rigorously derive the formalism of the theory. We consider only the case in which the number of distinguishable states is finite. The following three axioms separate classical probability theory and quantum theory from all other probabilistic theories:

**Axiom 1** *(Information Capacity). An elementary system has the information-carrying capacity of at most one bit. All systems of the same information carrying-capacity are equivalent.*

**Axiom 2** *(Locality). The state of a composite system is completely determined by local measurements on its subsystems and their correlations.*

**Axiom 3** *(Reversibility). Between any two pure states, there exists a reversible transformation.*

A few comments on these axioms are appropriate here. The most elementary system in the theory is a two-dimensional system. All higher-dimensional systems will be built from two-dimensional ones. Recall that the *dimension* is defined as the maximal number of states that can be reliably distinguished from one another in a single-shot experiment. Under the clause "an elementary system has an information capacity of at most one bit," we precisely assume that for any state (pure or mixed) of a two-dimensional system there is a measurement such that the state is a mixture of *two* states that are distinguished reliably in the measurement. An alternative formulation could be that *any state of a two-dimensional system can be prepared by mixing at most two basis (i.e., perfectly distinguishable in a measurement) states* (Fig. 9.2). Roughly speaking, Axiom 1 assumes that a state of an elementary system can always be represented as a mixture of *two classical bit values*. This part of the axiom is inspired by Zeilinger's proposal for a foundation principle for quantum theory [24].

The second statement in Axiom 1 is motivated by the intuition that at the fundamental level there should be no difference between systems of the same information-carrying capacity. All elementary systems— whether or not they are part of higher-dimensional systems—should have equivalent state spaces and equivalent sets of transformations and measurements. This seems to be a natural assumption if one makes no prior restrictions to the theory and preserves the full symmetry between all possible elementary systems. This is why we have decided to put the statement as a part of

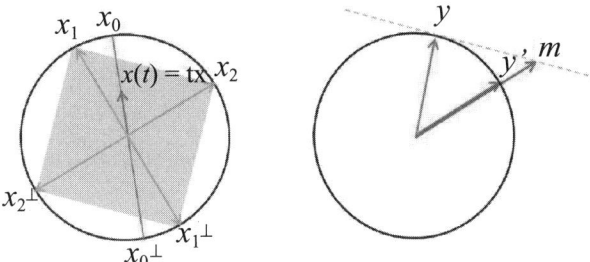

**Figure 9.2.** Illustration of the assumption stated in Axiom 1. Consider a toy-world of a two-dimensional system in which the set of pure states consists of only $\mathbf{x}_1$ and $\mathbf{x}_2$ and their orthogonal states $\mathbf{x}_1^\perp$ and $\mathbf{x}_2^\perp$, respectively, and where only two measurements exist, which distinguish $\{\mathbf{x}_1, \mathbf{x}_1^\perp\}$ and $\{\mathbf{x}_2, \mathbf{x}_2^\perp\}$. The convex set (represented by the gray area within the circle), whose vertices are the four states, contains all physical (pure or mixed) states in the toy-world. Now choose a point in the set, say $\mathbf{y} = \lambda\mathbf{x}_1 + (1-\lambda)\mathbf{x}_2$. Axiom 1 states that any physical state can be represented as a mixture of two orthogonal states (i.e., states perfectly distinguishable in a single-shot experiment); for example, $\mathbf{y} = \eta\mathbf{x} + (1-\eta)\mathbf{x}^\perp$. This is not fulfilled in the toy-world but is satisfied in a theory in which the entire circle represents the pure states and where measurements can distinguish all pairs of orthogonal states.

Axiom 1 rather than as a separate axiom. The particular formulation used here is from Grinbaum [53], who suggested rephrasing the "subspace axiom" of Hardy's reconstruction using physical rather than mathematical language. The subspace axiom states that a system whose state is constrained to belong to an $M$ dimensional subspace (i.e., have support on only $M$ of a set of $N$ possible distinguishable states) behaves like a system of dimension $M$.

In logical terms, Axiom 1 means the following. We can think of two basis states as two binary propositions about an individual system, such as (1) "The outcome of measurement $A$ is $+1$," and (2) "The outcome of measurement $A$ is $-1$." An alternative choice for the pair of propositions can be propositions about joint properties of two systems, such as (1') "The outcomes of measurement $A$ on the first system and of $B$ on the second system are correlated" (i.e., either both $+1$ or both $-1$), and (2') "The outcomes of measurement $A$ on the first system and of $B$ on the second system are anticorrelated." The two choices for the pair of propositions correspond to two choices of basis states, each of which can be used to span the full state space of an abstract elementary system (also called "generalized bit"). As we see later, taking the latter choice, it follows from Axiom 1 alone that the state space must contain entangled states.

Axiom 2 assumes that a specification of the probabilities for a complete set of local measurements for each of the subsystems plus the joint probabilities for correlations between these measurements is sufficient to determine completely the global state. Note that this property holds in both quantum theory and classical probability theory but not in quantum theory formulated on the basis of real or quaternionic instead of complex amplitudes. A closely related formulation of the axiom was given by Barrett [17].

Finally, Axiom 3 requires that transformations are reversible. This is assumed alone for the purposes that the set of transformations builds a group structure. It is natural to assume that a composition of two physical transformations is again a physical transformation. It should be noted that this axiom could be used to exclude the theories in which "nonlocal boxes" occur because there, the dynamical group is trivial, in the sense that it is generated solely by local operations and permutations of systems with no entangling reversible transformations (i.e., non-local boxes cannot be prepared from product states) [54].

If one requires the reversible transformation from Axiom 3 to be continuous:

**Axiom 4** *(Continuity). Between any two pure states, there exists a continuous reversible transformation,*

which separates quantum theory from classical probability theory. The same axiom is also present in Hardy's reconstruction. By a *continuous transformation* is here meant that every transformation can be made up from a sequence of transformations only infinitesimally different from the identity.

A remarkable result following from our reconstruction is that *quantum theory is the only probabilistic theory in which one can construct entangled states and fulfill the three axioms*. In particular, in the higher-order theories [20, 46, 47] composite systems can only enjoy trivial separable states. Conversely, we see that Axiom 1 alone requires entangled states to exist in all non-classical theories. This will allow us to discard the higher-order theories in our reconstruction, scheme without invoking the simplicity argument.

As a by-product of our reconstruction, we will be able to answer why in nature only "odd" correlations (i.e., $(1, 1, -1)$, $(1, -1, 1)$, $(-1, 1, 1)$, and $(-1, -1, -1)$) are observed when two maximally entangled qubits (spin-1/2 particles) are both measured along direction $x$, $y$, and $z$, respectively. The most familiar example is of the singlet state $|\psi^-\rangle = \frac{1}{2}(|0\rangle_1|1\rangle_2 - |0\rangle_1|1\rangle_2)$ with anticorrelated results whenever the qubits are measured along the same direction. We will show that the "mirror quantum mechanics" in which only "even" correlations appear cannot be extended consistently to composite systems of three bits.

Our reconstruction is given in the framework of a typical experimental situation that an observer faces in the laboratory. Although this instrumentalist approach is a useful paradigm to work with, it might not be necessary. One could think about Axioms 1 and 3 as referring to objective features of elementary constituents of the world, which need not necessarily be related to laboratory actions. In contrast, Axiom 2 seems to acquire a meaning only within the instrumentalist approach because it involves the word *measurement*. Even here one could follow a suggestion of Grinbaum [53] and rephrase the axiom to the assumption of "multiplicability of the information carrying capacity of subsystems."

Concluding this section, we note that the conceptual groundwork for the ideas presented here has been prepared most notably by Weizsäcker [55], Wheeler [56], and Zeilinger [24], who proposed that the notion of the elementary yes–no alternative, or the "Ur," should play a pivotal role when reconstructing quantum physics.

## 9.3 Basic Notions

Following Hardy [20], we distinguish three types of devices in a typical laboratory. The preparation device prepares systems in some state. It has a set of switches on it for varying the state produced. After state preparation, the system passes through a transformation device. It also has a set of switches on it for varying the transformation applied on the state. Finally, the system is measured in a measurement apparatus. It again has switches on it, which help an experimenter choose different measurement settings. This device outputs classical data (e.g., a click in a detector or a spot on a observation screen).

We define the *state* of a system as that mathematical object from which one can determine the probability for any conceivable measurement. Physical theories can have enough structure that it is not necessary to give an exhaustive list of all probabilities for all possible measurements but rather only a list of probabilities for some minimal subset of them. We refer to this subset as a *fiducial* set. Therefore, the state is specified by a list of $d$ (where $d$ depends on dimension $N$) probabilities for a set of fiducial measurements: $\mathbf{p} = (p_1, \ldots, p_d)$. The state is *pure* if it is not a (convex) mixture of other states. The state is mixed if it is not pure. For example, the mixed state $\mathbf{p}$ generated by preparing state $\mathbf{p}_1$ with probability $\lambda$ and $\mathbf{p}_2$ with probability $1 - \lambda$ is $\mathbf{p} = \lambda \mathbf{p}_1 + (1 - \lambda)\mathbf{p}_2$.

When we refer to an $N$-dimensional system, we assume that there are $N$ states, each of which identifies a different outcome of some measurement setting, in the sense that they return probability one for the outcome. We call this set a set of *basis* or *orthogonal states*. Basis states can be chosen to be pure. To see this, assume that some mixed state identifies one outcome. We can decompose the state into a mixture of pure states, each of which has to return probability one; thus, we can use one of them to be a basis state. We show later that each pure state corresponds to a unique measurement outcome.

If the system in state $\mathbf{p}$ is incident on a transformation device, its state will be transformed to some new state $U(\mathbf{p})$. The transformation $U$ is a linear function of the state $\mathbf{p}$ because it needs to preserve the linear structure of mixtures. For example, consider the mixed state $\mathbf{p}$ that is generated by preparing state $\mathbf{p}_1$ with probability $\lambda$ and $\mathbf{p}_2$ with probability $1 - \lambda$. Then, in each single run, either $\mathbf{p}_1$ or $\mathbf{p}_2$ is transformed and thus one has

$$U(\lambda \mathbf{p}_1 + (1 - \lambda)\mathbf{p}_2) = \lambda U(\mathbf{p}_1) + (1 - \lambda)U(\mathbf{p}_2). \tag{9.1}$$

It is natural to assume that a composition of two or more transformations is again from a set of (reversible) transformations. This set forms some abstract group. Axiom 3 states that the transformations are reversible (i.e., for every $U$ there is an inverse group element $U^{-1}$). Here, we assume that every transformation has its matrix representation $U$ and that there is an orthogonal representation of the group: there exists an invertible matrix $S$ such that $O = SUS^{-1}$ is an orthogonal matrix; that is, $O^T O = \mathbb{1}$, for every $U$ (we use the same notation both for the group element and for its matrix representation). This does not put severe restrictions on the group of transformations because it is known that all compact groups have such a representation (the Schur–Auerbach lemma) [60]. Because the transformation keeps the probabilities in the range $[0, 1]$, it has to be a compact group [20]. All finite groups and all continuous Lie groups are therefore included in our consideration.

Given a measurement setting, the outcome probability $P_{\text{meas}}$ can be computed by some function $f$ of the state $\mathbf{p}$,

$$P_{\text{meas}} = f(\mathbf{p}). \tag{9.2}$$

Like a transformation, the measurement cannot change the mixing coefficients in a mixture; therefore, the measured probability is a linear function of the state **p**:

$$f(\lambda \mathbf{p}_1 + (1-\lambda)\mathbf{p}_2) = \lambda f(\mathbf{p}_1) + (1-\lambda) f(\mathbf{p}_2). \tag{9.3}$$

## 9.4 Elementary System: System of Information Capacity of One Bit

A two-dimensional system has two distinguishable outcomes that can be identified by a pair of basis states $\{\mathbf{p}, \mathbf{p}^\perp\}$. The state is specified by $d$ probabilities $\mathbf{p} = (p_1, ..., p_d)$ for $d$ fiducial measurements, where $p_i$ is probability for a particular outcome of the $i$-th fiducial measurement (the dependent probabilities $1 - p_i$ for the opposite outcomes are omitted in the state description). Instead of using the probability vector **p**, we will specify the state by its *Bloch representation* **x** defined as a vector with $d$ components:

$$x_i = 2p_i - 1. \tag{9.4}$$

The mapping between the two different representations is an invertible linear map and therefore preserves the structure of the mixture $\lambda \mathbf{p}_1 + (1-\lambda)\mathbf{p}_2 \mapsto \lambda \mathbf{x}_1 + (1-\lambda)\mathbf{x}_2$.

It is convenient to define a *totally mixed state* $\mathbf{E} = \frac{1}{\mathcal{N}} \sum_{x \in \mathcal{S}_{\text{pure}}} \mathbf{x}$, where $\mathcal{S}_{\text{pure}}$ denotes the set of pure states and $\mathcal{N}$ is the normalization constant. In the case of a continuous set of pure states, the summation has to be replaced by a proper integral. It is easy to verify that **E** is a totally invariant state. This implies that every measurement and, in particular, the fiducial ones, will return the same probability for all outcomes. In the case of a two-dimensional system, this probability is $1/2$. Therefore, the Bloch vector of the totally mixed state is the zero-vector $\mathbf{E} = \vec{0}$.

The transformation $U$ does not change the totally mixed state; hence, $U(\vec{0}) = \vec{0}$. The last condition together with the linearity condition (Equation 9.1) implies that any transformation is represented by some $d \times d$ real invertible matrix $U$. The same reasoning holds for measurements. Therefore, the measured probability is given by the formula:

$$P_{\text{meas}} = \frac{1}{2}(1 + \mathbf{r}^T \mathbf{x}). \tag{9.5}$$

The vector **r** represents the outcome for the given measurement setting. For example, the vector $(1, 0, 0, ...)$ represents one of the outcomes for the first fiducial measurement.

According to Axiom 1, any state is a classical mixture of some pair of orthogonal states. For example, the totally mixed state is an equally weighted mixture of some orthogonal states $\vec{0} = \frac{1}{2}\mathbf{x} + \frac{1}{2}\mathbf{x}^\perp$. Take **x** to be the reference state. According to Axiom 3, we can generate the full set of states by applying all possible transformations to the reference state. Because the totally mixed state is invariant under the transformations, the pair of orthogonal states is represented by a pair of antiparallel vectors $\mathbf{x}^\perp = -\mathbf{x}$. Consider the set $\mathcal{S}_{\text{pure}} = \{U\mathbf{x} \mid \forall U\}$ of all pure states generated by applying all transformations to the reference state. If one uses the orthogonal representation of the transformations, $U = S^{-1}OS$, which was introduced already herein, one maps $\mathbf{x} \mapsto S\mathbf{x}$ and $U \mapsto O$. Hence, the transformation $U\mathbf{x} \mapsto SU\mathbf{x} = OS\mathbf{x}$ is norm-preserving. We conclude that all pure states are points on a $d$-dimensional ellipsoid described by $\|S\mathbf{x}\| = c$ with $c > 0$.

# ELEMENTARY SYSTEM: SYSTEM OF INFORMATION CAPACITY OF ONE BIT

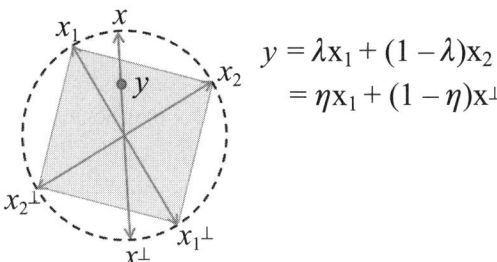

**Figure 9.3.** (*Left*) Illustration to the proof that the entire $d$-dimensional ellipsoid (here, represented by a circle; $d = 2$) contains physical states. Consider a line $\mathbf{x}(t) = t\mathbf{x}$ through the origin. A point on the line can be expanded into a set of linearly independent vectors $\mathbf{x}_i$ (here, $\mathbf{x}_1$ and $\mathbf{x}_2$). For sufficiently small $t$ (i.e., when the line is within the gray square), the point $\mathbf{x}(t)$ can be represented as a convex mixture over $\mathbf{x}_i$ and their orthogonal vectors $\mathbf{x}_i^\perp$ and thus is a physical state. According to Axiom 1, $\mathbf{x}(t)$ can be represented as a convex mixture of two orthogonal pure states $\mathbf{x}_0$ and $\mathbf{x}_0^\perp$: $\mathbf{x}(t) = t\mathbf{x} = \alpha \mathbf{x}_0 + (1-\alpha)(-\mathbf{x}_0)$, where $\mathbf{x} = \mathbf{x}_0$ (see text for details). This implies that every point in the ellipsoid is a physical state. (*Right*) Illustration to the proof that in the orthogonal representation the measurement vector $\mathbf{m}$ that identifies the state $\mathbf{x}$—that is, for which the probability $P_{\text{meas}} = \frac{1}{2}(1 + \mathbf{m}^T\mathbf{y}) = 1$—is identical to the state vector, $\mathbf{m} = \mathbf{y}$. Suppose that $\mathbf{m} \neq \mathbf{y}$, then $||\mathbf{m}|| > 1$ because the state vector is normalized. But then the same measurement for state $\mathbf{y}'$ parallel to $\mathbf{m}$ would return a probability larger than 1, which is nonphysical. Thus, $\mathbf{m} = \mathbf{y}$.

Now we want to show that any vector $\mathbf{x}$ satisfying $||S\mathbf{x}|| = c$ is a physical state; therefore, the set of states has to be the whole ellipsoid. Let $\mathbf{x}$ be some vector satisfying $||S\mathbf{x}|| = c$ and $\mathbf{x}(t) = t\mathbf{x}$ a line through the origin (totally mixed state) as given in Fig 9.3 (left). Within the set of pure states, we can always find $d$ linearly independent vectors $\{\mathbf{x}_1, \ldots, \mathbf{x}_d\}$. For each state $\mathbf{x}_i$, there is a corresponding orthogonal state $\mathbf{x}_i^\perp = -\mathbf{x}_i$ in a set of states. We can expand a point on the line into a linearly independent set of vectors: $\mathbf{x}(t) = t\sum_{i=1}^{d} c_i \mathbf{x}_i$. For sufficiently small $t$, we can define a pair of non-negative numbers $\lambda_i(t) = \frac{1}{2}(\frac{1}{d} + tc_i)$ and $\lambda_i^\perp(t) = \frac{1}{2}(\frac{1}{d} - tc_i)$ with $\sum_i(\lambda_i(t) + \lambda_i^\perp(t)) = 1$ such that $\mathbf{x}(t)$ is a mixture $\mathbf{x}(t) = \sum_{i=1}^{d} \lambda_i(t)\mathbf{x}_i + \lambda_i^\perp(t)\mathbf{x}_i^\perp$ and therefore is a physical state. Then, according to Axiom 1, there exists a pair of basis states $\{\mathbf{x}_0, -\mathbf{x}_0\}$ such that $\mathbf{x}(t)$ is a mixture of them

$$\mathbf{x}(t) = t\mathbf{x} = \alpha \mathbf{x}_0 + (1-\alpha)(-\mathbf{x}_0), \tag{9.6}$$

where $\alpha = \frac{1+t}{2}$ and $\mathbf{x} = \mathbf{x}_0$. This implies that $\mathbf{x}$ is a pure state and, therefore, all points of the ellipsoid are physical states.

For every pure state $\mathbf{x}$, there exists at least one measurement setting with the outcome $\mathbf{r}$ such that the outcome probability is one; hence, $\mathbf{r}^T\mathbf{x} = 1$. Let us define new coordinates $\mathbf{y} = \frac{1}{c}S\mathbf{x}$ and $\mathbf{m} = cS^{-1T}\mathbf{r}$ in the orthogonal representation. The set of pure states in the new coordinates is a $(d-1)$-sphere $\mathcal{S}^{d-1} = \{\mathbf{y} \mid ||\mathbf{y}|| = 1\}$ of the unit radius. The probability rule (Equation 9.5) remains unchanged in the new coordinates:

$$P_{\text{meas}} = \frac{1}{2}(1 + \mathbf{m}^T\mathbf{y}). \tag{9.7}$$

Thus, one has $\mathbf{m}^T\mathbf{y} = 1$. Now, assume that $\mathbf{m} \neq \mathbf{y}$. Then, $||\mathbf{m}|| > 1$ and the vectors $\mathbf{m}$ and $\mathbf{y}$ span a two-dimensional plane as illustrated in Fig 9.3 (right). The set of pure states within this plane is a unit circle. Choose the pure state $\mathbf{y}'$ to be parallel to $\mathbf{m}$. Then, the outcome probability is $P_{\text{measur}} = \frac{1}{2}(1 + ||\mathbf{m}||||\mathbf{y}'||) > 1$, which is nonphysical; hence, $\mathbf{m} = \mathbf{y}$. Therefore, to each pure state $\mathbf{y}$, we associate a measurement vector $\mathbf{m} = \mathbf{y}$, which identifies it. Equivalently, in the original coordinates, to each $\mathbf{x}$ we associate a measurement vector $\mathbf{r} = D\mathbf{x}$, where $D = \frac{1}{c^2}S^T S$ is a positive, symmetric matrix. A proof of this relation for the restricted case of $d = 3$ can be found in [20].

From now on, instead of the measurement vector $\mathbf{r}$, we use the pure state $\mathbf{x}$, which identifies it. When we say that the measurement along the state $\mathbf{x}$ is performed, we mean the measurement given by $\mathbf{r} = D\mathbf{x}$. The measurement setting is given by a pair of measurement vectors $\mathbf{r}$ and $-\mathbf{r}$. The measured probability when the state $\mathbf{x}_1$ is measured along the state $\mathbf{x}_2$ follows from formula (9.5):

$$P(\mathbf{x}_1, \mathbf{x}_2) = \frac{1}{2}(1 + \mathbf{x}_1^T D \mathbf{x}_2). \tag{9.8}$$

We can choose orthogonal eigenvectors of the matrix $D$ as the fiducial set of states (measurements):

$$D\mathbf{x}_i = a_i \mathbf{x}_i, \tag{9.9}$$

where $a_i$ are eigenvalues of $D$. Because $\mathbf{x}_i$ are pure states, they satisfy $\mathbf{x}_i^T D \mathbf{x}_j = \delta_{ij}$. The set of pure states becomes a unit sphere $\mathcal{S}^{d-1} = \{\mathbf{x} \mid ||\mathbf{x}|| = 1\}$ and the probability formula is reduced to

$$P(\mathbf{x}_1, \mathbf{x}_2) = \frac{1}{2}(1 + \mathbf{x}_1^T \mathbf{x}_2). \tag{9.10}$$

This corresponds to a choice of a complete set of mutually complementary measurements (i.e., mutually unbiased basis sets [61, 62]) for the fiducial measurements. The states identifying outcomes of complementary measurements satisfy $P(\mathbf{x}_i, \mathbf{x}_j) = \frac{1}{2}$ for $i \neq j$. Two observables are said to be mutually complementary if complete certainty about one of the observables (i.e., one of two outcomes occurs with probability one) precludes any knowledge about the others (i.e., the probability for both outcomes is $1/2$). Given some state $\mathbf{x}$, the $i$-th fiducial measurement returns probability $p_i = \frac{1}{2}(1 + x_i)$. Therefore, $x_i$ is a mean value of a dichotomic observable $\mathbf{b}_i = +1\mathbf{x}_i - 1\mathbf{x}_i^{\perp}$ with two possible outcomes $b_i = \pm 1$.

A theory in which the state space of the generalized bit is represented by a $(d-1)$-sphere has $d$ mutually complementary observables. This is a characteristic feature of the theories, and they can be ordered according to their number. For example, classical physics has no complementary observables, real quantum mechanics has two, complex (standard) quantum mechanics has three (e.g., the spin projections of a spin-1/2 system along three orthogonal directions), and the one based on quaternions has five mutually complementary observables. Note that higher-order theories of a single generalized bit are such that the qubit theory can be embedded in them in the same way in which classical theory of a bit can be embedded in qubit theory itself.

Higher-order theories can have even better information-processing capacity than quantum theory. For example, the computational abilities of the theories with $d = 2^r$ and $r \in \mathbb{N}$ in solving the Deutsch–Josza type of problems increases with the number of mutually complementary measurements [47]. It is likely that the larger this number is, the larger the error rate would be in secret key distribution in these theories, in a similar manner in which the 6-state is advantageous over the 4-state protocol in (standard) quantum mechanics. In the first case, one uses all three mutually complementary observables and in the second one only two of them. (See [57] for a review on characterizing generalized probabilistic theories in terms of their information-processing power and [58] for investigating the same question in much more general framework of compact closed categories.)

A final remark on higher-order theories is of a more speculative nature. In various approaches to quantum theory of gravity, one predicts at the Planck scale the dimension of spacetime to be different from $3 + 1$ [59]. If one considers directional degrees of freedom (spin), then the $d - 1$-sphere (Bloch sphere) might be interpreted as the state space of a spin system embedded in real (ordinary) space of dimension $d$, in general different from 3, which is the special case of quantum theory.

The reversible transformation $R$ preserves the purity of state $||R\mathbf{x}|| = ||x||$; therefore, $R$ is an orthogonal matrix. We have shown that the state space is the full $(d - 1)$-sphere. According to Axiom 3, the set of transformations must be rich enough to generate the full sphere. If $d = 1$ (classical bit), the group of transformations is discrete and contains only the identity and the bit-flip. If $d > 1$, the group is continuous and is some subgroup of the orthogonal group $O(d)$. Every orthogonal matrix has the determinant either 1 or $-1$. The orthogonal matrices with determinant 1 form a normal subgroup of $O(d)$, known as the *special orthogonal* group $SO(d)$. The group $O(d)$ has two connected components: the identity component, which is the $SO(d)$ group, and the component formed by orthogonal matrices with determinant $-1$. Because every two points on the $(d - 1)$-sphere are connected by some transformation, the group of transformations is at least the $SO(d)$ group. If we include even a single transformation with determinant $-1$, the set of transformations becomes the entire $O(d)$ group. However, we show later that any transformation of determinant $-1$ cannot represent a physical transformation; thus, $SO(d)$ is the complete set of physical transformations.

## 9.5 Composite System and the Notion of Locality

We now introduce a description of composite systems. We assume that when one combines two systems of dimension $N_1$ and $N_2$ into a composite one, one obtains a system of dimension $N_1 N_2$ [20]. Consider a composite system consisting of two generalized bits, and choose a set of $d$ complementary measurements on each subsystem as fiducial measurements. According to Axiom 2, the state of the composite system is completely determined by a set of real parameters obtainable from local measurements on the two generalized bits and their correlations. We obtain $2d$ independent real parameters from the set of local fiducial measurements and additional $d^2$ parameters from correlations between them. This gives altogether $d^2 + 2d = (d + 1)^2 - 1$ parameters. They are the components $x_i$, $y_i$, $i \in \{1, ..., d\}$ of the local Bloch vectors and $T_{ij}$ of the correlation tensor:

$$x_i = p^{(i)}(A = 1) - p^{(i)}(A = -1), \tag{9.11}$$

$$y_j = p^{(j)}(B = 1) - p^{(j)}(B = -1), \tag{9.12}$$

$$T_{ij} = p^{(ij)}(AB = 1) - p^{(ij)}(AB = -1). \tag{9.13}$$

Here, for example, $p^{(i)}(A = 1)$ is the probability to obtain outcome $A = 1$ when the $i$-th measurement is performed on the first subsystem, and $p^{(ij)}(AB = 1)$ is the joint probability to obtain correlated results (i.e., either $A = B = +1$ or $A = B = -1$) when the $i$-th measurement is performed on the first subsystem and the $j$-th measurement on the second one.

Note that Axiom 2, "The state of a composite system is completely determined by local measurements on its subsystems and their correlations," is formulated in a way that the non-signaling condition is implicitly assumed to hold. This is because it is sufficient to speak about "local measurements" alone without specifying the choice of measurement setting on the other potentially distant subsystem. Therefore, $x_i$ does not depend on $j$, and $y_j$ does not depend on $i$.

We represent a state by the triple $\psi = (\mathbf{x}, \mathbf{y}, T)$, where $\mathbf{x}$ and $\mathbf{y}$ are the local Bloch vectors and $T$ is a $d \times d$ real matrix representing the correlation tensor. The product (separable) state is represented by $\psi_p = (\mathbf{x}, \mathbf{y}, T)$, where $T = \mathbf{x}\mathbf{y}^T$ is of product form because the correlations are just products of the components of the local Bloch vectors. We call the pure state *entangled* if it is not a product state.

The measured probability is a linear function of the state $\psi$. If we prepare totally mixed states of the subsystems $(0, 0, 0)$, the probability for any outcome of an arbitrary measurement will be $1/4$. Therefore, the outcome probability can be written as

$$P_{\text{measur}} = \frac{1}{4}(1 + (r, \psi)), \tag{9.14}$$

where $r = (\mathbf{r}_1, \mathbf{r}_2, K)$ is a measurement vector associated to the observed outcome and $(\dots, \dots)$ denotes the scalar product:

$$(r, \psi) = \mathbf{r}_1^T \mathbf{x} + \mathbf{r}_2^T \mathbf{y} + \text{Tr}(K^T T). \tag{9.15}$$

Now, assume that $r = (\mathbf{r}_1, \mathbf{r}_2, K)$ is associated to the outcome, which is identified by some product state $\psi_p = (\mathbf{x}_0, \mathbf{y}_0, T_0)$. If we preform a measurement on the arbitrary product state $\psi = (\mathbf{x}, \mathbf{y}, T)$, the outcome probability has to factorize into the product of the local outcome probabilities of the form (Equation 9.10):

$$P_{\text{measur}} = \frac{1}{4}(1 + \mathbf{r}_1^T \mathbf{x} + \mathbf{r}_2^T \mathbf{y} + \mathbf{x}^T K \mathbf{y}) \tag{9.16}$$

$$= P_1(\mathbf{x}_0, \mathbf{x}) P_2(\mathbf{y}_0, \mathbf{y}) \tag{9.17}$$

$$= \frac{1}{2}(1 + \mathbf{x}_0^T \mathbf{x}) \frac{1}{2}(1 + \mathbf{y}_0^T \mathbf{y}) \tag{9.18}$$

$$= \frac{1}{4}(1 + \mathbf{x}_0^T \mathbf{x} + \mathbf{y}_0^T \mathbf{y} + \mathbf{x}^T \mathbf{x}_0 \mathbf{y}_0^T \mathbf{y}), \tag{9.19}$$

which holds for all $\mathbf{x}, \mathbf{y}$. Therefore, we have $r = \psi_p$. For each product state $\psi_p$, there is a unique outcome $r = \psi_p$ that identifies it. We later show that correspondence $r = \psi$ holds for *all* pure states $\psi$.

If we perform local transformations $R_1$ and $R_2$ on the subsystems, the global state $\psi = (\mathbf{x}, \mathbf{y}, T)$ is transformed to

$$(R_1, R_2)\psi = (R_1 \mathbf{x}, R_2 \mathbf{y}, R_1 T R_2^T). \tag{9.20}$$

$T$ is a real matrix and we can find its singular value decomposition $\text{diag}[t_1, \dots, t_d] = R_1 T R_2^T$, where $R_1, R_2$ are orthogonal matrices that can be chosen to have determinant 1. Therefore, we can choose the local bases such that correlation tensor $T$ is a diagonal matrix:

$$(R_1, R_2)(\mathbf{x}, \mathbf{y}, T) = (R_1\mathbf{x}, R_2\mathbf{y}, \text{diag}[t_1, \ldots, t_d]). \tag{9.21}$$

The last expression is called *Schmidt decomposition* of the state.

The local Bloch vectors satisfy $||\mathbf{x}||, ||\mathbf{y}|| \leq 1$, which implies a bound on the correlation $||T|| \geq 1$ for all pure states. The following lemma identifies a simple entanglement witness for pure states. The proof of this and all subsequent lemmas is given in the Appendix.

**Lemma 1.** The lower bound $||T|| = 1$ is saturated, if and only if the state is a product state $T = \mathbf{x}\mathbf{y}^T$.

Recall that for every transformation $U$, we can find its orthogonal representation $U = SOS^{-1}$ (the Schur–Auerbach lemma), where $S$ is an invertible matrix and $O^T O = \mathbb{1}$. The matrix $S$ is characteristic of the representation and should be the same for all transformations $U$. If we choose some local transformation $U = (R_1, R_2)$, $U$ will be orthogonal and thus we can choose to set $S = \mathbb{1}$. The representation of transformations is orthogonal; therefore, they are norm-preserving. By applying simultaneously all (local and non-local) transformations $U$ to some product state (the reference state) $\psi$ and to the measurement vector that identifies it, $r = \psi$, we generate the set of all pure states and corresponding measurement vectors. Because we have $1 = P(r = \psi, \psi) = P(Ur, U\psi)$, correspondence $r = \psi$ holds for any pure state $\psi$. Instead of the measurement vector $r$ in formula (9.14), we use the pure state that identifies it. If the state $\psi_1 = (\mathbf{x}_1, \mathbf{y}_1, T_1)$ is prepared and measurement along the state $\psi_2 = (\mathbf{x}_2, \mathbf{y}_2, T_2)$ is performed, the measured probability is given by

$$P_{12}(\psi_1, \psi_2) = \frac{1}{4}(1 + \mathbf{x}_1^T \mathbf{x}_2 + \mathbf{y}_1^T \mathbf{y}_2 + \text{Tr}(T_1^T T_2)). \tag{9.22}$$

The set of pure states obeys $P_{12}(\psi, \psi) = 1$. We can define the normalization condition for pure states $P_{12}(\psi, \psi) = \frac{1}{4}(1 + ||\mathbf{x}||^2 + ||\mathbf{y}||^2 + ||T||^2) = 1$, where $||T||^2 = \text{Tr}(T^T T)$. Therefore, we have:

$$||\mathbf{x}||^2 + ||\mathbf{y}||^2 + ||T||^2 = 3 \tag{9.23}$$

for all pure states.

An interesting observation can be made here. Although seemingly Axiom 2 does not imply any strong prior restrictions to $d$, we surprisingly have obtained the explicit number 3 in the normalization condition (Equation 9.23). As we see soon, this relation will play an important role in deriving $d = 3$ as the only non-classical solution consistent with the axioms.

## 9.6 The Main Proofs

We now show that only classical probability theory and quantum theory are in agreement with the three axioms.

### 9.6.1 Ruling Out the *d* Even Case

For $d$ even, the total inversion $E\mathbf{x} = -\mathbf{x}$ has determinant 1 and thus is a physical transformation. Let $\psi = (\mathbf{x}, \mathbf{y}, T)$ be a pure state of composite system. We apply total inversion to one of the subsystems and obtain the state $\psi' = (E, \mathbb{1})(\mathbf{x}, \mathbf{y}, T) = (-\mathbf{x}, \mathbf{y}, -T)$. The probability

$$P_{12}(\psi, \psi') = \frac{1}{4}(1 - ||\mathbf{x}||^2 + ||\mathbf{y}||^2 - ||T||^2) \quad (9.24)$$

$$= \frac{1}{2}(||\mathbf{y}||^2 - 1) \quad (9.25)$$

has to be nonnegative and, therefore, we have $||\mathbf{y}|| = 1$. Similarly, we apply $(\mathbb{1}, E)$ to $\psi$ and obtain $||\mathbf{x}|| = 1$. Because the local vectors are of the unit norm, we have $||T|| = 1$; thus, according to lemma 1, the state $\psi$ is a product state. We conclude that no entangled states can exist for $d$ even. As we will soon see, according to Axiom 1, entangled states must exist; therefore, $d$ has to be an odd number.

As mentioned, the set of transformations contains the SO($d$) group. If we include even a single transformation with determinant $-1$, the set of transformations becomes the entire set of orthogonal matrices; hence, total inversion $E$ becomes a physical transformation. Therefore, the set of physical transformations cannot be larger than SO($d$) group.

### 9.6.2 Ruling Out the *d* > 3 Case

Let us define one basis set of two generalized bit product states:

$$\psi_1 = (\mathbf{e}_1, \mathbf{e}_1, T_0 = \mathbf{e}_1 \mathbf{e}_1^T) \quad (9.26)$$
$$\psi_2 = (-\mathbf{e}_1, -\mathbf{e}_1, T_0) \quad (9.27)$$
$$\psi_3 = (-\mathbf{e}_1, \mathbf{e}_1, -T_0) \quad (9.28)$$
$$\psi_4 = (\mathbf{e}_1, -\mathbf{e}_1, -T_0) \quad (9.29)$$

with $\mathbf{e}_1 = (1, 0, \ldots, 0)^T$. Now, we define two subspaces $S_{12}$ and $S_{34}$ spanned by the states $\psi_1, \psi_2$ and $\psi_3, \psi_4$, respectively. Axiom 1 states that these two subspaces behave like one-bit spaces; therefore, they are isomorphic to the $(d-1)$-sphere $S_{12} \cong S_{34} \cong S^{d-1}$. The state $\psi$ belongs to $S_{12}$ if and only if the following holds:

$$P_{12}(\psi, \psi_1) + P_{12}(\psi, \psi_2) = 1. \quad (9.30)$$

Because the $\psi_1, \ldots, \psi_4$ form a complete basis set, we have

$$P_{12}(\psi, \psi_3) = 0, \quad P_{12}(\psi, \psi_4) = 0. \quad (9.31)$$

A similar reasoning holds for states belonging to the $S_{34}$ subspace. The states $\psi \in S_{12}$ and $\psi' \in S_{34}$ are perfectly distinguishable in a single-shot experiment; therefore, we have $P_{12}(\psi, \psi') = 0$. Therefore, $S_{12}$ and $S_{34}$ are orthogonal subspaces.

Axiom 1 requires the existence of entangled states as is apparent from the following Lemma 2.

**Lemma 2.** *The only product states belonging to $S_{12}$ are $\psi_1$ and $\psi_2$.*

We define a local mapping between orthogonal subspaces $S_{12}$ and $S_{34}$. Let the state $\psi = (\mathbf{x}, \mathbf{y}, T) \in S_{12}$, with $\mathbf{x} = (x_1, x_2, \ldots, x_d)^{\mathrm{T}}$ and $\mathbf{y} = (y_1, y_2, \ldots, y_d)^{\mathrm{T}}$. Consider the one-bit transformation $R$ with the property $R\mathbf{e}_1 = -\mathbf{e}_1$. The local transformation of this type maps the state from $S_{12}$ to $S_{34}$ as shown by the following lemma:

**Lemma 3.** If the state $\psi \in S_{12}$, then $\psi' = (R, \mathbb{1})\psi \in S_{34}$ and $\psi'' = (\mathbb{1}, R)\psi \in S_{34}$.

Let us define $\mathbf{T}_i^{(x)} = (T_{i1}, \ldots, T_{id})$ and $\mathbf{T}_i^{(y)} = (T_{1i}, \ldots, T_{di})^{\mathrm{T}}$. The correlation tensor can be rewritten in two different ways:

$$T = \begin{pmatrix} \mathbf{T}_1^{(x)} \\ \mathbf{T}_2^{(x)} \\ \vdots \\ \mathbf{T}_d^{(x)} \end{pmatrix} \quad \text{or} \quad T = \begin{pmatrix} \mathbf{T}_1^{(y)} & \mathbf{T}_2^{(y)} & \cdots & \mathbf{T}_d^{(y)} \end{pmatrix}. \tag{9.32}$$

Consider now the case $d > 3$. We define local transformations $R_i$ flipping the first and $i$-th coordinate and $R_{jkl}$ flipping the first and $j$-th, $k$-th, and $l$-th coordinate with $j \neq k \neq l \neq 1$. Let $\psi = (\mathbf{x}, \mathbf{y}, (\mathbf{T}_1^{(x)}, \ldots, \mathbf{T}_d^{(x)})^{\mathrm{T}})$ belong to $S_{12}$. According to Lemma 2, the states $\psi_i = (R_i, \mathbb{1})\psi$ and $\psi_{jkl} = (R_{jkl}, \mathbb{1})\psi$ belong to $S_{34}$; therefore, $P_{12}(\psi, \psi_i) = 0$ and $P_{12}(\psi, \psi_{jkl}) = 0$. We have:

$$0 = P_{12}(\psi, \psi_i) \tag{9.33}$$
$$1 - x_1^2 + x_2^2 + \cdots - x_i^2 + \cdots + x_d^2 + \|\mathbf{y}\|^2 \tag{9.34}$$
$$-\|\mathbf{T}_1^{(x)}\|^2 + \|\mathbf{T}_2^{(x)}\|^2 + \cdots - \|\mathbf{T}_i^{(x)}\|^2 + \cdots + \|\mathbf{T}_d^{(x)}\|^2 \tag{9.35}$$
$$= 1 - 2x_1^2 - 2x_i^2 - 2\|\mathbf{T}_1^{(x)}\|^2 - 2\|\mathbf{T}_i^{(x)}\|^2 + \|\mathbf{x}\|^2 + \|\mathbf{y}\|^2 + \|T\|^2$$
$$= 2(2 - x_1^2 - x_i^2 - \|\mathbf{T}_1^{(x)}\|^2 - \|\mathbf{T}_i^{(x)}\|^2). \tag{9.36}$$

Similarly, we expand $P_{12}(\psi, \psi_{jkl}) = 0$ and together with the last equation, we obtain

$$x_1^2 + x_i^2 + \|\mathbf{T}_1^{(x)}\|^2 + \|\mathbf{T}_i^{(x)}\|^2 = 2 \tag{9.37}$$
$$x_1^2 + x_j^2 + x_k^2 + x_l^2 + \|\mathbf{T}_1^{(x)}\|^2 + \|\mathbf{T}_j^{(x)}\|^2 + \|\mathbf{T}_k^{(x)}\|^2 + \|\mathbf{T}_l^{(x)}\|^2 = 2.$$

Because this has to hold for all $i, j, k, l$, we have

$$x_2 = x_3 = \cdots = x_d = 0 \tag{9.38}$$
$$\mathbf{T}_2^{(x)} = \mathbf{T}_3^{(x)} = \cdots = \mathbf{T}_d^{(x)} = 0. \tag{9.39}$$

We repeat this kind of reasoning for the transformations $(\mathbb{1}, R_i)$ and $(\mathbb{1}, R_{jkl})$ and obtain

$$y_1^2 + y_i^2 + \|\mathbf{T}_1^{(y)}\|^2 + \|\mathbf{T}_i^{(y)}\|^2 = 2 \tag{9.40}$$
$$y_1^2 + y_j^2 + y_k^2 + y_l^2 + \|\mathbf{T}_1^{(y)}\|^2 + \|\mathbf{T}_j^{(y)}\|^2 + \|\mathbf{T}_k^{(y)}\|^2 + \|\mathbf{T}_l^{(y)}\|^2 = 2.$$

Therefore, we have

$$y_2 = y_3 = \cdots = y_d = 0 \tag{9.41}$$
$$\mathbf{T}_2^{(y)} = \mathbf{T}_3^{(y)} = \cdots = \mathbf{T}_d^{(y)} = 0. \tag{9.42}$$

The only non-zero element of the correlation tensor is $T_{11}$ and it has to be exactly 1, because $||T|| \geq 1$. This implies that $\psi$ is a product state; furthermore, $\psi = \psi_1$ or $\psi = \psi_2$.

This concludes our proof that only the cases $d = 1$ and $d = 3$ are in agreement with our three axioms. To distinguish between the two cases, one can invoke the continuity Axiom 4 and proceed as in the reconstruction given by Hardy [20].

## 9.7 "Two" Quantum Mechanics

We now obtain two solutions for the theory of a composite system consisting of two bits in the case when $d = 3$. One of them corresponds to the standard quantum theory of two qubits, the other one to its "mirror" version in which the states are obtained from the ones from the standard theory by partial transposition. Both solutions are regular as far as one considers composite systems of two bits, but the "mirror" one cannot be consistently constructed already for systems of three bits.

Two conditions, (Equations 9.30) and (9.31), put the constraint to the form of $\psi$:

$$x_1 = -y_1, \quad T_{11} = 1. \tag{9.43}$$

The subspace $S_{12}$ is isomorphic to the sphere $\mathcal{S}^2$. Let us choose $\psi$ complementary to the one-bit basis $\{\psi_1, \psi_2\}$ in $S_{12}$. We have $P_{12}(\psi, \psi_1) = P_{12}(\psi, \psi_2) = 1/2$ and thus $x_1 = y_1 = 0$. For simplicity, we write $\psi$ in the form:

$$\psi = \left( \begin{pmatrix} 0 \\ \mathbf{x} \end{pmatrix}, \begin{pmatrix} 0 \\ \mathbf{y} \end{pmatrix}, \begin{pmatrix} 1 & \mathbf{T}_y^T \\ \mathbf{T}_x & T \end{pmatrix} \right), \tag{9.44}$$

with $\mathbf{x} = (x_2, x_3)^T$, $\mathbf{y} = (y_2, y_3)^T$, $\mathbf{T}_y = (T_{12}, T_{13})^T$, $\mathbf{T}_x = (T_{21}, T_{31})^T$, and $T = \begin{pmatrix} T_{22} & T_{23} \\ T_{32} & T_{33} \end{pmatrix}$.

Let $R(\phi)$ be a rotation around the $\mathbf{e}_1$ axis. This transformation keeps $S_{12}$ invariant. Now, we show that the state $\psi$ as given by Equation (9.44) cannot be invariant under local transformation $(\mathbb{1}, R(\phi))$. To prove this by *reductio ad absurdum*, suppose the opposite; that is, that $(\mathbb{1}, R(\phi))\psi = \psi$. We have three conditions:

$$R(\phi)\mathbf{y} = \mathbf{y}, \quad \mathbf{T}_y^T R^T(\phi) = \mathbf{T}_y^T, \quad T R^T(\phi) = T, \tag{9.45}$$

which implies $\mathbf{y} = 0$, $\mathbf{T}_y^T = 0$, and $T = 0$; thus

$$\psi = \left( \begin{pmatrix} 0 \\ x_2 \\ x_3 \end{pmatrix}, \begin{pmatrix} 0 \\ 0 \\ 0 \end{pmatrix}, \begin{pmatrix} 1 & 0 & 0 \\ T_2 & 0 & 0 \\ T_3 & 0 & 0 \end{pmatrix} \right). \tag{9.46}$$

According to Equations (9.37) and (9.40), we can easily check that $||\mathbf{x}|| = 1$; thus, $\psi$ is locally equivalent to the state:

$$\psi' = \left( \begin{pmatrix} 0 \\ 0 \\ 1 \end{pmatrix}, \begin{pmatrix} 0 \\ 0 \\ 0 \end{pmatrix}, \begin{pmatrix} 1 & 0 & 0 \\ T_2' & 0 & 0 \\ T_3' & 0 & 0 \end{pmatrix} \right). \tag{9.47}$$

Let $\chi_1 = (-\mathbf{e}_3, \mathbf{e}_1, -\mathbf{e}_3\mathbf{e}_1^T)$ and $\chi_2 = (-\mathbf{e}_3, -\mathbf{e}_1, \mathbf{e}_3\mathbf{e}_1^T)$. The two conditions $P(\psi', \chi_1) \geq 0$ and $P(\psi', \chi_2) \geq 0$ become

$$\frac{1}{4}(1 - 1 - T_3') = -\frac{1}{4}T_3' \geq 0 \tag{9.48}$$

$$\frac{1}{4}(1 - 1 + T_3') = \frac{1}{4}T_3' \geq 0; \tag{9.49}$$

thus, $T_3' = 0$. The normalization condition (Equation 9.23) gives $T_2' = \pm 1$. The state $\psi'$ is not physical. This can be seen when one performs the rotation $(R, \mathbb{1})$, where

$$R = \begin{pmatrix} \frac{1}{\sqrt{2}} & -\frac{1}{\sqrt{2}} & 0 \\ \frac{1}{\sqrt{2}} & \frac{1}{\sqrt{2}} & 0 \\ 0 & 0 & 1 \end{pmatrix}. \tag{9.50}$$

The transformed correlation tensor has a component $\sqrt{2}$, which is non-physical. Therefore, the transformation $(\mathbb{1}, R(\phi))\psi$ draws a full circle of pure states in a plane orthogonal to $\psi_1$ within the subspace $S_{12}$. Similarly, the transformation $(R(\phi), \mathbb{1})$ draws the same set of pure states when applied to $\psi$. Hence, for every transformation $(\mathbb{1}, R(\phi_1))$, there exists a transformation $(R(\phi_2), \mathbb{1})$ such that $(\mathbb{1}, R(\phi_1))\psi = (R(\phi_2), \mathbb{1})\psi$. This gives us a set of conditions

$$R(\phi_2)\mathbf{x} = \mathbf{x} \tag{9.51}$$

$$R(\phi_1)\mathbf{y} = \mathbf{y} \tag{9.52}$$

$$R(\phi_2)\mathbf{T}_x = \mathbf{T}_x \tag{9.53}$$

$$\mathbf{T}_y^T R^T(\phi_1) = \mathbf{T}_y^T \tag{9.54}$$

$$R(\phi_2)T = T R^T(\phi_1), \tag{9.55}$$

which are fulfilled if $\mathbf{x} = \mathbf{y} = \mathbf{T}_x = \mathbf{T}_y = 0$ and $T = \text{diag}[T_1, T_2]$. Equation (9.37) gives $T_2^2 = T_3^2 = 1$ and we finally end up with two different solutions:

$$\psi_{QM} = (0, 0, \text{diag}[1, -1, 1]) \ \lor \ \psi_{MQM} = (0, 0, \text{diag}[1, 1, 1]). \tag{9.56}$$

The first "M" in $\psi_{MQM}$ stands for "mirror." The two solutions are incompatible and cannot coexist within the same theory. The first solution corresponds to the triplet state $\phi^+$ of ordinary quantum mechanics. The second solution is a totally invariant state and has a negative overlap with, for example, the singlet state $\psi^-$ for which $T = \text{diag}[-1, -1, -1]$. That is, if the system were prepared in one of the two states and the other one were measured, the probability would be negative. Nevertheless, both solutions are regular at the level of two bits. The first belongs to ordinary quantum mechanics with the singlet in the "antiparallel" subspace $S_{34}$, and the second solution is "the singlet state in the parallel subspace" $S_{12}$. We will show that one can build the full state space, transformations, and measurements in both cases. The states from one quantum mechanics can be obtained from the other by partial transposition $\psi_{QM}^{PT} = \psi_{MQM}$. In particular, the four maximal entangled states (Bell states) from "mirror quantum mechanics" have correlations of the opposite sign of those from the standard quantum mechanics (Fig 9.4).

Now we show that the theory with "mirror states" is physically inconsistent when applied to a composite system of three bits. Let us first derive the full set of states

| QM | xx | yy | zz |
|---|---|---|---|
| $\Phi^+$ | +1 | -1 | +1 |
| $\Phi^-$ | -1 | +1 | +1 |
| $\Psi^+$ | +1 | +1 | -1 |
| $\Psi^-$ | -1 | -1 | -1 |

| MQM | xx | yy | zz |
|---|---|---|---|
| $\Phi^+_{PT}$ | -1 | +1 | -1 |
| $\Phi^-_{PT}$ | +1 | -1 | -1 |
| $\Psi^+_{PT}$ | -1 | -1 | +1 |
| $\Psi^-_{PT}$ | +1 | +1 | +1 |

**Figure 9.4.** Correlations between results obtained in measurements of two bits in a maximal entangled (Bell's) state in standard quantum mechanics (*left*) and "mirror quantum mechanics" (*right*) along x, y, and z directions. Why do we never see correlations as given in the table on the right? The opposite sign of correlations on the right and on the left is not a matter of convention or labeling of outcomes. If one can transport the two bits parallel to the same detector, one can distinguish operationally between the two types of correlations.

and transformations for two qubits in standard quantum mechanics. We have seen that the state $\psi_{QM}$ belongs to the subspace $S_{12}$ and, furthermore, that it is complementary (within $S_{12}$) to the product states $\psi_1$ and $\psi_2$. The totally mixed state within the $S_{12}$ subspace is $E_{12} = \frac{1}{2}\psi_1 + \frac{1}{2}\psi_2$. The states $\psi_1$ and $\psi_{QM}$ span one two-dimensional plane, and the set of pure states within this plane is a circle:

$$\psi(x) = E_{12} + \cos x \, (\psi_1 - E_{12}) + \sin x \, (\psi_{QM} - E_{12}) \tag{9.57}$$

$$= (\cos x \, \mathbf{e}_1, \cos x \, \mathbf{e}_1, \text{diag}[1, -\sin x, \sin x]). \tag{9.58}$$

We can apply a complete set of local transformations to the set $\psi(x)$ to obtain the set of all pure two-qubit states. Let us represent a pure state $\psi = (\mathbf{x}, \mathbf{y}, T)$ by the $4 \times 4$ Hermitian matrix $\rho$:

$$\rho = \frac{1}{4}(\mathbb{1} \otimes \mathbb{1} + \sum_{i=1}^{3} x_i \sigma_i \otimes \mathbb{1} + \sum_{i=1}^{3} y_i \mathbb{1} \otimes \sigma_i + \sum_{i,j=1}^{3} T_{ij} \sigma_i \otimes \sigma_j), \tag{9.59}$$

where $\sigma_i, i \in \{1, 2, 3\}$ are the three Pauli matrices. It is easy to show that the set of states (Equation 9.57) corresponds to the set of one-dimensional projectors $|\psi(x)\rangle\langle\psi(x)|$, where $|\psi(x)\rangle = \cos \frac{x}{2}|00\rangle + \sin \frac{x}{2}|11\rangle$. The action of local transformations $(R_1, R_2)\psi$ corresponds to local unitary transformation $U_1 \otimes U_2 |\psi\rangle\langle\psi| U_1^\dagger \otimes U_2^\dagger$, where the correspondence between $U$ and $R$ is given by the isomorphism between the groups SU(2) and SO(3):

$$U\rho U^\dagger = \frac{1}{2}\left(\mathbb{1} + \sum_{i=1}^{3}\left(\sum_{j=1}^{3} R_{ij} x_j\right) \sigma_i\right). \tag{9.60}$$

Here, $R_{ij} = \text{Tr}(\sigma_i U \sigma_j U^\dagger)$ and $x_i = \text{Tr}\sigma_i \rho$. When we apply a complete set of local transformations to the states $|\psi(x)\rangle$, we obtain the whole set of pure states for two qubits. The group of transformations is the set of unitary transformations SU(4).

The set of states from "mirror quantum mechanics" can be obtained by applying partial transposition to the set of quantum states. Formally, partial transposition with respect to subsystem 1 is defined by action on a set of product operators:

$$\text{PT}_1(\rho_1 \otimes \rho_2) = \rho_1^T \otimes \rho_2, \tag{9.61}$$

where $\rho_1$ and $\rho_2$ are arbitrary operators. Similarly, we can define the partial transposition with respect to subsystem 2 $\text{PT}_2$. To each unitary transformation $U$ in quantum mechanics, we define the corresponding transformation in "mirror mechanics"—for example, with respect to subsystem 1: $PT_1 U PT_1$. Therefore, the set of transformations is a conjugate group $PT_1 SU(4) PT_1 := \{PT_1 U PT_1 \mid U \in SU(4)\}$. Note that we could equally have chosen to apply partial transposition with respect to subsystem 2 and would obtain the same set of states. In fact, one can show that $PT_1 U PT_1 = PT_2 U^* PT_2$, where $U^*$ is a conjugate unitary transformation (see Lemma 4 in the Appendix). Therefore, the two conjugate groups are the same $PT_1 SU(4) PT_1 = PT_2 SU(4) PT_2$. We can generate the set of "mirror states" by applying all of the transformations $PTUPT$ to some product state, regardless of which particular partial transposition is used.

Now, we show that "mirror mechanics" cannot be consistently extended to composite systems consisting of three bits. Let $\psi_p = (\mathbf{x}, \mathbf{y}, \mathbf{z}, T_{12}, T_{13}, T_{23}, T_{123})$ be some product state of three bits, where $\mathbf{x}$, $\mathbf{y}$, and $\mathbf{z}$ are local Bloch vectors and $T_{12}$, $T_{13}$, $T_{23}$, and $T_{123}$ are two- and three-body correlation tensors, respectively. We can apply the transformations $PT U_{ij} PT$ to a composite system of $i$ and $j$, and we are free to choose with respect to which subsystem ($i$ or $j$) to take the partial transposition. Furthermore, we can combine transformations in 12 and 13 subsystems such that the resulting state is genuine three-partite entangled, and we can choose to partially transpose subsystem 2 in both cases. We obtain the transformation

$$U_{123} = PT_2 U_{12} PT_2 PT_2 U_{23} PT_2 \qquad (9.62)$$
$$= PT_2 U_{12} U_{23} PT_2. \qquad (9.63)$$

When we apply $U_{123}$ to $\psi_p$, we obtain the state $PT_2 U_{12} U_{23} \phi_p$, where $\phi_p = PT_2 \psi_p$ is again some product state. The state $U_{12} U_{23} \phi_p$ is a quantum three-qubit state. Because states $\psi_p$ and $\phi_p$ are product states and do belong to standard quantum states, we can use the formalism of quantum mechanics and denote them as $|\psi_p\rangle$ and $|\phi_p\rangle$. Furthermore, because the state $|\psi_p\rangle$ is an arbitrary product state, without loss of generality, we set $|\phi_p\rangle = |0\rangle|0\rangle|0\rangle$. We can choose $U_{12}$ and $U_{23}$ such that:

$$U_{12}|0\rangle|0\rangle = |0\rangle|0\rangle \qquad (9.64)$$

$$U_{12}|0\rangle|1\rangle = \frac{1}{\sqrt{2}}(|0\rangle|1\rangle + |1\rangle|0\rangle) \qquad (9.65)$$

$$U_{23}|0\rangle|0\rangle = \frac{1}{\sqrt{3}}|0\rangle|1\rangle + \sqrt{\frac{2}{3}}|1\rangle|0\rangle. \qquad (9.66)$$

This way, we can generate the $W$-state

$$|W\rangle = U_{12} U_{23} |0\rangle|0\rangle|0\rangle \qquad (9.67)$$
$$= \frac{1}{\sqrt{3}}(|0\rangle|0\rangle|1\rangle + |0\rangle|1\rangle|0\rangle + |1\rangle|0\rangle|0\rangle). \qquad (9.68)$$

When we apply partial transposition with respect to subsystem 2, we obtain the corresponding "mirror W-state," which we denote as $W_M$-state, $W_M = PT_2 W$. The local Bloch vectors and two-body correlation tensors for the $W$ state are

$$\mathbf{x} = \mathbf{y} = \mathbf{z} = (0, 0, \tfrac{1}{3})^T, \qquad (9.69)$$

$$T_{12} = T_{13} = T_{23} = \mathrm{diag}[\tfrac{2}{3}, \tfrac{2}{3}, -\tfrac{1}{3}], \qquad (9.70)$$

where $|0\rangle$ corresponds to result $+1$. Consequently, the local Bloch vectors and the correlation tensor for $W_M$-state are

$$\mathbf{x} = \mathbf{y} = \mathbf{z} = (0, 0, \tfrac{1}{3})^T, \tag{9.71}$$

$$T_{12} = T_{23} = \text{diag}[\tfrac{2}{3}, -\tfrac{2}{3}, -\tfrac{1}{3}], \tag{9.72}$$

$$T_{13} = \text{diag}[\tfrac{2}{3}, \tfrac{2}{3}, -\tfrac{1}{3}]. \tag{9.73}$$

The asymmetry in the signs of correlations in the tensors $T_{12}$, $T_{23}$, and $T_{13}$ leads to inconsistencies because they define three different reduced states $\psi_{ij} = (\mathbf{x}_i, \mathbf{x}_j, T_{ij})$, $ij \in \{12, 23, 13\}$, which cannot coexist within a single theory. The states $\psi_{12}$ and $\psi_{23}$ belong to "mirror quantum mechanics," whereas the state $\psi_{13}$ belongs to ordinary quantum mechanics. To see this, take the state $\psi = (0, 0, \text{diag}[-1, -1, 1])$, which is locally equivalent to state $\psi_{MQM} = (0, 0, \mathbb{1})$. The overlap (measured probability) between the states $\psi_{13}$ and $\psi$ is negative

$$P(\psi, \psi_{13}) = \frac{1}{4}(1 - \frac{2}{3} - \frac{2}{3} - \frac{1}{3}) = -\frac{1}{6}. \tag{9.74}$$

We conclude that "mirror quantum mechanics"—although being a perfectly regular solution for a theory of two bits—cannot be consistently extended to also describe systems consisting of many bits. This also answers the question of why we find in nature only four types of correlations as given in the table (see Figure 9.4) on the left rather than all eight logically possible ones.

## 9.8 Higher-Dimensional Systems and State Update Rule in Measurement

Having obtained $d = 3$ for a two-dimensional system, we have derived quantum theory of this system. We have also reconstructed quantum mechanics of a composite system consisting of two qubits. Further reconstruction of quantum mechanics can be proceeded as in Hardy's work [20]. In particular, the reconstruction of higher-dimensional systems from the two-dimensional ones and the general transformations of the state after measurement are explicitly given there. We only briefly comment on them here.

To derive the state space, measurements, and transformations for a higher-dimensional system, we can use quantum theory of a two-dimensional system in conjunction with Axiom 1. The axiom requires that upon any two linearly independent states, one can construct a two-dimensional subspace that is isomorphic to the state space of a qubit (2-sphere). The state space of a higher-dimensional system can be characterized such that if the state is restricted to any given two dimensional subspace, then it behaves like a qubit. The fact that all other (higher-dimensional) systems can be built out of two-dimensional ones suggests that the latter can be considered as fundamental constituents of the world and gives a justification for the usage of the term *elementary system* in the formulations of the axioms.

When a measurement is performed and an outcome is obtained, our knowledge about the state of the system changes and its representation in form of the probabilities must be updated to be in agreement with the new knowledge acquired in the measurement. This is the most natural update rule present in any probability theory. Only if one views this change as a real physical process conceptual problems arise related to discontinuous and abrupt "collapse of the wave function." There is no basis for any such assumption. Associated with each outcome is the measurement vector **r**. When the outcome is observed, the state after the measurement is updated to **r** and the measurement will be a certain transformation on the initial state. Update rules for more general measurements can accordingly be given.

## 9.9 What the Present Reconstruction Tells Us about Quantum Mechanics

It is often said that reconstructions of quantum theory within an operational approach are devoid of ontological commitments and that nothing can be generally said about the ontological content that arises from the first principles or about the status of the notion of realism. As a supporting argument, one usually notes that within a realistic worldview, one would anyway expect quantum theory at the operational level to be deducible from some underlying theory of "deeper reality." After all, we have the Broglie–Bohm theory [63,64], which is a nonlocal realistic theory in full agreement with the predictions of (nonrelativistic) quantum theory. Having said this, we cannot but emphasize that realism does stay "orthogonal" to the basic idea behind our reconstruction.

Whether local or nonlocal, realism asserts that outcomes correspond to actualities objectively existing before and independent of measurements. Conversely, we have shown that the finiteness of information-carrying capacity of quantum systems is an important ingredient in deriving quantum theory. This capacity is not enough to allow assignment of definite values to outcomes of all possible measurements. The elementary system has the information-carrying capacity of one bit. This is signified by the possibility to decompose any state of an elementary system (qubit) in quantum mechanics in two orthogonal states. In a realistic theory based on hidden variables and an "epistemic constraint" on an observer's knowledge of the "variables" values, one can reproduce this feature at the level of the entire distribution of the hidden variables.[3] That this is possible is not surprising if one bears in mind that hidden-variable theories in the first place were introduced to *reproduce* quantum mechanics and yet give a more complete description.[4] But any realism of that kind at the same time assumes an *infinite information capacity* at the level of hidden variables. Even to reproduce measurements on a single qubit requires infinitely many orthogonal hidden-variable states [65–67]. It might be a matter of taste whether or not one is ready to work with this "ontological access baggage" [65] not doing any explanatory work at the operational level. But it is certainly conceptually distinctly different from the theory analyzed here, in which the information capacity of the most elementary systems— those that are by definition not reducible further—is fundamentally limited.

---

[3] See [40] for a local version of such hidden-variable theory in which quantum mechanical predictions are partially reproduced.

[4] That this cannot be done without allowing nonlocal influences from space-like distant regions is a valid point in itself, which we do not want to follow here further.

To clarify our position further, consider the Mach–Zehnder interferometer in which both the path information and interference observable are dichotomic (i.e., two-valued observables). It is meaningless to speak about "the path the particle took in the interferometer in the interference experiment" because this would already require the assignment of two bits of information to the system, which would exceed its information capacity of one bit [68]. The information capacity of the system is simply not enough to provide definite outcomes to all possible measurements. Then, by necessity, the outcome in some experiments must contain an element of randomness, and there must be observables that are complementary to each other. Entanglement and, consequently, the violation of Bell's inequality (and thus of local realism) arise from the possibility of defining an abstract elementary system carrying at most one bit such that correlations ("00" and "11" in a joint measurement of two subsystems) are basis states.

## 9.10 Conclusions

Quantum theory is our most accurate description of nature and is fundamental to our understanding of, for example, the stability of matter, the periodic table of chemical elements, and the energy of the sun. It has led to the development of great inventions like the electronic transistor, laser, and quantum cryptography. Given the enormous success of quantum theory, can we consider it as our final and ultimate theory? Quantum theory has caused much controversy in interpreting what its philosophical and epistemological implications are. At the heart of this controversy lies the fact that the theory makes only probabilistic predictions. In recent years it was shown, however, that some features of quantum theory that one might have expected to be uniquely quantum turned out to be highly generic for generalized probabilistic theories. Is there any reason why the universe should obey the laws of quantum theory, as opposed to any other possible probabilistic theory?

In this chapter, we have shown that classical probability theory and quantum theory—the only two probability theories for which we have empirical evidences—are special in a way that they fulfill three reasonable axioms on the systems' information-carrying capacity, on the notion of locality, and on the reversibility of transformations. The two theories can be separated if one restricts the transformations between the pure states to be continuous [20]. An interesting finding is that quantum theory is the *only* non-classical probability theory that can exhibit entanglement without conflicting one or more axioms. Therefore—to use Schrödinger's words [69]— entanglement is not only "*the* characteristic trait of quantum mechanics, the one that enforces its entire departure from classical lines of thought," but also the one that enforces the departure from a broad class of more general probabilistic theories.

## Acknowledgments

We thank M. Aspelmeyer, J. Kofler, T. Paterek, and A. Zeilinger for discussions. We acknowledge support from the Austrian Science Foundation FWF within Project No. P19570-N16, SFB and CoQuS No. W1210-N16, the European Commission Project QAP (No. 015848), and the Foundational Question Institute (FQXi).

# Appendix

In this Appendix, we give the proofs of the lemmas from the main text.

**Lemma 1.** The lower bound $||T|| = 1$ is saturated if and only if the state is a product state $T = \mathbf{xy}^T$.

**Proof.** If the state is a product state, then $||T||^2 = ||\mathbf{x}||^2 ||\mathbf{y}||^2 = 1$. Conversely, assume that the state $\psi = (\mathbf{x}, \mathbf{y}, T)$ satisfies $||T|| = 1$. Normalization (Equation 9.23) gives $||\mathbf{x}|| = ||\mathbf{y}|| = 1$. Let $\phi_p = (-\mathbf{x}, -\mathbf{y}, T_0 = \mathbf{xy}^T)$ be a product state. We have $P(\psi, \phi_p) \geq 0$ and, therefore,

$$1 - ||\mathbf{x}||^2 - ||\mathbf{y}||^2 + \text{Tr}(T^T T_0) = -1 + \text{Tr}(T^T T_0) \geq 0. \tag{9.75}$$

The last inequality $\text{Tr}(T^T T_0) \geq 1$ can be seen as $(T, T_0) \geq 1$, where $(,)$ is the scalar product in Hilbert–Schmidt space. Because the vectors $T, T_0$ are normalized, $||T|| = ||T_0|| = 1$, the scalar product between them is always $(T, T_0) \leq 1$. Therefore, we have $(T, T_0) = 1$, which is equivalent to $T = T_0 = \mathbf{xy}^T$.

QED

**Lemma 2.** The only product states belonging to $S_{12}$ are $\psi_1$ and $\psi_2$.

**Proof.** Let $\psi_p = (\mathbf{x}, \mathbf{y}, \mathbf{xy}^T) \in S_{12}$. We have

$$1 = P_{12}(\psi_p, \psi_1) + P_{12}(\psi_p, \psi_2) \tag{9.76}$$

$$= \frac{1}{4}(1 + \mathbf{xe}_1 + \mathbf{ye}_1 + (\mathbf{xe}_1)(\mathbf{ye}_1)) \tag{9.77}$$

$$+ \frac{1}{4}(1 - \mathbf{xe}_1 - \mathbf{ye}_1 + (\mathbf{xe}_1)(\mathbf{ye}_1)) \tag{9.78}$$

$$= \frac{1}{2}(1 + (\mathbf{xe}_1)(\mathbf{ye}_1)) \tag{9.79}$$

$$\Rightarrow \mathbf{xe}_1 = \mathbf{ye}_1 = 1 \ \vee \ \mathbf{xe}_1 = \mathbf{ye}_1 = -1 \tag{9.80}$$

$$\Leftrightarrow \mathbf{x} = \mathbf{y} = \mathbf{e}_1 \ \vee \ \mathbf{x} = \mathbf{y} = -\mathbf{e}_1. \tag{9.81}$$

QED

**Lemma 3.** If the state $\psi \in S_{12}$, then $\psi' = (R, \mathbb{1})\psi \in S_{34}$ and $\psi'' = (\mathbb{1}, R)\psi \in S_{34}$.

**Proof.** If $\psi \in S_{12}$, we have

$$1 = P_{12}(\psi, \psi_1) + P_{12}(\psi, \psi_2) \tag{9.82}$$

$$= P_{12}((R, \mathbb{1})\psi, (R, \mathbb{1})\psi_1) + P_{12}((R, \mathbb{1})\psi, (R, \mathbb{1})\psi_2) \tag{9.83}$$

$$= P_{12}(\psi', \psi_3) + P_{12}(\psi', \psi_4). \tag{9.84}$$

Similarly, one can show that $(\mathbb{1}, R)\psi \in S_{34}$.

QED

**Lemma 4.** Let $U$ be some operator with the following action in the Hilbert–Schmidt space; $U(\rho) = U\rho U^\dagger$, and $PT_1$ and $PT_2$ are partial transpositions with respect to subsystems 1 and 2, respectively. The following identity holds: $\mathrm{PT}_1 U \mathrm{PT}_1 = \mathrm{PT}_2 U^* \mathrm{PT}_2$, where $U^*$ is the complex-conjugate operator.

**Proof.** We can expand $U$ into some product basis in the Hilbert–Schmidt space $U = \sum_{ij} u_{ij} A_i \otimes B_j$. We have

$$\mathrm{PT}_1 U \mathrm{PT}_1 (\rho_1 \otimes \rho_2) = \mathrm{PT}_1 \{ U \rho_1^\mathrm{T} \otimes \rho_2 U^\dagger \} \qquad (9.85)$$

$$= \sum_{ijkl} u_{ij} u_{kl}^* (A_k^* \rho_1 A_i^\mathrm{T}) \otimes (B_j \rho_2 B_l^\dagger)$$

$$= \mathrm{PT}_2 \{ \sum_{ijkl} u_{ij} u_{kl}^* (A_k^* \rho_1 A_i^\mathrm{T}) \otimes (B_l^* {}^\mathrm{T} \rho_2^\mathrm{T} B_j^\mathrm{T}) \}$$

$$= \mathrm{PT}_2 \{ \sum_{ijkl} u_{kl}^* u_{ij} (A_k^* \otimes B_l^*)(\rho_1 \otimes \rho_2^\mathrm{T})(A_i^T \otimes B_j^T) \}$$

$$= \mathrm{PT}_2 U^* \mathrm{PT}_2 (\rho_1 \otimes \rho_2)$$

for arbitrary operators $\rho_1$ and $\rho_2$.

QED

## References

[1] S. Kochen and E. P. Specker. The problem of hidden variables in quantum mechanics. *Journal of Mathematics Mechanics* **17** (1967), 59.

[2] J. S. Bell. *Speakable and Unspeakable in Quantum Mechanics*. Cambridge: Cambridge University Press, 1987.

[3] A. J. Leggett. Nonlocal hidden-variable theories and quantum mechanics: An incompatibility theorem. *Foundations of Physics* **33** (2003), 1469.

[4] S. Gröblacher, T. Paterek, R. Kaltenbaek, et al. An experimental test of non-local realism. *Nature* **446** (2007), 871.

[5] I. Biaynicki-Birula and J. Mycielski. Nonlinear wave mechanics. *Annals of Physics* **100** (1976), 62.

[6] A. Shimony. Proposed neutron interferometer test of some nonlinear variants of wave mechanics. *Physical Review A* **20** (1979), 394.

[7] C. G. Shull, D. K. Atwood, J. Arthur, et al. Search for a nonlinear variant of the Schrödinger equation by neutron interferometry. *Physical Review Letters* **44** (1980), 765.

[8] R. Gähler, A. G. Klein, and A. Zeilinger. Neutron optical tests of nonlinear wave mechanics. *Physical Review A* **23** (1981), 1611.

[9] G. C. Ghirardi, A. Rimini, and T. Weber. Unified dynamics for microscopic and macroscopic systems. *Physical Review D* **34** (1986), 470.

[10] F. Károlyházy. *Nuovo Cimento* **42** (1966), 390.

[11] F. Károlyházy. Gravitation and quantum mechanics of macroscopic bodies (Thesis, in Hungarian). *Magyar Fizikai Folyóirat* **22** (1974), 23.

[12] L. Diosi. Models for universal reduction of macroscopic quantum fluctuations. *Physical Review A* **40** (1989), 1165.

[13] R. Penrose. On gravity's role in quantum state reduction. *General Relativity and Gravitation* **28** (1996), 581.

[14] P. Pearle. Reduction of the state vector by a nonlinear Schrödinger equation. *Physical Review D* **13** (1976), 857.

# REFERENCES

[15] H. Barnum, J. Barrett, M. Leifer, et al. Cloning and broadcasting in generic probabilistic models (2006). arXiv:quant-ph/061129.

[16] H. Barnum, J. Barrett, M. Leifer, et al. A general no-cloning theorem. *Physical Review Letters* **99** (2007), 240501.

[17] J. Barrett. Information processing in general probabilistic theories. *Physical Review A.* **75** (2007), 032304.

[18] S. Popescu and D. Rohrlich. Quantum nonlocality as an axiom. *Foundations of Physics* **24** (1994), 379.

[19] G. W. Mackey. Quantum mechanics and Hilbert space. *American Mathematical Monthly* **64** (1957), 45.

[20] L. Hardy. Quantum theory from five reasonable axioms (2001). arXiv:quant-ph/0101012.

[21] G. M. Dariano. Probabilistic theories: What is special about quantum mechanics? In *Philosophy of Quantum Information and Entanglement*, eds. A. Bokulich and G. Jaeger. Cambridge: Cambridge University Press, 2010.

[22] P. Goyal, K. H. Knuth, and J. Skilling. Origin of complex quantum amplitudes and Feynman's rules (2009). arXiv:quant-ph/0907.0909.

[23] C. Rovelli. Relational quantum mechanics. *International Journal of Theoretical Physics* **35** (1996), 1637.

[24] A. Zeilinger. A foundational principle for quantum mechanics. *Foundations of Physics* **29** (1999), 631.

[25] Č. Brukner and A. Zeilinger. Information and fundamental elements of the structure of quantum theory. In *Time, Quantum, Information*, eds. L. Castell and O. Ischebeck. Springer, 2003.

[26] Č. Brukner and A. Zeilinger. Information invariance and quantum probabilities. *Foundations of Physics* **39** (2009), 677.

[27] R. Clifton, J. Bub, and H. Halvorson. Characterizing quantum theory in terms of information-theoretic constraints. *Foundations of Physics* **33**(11) (2003), 1561.

[28] A. Grinbaum. Elements of information-theoretic derivation of the formalism of quantum theory. *International Jouranal Quantum Information* **1**(3) (2003), 289.

[29] W. K. Wootters. Statistical distance and Hilbert space. *Physical Review D* **23** (1981), 357.

[30] D. I. Fivel. How interference effects in mixtures determine the rules of quantum mechanics. *Physical Review A* **59** (1994), 2108.

[31] J. Summhammer. Maximum predictive power and the superposition principle. *International Journal of Theoretical Physics* **33** (1994), 171.

[32] J. Summhammer. Quantum theory as efficient representation of probabilistic information (2007). arXiv:quant-ph/0701181.

[33] A. Bohr and O. Ulfbeck. Primary manifestation of symmetry: Origin of quantal indeterminacy. *Reviews of Modern Physics* **67** (1995), 1.

[34] A. Caticha. Consistency, amplitudes and probabilities in quantum theory. *Physical Review A* **57** (1998), 1572.

[35] C. A. Fuchs. Quantum mechanics as quantum information (and only a little more). In *Quantum Theory: Reconstruction of Foundations*, ed. A. Khrenikov. Växjo: Växjo, Sweden University Press, 2002.

[36] C. A. Fuchs and R. Schack. Quantum-Bayesian coherence (2009). arXiv:quant-ph/0906.2187.

[37] P. Grangier. Contextual objectivity: A realistic interpretation of quantum mechanics. *European Journal of Physics* **23** (2002), 331.

[38] P. Grangier. Contextual objectivity and the quantum formalism. *International Journal Quantum Information* **3** (2005), 17–22.

[39] S. Luo. Maximum Shannon entropy, minimum Fisher information, and an elementary game. *Foundations of Physics* **32** (2002), 1757.

[40] R. Spekkens. Evidence for the epistemic view of quantum states: A toy theory. *Physical Review A* **75** (2007), 032110.

[41] P. Goyal. Information-geometric reconstruction of quantum theory. *Physical Review A* **78** (2008), 052120.

[42] W. van Dam. Implausible consequences of superstrong nonlocality (2005). arXiv:quant-ph/0501159.

[43] G. Brassard, H. Buhrman, N. Linden, et al. A limit on nonlocality in any world in which communication complexity is not trivial. *Physical Review Letters* **96** (2006), 250401.

[44] M. Pawlowski, T. Paterek, D. Kaszlikowski, et al. A new physical principle: Information causality. *Nature* **461** (2009), 1101.

[45] M. Navascues and H. Wunderlich. A glance beyond the quantum model. *Proceedings of the Royal Society A* **466** (2009), 2115.

[46] K. Zyczkowski. Quartic quantum theory: An extension of the standard quantum mechanics. *Journal of Physics A* **41** (2008), 355302.

[47] T. Paterek, B. Dakić, and Č. Brukner. Theories of systems with limited information content. *New Journal of Physics* **12** (2010), 053037.

[48] A. Peres. Proposed test for complex versus quaternion quantum theory. *Physical Review Letters* **42** (1979), 683.

[49] H. Kaiser, E. A. George, and S. A. Werner. Neutron interferometric search for quaternions in quantum mechanics. *Physical Review A* **29** (1984), 2276.

[50] A. Peres. Quaternionic quantum interferometry. In *Quantum Interferometry*, eds. F. De Martini et al. VCH Publication, 1996, pp 431–437.

[51] R. D. Sorkin. Quantum mechanics as quantum measure theory. *Modern Physics Letters A* **9** (1994), 3119.

[52] U. Sinha, C. Couteau, Z. Medendorp, et al. Testing Born's Rule in quantum mechanics with a triple slit experiment. *AIP Conference Proceedings* **1101** (2009), 200–207.

[53] A. Grinbaum. Reconstruction of quantum theory. *British Journal for the Philosophy of Science* **8** (2007), 387.

[54] D. Gross, M. Mueller, R. Colbeck, et al. All reversible dynamics in maximally non-local theories are trivial (2009). arXiv:quant-ph/0910.1840.

[55] C. F. von Weizsäcker. *Aufbau der Physik*. Carl Hanser, München, 1958.

[56] J. A. Wheeler. Law without law in quantum theory and measurement. In *Quantum Theory and Measurement*, eds. J. A. Wheeler and W. H. Zurek. Princeton, NJ: Princeton University Press, 1983, p. 182.

[57] H. Barnum and A. Wilce. Information processing in convex operational theories. arXiv:quant-ph/0908.2352.

[58] S. Abramsky and B. Coecke. A categorical semantics of quantum protocols. *Logic in Computer Science, Symposium on* **0** (2004), 415–25.

[59] J. Ambjorn, J. Jurkiewicz, and R. Loll. Reconstructing the universe. *Physical Review D* **72** (2005), 064014.

[60] H. Boerner. *Representations of groups*. Amsterdam: North-Holland Publishing Company, 1963.

[61] W. K. Wooters and B.D. Fields. Optimal state-determination by mutually unbiased measurements. *Annals of Physics (N.Y.)* **191** (1989), 363.

[62] I. D. Ivanovic. Geometrical description of quantal state determination. *Journal of Physics A* **14** (1981), 3241.

[63] D. Bohm. A suggested interpretation of the quantum theory in terms of "hidden variables" I. *Physical Review* **85** (1952), 166.

[64] D. Bohm. A suggested interpretation of the quantum theory in terms of "hidden variables" II. *Physical Review* **85** (1952), 180.

[65] L. Hardy. Quantum ontological excess baggage. *Studies in History and Philosophy of Modern Physics* **35** (2004), 267.

[66] A. Montina. Exponential complexity and ontological theories of quantum mechanics. *Physical Review A* **77** (2008), 022104.

[67] B. Dakić, M. Suvakov, T. Paterek, et al. efficient hidden-variable simulation of measurements in quantum experiments. *Physical Review Letters* **101** (2008), 190402.

[68] Č. Brukner and A. Zeilinger. Young's experiment and the finiteness of information. *Philosophical Transactions of the Royal Society of London A* **360** (2002), 1061.

[69] E. Schrödinger. Discussion of probability relations between separated systems, *Proceedings of the Cambridge Philosophical Society* **31** (1935), 555–63; **32** (1936), 446–51.

# CHAPTER 10

# Is Von Neumann's "No Hidden Variables" Proof Silly?

Jeffrey Bub

## 10.1 Introduction

In his *Mathematical Foundations of Quantum Mechanics* [11, Chapter 4], John von Neumann presented a proof that the quantum statistics cannot be recovered from probability distributions over "hidden" deterministic states that assign definite premeasurement values to all physical quantities.[1] Von Neumann's proof has been dismissed as "silly" by John Bell and by David Mermin, who writes [8, pp. 805–6]:

> Many generations of graduate students who might have been tempted to try to construct hidden-variables theories were beaten into submission by the claim that von Neumann, 1932, had proved that it could not be done. A few years later (see Jammer, 1974, p. 273) Grete Hermann, 1935, pointed out a glaring deficiency in the argument, but she seems to have been entirely ignored. Everybody continued to cite the von Neumann proof. A third of a century passed before John Bell, 1966, rediscovered the fact that von Neumann's no-hidden-variables proof was based on an assumption that can only be described as silly—so silly, in fact, that one is led to wonder whether the proof was ever studied by either the students or those who appealed to it to rescue them from speculative adventures.

A footnote refers to Bell's comments in an interview with *Omni*, May 1988, p. 88:

> Yet the von Neumann proof, if you actually come to grips with it, falls apart in your hands! There is *nothing* to it. It's not just flawed, it's *silly*! ... When you translate [his assumptions] into terms of physical disposition, they're nonsense. You may quote me on that: The proof of von Neumann is not merely false but *foolish*!

In this chapter, I reexamine von Neumann's proof and argue that it has been misunderstood by detractors; I look at the proof in more detail in Section 10.4. Here, I

---

[1] Von Neumann uses the term *physical quantity* or sometimes *measurable quantity*. The term in common use is *observable*. Observables are usually identified with their representative Hermitian operators, but in the following it is important to keep the two notions distinct. So I use von Neumann's term *physical quantity* and distinguish physical quantities (denoted by $\mathcal{A}, \mathcal{B}, \mathcal{S}_x$, and so on) from Hermitian operators (denoted by $A, B, \sigma_x$, and so on).

simply state the allegedly silly assumption: von Neumann assumed that the expectation value of a sum of physical quantities, $\text{Exp}(\mathcal{A} + \mathcal{B})$, is equal to the sum of the expectation values, $\text{Exp}(\mathcal{A}) + \text{Exp}(\mathcal{B})$, for all statistical ensembles—that is, for all expectation value functions. This is the case for simultaneously measurable physical quantities, represented by commuting Hermitian operators. As von Neumann remarks [11, p. 308], the relation "depends on this theorem on probability: the expectation value of a sum is always the sum of the expectation values of the individual terms, independent of whether probability dependencies exist between these or not (in contrast, for example, to the probability of the product)." For physical quantities that are not simultaneously measurable, represented by noncommuting operators, von Neumann justified the assumption on the grounds that the relation holds for quantum states. The objection is that it is unreasonable—in fact, "silly"—to impose the condition on hypothetical deterministic states, which assign probabilities 0 or 1 to the possible values of a physical quantity (the eigenvalues of the corresponding Hermitian operator) because for such states, the expectation value of a physical quantity is equal to the value with probability 1, and the eigenvalues of a sum of noncommuting operators are *not* equal to the sums of the eigenvalues of the summed operators. For a spin-1/2 particle, for example, the eigenvalues of $\sigma_x$ and $\sigma_z$ are both $\pm 1$, while the eigenvalues of $\sigma_x + \sigma_z$ are $\pm\sqrt{2}$, so the relation cannot hold.

Now, von Neumann was surely aware of this. In [11, pp. 309–10, footnote 164], he refers to the kinetic energy of an electron in an atom, which is a sum of two physical quantities represented by noncommuting operators, a function of momentum, $\mathcal{P}$, and a function of position, $\mathcal{Q}$:

$$\mathcal{E} = \frac{\mathcal{P}^2}{2m} + V(\mathcal{Q}),$$

and he observes that we measure the energy by measuring the frequency of the spectral lines in the radiation emitted by the electron, not by measuring the electron's position and momentum, computing the values of $\mathcal{P}^2/2m$ and $V(\mathcal{Q})$, and adding the results.

Before turning to von Neumann's proof, I first consider—from a slightly different perspective than usual (a variation on the analysis in [3])—the two classic "no hidden variables" proofs that followed von Neumann's proof: the Kochen–Specker "no noncontextual hidden variables" proof and Bell's "no local hidden variables" proof. Following that, we consider whether von Neumann's proof has been misunderstood or whether it is silly, as Bell and Mermin claim.

## 10.2 The Kochen–Specker Proof: Klyachko Version

Can the quantum statistics for all quantum states and all physical quantities, including all correlations among physical quantities, be recovered from probability distributions over "hidden" deterministic states that assign definite premeasurement values to all physical quantitities? If the question concerns simply the existence, in principle, of such hypothetical deterministic states, and no constraints are imposed on the value assignments (e.g., that functional relations among physical quantities should be preserved by the value assignments), then the answer is "yes." As Werner and Wolf observe [12, p. 5], deterministic states are defined—trivially—by the correlation data for any experiment involving any number of physical quantities. Of course, such deterministic states are wildly context-dependent in the sense that they are defined relative to the physical quantities associated with an experimental context (so one could take the full probability space as the product space of the probability spaces for each experimental context, treating the probability spaces for different experimental contexts as independent), and these hypothetical deterministic states are indeed "hidden" until we perform the measurements, but this is irrelevant to the question of the existence of such states as a mathematical possibility.

The question about hidden deterministic states becomes interesting only if some constraint is imposed on the representation of physical quantities. Kochen and Specker require each physical quantity to be assigned a definite value by a deterministic state, equal to an eigenvalue of the representative Hermitian operator, and impose the constraint that any functional relation satisfied by a set of physical quantities represented by *commuting* operators is also satisfied by the eigenvalues assigned to these physical quantities by a deterministic state.

To see the implications of this constraint, which is significant for physical quantities represented by Hermitian operators on a Hilbert space of more than two dimensions, consider the physical quantity $\mathcal{S}_x$, spin in the $x$-direction of a spin-1 particle, with three possible values: $-1, 0, +1$, represented by the Pauli operator $\sigma_x$. The eigenvectors of $\sigma_x$, $|-1\rangle_x, |0\rangle_x, |+1\rangle_x$ form a basis or orthogonal triple in a three-dimensional Hilbert space. The 0-eigenvectors of the spin operators in any three orthogonal directions $x, y, z$ form an orthogonal triple and, hence, correspond to a Hermitian operator $R$. The operator $R$, which does not commute with $\sigma_x$, represents a maximal physical quantity $\mathcal{R}$ with three possible values. Now, even though $\sigma_x, \sigma_y, \sigma_z$ do not commute, the squares of these operators commute pairwise, and they all commute with $R$. It follows that the nonmaximal physical quantity $\mathcal{S}_x^2$, represented by the operator $\sigma_x^2$ with eigenvalues 0 and 1, is a function of $\mathcal{S}_x$ and of $\mathcal{R}$. In fact, $\sigma_x^2 = 1 - P_{|0\rangle_x}$, where $P_{|0\rangle_x}$ is the projection operator onto $|0\rangle_x$, the 0-eigenvector of $\sigma_x$, and 1 here represents the identity operator. For quantum states, measuring $\mathcal{S}_x^2$ via $\mathcal{S}_x$ will yield the same probabilities for the values of $\mathcal{S}_x^2$ as measuring $\mathcal{S}_x^2$ via $\mathcal{R}$, even though there is no measurement context in which all three physical quantities can be measured simultaneously (i.e., $\sigma_x^2$ commutes with $\sigma_x$ and with $R$, but $\sigma_x$ does not commute with $R$). The Kochen–Specker functional relation constraint requires that the value assigned by a deterministic state to a nonmaximal physical quantity such as $\mathcal{S}_x^2$ should be the same value whether $\mathcal{S}_x^2$ is considered to be a function of $\mathcal{S}_x$ or a function of $\mathcal{R}$. Because the value is the *measured* value, the constraint requires that the measured value of $\mathcal{S}_x^2$ should be the same whether $\mathcal{S}_x^2$ is assigned a value via a measurement of $\mathcal{S}_x$ or via a very different measurement of $\mathcal{R}$. That is, the constraint requires that $\mathcal{S}_x^2$ has a definite value independent of the measurement context. Note that this is a counterfactual claim, because the two measurements cannot be performed simultaneously and is a claim that is rejected by the standard interpretation of quantum mechanics. The question posed by Kochen and Specker is whether a hidden variable theory subject to this "noncontextuality" constraint can be excluded on the basis of a mathematical argument, not merely as a matter of accepted doctrine.

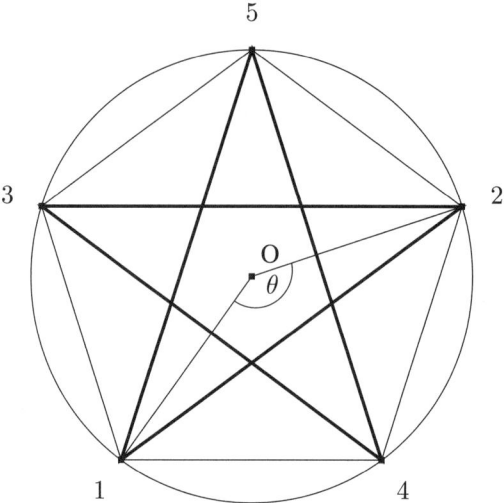

**Figure 10.1.** Circle $\Sigma_1$ with inscribed pentagram.

Kochen and Specker identified a finite set of 2-valued physical quantities represented by noncommuting one-dimensional projection operators on a three-dimensional Hilbert space, in which an individual projection operator can belong to different orthogonal triples of projection operators representing different bases or contexts, such that no assignment of 0 and 1 values to the physical quantities is possible that is both (1) noncontextual (i.e., each physical quantity is assigned one and only one value, independent of context); and (2) satisfies the functional relations for simultaneously measurable physical quantities, which amounts to respecting the orthogonality relations for commuting projection operators (so that, e.g., the assignment of the value 1 to a projection operator $P$ requires the assignment of 0 to any projection operator orthogonal to $P$).

This result is actually stronger than we need here. In fact, a much simpler proof suffices if all we want to show is that the quantum probabilities of at least one quantum state cannot be recovered from probability distributions over hypothetical deterministic states that assign values to all relevant physical quantities noncontextually. Here is a proof first proposed by Klyachko [6], involving just five one-dimensional projection operators on a three-dimensional Hilbert space.

Consider a unit sphere and imagine a circle $\Sigma_1$ on the equator of the sphere with an inscribed pentagon and pentagram, with the vertices of the pentagram labeled in order 1, 2, 3, 4, 5 (Figure 10.1).[2] Note that the angle subtended at the center $O$ by adjacent vertices of the pentagram defining an edge (e.g., 1 and 2) is $\theta = 4\pi/5$, which is greater than $\pi/2$. It follows that if the radii linking $O$ to the vertices are pulled upward toward the north pole of the sphere, the circle with the inscribed pentagon and pentagram will move up on the sphere toward the north pole. Because $\theta = 0$ when the radii point to the north pole (and the circle vanishes), $\theta$ must pass through $\pi/2$ before the radii point

---

[2] The following formulation of Klyachko's proof owes much to a discussion with Ben Toner and differs from the analysis in [6, 7].

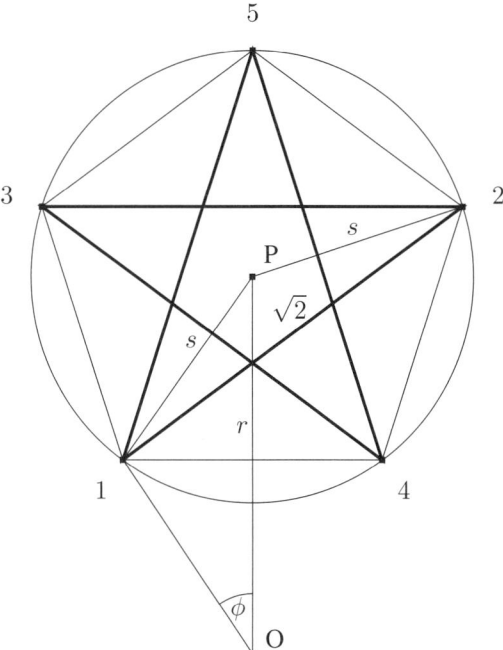

**Figure 10.2.** Circle $\Sigma_2$ with inscribed pentagram.

to the north pole, which means that it is possible to draw a circle $\Sigma_2$ with an inscribed pentagon and pentagram on the sphere at some point between the equator and the north pole *such that the angle subtended at O by an edge of the pentagram is $\pi/2$*. Label the center of this circle $P$ (Figure 10.2; note that the line OP is orthogonal to the circle $\Sigma_2$ and is not in the plane of the pentagram).

Now consider measuring the spins in the directions corresponding to the vertices of the pentagram/pentagon on the circle $\Sigma_2$ and assigning 1's and 0's to these vertices noncontextually, according to whether a "yes" (1) or "no" (0) answer is given to the question $\mathcal{Q}$: "Is the value of the spin in the direction $i$ equal to zero?" Note that at most one 1 can be assigned to two vertices defining an edge of the pentagram noncontextually because these vertices are orthogonal. As we saw previously, the spin-0 eigenvectors for any three orthogonal directions in real space form an orthogonal triple in Hilbert space and hence define a physical quantity $\mathcal{R}$. Because a deterministic state assigns these physical quantities definite values, it follows that the answer to the question for a given deterministic state cannot be "yes" for two orthogonal directions. Here, noncontextuality amounts to the requirement that the assignment of a 1 or a 0 to a vertex is independent of the pentagram edge to which the vertex belongs.

It is obvious by inspection that the constraint on 0, 1 assignments for pentagram edges can be satisfied noncontextually by deterministic states that assign zero 1's, one 1, or two 1's to all five vertices (but not by three 1's, four 1's, or five 1's). It follows that for such deterministic states, where $v(i)$ is the value assigned to the vertex $i$:

$$\sum_{i=1}^{5} v(i) \leq 2. \tag{10.1}$$

If we label the possible deterministic states with a hidden variable $\lambda \in \Lambda$ and average over $\Lambda$, then the probability that a vertex is assigned the value 1 is given by

$$p(v(i) = 1) = \sum_{\Lambda} v(i|\lambda) p(\lambda) \tag{10.2}$$

so

$$\sum_{i=1}^{5} p(v(i) = 1) \leq 2. \tag{10.3}$$

This is the Klyachko inequality: the sum of the probabilities assigned to the vertices of the pentagram on the circle $\Sigma_2$ must be less than or equal to 2 if values are assigned noncontextually by deterministic states in such a way as to satisfy the constraint that two 1's cannot be assigned to the vertices defining a pentagram edge. Note that the inequality follows without any assumption about the relative weighting of the deterministic states.

Now note that the spin measurement directions defined by the vertices of the pentagram/pentagon in real space correspond to spin states on the unit sphere in a three-dimensional Hilbert space. Consider a quantum system in the state defined by a unit vector that passes through the north pole of this unit sphere, which we label similarly to the sphere in real space. This vector passes through the point $P$ in the center of the circle $\Sigma_2$. Call this state $|\psi\rangle$. A simple geometric argument shows that if probabilities are assigned to the answer "yes" to the question $Q$ for the directions defined by the vertices of the pentagram on $\Sigma_2$ by the state $|\psi\rangle$, then the sum of the probabilities is greater than 2!

To see this, note that the probability assigned to a vertex (e.g., the vertex 1) is

$$|\langle 1|\psi\rangle|^2 = \cos^2 \phi, \tag{10.4}$$

where $|1\rangle$ is the unit vector defined by the radius from $O$ to the vertex 1. Because the lines from the center $O$ of the sphere to the vertices of an edge of the pentagram on $\Sigma_2$ are radii of length 1 subtending a right angle, each edge of the pentagram has length $\sqrt{2}$. The angle subtended at $P$ by the lines joining $P$ to the two vertices of an edge is $4\pi/5$, so the length, $s$, of the line joining $P$ to a vertex of the pentagram is

$$s = \frac{1}{\sqrt{2} \cos \frac{\pi}{10}}. \tag{10.5}$$

Now, $\cos \phi = r$, where $r$ is the length of the line $OP$, and $r^2 + s^2 = 1$, so

$$\cos^2 \phi = r^2 = 1 - s^2 = \frac{\cos \frac{\pi}{5}}{1 + \cos \frac{\pi}{5}} = 1/\sqrt{5} \tag{10.6}$$

(because $\cos \pi/5 = \frac{1}{4}(1 + \sqrt{5})$), and so

$$\sum_{i=1}^{5} p(v(i) = 1) = 5 \times 1/\sqrt{5} = \sqrt{5} > 2. \tag{10.7}$$

So, the probabilities defined by the quantum state $|\psi\rangle$ cannot be recovered from probability distributions over hypothetical deterministic states that assign definite premeasurement values noncontextually to the spin physical quantities associated with the directions defined by the vertices of the pentagram/pentagon.[3]

## 10.3 Bell's Proof

Bell proposed weakening the noncontextuality constraint to allow a certain constrained contextuality, where the constraint is physically motivated by a locality assumption for bipartite systems **AB**. For deterministic states, the locality constraint is the requirement that the value assigned to an **A**-quantity $\mathcal{X}$ by a deterministic state is independent of the remote measurement context defined by the measurement of a **B**-quantity, $\mathcal{Y}$.[4]

Note that there is no loss of generality in considering deterministic states rather than states that assign values stochastically. Werner and Wolf [12, p. 7] point out that one can always "upgrade" a stochastic hidden variable theory to a deterministic theory by the following trick. Suppose $\lambda$ is the hidden variable parametrizing the hidden stochastic states, which satisfy Bell's locality condition,

$$p(x, y | \mathcal{X}, \mathcal{Y}) = \int p_\lambda(x|\mathcal{X}) p_\lambda(y|\mathcal{Y}) \rho(\lambda) d\lambda; \qquad (10.8)$$

that is, $p_\lambda(x|\mathcal{X})$ is the probability, in the hidden state $\lambda$, of obtaining the outcome $x$ in a measurement of the physical quantity $\mathcal{X}$ on system **A** and similarly for $\mathcal{Y}$ and system **B**. Introduce a new hidden variable $\tilde{\lambda} = (\lambda, \lambda_\mathbf{A}, \lambda_\mathbf{B})$, where $\lambda, \lambda_\mathbf{A}, \lambda_\mathbf{B}$ are independent random variables, and $\lambda_\mathbf{A}, \lambda_\mathbf{B}$ are uniformly distributed over $[0, 1]$.[5] One can then define a deterministic state $\tilde{\lambda}$ via

$$p_{\tilde{\lambda}}(x|\mathcal{X}) \equiv p_{\lambda,\lambda_\mathbf{A},\lambda_\mathbf{B}}(x|\mathcal{X}) = \begin{cases} 1 & \lambda_\mathbf{A} \leq p_\lambda(x|\mathcal{X}) \\ 0 & \text{otherwise} \end{cases} \qquad (10.9)$$

and similarly for **Y**. This deterministic theory produces the same correlations as the stochastic theory satisfying Equation (10.8).

Bell's locality condition is a probabilistic noncontextuality constraint. As shown by Jarrett [5], Equation (10.8) is equivalent to the conjunction of two independent conditions: the probability, conditional on $\lambda$, that an **A**-quantity, $\mathcal{X}$, takes the value $x$ is (I) independent of the quantity $\mathcal{Y}$ measured at **B** and also (II) independent of the outcome of a $\mathcal{Y}$-measurement at **B** (and conversely). Note that (I) is not the same as the "no signaling" condition and that the marginal probabilities of measurement outcomes at **A** do not depend on the quantity measured at **B** (and conversely). The "no signaling" condition refers to "surface probabilities," whereas (I) refers to "hidden probabilities"

---

[3] More precisely, the relevant physical quantities are the five $\mathcal{S}_i^2$-quantities associated with the five vertices and the five $\mathcal{R}$-quantities associated with the five edges of the pentagram (i.e., the physical quantities represented by Hermitian operators whose eigenvectors are the spin-0 eigenvectors of $\sigma_x, \sigma_y, \sigma_z$, where $x, y$, and $z$ are orthogonal and two of these directions correspond to two vertices defining an edge of the pentagram).

[4] This is the condition of "parameter independence," introduced herein, for deterministic states. The condition of "outcome independence" is automatically satisfied for deterministic states; 0, 1 probabilities cannot be further refined by additional information.

[5] We can think of $\lambda_\mathbf{A}$ and $\lambda_\mathbf{B}$ as local random variables associated with the measuring devices for **A** and **B**.

(to use a terminology due to van Fraassen [10]). Shimony [9] calls conditions (I) and (II) "parameter independence" and "outcome independence," respectively.

Consider now a pair of spin-1 systems in a maximally entangled correlated state of the form:

$$|\Psi\rangle = \frac{1}{\sqrt{3}} \sum_{i=1}^{3} |a_i\rangle|b_i\rangle. \qquad (10.10)$$

The state $|\Psi\rangle$ has the form of a biorthogonal representation with equal coefficients, which is the same for any basis. In a local hidden variable simulation of the correlations, local noncontextuality is forced by the requirement of perfect correlation for the outcomes of spin measurements in the same direction on **A** and **B**. That is, the answers to the question $Q$ for a pair of spin directions, one for **A** and one for **B**, cannot be locally contextual: the **A**-answer cannot depend on the edge to which the vertex defining the direction for the **A**-question belongs (and similarly for the **B**-question) because the **A**-answer (e.g., for vertex 1) must be the same for two **B**-directions associated with vertices defining two different edges of the pentagram (1-2 and 1-5), and the answers for the maximal contexts defined by these edges are perfectly correlated for **A** and for **B**. Note that local noncontextuality for **A** and for **B** follows from locality and perfect correlation while this is an assumption in the Klyachko version of the Kochen–Specker proof.

What is the probability that the spins are not both 0 for two directions corresponding to an edge of the *pentagon*?

Suppose that this probability is generated by a probability distribution over deterministic states that generates the correct quantum mechanical marginal probabilities of 1/3 for the possible spin values in any direction for **A** and for **B**. As we saw, a deterministic state can assign at most two 1's to the five vertices without violating the constraint that two vertices defining an edge of the pentagram cannot both be assigned a 1. For deterministic states $D_1$ that assign one 1 to the vertices, the probability of a vertex is 1/5, which is less than 1/3. So, any mixture of deterministic states yielding the marginal 1/3 would have to include deterministic states $D_2$ that assign two 1's to the vertices. It is easy to see that there is only one mixture, $M_{21}$, of deterministic states $D_2$ and $D_1$, and one mixture, $M_{20}$, of deterministic states $D_2$ and $D_0$ yielding a probability of 1/3 for each vertex[6]:

$M_{21}$: 2/3 $C_2$, 1/3 $C_1$
$M_{20}$: 5/6 $C_2$, 1/6 $C_0$

To obtain the probability 1/3, each particle pair in the sequence is labeled with a shared hidden variable whose value selects a particular deterministic state in the appropriate mixture, either $D_2$ or $D_1$ in the case of mixture $M_{21}$, or $D_2$ or $D_0$ in the case of mixture $M_{20}$, where the values occur in the sequence with probabilities corresponding to the mixture weights. Other mixtures yielding the marginal 1/3 involve appropriate mixtures of deterministic states of all three types: $D_2$, $D_1$, and $D_0$.

---

[6] The weights are the solutions to the linear equations $x \cdot \frac{2}{5} + y \cdot \frac{1}{5} = \frac{1}{3}$, $x + y = 1$ for the mixture $M_{21}$, and $x \cdot \frac{2}{5} + y \cdot 0 = \frac{1}{3}$, $x + y = 1$ for the mixture $M_{20}$.

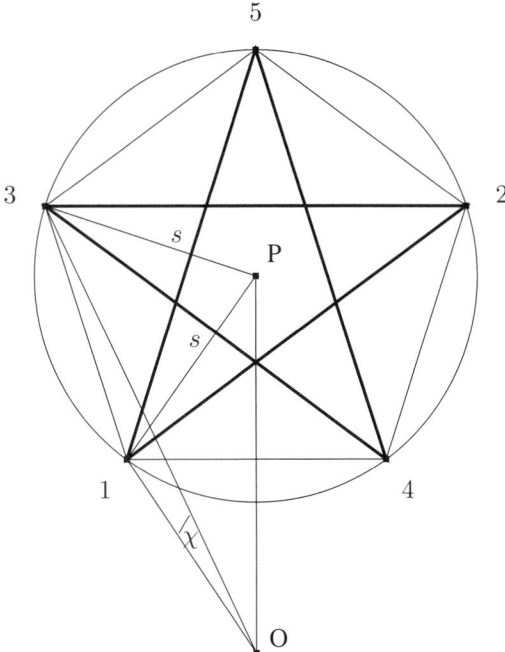

**Figure 10.3.** Pentagram on $\Sigma_2$ showing angle $\chi$ between states $|1\rangle$ and $|3\rangle$.

For the mixture $M_{21}$:

$$\text{prob(spins not both 0| pentagon edge directions)}_{M_1} = 1 - (\frac{2}{3} \times \frac{1}{5} + \frac{1}{3} \times 0)$$
$$\approx 0.8667 \qquad (10.11)$$

For the mixture $M_{20}$:

$$\text{prob(spins not both 0| pentagon edge directions)}_{M_2} = 1 - (\frac{5}{6} \times \frac{1}{5} + \frac{1}{6} \times 0)$$
$$\approx 0.8333 \qquad (10.12)$$

Because any mixture of deterministic states $D_2$, $D_1$, $D_0$ will yield probabilities between these values, it follows that the *optimal* probability for a local hidden variable theory is 0.8667.

Now consider the probability defined by the state $|\Psi\rangle$. We saw that the five directions defined by the vertices lie at an angle $\phi$ from the north pole, where $\cos^2 \phi = 1/\sqrt{5}$, and are symmetrically placed around the north pole, or around the point $P$ in the center of the pentagram/pentagon. (Therefore, $\phi \approx 48.03°$.) The angle between two vertices defining an edge of the pentagon (Figure 10.3) is $\chi$, where $\cos \chi = (\sqrt{5} - 1)/2$.[7] (Therefore, $\chi \approx 51.82°$.)

---

[7] This is the inverse of the golden ratio, the limit of the ratio of successive terms in the Fibonacci series: $\tau = \frac{\sqrt{5}+1}{2}$: $1/\tau = \tau - 1$.

To see this, note that:

$$\sin \frac{\chi}{2} = s \sin \frac{\pi}{5} = \frac{\sin \frac{\pi}{5}}{\sqrt{2} \cos \frac{\pi}{10}} = \sqrt{2} \sin \frac{\pi}{10} = \sqrt{2} \frac{\sqrt{5}-1}{4}. \quad (10.13)$$

It follows that the probability, in the state $|\Psi\rangle$, that the spins are not both zero for two directions corresponding to an edge of the pentagon is given by:

$$\text{prob(spins not both 0| pentagon edge directions)}_\Psi = 1 - \frac{1}{3} \cos^2 \chi$$

$$= 1 - \frac{1}{3}(\frac{\sqrt{5}-1}{2})^2$$

$$= \approx 0.8727. \quad (10.14)$$

Because this probability is greater than the optimal probability calculated on the basis of the locality assumption for probability distributions over hypothetical deterministic states, it follows that the quantum correlations defined by $|\Psi\rangle$ cannot be reproduced by such probability distributions.

## 10.4 Von Neumann's Proof

Von Neumann begins the discussion of the hidden variable question on p. 297 of [11]. He writes as follows:

> Let us forget the whole of quantum mechanics but retain the following. Suppose a system **S** is given, which is characterized for the experimenter by the enumeration of all the effectively measurable quantities in it and their functional relations with one another. With each quantity we include the directions as to how it is to be measured—and how its value is to be read or calculated from the indicator positions on the measuring instruments. If $\mathcal{R}$ is a quantity and $f(x)$ any function, then the quantity $f(\mathcal{R})$ is defined as follows: To measure $f(\mathcal{R})$, we measure $\mathcal{R}$ and find the value $a$ (for $\mathcal{R}$). Then $f(\mathcal{R})$ has the value $f(a)$. As we see, all quantities $f(\mathcal{R})$ ($\mathcal{R}$ fixed, $f(x)$ an arbitrary function) are measured simultaneously with $\mathcal{R}$. ... But it should be realized that it is completely meaningless to try to form $f(\mathcal{R}, \mathcal{S})$ if $\mathcal{R}, \mathcal{S}$ are not simultaneously measurable: there is no way of giving the corresponding measuring arrangement.

Note that von Neumann does not begin with quantum mechanics but rather with physical or measurable quantities satisfying certain functional relations abstracted from the statistics of measurements. Note also that von Neumann is, from the start, careful to distinguish simultaneously measurable quantities from quantities that are not simultaneously measurable.

Von Neumann's proof proceeds from six assumptions. Two assumptions [11, p. 312], $\alpha'$, $\beta'$, define the notion of a deterministic or "dispersion-free" expectation value function and a homogeneous or pure expectation value function, where an expectation value function, defined for all physical quantities of a system **S**, is understood as completely characterizing the statistical properties of a statistical ensemble of copies of **S**. An ensemble is dispersion free if, for each quantity $\mathcal{R}$, $\text{Exp}(\mathcal{R}^2) = (\text{Exp}(\mathcal{R}))^2$, so the probabilities are all 0 or 1, and homogeneous or pure if $\text{Exp}(\mathcal{R})$ cannot be expressed as a convex combination of two different expectation value functions, neither of them equal to Exp($\mathcal{R}$).

Two assumptions [11, p. 311], **A′**, **B′**, concern *physical quantities*:

**A′**: If the quantity $\mathcal{R}$ is by nature nonnegative, for example, if it is the square of another quantity $\mathcal{S}$, then Exp($\mathcal{R}$) ≥ 0.
**B′**: If $\mathcal{R}, \mathcal{S}, \ldots$ are arbitrary quantities and $a, b, \ldots$ are real numbers, then Exp($a\mathcal{R} + b\mathcal{S} + \cdots$) = $a$Exp($\mathcal{R}$) + $b$Exp($\mathcal{S}$)+.

Von Neumann emphasizes [11, p. 309] that if the quantities $\mathcal{R}, \mathcal{S}, \ldots$ are simultaneously measurable, then $a\mathcal{R} + b\mathcal{S} + \ldots$ is the ordinary sum, but if they are not simultaneously measurable, then "the sum is characterized... only in an implicit way." It is at this point that von Neumann mentions the example of energy as the sum of a momentum function and a position function. Clearly, assumption **B′** is to be taken as a *definition* of the quantity $a\mathcal{R} + b\mathcal{S} + \ldots$ when the quantities $\mathcal{R}, \mathcal{S}, \ldots$ are not simultaneously measurable—that is, as defining a quantity, $\mathcal{X} \equiv a\mathcal{R} + b\mathcal{S} + \ldots$, whose expectation value is the sum $a$Exp($\mathcal{R}$) + $b$Exp($\mathcal{S}$) + $\ldots$ in all statistical ensembles (i.e., for all expectation value functions).

The two remaining assumptions [11, pp. 313–14], **I**, **II**, relate *physical quantities* to Hilbert space *operators*. It is assumed that each physical quantity of a quantum mechanical system is represented by a hypermaximal Hermitian operator in a Hilbert space and that:

**I**: If the quantity $\mathcal{R}$ has the operator $R$, then the quantity $f(\mathcal{R})$ has the operator $f(R)$.
**II**: If the quantities $\mathcal{R}, \mathcal{S}, \ldots$ have the operators $R, S, \ldots$, then the quantity $\mathcal{R} + \mathcal{S} + \ldots$ has the operator $R + S + \ldots$.

On the basis of these assumptions, von Neumann proved that the expectation value function is uniquely defined by a trace function:

$$\text{Exp}(\mathcal{R}) = \text{Tr}(WR) \tag{10.15}$$

where $W$ is a Hermitian operator independent of $R$ that characterizes the ensemble. It follows that there are no dispersion-free ensembles and that there are homogeneous or pure ensembles corresponding to the pure states of quantum mechanics.

Von Neumann's proof characterizes the convex set of quantum probability distributions.[8] As von Neumann asserts, the existence of dispersion-free or deterministic states is inconsistent with assumption **II** for physical quantities that are not simultaneously measurable. For example, if the physical quantities $\mathcal{R} = \mathcal{P}^2/2m$ and $\mathcal{S} = V(\mathcal{Q})$ are represented by the noncommuting operators $R = P^2/2m$ and $S = V(Q)$, then the physical quantity $\mathcal{E} = \mathcal{R} + \mathcal{S} = \mathcal{P}^2/2m + V(\mathcal{Q})$ representing the energy of the system (whose value in a deterministic state is the sum of the values of $\mathcal{R}$ and $\mathcal{S}$) could not be represented by the operator $R + S = P^2/2m + V(Q)$.

In his classic article on hidden variables, Bell characterizes von Neumann's proof as follows [1, p. 449]:

---

[8] Gleason's theorem [4] establishes the same result on the basis of weaker assumptions: in effect, **II** is required to hold only for simultaneously measurable quantities.

Thus the formal proof of von Neumann does not justify his informal conclusion: "It is therefore not, as is often assumed, a question of reinterpretation of quantum mechanics—the present system of quantum mechanics would have to be objectively false, in order that another description of the elementary processes than the statistical one be possible." It was not the objective measurable predictions of quantum mechanics which ruled out hidden variables. It was the arbitrary assumption of a particular (and impossible) relation between the results of incompatible measurements either of which might be made on a given occasion but only one of which can in fact be made.

The "arbitrary assumption of a particular (and impossible) relation" is the assumption that the expectation value of a sum of physical quantities is the sum of the expectation values of the summed quantities, for quantities that are not simultaneously measurable, represented by noncommuting operators, as well as simultaneously measurable quantities represented by commuting operators (i.e., **B′** together with **II**). The quotation suggests that von Neumann concluded that "the objective measurable predictions of quantum mechanics"—that is, the quantum statistics—rule out hidden variables. This is misleading. In the comments immediately preceding this quotation, von Neumann writes [11, pp. 324–5]:

It should be noted that we need not go any further into the mechanism of the "hidden parameters," since we now know that the established results of quantum mechanics can never be re-derived with their help. In fact, we have even ascertained that it is impossible that the same physical quantities exist with the same function connections (i.e., that **I, II** hold) if other variables (i.e., "hidden parameters") should exist in addition to the wave function.

Nor would it help if there existed other, as yet undiscovered, physical quantities, in addition to those represented by the operators in quantum mechanics, because the relations assumed by quantum mechanics (i.e., **I, II**) would have to fail already for the by now known quantities, those that we discussed above. It is therefore not, as is often assumed, . . .

So the sense in which "the present system of quantum mechanics would have to be objectively false" is that *the representation of known physical quantities—like energy, position, momentum—in terms of Hermitian operators in Hilbert space would have to fail.*

According to Bell and Mermin, von Neumann proved only the impossibility of hidden deterministic states that assign values to a sum of physical quantities, $\mathcal{R} + \mathcal{S}$, that are the sums of the values assigned to the quantities $\mathcal{R}$ and $\mathcal{S}$, even when $\mathcal{R}$ and $\mathcal{S}$ cannot be measured simultaneously. As we saw, von Neumann regarded a sum of physical quantities that cannot be measured simultaneously as implicitly defined by the statistics, and he drew the conclusion that such an implicitly defined physical quantity cannot be represented by the operator sum in a hidden variable theory.

From Kochen and Specker, we learn that a hidden variable theory must be contextual, and from Bell that it must be nonlocal—both features of Bohm's hidden variable theory. It follows from von Neumann's argument too that a hidden variable theory must be like Bohm's theory in the sense that the physical quantities of the theory cannot be identified with the Hilbert space operators representing quantum "observables." For example, the momentum of a Bohmian particle is the rate of change of position, but the expectation value of momentum in a quantum ensemble is not derived by averaging over the particle momenta. Rather, the expectation value is derived via a theory of measurement, which yields the trace formula involving the momentum *operator* if we assume, as a contingent fact, that the probability distribution of hidden variables—particle positions—has reached equilibrium. Bohm [2, p. 387] gives an example of a free particle in a box of length $L$ with perfectly reflecting walls. Because the wave function is real, the particle is at rest. The kinetic energy of the particle is $E = P^2/2m = (nh/L)^2/2m$. Bohm asks: How can a particle with high energy be at rest in the empty space of the box? The solution to the puzzle is that a measurement of the particle's momentum changes the wave function, which plays the role of a guiding field for the particle's motion, in such a way that the *measured* momentum values will be $\pm nh/L$ with equal probability. Bohm comments [2, pp. 386–7, my italics]:

> This means that the measurement of an "observable" *is not really a measurement of any physical property belonging to the observed system alone.* Instead, the value of an "observable" measures only an incompletely predictable and controllable potentiality belonging just as much to the measuring apparatus as to the observed system itself.... We conclude then that this measurement of the momentum "observable" leads to the same result as is predicted in the usual interpretation. However, the actual particle momentum existing before the measurement took place is quite different from the numerical value obtained for the momentum "observable," which, in the usual interpretation, is called the "momentum."

What of Bell's counterexample to von Neumann's proof? Bell [1, p. 448] constructed a hidden variable schema for spin-1/2 systems—that is, for quantum systems whose states are represented on a two-dimensional Hilbert space, in which eigenvalues are assigned to the spins in all directions, given a quantum pure state $|\psi\rangle$ and a hidden variable. Averaging over the hidden variable yields the quantum statistics for the spins, as defined by the state $|\psi\rangle$.

One might imagine that von Neumann's response would have been something like the following:

> Your construction is clever, but as a counterexample to my analysis of the hidden variable question it is rather silly. You show that, for a quantum pure state and the value of a hidden variable, definite values can be assigned to spin physical quantities $S_x, S_z$ equal to the eigenvalues of the noncommuting Hermitian operators $\sigma_x, \sigma_z$ representing these quantities. There is then an implicitly defined physical quantity in my sense, $S = S_x + S_z$, whose value is simply the sum of the values of $S_x$ and $S_y$ so, trivially, $\text{Exp}(S) = \text{Exp}(S_x) + \text{Exp}(S_z)$ for all statistical ensembles—that is, for all expectation value functions calculated by taking the deterministic states as extremal (not via the trace formula for the representative operators, which takes the quantum pure states as extremal). In this case, it would seem quite unmotivated to take the sum of the Hermitian operators $\sigma_x, \sigma_z$, with eigenvalues $\pm\sqrt{2}$, as representing the physical quantity $S$. Rather, it would seem necessary to account for the applicability of the trace formula to the Hermitian operator $\sigma_x + \sigma_z$ in generating the quantum statistics of the physical quantity $S$, that is, the quantum statistics of the implicitly defined sum of the physical quantities $S_x$ and $S_z$ that are not simultaneously measurable—perhaps on the basis of speculations about the role of measurement in generating the quantum statistics.

Here, one imagines, von Neumann sketches a possible account along the lines of Bohm's theory, and continues:

But then the quantum statistics would be nothing more than an artefact of measurement for physical quantities introduced in some *a priori* way. This hardly makes sense if you understand the physical quantities and their functional relations as abstracted from the measurement statistics, and the quantum theory, including the quantum theory of measurement, as a theory about these physical quantities.

In fact, your counterexample implicitly adopts a position analogous to Lorentz's objection to Einstein's special theory of relativity. Lorentz wanted to maintain Euclidean geometry and Newtonian kinematics and explain the anomalous behavior of light in terms of a dynamical theory about how rods contract as they move through the ether. His dynamical theory plays a similar role, in showing how Euclidean geometry can be preserved, as Bohm's dynamical theory of measurement in explaining how the quantum statistics can be generated in a classical or Boolean theory of probability. My claim is that the quantum statistics precludes "hidden variables" in a similar sense to which special relativity precludes "hidden forces": they are both inconsistent with the pre-dynamic structure abstracted from measurement. In quantum mechanics, this pre-dynamic structure is given by the Hilbert space representation of physical quantities as Hermitian operators and by the trace formula for expectation values, which I showed follows from certain assumptions about the representation of physical quantities; in special relativity, it is given by Minkowski spacetime.

## Acknowledgments

Research was supported by the University of Maryland Institute for Physical Science and Technology.

## References

[1] J. S. Bell. On the problem of hidden variables in quantum mechanics. *Reviews of Modern Physics*, **38** (1966), 447–2. Reprinted in J. S. Bell. *Speakable and Unspeakable in Quantum Mechanics*. Cambridge: Cambridge University Press, 1989.

[2] D. Bohm. A suggested interpretation of quantum theory in terms of "hidden" variables. i and ii. *Physical Review*, **85** (1952), 166–93.

[3] J. Bub and A. Stairs. Contextuality and nonlocality in "no-signaling" theories. *Foundations of Physics*, **39** (2009), 690–711. arXiv:0903.1462v2 [quant-ph].

[4] A. N. Gleason. Measures on the closed sub-spaces of Hilbert spaces. *Journal of Mathematics and Mechanics*, **6** (1957), 885–93.

[5] J. Jarrett. On the physical significance of the locality conditions in the Bell arguments. *Noûs*, **18** (1984), 569–89.

[6] A. A. Klyachko. Coherent states, entanglement, and geometric invariant theory. arXiv quant-ph/0206012 (2002).

[7] A. A. Klyachko, M. A. Can, S. Binicioğlu, and A. S. Shumovsky. A simple test for hidden variables in spin-1 system. *Physical Review Letters*, **101** (2008), 020403–6.

[8] N. D. Mermin. Hidden variables and the two theorems of John Bell. *Reviews of Modern Physics*, **65** (1993), 803–15.

[9] A. Shimony. Aspects of nonlocality in quantum mechanics. In *Quantum Mechanics at the Crossroads*. Berlin: Springer, 2006.

[10] B. van Fraassen. The Charybdis of realism: Epistemological implications of Bell's inequality. *Synthese*, **52** (1982), 25–38.

[11] J. von Neumann. *Mathematical Foundations of Quantum Mechanics*. Princeton, NJ: Princeton University Press, 1955.

[12] R. F. Werner and M. M. Wolf. Bell inequalities and entanglement. *Quantum Information and Computation*, **1** (2001), 1–25.

CHAPTER 11

# Foliable Operational Structures for General Probabilistic Theories

Lucien Hardy

## 11.1 Preamble

In this chapter, a general mathematical framework for probabilistic theories of operationally understood circuits is laid out. Circuits consist of operations and wires. An *operation* is one use of an apparatus, and a *wire* is a diagrammatic device for showing how apertures on the apparatuses are placed next to each other. Mathematical objects are defined in terms of the circuit understood graphically. In particular, we do not think of the circuit as sitting in a background time. Circuits can be foliated by hypersurfaces composed of sets of wires. Systems are defined to be associated with wires. A closable set of operations is defined to be one for which the probability associated with any circuit built from this set is independent both of choices on other circuits and of extra circuitry that may be added to outputs from this circuit. States can be associated with circuit fragments corresponding to preparations. These states evolve on passing through circuit fragments corresponding to transformations. The composition of transformations is treated. A number of theorems are proven, including one that rules out quaternionic quantum theory. The case of locally tomographic theories (where local measurements on a systems components suffice to determine the global state) is considered. For such theories, the probability can be calculated for a circuit from matrices pertaining to the operations that constitute that circuit. Classical probability theory and quantum theory are exhibited as examples in this framework.

## 11.2 Introduction

Prior to Einstein's 1905 article [14] laying the foundations of special relativity, it was known that Maxwell's equations are invariant under the Lorentz transformations. Mathematically, the Lorentz transformations are rather complicated, and it must have been unclear why nature would choose these transformations over the rather more natural-looking Galilean transformations. Further, there was an understanding of the physical reasons for Galilean transformations in terms of boosts and the additive nature of

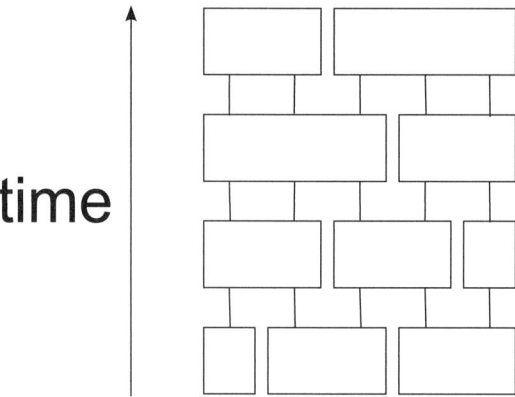

**Figure 11.1.** A naive picture of operationalism. Systems pass through boxes with respect to a background time.

velocities. We find ourselves in a similar situation today with respect to quantum theory. Regarded as a probabilistic theory, it is much more complicated from a mathematical point of view than the rather natural equations of classical probability theory. And further, we can motivate classical probability by ordinary reasoning by imagining that the probabilities pertain to some underlying mutually exclusive set of possibilities. The situation with respect to the Lorentz transformations was resolved by Einstein when he showed that they follow from two very reasonable postulates: that the laws of physics are the same in all reference frames and that the speed of light in vacuum is independent of the motion of the source. Once we see Einstein's reconstruction of the Lorentz transformations, we have a sense that we understand why, at a fundamental level, nature prefers these over the mathematically simpler Galilean transformations. We need something similar for quantum theory [16].

The subject of reconstructing quantum theory has seen something of a revival in the last decade [36]. Generally, to reconstruct quantum theory, we write down a set of basic axioms or postulates that are supposed to be well motivated. They should not appear unduly mathematical. Then, we apply these in the context of some framework for physical theories and show that we obtain quantum theory. This framework itself should be well motivated and may even follow from one or more of the given postulates.

The purpose of this chapter is to set up one such framework. This will be a framework for general probabilistic operational theories. There is a large body of literature on this (see Section 11.3). To construct the mathematics of such a framework, we must first specify what we mean by our *operational structure*. Only then can we add probability. The mathematics associated with the part of this where we add probability has become fairly sophisticated. However, a fairly simple-minded point of view is usually taken with respect to setting up the operational structure on which the whole endeavor is founded. The picture normally adopted is of a system passing sequentially through various boxes representing operations or, more generally, of many systems where, at any given time, each system passes through a box with, possibly, the same box acting on more than one system at once (Figure 11.1). This simple picture is problematic for various reasons. First, there is no reason why the types and number of systems going into a box are equal to the types and number emerging. Second, the notion of system itself is not fully operational. Third (and most significantly), this circuit is understood as being embedded in a background Newtonian time, and this constitutes structure in addition to a purely graphical interpretation of the diagram (it matters how high on the page the box is placed because this corresponds to the time at which the operation happens).

To deal with these three points, we set up a more general framework where (1) we allow the number and types of systems going into a box to be different from the number and types of systems going out, (2) give an operational definition of the notion of system, and (3) define our temporal concepts entirely in terms of the graphical information in the diagram. This third point gives rise to a natural notion of spacelike hypersurface in such a way that multiple foliations are possible. Hence, we call this a *foliable operational structure*.

It is worth being careful to formulate the operational structure well because such structures form a foundation for general probabilistic theories. Different operational structures can lead to different probabilistic frameworks. Once we have an operational structure, we can introduce probabilities. We then proceed along a fairly clear path introducing the notions of preparation, transformation, measurement, and associating mathematical objects with these that allow probabilities to be calculated. This gives a example of how an operational framework can be a foundation for a general probabilistic theory. The foliable framework presented here is sufficient for the formulation of classical probability theory, quantum theory, and potentially many theories beyond.

However, the foliable operational structure still, necessarily, has a notion of definite causal structure—when a system passes between two boxes that corresponds to a timelike separation (or null in the case of photons). We anticipate that a theory of quantum gravity will be a probabilistic theory with indefinite causal structure. If this is true, then we need a more general framework than the one presented here for quantum gravity. Preliminary ideas along this line have been presented in [24]. In future work, it will be shown how the approach taken in this chapter can be generalized to theories without definite causal structure—that is, nonfoliable theories. First, we must specify a sufficiently general nonfoliable operational structure and then add probabilities (see [26] for an outline of these ideas).

## 11.3 Related Work

The work presented here is a continuation of work initiated by the author in [23] in which a general probabilistic framework, sometimes called the **r-p** framework (because these vectors represent effects and states), was obtained for the purpose of reconstructing quantum theory from some simple postulates. In [24, 25], the author adapted the **r-p** for the purpose of describing a situation with indefinite causal structure to obtain a general probabilistic framework that might be suitable for a theory of quantum gravity. The idea that states should be represented by joint rather than conditional probabilities used in these papers is also adopted here. Preliminary versions of the work presented here can be seen in [26, 27].

In this chapter, we consider arbitrary foliations of circuits. Thus, we take a more spacetime-based approach. There are many other spacetime-based approaches in the context of a discrete setting in the literature (particularly work on quantum gravity). Sorkin builds a discrete model of spacetime based on causal sets [38]. Work in the consistent (or decoherent) histories tradition [17, 18, 28, 33] takes whole histories as the basic objects of study. A particularly important and influential piece of work is the quantum causal histories approach developed by Markopoulou [32]. In this, completely positive maps are associated with the edges of a graph with Hilbert spaces living on the vertices. Blute, Ivanov, and Panangaden [9] (see also [34]), motivated by Markopoulou's work, took the dual point of view with systems living on the edges (wires) and completely positive maps on the vertices. The work of Blute et al., although restricted to quantum theory rather than general probabilistic theories, bears much similarity with the present work. In particular, similar notions of foliating circuits are to be found in that paper. Leifer [29] has also done interesting work concerning the evolution of quantum systems on a causal circuit.

Abramsky and Coecke showed how to formulate quantum theory in a category theoretic framework [1] (see also [11]). This leads to a rich and beautiful diagramatic theory in which many essential aspects of quantum theory can be understood in terms of simple manipulations of diagrams. The diagrams can be understood operationally. Ideas from that work are infused into the present approach. Indeed, in category theoretic terms, the diagrams in this work can be understood as symmetric monoidal categories.

The **r-p** framework in [23] is actually a simple example of a framework for general probability theories going back originally to Mackey [31] and has been worked on (and often rediscovered) by many others since including Ludwig [30], Davies and Lewis [13], Araki [2], Gudder et al. [21], and Foulis and Randall [15].

Barrett elaborated on **r-p** framework in [8]. He makes two assumptions: that local operations commute and that local tomography is possible (whereby the state of a composite system can be determined by local measurements). In this chapter, we do not make either assumption. The first assumption, in any case, would have no content because we are interested in the graphical information in a circuit diagram and interchanging the relative height of operations does not change the graph. Under these assumptions, Barrett showed that composite systems can be associated with a tensor product structure. We recover this here for the special case when we have local tomography, but the more general case is also studied. In his article, Barrett shows that some properties that are thought to be specific to quantum theory are actually properties of any nonclassical probability theory.

More examples of this nature are discussed in various papers by Barrett, Barnum, Leifer, and Wilce in [3–5] and in [6, 7], the general probability framework is further developed.

The assumption of local tomography is equivalent to the assumption that $K_{ab} = K_a K_b$, where $K_{ab}$ is the number of probabilities needed to specify the state of the composite system $ab$ and $K_a$ ($K_b$) is the number needed to describe system $a$ ($b$) alone (this is the content of Theorem 5 herein). Theories having this property were investigated by Wootters [39] (see also [40]) in 1990, who showed that they are consistent with the relation $K_a = N_a^r$, where $N_a$ is the number of states that can be distinguished in a single-shot measurement (this was used in [23] as part of the axiomatic structure).

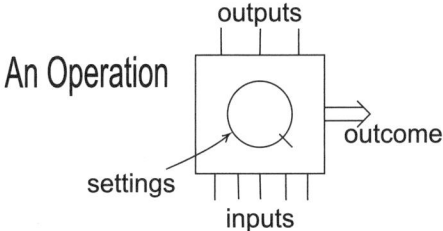

**Figure 11.2.** An operation having knob settings, measurement outcomes, and inputs (at the bottom of box) and outputs (at the top).

In 1994, Popescu and Rohrlich [37] exhibited correlations that maximally violate Bell's inequality but do not permit signalling. These correlations are more nonlocal than quantum correlations. Barrett asked which principles would be required to restrict such no-signalling correlations to the quantum limit [8]. Pawlowski et al. [35] have shown that the Popescu–Rohrlich correlations (and, indeed, any correlations more nonlocal than quantum theory) violate a natural principle they call the *information causality principle*. And Gross et al. [20] have shown, as speculated by Barrett [8], that the dynamics in any theory allowing Popescu–Rohrlich correlations are trivial.

Another line of work in this type of framework has been initiated by D'Ariano in [12], who has a set of axioms from which he obtains quantum theory. A recent report by Chiribella, D'Ariano, and Perinotti [10] set up a general probabilistic framework also having the local tomography property. Like Abramsky, Coecke, and co-workers, Chiribella et al. develop a diagrammatic notation with which calculations can be performed. They show that theories having the property that every mixed state has a purification have many properties in common with quantum theory.

There have been many attempts at reconstructing quantum theory, not all in the probabilistic framework of the sort considered in the preceding works. A recent conference on the general problem of reconstructing quantum theory can be seen at [36].

## 11.4 Essential Concepts

### 11.4.1 Operations and Circuits

The basic building block will be an *operation*. An operation is *one use of an apparatus*. An operation has *inputs* and *outputs*, and it also has *settings* and *outcomes* (Figure 11.2). The inputs and outputs are apertures through we imagine a system can pass. Each input or output can be open or closed. For example, we may close an output by blocking the aperture (we will explain the significance of this later). The settings can be adjusted by *knobs*. The outcomes may be read off a meter or digital display or correspond to a detector clicking or lights flashing, for example. It is possible that there is no outcome readout on the apparatus, in which case we can simply say that the set of possible outcomes has only one member. The same apparatus may be used multiple times in a given experiment. Each separate use constitutes an operation.

Each input or output is of a given *type*. We can think of the type as being associated with the type of system that we imagine passes through. The type associated with an electron is different from that associated with a photon. However, from an operational point of view, talk of electrons or photons is a linguistic shortcut for certain operational procedures. We might better say that the type corresponds to the nature and intended use of the aperture. Operations can be connected by wires between outputs and inputs of the same type. These wires do not represent actual wires but rather are a diagrammatic device to show how the apertures on the operations are placed next to one another; this is something an experimentalist would be aware of and so constitutes part of the operational structure. If we actually have a wire (e.g., an optical fiber), this wire should be thought of as an operation itself and be represented by a box rather than a wire. Likewise, passage through a vacuum also should be thought of as an operation. The wires show how the experiment is assembled. Often, a piece of self-assembly furniture (from Ikea, e.g.) comes with a diagram showing an exploded view with lines connecting the places on the different parts of the furniture that should be connected. The wires in our diagrams are similar in some respects to the lines in these diagrams (although an experiment is something that changes in time, so the wires represent connections that may be transient, whereas the connections in a piece of furniture are static).

There is nothing to stop us from trying to match an electron output with a photon input or even a small rock output with an atom input (this would amount to firing small rocks at an aperture intended for individual atoms). However, this would fall outside the intended use of the apparatus, so we would not expect our theory to be applicable (and the apparatuses may even get damaged).

We often refer to *tracing forward* through a circuit. By tracing forward, we mean following a path through the circuit from the output of one operation, along the wire attached, to the input of another operation and then from an output of that operation, along the wire attached, to the input of another operation, and so on. Such paths are analogous to future-directed time-like trajectories in spacetime physics.

We require that there can be *no closed loops* as we trace forward (i.e., that we cannot get back to the same operation by tracing forward). This is a natural requirement given that an operation corresponds to a single use of an apparatus (as long as there are no backward-in-time influences). It is this requirement of no closed loops that will enable us to foliate.

In the case that we have a bunch of operations wired together with no open inputs or outputs left over, then we will say we have a *circuit* (we allow circuits to consist of disconnected parts). An example of a circuit is shown in Figure 11.3.

As mentioned, we assume that any input or output can be closed. This means that if we have a circuit fragment with open inputs and outputs left over, we can simply close them to create a circuit. This is useful because the mathematical machinery that we will set up starts with circuits. Closing an output can be thought of as simply blocking it off. The usefulness of the notion of closing an output relates to the possibility of having no influences from the future. This is discussed in Section 11.4.3. Closing an input can be thought of as sending in a system corresponding to the type of input in some fiducial state. We do not make particular use of the notion of closing an input (beyond that it allows us to get circuits from circuit fragments), and so we need not be more specific than this. We could set up the mathematical machinery in this chapter without assuming that we can close inputs and outputs without much more effort, but the present approach has certain pedagogical advantages.

# ESSENTIAL CONCEPTS

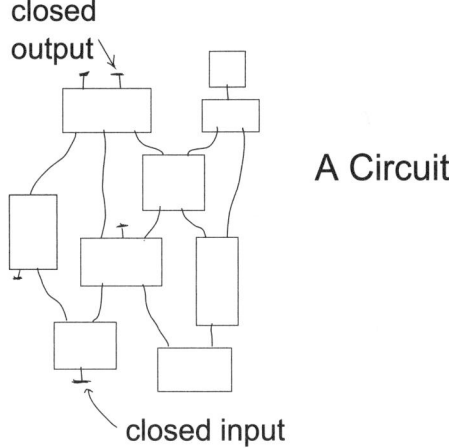

**Figure 11.3.** A bunch of operations wired together form a circuit if there are no open inputs or outputs left over. We require that there be no closed loops as we trace forward. We have not drawn in the settings or the outcomes (these will usually be taken to be implicit in these circuit diagrams). There are some closed inputs and outputs.

### 11.4.2 Time in a Circuit

We do not think of time as something in the background but rather define it in terms of the circuit itself. We take the attitude that two circuits having the same circuit diagram (in the graphical sense) are equivalent. Hence, there is no physical meaning to sliding operations along wires with respect to some background time. This is a natural attitude given the interpretation of wires as showing how apertures are placed next to each other rather than as actual wires.

We define *a synchronous set of wires* to be any set of wires for which there does not exist a path from one wire to another in the set if we trace forward along wires from output to input. See Figure 11.4 for examples.

We call a synchronous set of wires a *hypersurface*, $H$, if it partitions the circuit into two parts, $\gamma_H^-$ and $\gamma_H^+$, that are not connected other than by wires in the hypersurface. Each of the wires in the hypersurface has an end connected to an output (the "past") and

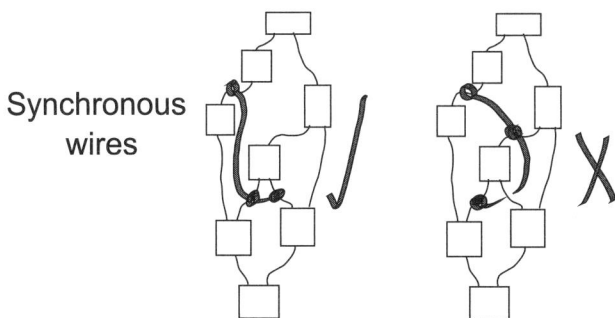

**Figure 11.4.** A set of wires is synchronous if it is not possible to get from one to another by tracing forward. On the left we see an example of a set of wires that are synchronous and on the right an example of a set that is not.

an end connected to an input (the "future"). $\gamma_H^-$ is the part of the circuit to the "past" and $\gamma_H^+$ is the part of the circuit to the "future" of the hypersurface. A *hypersurface*, as defined here, is the circuit analogue of a spacelike hypersurface in spacetime physics.

We say two hypersurfaces are distinct if at least some of the wires are different. We say that hypersurface $H_2$ is after hypersurface $H_1$ if the intersection of the past of $H_1$ (this is $\gamma_{H_1}^-$) and the future of $H_2$ (this is $\gamma_{H_2}^+$) has no operations in common. If we can get from every wire in $H_2$ by tracing forward from a wire in $H_1$, then $H_2$ is *after* $H_1$ (there are, however, examples of $H_2$ after $H_1$ that are not like this).

A *foliation* is an ordered set of hypersurfaces $\{H_t\}$ such that $H_{t+1}$ is after $H_t$. A *complete foliation* is a foliation that includes every wire in the circuit. It is easy to prove that complete foliations exist for every circuit. Define an *initial wire* to be one connected to an output of an operation having no open inputs. Take the set of all initial wires (this cannot be the null set as long as we have at least one connecting wire in this circuit). These wires form a hypersurface $H_1$. Consider the set of operations for which these wires form inputs. According to the wiring constraints, there can be no closed loops, so there must exist at least one operation in this set that has no inputs from wires connected to outputs of other operations in this set. Substitute the wires coming out of one such operation for the wires going into the operation in $H_1$ to form a new spacelike hypersurface $H_2$ (this is after $H_1$). This can be repeated until all wires have been included, forming a complete foliation. This proves that complete foliations always exist. There can, of course, exist other complete foliations that are not obtainable by this technique.

Although we do not use a notion of a background time to time-order our operations, it is the case that these foliable structures are consistent with a Newtonian notion of an absolute background time. Simply choose one complete foliation and take that as corresponding to our Newtonian time. They are, however, more naturally consistent with relativistic ideas because, for a general circuit, there exist multiple foliations.

### 11.4.3 Probability

Now we are in a position to introduce probability into the picture. Probability is a deeply problematic notion from a philosophical point of view [19]. There are various competing interpretations. All of these interpretations attempt to account for the empirical fact that, in the long run, relative frequencies are stable—that if you toss a coin a million times and get 40% heads, then if you toss the same coin a million times again, you will get 40% heads again (more or less). It is not the purpose of this chapter to solve the interpretational problems of probability, and so we adopt the point of view that probability is a limiting relative frequency. This gives us the basic mathematical properties of probability:

1. Probabilities are nonnegative.
2. Probabilities sum to 1 over a complete set of mutually exclusive events.
3. Bayes rule, $\text{Prob}(A\&B) = \text{Prob}(A|B)\text{Prob}(B)$, applies.

We could equally adopt any other interpretation of probability that gives us these mathematical properties and set up the same theoretical framework.

# ESSENTIAL CONCEPTS

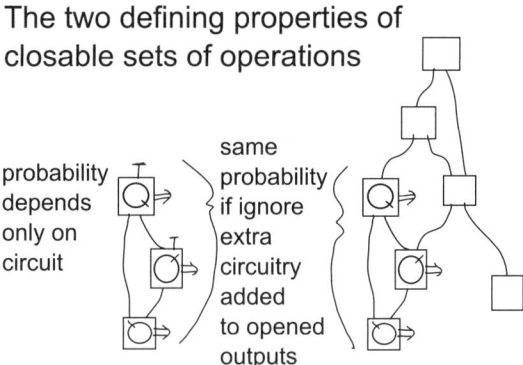

**Figure 11.5.** A set of operations is closable if any circuit built from it has a probability associated with it depending only on that circuit and if that probability is independent of any extra circuitry.

Typically, an experimentalist will have available to him some set of operations, $\mathcal{O}$, he can use to build circuits. On each operation in the circuit are various possible settings (among which the experimentalist can choose) and various outcomes, one of which will happen. We say the circuit is setting specified if each operation is given. We say the circuit is setting-outcome specified if the setting and outcome on each operation are specified. A setting-outcome–specified circuit corresponds to what happens in single run of the experiment. We define the following:

**Definition. Closable sets of operations.** A set of operations, $\mathcal{O}$, is said to be closable if, for every setting-outcome–specified circuit that can be built from $\mathcal{O}$,

(i) there is a probability depending only on the particulars of this circuit that is independent of choices made elsewhere; and
(ii) if we open closed outputs on this circuit and add on extra circuitry, then the probability associated with the original bit of setting-outcome specified circuitry (ignoring outcomes associated with the extra bit) is unchanged for any such extra circuitry we can add (Figure 11.5).

Part (i) of this definition concerns choices made elsewhere. These could be choices of settings on operations in other circuits (disjoint from this one), choices of what circuits to build elsewhere, or choices not even having to do with circuits built from the given set of operations. We might also have said that the probability associated with a setting-outcome–specified circuit is independent of the outcomes seen in other circuits. This is a natural assumption because otherwise the probability attached to a particular circuit could be different if we restrict our attention to the case where we had seen some particular outcomes on other circuits. In the case in which (i) and (ii) hold and also the probability is independent of the outcomes in other circuits, we say the set of operations is *fully closable*. It turns out we can go a very long way without assuming this. Further, we prove that in Section 11.7.6 that if a very natural condition holds, then closable sets are, in any case, fully closable.

Part (ii) imposes a kind of closure from the future; that is, that choices on operations connected only to a part of a circuit by outgoing wires (or even choices of what circuitry to place after outgoing wires) do not affect the probabilities for this circuit part. This could almost be regarded as a definition of what we mean by wires going from output to input. We do not regard it as a definition, however, because it corresponds to a rather global property of circuits rather than a property specific to a given wire.

It is interesting to consider examples of sets of operations that are not closable. Imagine, for example, that among his operations, an experimentalist has an apparatus that he specifies as implementing an operation on two qubits; but, actually, it implements an operation on three qubits, and there is an extra input aperture of which he is unaware. If he builds a circuit using this gate, then an adversary can send a qubit into the extra input, which will affect the probability for the circuit. Thus, the probability would depend on a choice made elsewhere. In this case, we could fix the situation by properly specifying the operation to include the extra input. The notion of closability is important because it ensures that the experimentalist has full control of his apparatuses. It is possible, at least in principle, that a set of operations cannot be closed by discovering extra apertures. By restricting ourselves to physical theories that admit closability, we are considering a subset of all possible theories (although a rather important one).

### 11.4.4 Can No Signalling Be an Axiom?

Many authors have promoted no signalling as an axiom for quantum theory. It may appear that part (ii) of the definition of closability sneaks in a no-signalling assumption here. In fact, this is not the case. Indeed, a no-signalling axiom would actually have limited content in this circuit framework. Consider the circuit shown in Figure 11.6(a). A no-signalling axiom would assert that a choice at one end, $B$ say, cannot affect the probability of outcomes at the other end, $A$. However, this is actually implied by part (ii) because $B$ is connected to the $AC$ part of the circuit by an outgoing wire. Hence, it looks like we are assuming no signalling. However, this is not an assumption for the framework but only a consequence of asserting that the correct circuit is the one in Figure 11.6(a). Imagine that there actually was signalling such that probabilities for the $AC$ part of the circuit depended on a choice at $B$ and then, under the assumption that the operations are drawn from a closable set, it is clear that the situation cannot be described by Figure 11.6(a). Rather, the situation would have to be that shown in Figure 11.6(b), where there is an extra wire going from $B$ to $A$ (or something like this but with more structure). This framework is perfectly capable of accommodating both signalling and no-signalling situations by appropriate circuits and, consequently, we are not sneaking in a no-signalling assumption.

This reasoning leads us to question whether a no-signalling axiom could do any work at all. Indeed, it is often unappreciated in such axiomatic discussions that the usual framework of quantum theory does allow signalling. One can write down nonlocal Hamiltonians that will, for example, entangle product states. Of course, in quantum field theory, one incorporates a no-signalling property by demanding that field operators for spacelike separated regions commute so that such nonlocal Hamiltonians are ruled out. However, this is an example where we have a given background. In general, a no-signalling axiom with respect to some particular given background would restrict the type of circuits we allow. For example, the circuit in Figure 11.6(b) would not be allowed unless $A$ was in the future light cone of $B$. In quantum field theory, we have an

# ESSENTIAL CONCEPTS

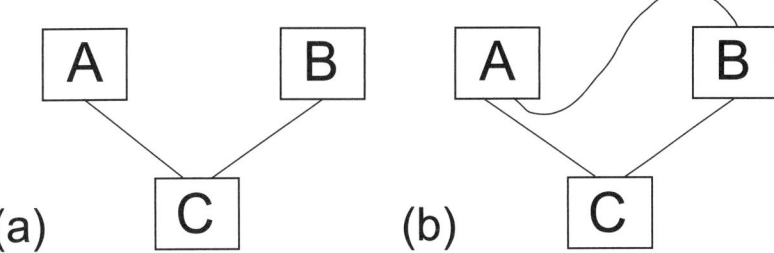

**Figure 11.6.** If there is no signalling between *A* and *B*, then the correct circuit is that shown in (a). However, if there is signalling from *B* to *A*, then (a) could not be the correct circuit. Instead, the correct circuit would have to be something like that shown in (b). The framework is perfectly capable of incorporating signalling. Hence, the assumption of no signalling is not an assumption of the framework but rather corresponds to asserting that the correct circuit for the given no-signalling situation is the one in (a).

example where a no-signalling axiom with respect to a Minkowski background restricts the types of unitary evolution and measurement that are allowed. However, it is often claimed that the abstract Hilbert space framework itself (which makes no mention of Minkowski spacetime) can be derived using no signalling as one of the axioms. It is this more ambitious claim we question. In fact, we will see that we can define this abstract framework of quantum theory for any circuit (as long as there are no closed loops), including no-signalling and signalling situations, as in Figure 11.6(a) and (b). Hence, a no-signalling axiom clearly cannot be regarded as a constraint on this abstract framework.

This criticism of the usefulness of no signalling as an axiom does not apply to a recently proposed generalization of this principle in an extraordinary article by Pawlowski et al. [35]. They introduced the *information causality principle*. It was shown that this compelling principle limits violations of the Bell inequality to the quantum limit. Imagine that Alice receives $n$ classical bits of information. She communicates $m$ classical bits to Bob. Bob is expected to reveal the value of one of the $n$ classical bits, although neither Alice nor Bob know which one this will be in advance. The information causality principle is that Alice and Bob can be successful only when $n$ is less than or equal to $m$. For $m = 0$, this is the no-signalling assumption we have criticized. The information causality principle can be read as implying that if the task cannot be achieved for $m = 0$, then it cannot be achieved for any other value of $m$. This principle would be useful in prescribing what is possible in the framework described in this chapter. Consider two fragments of a circuit that cannot be connected by tracing forward (these fragments are analogous to spacelike separated regions). The information causality principle implies that there is no way of accomplishing this task for any $m$ with $n > m$ between these two circuit fragments. That we cannot do this for $m = 0$ is already implied (assuming that we have the correct circuit).

Although a simple no-signalling "across space" principle is of limited use for the aforementioned reasons, we do employ what might be regarded as a no-signalling backward-in-time principle because we do not allow closed loops in a circuit.

### 11.4.5 Systems and Composite Systems

We wish to give an operational definition of what we mean by the notion of a *system*. We may find that whenever we press a button on one box, a light goes on on another box. We can interpret this in terms of a system passing between the two boxes. We find this happens only when we place the boxes in a certain arrangement with respect to one another (which we think of as *aligning apertures*). Given this, we clearly want to associate systems with wires. Hence, we adopt the following definition:

> **Definition.** A *system* of type $abc\ldots$ is, by definition, associated with any set of synchronous wires of type $a, b, c, \ldots$ in any circuit formed from a closable set of operations.

We may refer to a system type corresponding to more than one wire by a single letter. Thus, we may denote the system type $abc$ by $d$.

The usefulness of the notion of closable sets of operations is that it leads to wires being associated with the sort of correlation we expect for systems, given our usual intuitions about what systems are. Nevertheless, our definition of system is entirely operational because wires are defined operationally.

It is common to speak of composite systems. We define a composite system as follows:

> **Definition.** A *composite system*, $AB$, is associated with any two systems (each associated with disjoint sets of wires) if the union of the sets of wires associated with system $A$ and system $B$ forms a synchronous set.

This definition generalizes in the obvious way for more than two systems. A system of type $aabc$ can be regarded as a composite of systems of type $aa$ and $bc$; or a composite of systems of type $aac$ and $b$; or a composite of systems of type $ac$, $a$ and $b$—to list just a few possibilities. Systems associated with a single wire cannot be regarded as composite.

A hypersurface consists of synchronous wires and so can be associated with a system (or composite system). A complete foliation can therefore be associated with the evolution of a system through the circuit (although the system type can change after each step). This evolution can also be viewed as the evolution of a composite system.

## 11.5 Preparations, Transformations, and Effects

### 11.5.1 Circuit Fragments

We can divide up a circuit into fragments corresponding to preparations, transformations, and effects as shown in Figure 11.7. By the term *circuit fragment*, we mean a part of a circuit (i.e., a subset of the operations in the circuit along with the wires connecting them) having inputs coming from a synchronous set of wires and outputs going into a synchronous set of wires. We allow lone wires in a circuit fragment (i.e., wires not connected to any operations in the fragment). An example of a lone wire is seen in the circuit fragment in the rectangle on the left in Figure 11.7. The lone wire

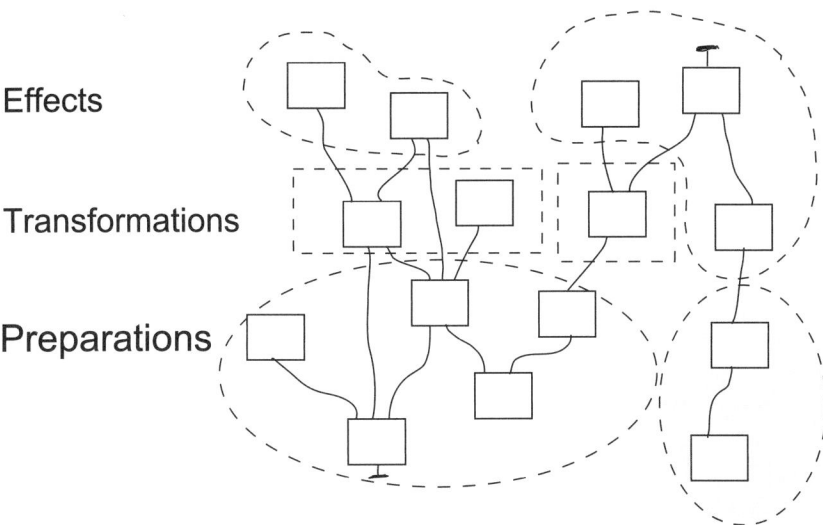

**Figure 11.7.** A circuit fragment is a part of the circuit having inputs wires in a synchronous set and output wires also in a synchronous set. A circuit can be divided up many ways into circuit fragments corresponding to preparations (no open inputs), transformations, and effects (no open outputs).

corresponds to the identity transformation on that system and contributes an input and output to the circuit fragment. Generally, we take the term *circuit fragment* to imply that the settings and outcomes at each operation associated with these circuit fragments have been specified. A circuit fragment is, essentially, an operation at a course-grained level. *Preparations* correspond to a circuit fragment having outputs but no open inputs. *Transformations* have inputs and outputs. *Effects* have inputs but no open outputs. Note that *preparations* and *effects* are special cases of transformations.

### 11.5.2 States

A preparation prepares a system. For a given type of system, there will be many possible preparations. We will label them with $\alpha$. This label tells us which circuit fragment is being used to accomplish the preparation (including the specification of the knob settings and outcomes on each operation). Associated with each preparation for a system of type $a$ will be a state (labeled by $\alpha \in \text{Prep}_a$). We can build a circuit having this preparation by adding an effect for a system of type $a$. There are many possible effects labeled by $\beta \in \text{Eff}_a$. The label $\beta$ tells us which circuit fragment is used to accomplish the effect along with the knob settings and outcomes at each operation. Associated with the circuit is a probability $p^{\alpha\beta}$. We define the *state* associated with preparation $\alpha$ to be that thing represented by any mathematical object that can be used to calculate $p^{\alpha\beta}$ for all $\beta$. We could represent the state by the long list

$$\mathbf{P}_a^\alpha = \begin{pmatrix} \vdots \\ p^{\alpha\beta} \\ \vdots \end{pmatrix} \quad \beta \in \text{Eff}_a. \tag{11.1}$$

However, in general, physical theories relate physical quantities. Hence, it is necessary only to list a subset of these quantities where the remaining quantities can be calculated by the equations of the physical theory. We call the forming of this subset of quantities *physical compression*. In the current case, we expect the probabilities in this list to be related. We consider the maximum amount of physical compression that is possible by linear means. Thus, we write the state as

$$\mathbf{p}_a^\alpha = \begin{pmatrix} \vdots \\ p^{\alpha\beta} \\ \vdots \end{pmatrix} \quad \beta \in \Omega_a \subseteq \text{Eff}_a, \tag{11.2}$$

where there exist vectors $\mathbf{r}_a^\alpha$ such that

$$p^{\alpha\beta} = \mathbf{r}_a^\beta \cdot \mathbf{p}_a^\alpha \quad \text{for all} \quad \alpha \in \text{Prep}_a, \quad \beta \in \text{Eff}_b. \tag{11.3}$$

We call the set $\Omega_a$ the fiducial set of effects for a system of type $a$. The choice of fiducial set need not be unique; we simply make one choice and stay with it. Because we have applied as much linear physical compression as possible, $|\Omega_a|$ is the minimum number of probabilities required to calculate all the other probabilities by linear equations. The vectors $\mathbf{r}_a^\beta$ are associated with effects on a system of type $a$. For the fiducial effects, they consist of a 1 in the $\beta$ position and 0's elsewhere.

An important subtlety here is that we define states in terms of joint rather than conditional probabilities. This makes more sense for the circuit model because, generally, we want to calculate a probability for a circuit. If we want to calculate conditional probabilities, we can use Bayes's rule in the standard way.

### 11.5.3 Transformations

Now, consider a preparation $\alpha$ that prepares a system of type $a$ in state $\mathbf{p}_a^\alpha$ followed by a transformation $\beta$ that outputs a system of type $b$. We can regard the preparation and transformation, taken together, as a new preparation $\alpha\beta$ for a system of type $b$ with state $\mathbf{p}_b^{\alpha\beta}$. Now, follow this by a fiducial effect $\gamma \in \Omega_b$. The probability for the circuit $\alpha\beta\gamma$ can be written

$$p^{\alpha\beta\gamma} = \mathbf{r}_b^\gamma \cdot \mathbf{p}_b^{\alpha\beta} = \mathbf{r}_a^{\beta\gamma} \cdot \mathbf{p}_a^\alpha, \tag{11.4}$$

where $\mathbf{r}_a^{\beta\gamma}$ is the effect vector associated with the measurement consisting of the transformation $\beta$ followed by the effect $\gamma$. Given the special form of the fiducial effect vectors, it follows that the state transforms as

$$\mathbf{p}_b^{\alpha\beta} = {}^b Z_a^\beta \mathbf{p}_a^\alpha, \tag{11.5}$$

where ${}^b Z_a^\beta$ is a $|\Omega_b| \times |\Omega_a|$ matrix such that its $\beta$ row is given by the components of the effect vectors $\mathbf{r}^{\beta\gamma}$. We use a subscript, $a$, for the inputted system type and a pre-superscript, $b$, for the outputted system type. Hence, a general transformation is given by a matrix acting on the state. If the matrix transforms from one type of system to another type of system with a different number of fiducial effects, then it will be rectangular.

The general equation for calculating probabilities is

$$p_{\alpha\beta\gamma} = (\mathbf{r}_b^\gamma)^T {}^b Z_a^\beta \mathbf{p}_a^\alpha, \tag{11.6}$$

where $T$ denotes transpose. If we have more than one transformation, then, by a clear extrapolation of this reasoning, we can write

$$p^{\alpha\beta\gamma\delta} = (\mathbf{r}_c^\delta)^T {}^c Z_b^\gamma {}^b Z_a^\beta \mathbf{p}_a^\alpha, \tag{11.7}$$

and so on. Now, because the $Z$ matrices can be rectangular, we can think of $\mathbf{p}_a^\alpha$ and $(\mathbf{r}_a^\gamma)^T$ as instances of a transformation matrix. The state $\mathbf{p}_a^\alpha$ can be thought of as corresponding to the transformation that turns a null system (no system at all) into a system outputted by this preparation, and we can change our notation to ${}^a Z^\alpha$ instead (a column vector being a special case of a rectangular matrix). Likewise, we can change our notation for the row vector $(\mathbf{r}_a^\delta)^T$ to $Z_a^\delta$ (a row vector being a special case of a rectangular matrix). Then, we can write

$$p^{\alpha\beta\gamma\delta} = Z_c^\delta {}^c Z_b^\gamma {}^b Z_a^\beta {}^a Z^\alpha. \tag{11.8}$$

The agreement of output and input system types is clear (by matching pre-superscript with subscripts between the $Z$'s).

The label $\alpha$ labels the circuit fragment along with the knob settings and the outcomes. Sometimes it is useful to break these up into separate labels. Thus, we write

$$\alpha \equiv (\mathcal{F}, \varphi, l), \tag{11.9}$$

where $\mathcal{F}$ denotes the circuit fragment before the settings and outcomes are specified, $\varphi$ denotes the settings on the operations, and $l$ denotes the outcomes. In particular, this means that we can notate the effects associated with the different outcomes of a measurement as $\alpha_i = (\mathcal{F}, \varphi, l_i)$.

Let $L$ be the set of possible outcomes $l_i$ for a given measurement (with fixed $\mathcal{F}$ and $\varphi$). We can subdivide the set $L$ into disjoint sets $L_k$ where $\cup_k L_k = L$. We could choose to be ignorant of the actual outcome $l_i$ and rather record only which set $L_k$ to which it belongs. In this case, the transformation effected can be denoted $\overline{\alpha}_k$. Because we have used linear compression, we must have

$$^b Z_a^{\overline{\alpha}_k} = \sum_{i \in I_k} {}^b Z_a^{\alpha_i}, \tag{11.10}$$

where $I_i$ is the set of $i$'s corresponding to the $l_i$'s in $L_k$. Because we can always choose to be ignorant in this way, we must include such transformations in the set of allowed transformations.

The matrices corresponding to the set of allowed transformations must be such that when closed expressions such as Equation (11.8) are calculated, they always give probabilities between 0 and 1. This is an important constraint on this framework.

### 11.5.4 The Identity Transformation

One transformation we can consider is one in which we do nothing. The wires coming in are the same as the wires coming out and no operation has intervened. We will denote this transformation by 0. Then, we have, for example,

$$^a Z_a^0. \tag{11.11}$$

This is a $|\Omega_a| \times |\Omega_a|$ matrix and must be equal to the identity because as long as it is type matched, it can be inserted as many times as we like into any expression where nontrivial transformations act also.

### 11.5.5 The Trace Measurement

One effect we can perform on a preparation $\alpha$ is to close all outputs. This forms a circuit and, hence, there is an associated probability, $p^{\alpha-}$ (where $-$ denotes that the outputs have been closed). This is an effect; hence, we must have

$$p^{\alpha-} = \mathbf{r}_a^- \cdot \mathbf{p}_a^\alpha, \tag{11.12}$$

where the vector $\mathbf{r}_a^-$ corresponds to this effect. We call this the *trace measurement* (terminology borrowed from quantum theory where this effect corresponds to taking the trace of the density matrix). It follows from part (ii) of the condition for a closable set of operations that this is the probability associated with the preparation part of the circuit even if the outputs are open and more circuitry is added.

In the case that $\mathbf{r}_a^- \cdot \mathbf{p}_a^\alpha = 1$, we say that the state is of *norm one*. In general, we do not expect states to be of norm one because they consist of joint rather than conditional probabilities; hence, we require only that $0 \leq \mathbf{r}_a^- \cdot \mathbf{p}_a^\alpha \leq 1$.

We can *normalize* a state by dividing by $\mathbf{r}_a^- \cdot \mathbf{p}_a^\alpha$. We denote the normalized state by

$$\bar{\mathbf{p}}_a^\alpha \equiv \frac{\mathbf{p}_a^\alpha}{\mathbf{r}_a^- \cdot \mathbf{p}_a^\alpha}. \tag{11.13}$$

We cannot guarantee that $\bar{\mathbf{p}}_a^\alpha$ belongs to the set of allowed states (i.e., that there exists a state for which $\mathbf{p}_a^\alpha = \bar{\mathbf{p}}_a^\alpha$) because preparations for all states parallel to $\bar{\mathbf{p}}_a^\alpha$ may be intrinsically probabilistic.

## 11.6 Mixtures

### 11.6.1 Forming Mixtures

Imagine that we have a box with a light on it that can flash and an aperture out of which a system of type $a$ can emerge. With probability $\lambda_i$, we place preparation $\alpha_i$ in the box such that the system (which we take to be of type $a$) will emerge out of the aperture and such that the light will flash if the outcomes corresponding to this preparation are seen. The state prepared for this one $i$ is $\lambda_i \mathbf{p}_a^{\alpha_i}$. If $\lambda_i = 0$, then the state prepared is the *null state* $\mathbf{0}_a$, which has all 0's as entries. If we use this box for a set $\alpha_i$ with $i \in I$ such that $\sum_{i \in I} \lambda_i \leq 1$, then the state prepared is

$$\sum_{i \in I} \lambda_i \mathbf{p}_a^{\alpha_i}. \tag{11.14}$$

This is a linear sum of terms because we have linear compression. This process of using a box may be beyond the experimental capacities of a given experimentalist. It certainly takes us outside the circuit model as previously described. However, we can always consider taking mixtures like this at a mathematical level.

A technique that can be described in the circuit model is the following. Consider placing a single preparation circuit into the box described herein, where $\alpha_j = (\mathcal{F}, \varphi, l_j)$ and where $l_j$ labels the outcomes. We can arrange things so that the light flashes only if $j \in J$ (where $J$ is some subset of the $j$'s). The state is then given by

$$\sum_{j \in J} \mathbf{p}_a^{\alpha_j}. \tag{11.15}$$

This technique does not require having a coin to generate probabilities $\lambda_i$, and neither does it require the placing of different circuit fragments into a box depending on the outcome of the coin toss.

The most general thing we can do is a mixture of these two techniques. With probability $\lambda_i$, we place a circuit $\mathcal{F}_i$ with settings $\varphi_i$ and outcomes $l_{ij}$ in the box for $i \in I$ and $j \in J_i$. The state we obtain is

$$\sum_{i \in I, j \in J_i} \lambda_i \mathbf{p}_a^{\alpha_{ij}}. \tag{11.16}$$

We can absorb the $\lambda$'s into the $\mathbf{p}$'s and relabel so we obtain that a general mixture is given by

$$\mathbf{p}^\alpha = \sum_i \mathbf{p}_a^{\alpha_i}. \tag{11.17}$$

We have the constraints that $\mathbf{r}_a^- \cdot \mathbf{p}_a^{\alpha_i} \geq 0$ and $\sum_i \mathbf{r}_a^- \cdot \mathbf{p}_a^{\alpha_i} \leq 1$.

Note that if we write $\mathbf{p}_a^{\alpha_i} \equiv \mu_i \bar{\mathbf{p}}_a^{\alpha_i}$ where $\bar{\mathbf{p}}_a^{\alpha_i}$ is normalized and include an extra state $\mathbf{p}_a^0 \equiv \mathbf{0}$ (the null state), then we have

$$\mathbf{p}^\alpha = \sum_i \mu_i \bar{\mathbf{p}}_a^{\alpha_i} \quad \text{where} \quad \mu_i \geq 0, \quad \sum_i \mu_i = 1, \tag{11.18}$$

where the sum now includes the null state. Hence, we can interpret a general mixture as a convex combination of normalized states and the null state.

### 11.6.2 Homogeneous, Pure, Mixed, and Extremal States

If two states are parallel, then they give rise to the same statistics up to an overall weighting, and if we condition on the preparation, then they have exactly the same statistics. We define a *homogeneous state* as one that can be written as a sum of parallel states. Thus, $\mathbf{p}_a^\alpha$ is homogeneous if, for any sum,

$$\mathbf{p}_a^\alpha = \sum_{i \in I} \mathbf{p}_a^{\alpha_i}, \tag{11.19}$$

we have $\mathbf{p}_a^{\alpha_i} = \eta_{ii'} \mathbf{p}_a^{\alpha_{i'}}$ for all $i, i' \in I$. A state that is not homogeneous (i.e., which can be written as a sum of at least two nonparallel states) is called a *heterogeneous state*.

Given a particular homogeneous state, there will, in general, be many others that are parallel to it but of different lengths. We call the longest among these a *pure state*.

A *mixed state* is defined to be any state that can be simulated by a probabilistic mixture of distinct states in the form $\sum_j \lambda_j \mathbf{p}_a^{\alpha_j}$, where $\lambda_j \geq 0$ and $\sum_j \lambda_j = 1$. A pure state is not a mixture (because the $\mathbf{p}_a^{\alpha_j}$'s would have to be parallel to the given pure state and, therefore, given the $\sum_j \lambda_j = 1$ condition, equal to the given pure state). Homogeneous states that are not *pure* are *mixtures*. Heterogeneous states are also mixtures. *Extremal states* are defined to be states that are not mixtures. Pure states are extremal. The null state is also extremal. If all pure states have norm equal to one (i.e., $\mathbf{r}_a^- \cdot \mathbf{p}_a^\alpha = 1$), then there are no more extremal states beyond the pure states and null state. However, if some pure states have $\mathbf{r}_a^- \cdot \mathbf{p}_a^\alpha < 1$, then there may be additional extremal states.

Usually, treatments of convex sets of states do not make these distinctions. More care is necessitated here because states are based on joint rather than conditional probabilities.

Any state, extremal or mixed, can be written as the sum of homogeneous states. This means that there must exist at least one set of $|\Omega_a|$ linearly independent homogeneous states, all of which can be pure. There cannot exist sets with more linearly independent states than this.

### 11.6.3 Optimality of Linear Compression

We state the following theorem:

**Theorem 1.** If we allow arbitrary mixtures of preparations, then (1) linear compression is optimal, and (2) optimal compression is necessarily linear.

The first point follows because there must exist $|\Omega_a|$ linearly independent states (otherwise, we could have implemented further linear compression). We can take an arbitrary mixture with $\sum_i \lambda_i \leq 1$; then, these $\lambda$'s are all independent and, hence, we need $|\Omega_a|$ parameters and the compression is optimal. To prove the second point, consider representing the state by a list of $|\Omega_a|$ probabilities $p_a^{\alpha\beta'}$ with $\beta' \in \Omega_a'$, where we do not demand that a general probability is given by a linear function of these probabilities. Represent this list by the vector $\mathbf{q}_a^\alpha$. Now,

$$p^{\alpha\beta'} = \mathbf{r}_a^{\beta'} \cdot \mathbf{p}_a^\alpha \quad \text{for all} \quad \beta' \in \Omega_a'. \tag{11.20}$$

Hence, $\mathbf{q}_a^\alpha = C\mathbf{p}_a^\alpha$, where $C$ is a square matrix with real entries. $C$ must be invertible; otherwise, we could specify $\mathbf{q}_a^\alpha$ with fewer than $|\Omega_a|$ probabilities. Hence,

$$p^{\alpha\beta} = \mathbf{r}_a^\beta \cdot C^{-1} \mathbf{p}_a^\alpha, \tag{11.21}$$

for a general $\beta$. Hence, the probability is linear in $\mathbf{q}_a^\alpha$ and so the compression is linear.

## 11.7 Composition

### 11.7.1 Preliminaries and Notation

As we discussed, systems associated with more than one wire can be thought of as *composite*. The **p**, **r**, $Z$ framework just discussed can be enriched by adapting it to deal separately with the components of composite systems (rather than regarding all of the wires at each time step as constituting a single system). The advantage of this is that we can break up the calculation into smaller parts and thereby define a theory by associating matrices to smaller transformations. Ultimately, we would like to have a matrix associated with each operation in the set of allowed operations that can be used to calculate the probability for any circuit. A transformation is now associated with a matrix such as

$$^{bacd}Z^{\alpha}_{acb}. \tag{11.22}$$

This transformation inputs a system of type $acb$ and outputs a system of type $bacd$. The label $\alpha$ denotes the circuit fragment used to do this, including the knob settings and the outputs at each operation in the fragment. The ordering of the symbols representing the system (e.g., $bacd$) is significant in that it is preserved between transformations to indicate how the wires are connected. Thus, the matrix for a transformation consisting of two successive transformations is written

$$^{d}Z^{\beta}_{bacd}\,^{bacd}Z^{\alpha}_{acb}. \tag{11.23}$$

The system types in between the two transformations must match (as in this example) because we employ the convention that the wires are in the same order from the output of one transformation to the input of the next. We can allow that the symbols for the systems (e.g., $c$) actually correspond to composite systems (so that they correspond to a cluster of wires). For example, it might be that $c = aabb$. Some transformations consist of disjoint circuit fragments, and it is useful to have notations for these. We write

$$^{(b)(ac)(-)}Z^{(\alpha)(\beta)(\gamma)}_{(a)(c)(b)}, \tag{11.24}$$

to indicate that the transformation consists of three disjoint parts: one transforming from system type $a$ to $b$ and labeled by $\alpha$, one transforming from $c$ to $ac$ and labeled by $\beta$, and one that inputs $b$ and has no open outputs (which we denote by $-$ when necessary for disambiguation) and labeled by $\gamma$. If it is clear from the context, we will sometimes use the less cumbersome notation

$$^{cd}Z^{\alpha\beta}_{ab} \equiv\,^{(c)(d)}Z^{(\alpha)(\beta)}_{(a)(b)}. \tag{11.25}$$

We may sometimes want to depart from the convention that the wires are in the same order from one transformation to the next, in which case we label the wires (and wire clusters, as appropriate) using integers $1, 2, \ldots$ as follows:

$$^{(b)_5(ac)_{64}(d)_7}Z^{(\alpha)(\beta)(\gamma)}_{(a)_1(c)_3(b)_2}. \tag{11.26}$$

In this example, wire 6 is an output wire of type $a$. We can then rewrite Equation (11.23) as

$$^{(d)_8}Z^{(\beta)}_{(dcab)_{7654}}\,^{(bacd)_{4567}}Z^{(\alpha)}_{(acb)_{123}}. \tag{11.27}$$

We see that the wires match (e.g., wire 6 is of type $c$ as an output from the first transformation and an input into the second transformation). The integers labeling the wires are of no significance, and any expression is invariant under any reassignment of these labels (this is a kind of discrete general covariance).

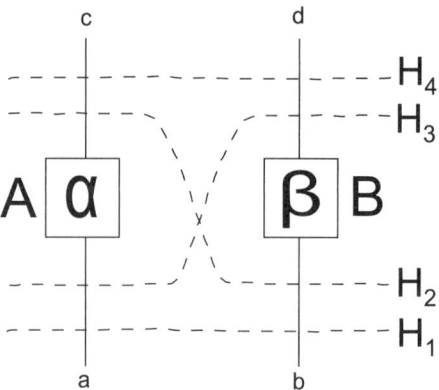

**Figure 11.8.** We can consider the evolution of this circuit with respect to different foliations.

### 11.7.2 Commutation

Consider the situation shown in Figure 11.8. By inspection of this diagram, we can write

$$^{cd}Z^{\alpha\beta}_{ab} = {}^{cd}Z^{0\beta}_{cb}\,{}^{cb}Z^{\alpha 0}_{ab} = {}^{cd}Z^{\alpha 0}_{ad}\,{}^{ad}Z^{0\beta}_{ab}, \tag{11.28}$$

where the 0 denotes that we do nothing (the identity transformation). The first equation here is obtained by first transforming from hyperplane $H_1$ to $H_2$ (past operation $A$) and then from $H_2$ to $H_4$ (past operation $B$). To get the second equation, we evolve from $H_1$ to $H_3$ first (past $B$), then from $H_3$ to $H_4$ (past $A$). There are a few points of interest here. First, note that we can break down a compound transformation into its parts. Second, we see that there is a kind of commutation; it does not matter whether we update at $A$ or $B$ first. This property is not assumed but derived from more basic assumptions and definitions. In the special case where wires $c$ and $d$ are of type $a$ and $b$, respectively (so the transformations do not change the system type), we have the commutation property

$$[^{ab}Z^{0\beta}_{ab}, {}^{ab}Z^{\alpha 0}_{ab}] \equiv {}^{ab}Z^{0\beta}_{ab}\,{}^{ab}Z^{\alpha 0}_{ab} - {}^{ab}Z^{\alpha 0}_{ab}\,{}^{ab}Z^{0\beta}_{ab} = 0. \tag{11.29}$$

The usual commutation relation is, then, a special case of the more general relation in Equation (11.28) where the local transformations may change the system type.

The fact that we can break up a compound transformation into smaller parts is potentially useful, but there is a stumbling block. The matrix $^{cb}Z^{\alpha 0}_{ab}$ transforms past the $A$ operation. However, it is a $|\Omega_{cb}| \times |\Omega_{ab}|$ matrix. That is, we still have to incorporate some baggage because we include wire $b$. It would be good if we could write

$$^{cb}Z^{\alpha 0}_{ab} = {}^{c}Z^{\alpha}_{a} \otimes {}^{b}Z^{0}_{b} \tag{11.30}$$

where $^{b}Z^{0}_{b}$ is just the identity matrix (as discussed previously). By considering the sizes of the matrices, it is clear that (11.30) implies $|\Omega_{ab}| = |\Omega_a||\Omega_b|$. In Section 11.8.1, we show that Equation (11.30) holds in general if $|\Omega_{ab}| = |\Omega_a||\Omega_b|$ is true (for any system types $a$ and $b$). We also see that this condition corresponds to a very natural class of physical theories. If Equation (11.30) holds true, we can break any circuit down into its basic operations appending the identity transformation as necessary. Then, we can

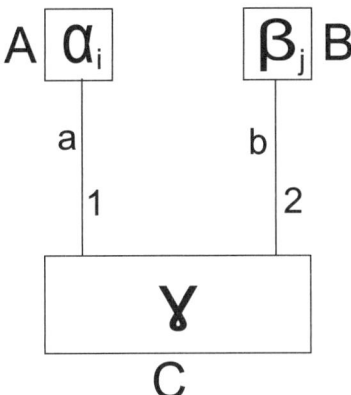

**Figure 11.9.** A bipartite system 12 of type $ab$ is prepared by some preparation $\gamma$ and subjected to effects $\alpha_i$ and $\beta_j$ on subsystem 1 and subsystem 2, respectively.

calculate the probability associated with any circuit from the transformation matrices associated with the operations.

### 11.7.3 Homogeneous States and Composite Systems

Consider a composite system 12 consisting of systems of types $a$ and $b$ with preparation $C$ labeled $\gamma$. The state prepared by this is $\mathbf{p}^\gamma_{ab}$. If we block the output 2 of the preparation, then we have a preparation for a system of type $a$. Let the state so prepared be $\mathbf{p}^\gamma_a$. Even if we do not block output 2, it follows from part (ii) of the condition for closable sets of operations that this state $\mathbf{p}^\gamma_a$ gives us the correct probabilities for all measurements on system 1 alone (that do not involve system 2). We call $\mathbf{p}^\gamma_a$ the *reduced state* for system 1. It is, effectively, the state of system 1 taken by itself.

We prove the following theorem.

**Theorem 2.** If one component of a bipartite system is in a homogeneous state (i.e., the reduced state for this system is homogeneous), all joint probabilities for separate effects measured on the two systems factorize.

Systems 1 and 2 can be subjected to measurements $A$ and $B$, respectively (see Figure 11.9). We also consider the possibility of closing either or both outputs from $C$. Let measurement $A$ have outcomes $l_i$, $i \in I$, the effects for which are labeled $\alpha_i \equiv (\mathcal{F}^A, \varphi^A, l^A_j)$. Similarly, for $B$ we have outcomes $l^B_j$ for $j \in J$ and effects labeled by $\beta_i \equiv (\mathcal{F}^B, \varphi^B, l^B_j)$. By part (ii) of the condition for a closable set of operations, we have

$$p^{\gamma--} = \sum_{i \in I} p^{\gamma \alpha_i -} = \sum_{j \in J} p^{\gamma - \beta_j}, \tag{11.31}$$

and

$$p^{\gamma - \beta_j} = \sum_{i \in I} p^{\gamma \alpha_i \beta_j}. \tag{11.32}$$

If the output 2 from $C$ is closed, then we say that the state prepared (for the system of type $a$) is $\mathbf{p}_a^\gamma$ (the reduced state). We say that the preparation due to $C$ and $B$ with outcome $l_j^B$ (this is a preparation circuit fragment) is $\mathbf{p}_a^{\gamma\beta_j}$. It follows from part (ii) of the condition for a closable set of operations that

$$\mathbf{p}_a^\gamma = \sum_{j \in J} \mathbf{p}_a^{\gamma\beta_j}. \tag{11.33}$$

If $\mathbf{p}_a^\gamma$ is homogeneous, then all the $\mathbf{p}_a^{\gamma\beta_j}$ states must be parallel to it. Hence, we can write $\mathbf{p}_a^{\gamma\beta_j} = \eta_j \mathbf{p}_a^\gamma$. We must have

$$p^{\gamma\alpha_i\beta_j} = \mathbf{r}_a^{\alpha_i} \cdot \mathbf{p}_a^{\gamma\beta_j} = \eta_j \mathbf{r}_a^{\alpha_i} \cdot \mathbf{p}_a^\gamma = \eta_j p^{\gamma\alpha_i -}. \tag{11.34}$$

Summing this over $i$ and using Equations (11.31) and (11.32), we obtain $p^{\gamma-\beta_j} = \eta_j p^{\gamma--}$. Hence,

$$p^{\gamma\alpha_i\beta_j} p^{\gamma--} = p^{\gamma\alpha_i -} p^{\gamma-\beta_j}. \tag{11.35}$$

Here, $p^{\gamma--}$ is the probability of the preparation being successful. Dividing this through by $(p^{\gamma--})^2$ and using Bayes's rule, we obtain

$$\text{prob}(l_i^A l_j^B | \text{prep}) = \text{prob}(l_i^A | \text{prep}) \text{prob}(l_j^B | \text{prep}). \tag{11.36}$$

Hence, we see that if one system is in a homogeneous state, then joint probabilities factorize between the two ends (obviously, the result also holds if both components are in homogeneous states). An obvious corollary follows:

**Corollary 1.** If the state of a bipartite system 12 of type $ab$ with preparation $\gamma$ is of norm one and the reduced state of either or both components is pure, then

$$p^{\gamma\alpha\beta} = p^{\gamma\alpha} p^{\gamma\beta}, \tag{11.37}$$

where $\alpha$ is any effect on 1 alone and $\beta$ is any effect on 2 alone.

This is true because pure states are necessarily homogeneous and because $p^{\gamma--} = 1$ because the state is of norm 1.

Now, we consider a related but slightly different situation. Imagine that we have a preparation consisting of two disjoint parts one of which prepares a homogeneous state.

**Theorem 3.** If a preparation, $\gamma\delta$, consists of two disjoint circuit fragments $\gamma$ and $\delta$, which prepare systems of type $a$ and $b$, respectively, and one of these circuit fragments, $\gamma$, taken by itself prepares a homogeneous state, $\mathbf{p}_a^\gamma$, then the state prepared by closing the outputs of the second-circuit fragment of $\gamma\delta$ is parallel to $\mathbf{p}_a^\gamma$ (and therefore also homogeneous).

The proof of this theorem is based on the same idea as the previous theorem. We can put $\delta = (\mathcal{F}, \varphi, l)$, where $\mathcal{F}$ denotes the actual circuit fragment, $\varphi$ the settings, and $l$ the outcomes. Then, we can put $\bar{\delta} = (\mathcal{F}, \varphi, \bar{l})$. This is the circuit fragment associated with not seeing $l$. Either $l$ or $\bar{l}$ must happen; hence,

$$\mathbf{p}_a^\gamma = \mathbf{p}_a^{\gamma\delta} + \mathbf{p}_a^{\gamma\bar{\delta}}. \tag{11.38}$$

Because $\mathbf{p}_a^\gamma$ is homogeneous, both $\mathbf{p}_a^{\gamma\delta}$ and $\mathbf{p}_a^{\gamma\bar{\delta}}$ must be parallel to it and hence are also homogeneous.

### 11.7.4 Fiducial Measurements for Composite Systems

Assume that system 1 is prepared by some preparation $\gamma$ in a homogeneous state $\mathbf{p}_a^\gamma$ and, similarly, system 2 is prepared in homogeneous state $\mathbf{p}_b^\delta$. Assume that these are two separate preparations corresponding to two disjoint circuit fragments. Hence, we can consider the joint preparation $\gamma\delta$ giving rise to the state $\mathbf{p}_{ab}^{\gamma\delta}$.

A natural question is: What is the relationship between the states $\mathbf{p}_a^\gamma$, $\mathbf{p}_b^\delta$, and $\mathbf{p}_{ab}^{\gamma\delta}$? By virtue of Theorem 3, we know that the reduced state at either end is homogeneous and, hence, by virtue of Theorem 2 (Equation (11.35) in particular) and the fact that the subsystems are in homogeneous states, we can write

$$\mathbf{r}_{ab}^{\alpha\beta} \cdot \mathbf{p}_{ab}^{\gamma\delta} = p^{\gamma\delta\alpha\beta} = \mu_{\gamma\delta} p^{\gamma\alpha} p^{\delta\beta} = \nu_{\gamma\delta} (\mathbf{r}_a^\alpha \otimes \mathbf{r}_b^\beta) \cdot (\mathbf{p}_a^\gamma \otimes \mathbf{p}_b^\gamma), \tag{11.39}$$

where we obtain

$$\nu_{\gamma\delta} = \frac{p^{\gamma\delta--}}{p^{\gamma-} p^{\delta-}}, \tag{11.40}$$

by putting $\alpha = -$ and $\beta = -$.

We know that there must exist $K_a \equiv |\Omega_a|$ linearly independent homogeneous states for system 1 (they can all be chosen to be pure). Let $\gamma \in \overline{\Omega}_a$ be the preparations associated with one such set of linearly independent homogeneous states for system 1 (where $|\overline{\Omega}_a| = |\Omega_a|$). Likewise, we have $K_b \equiv |\Omega_b|$ linearly independent homogeneous states for system 2 with preparations $\delta \in \overline{\Omega}_b$ (where $|\overline{\Omega}_b| = |\Omega_b|$). The $K_a K_b$ vectors

$$\mathbf{p}_a^\gamma \otimes \mathbf{p}_b^\delta \quad \text{for } \gamma\delta \in \overline{\Omega}_a \times \overline{\Omega}_b \tag{11.41}$$

are linearly independent and, similarly, the $K_a K_b$ vectors

$$\mathbf{r}_a^\alpha \otimes \mathbf{r}_b^\beta \quad \text{for } \alpha\beta \in \Omega_a \times \Omega_b \tag{11.42}$$

are linearly independent. It follows from Equation (11.39) and some simple linear algebra that the $K_a K_b$ vectors

$$\mathbf{p}_{ab}^{\gamma\delta} \quad \text{for } \gamma\delta \in \overline{\Omega}_a \times \overline{\Omega}_b \tag{11.43}$$

are linearly independent, as are the $K_a K_b$ vectors

$$\mathbf{r}_{ab}^{\alpha\beta} \quad \text{for } \alpha\beta \in \Omega_a \times \Omega_b. \tag{11.44}$$

From this, we can prove the following theorem:

**Theorem 4.** *For composite systems, we can choose $\Omega_{ab}$ such that*

$$\Omega_a \times \Omega_b \subseteq \Omega_{ab}. \tag{11.45}$$

This theorem has an immediate corollary:

**Corollary 2.** *For composite systems, $K_{ab} \geq K_a K_b$.*

Here, $K_a = |\Omega_a|$. This inequality follows from fact that we have at least $K_a K_b$ linearly independent states in Equation (11.43). The set relation Equation (11.45) follows as the effects in Equation (11.44) are linearly independent and, hence, that we can choose $K_a K_b$ of the $K_{ab}$ fiducial effects in $\Omega_{ab}$ to correspond to local effects. By a local effect, we mean one composed of disjoint circuit fragments, one on system $a$ and one on system $b$. It follows from Equation (11.45) that we can write

$$\Omega_{ab} = \check{\Omega}_{ab} \cup \tilde{\Omega}_{ab} \quad \text{where} \quad \check{\Omega}_{ab} = \Omega_a \times \Omega_b. \tag{11.46}$$

The fiducial effects in $\check{\Omega}_{ab}$ are local. Hence, we can write a general bipartite state, with preparation $\varepsilon$, as

$$\mathbf{p}^\varepsilon_{ab} = \check{\mathbf{p}}^\varepsilon_{ab} \oplus \tilde{\mathbf{p}}^\varepsilon_{ab}, \tag{11.47}$$

where the elements of $\check{\mathbf{p}}^\varepsilon_{ab}$ are the probabilities corresponding to effects in $\check{\Omega}_{ab}$ and the elements of $\tilde{\mathbf{p}}^\varepsilon_{ab}$ are the probabilities corresponding to the effects in $\tilde{\Omega}_{ab}$. Note that $\check{\mathbf{p}}^\varepsilon_{ab}$ lives in the tensor product space of the vector spaces for component systems because $\check{\Omega}_{ab} = \Omega_a \times \Omega_b$. If systems 1 and 2 are both in homogeneous states, then it follows from Equation (11.39) that

$$\check{\mathbf{p}}^{\gamma\delta}_{ab} = \nu_{\gamma\delta} \mathbf{p}^\gamma_a \otimes \mathbf{p}^\delta_b. \tag{11.48}$$

A similar result holds even if only one system is in a homogeneous state (this follows from Theorem 1). These results for bipartite systems generalize in the obvious way to composites having more than two component systems (for three systems, we use $\Omega_{abc} = (\Omega_a \times \Omega_b \times \Omega_c) \cup \tilde{\Omega}_{abc}$).

It is easy to see that if each system is subject to its own local transformation (so the circuit fragments corresponding to the transformations are disjoint), then the state updates as

$$\mathbf{p}^{\varepsilon\alpha\beta}_{cd} = [(^cZ^\alpha_a \otimes {}^dZ^\beta_b)\check{\mathbf{p}}^\varepsilon_{ab}] \oplus {}^{cd}\tilde{Z}^{\alpha\beta}_{ab}\mathbf{p}^\varepsilon_{ab}, \tag{11.49}$$

where the form of the ${}^{cd}\tilde{Z}^{\alpha\beta}_{ab}$ matrix depends, in general, on the particular theory (it acts on $\mathbf{p}^\varepsilon_{ab}$ to give $\tilde{\mathbf{p}}^{\varepsilon\alpha\beta}_{cd}$). This equation follows by considering the case in which both systems are in homogeneous states. Then, we have $K_a K_b$ linearly independent vectors $\check{\mathbf{p}}^{\gamma\delta}_{ab} = \mu_{\gamma\delta}\mathbf{p}^\gamma_a \otimes \mathbf{p}^\delta_b$ with $(\gamma\delta \in \overline{\Omega}_a \times \overline{\Omega}_b)$. The $\check{\mathbf{p}}$ part of the state must remain as a tensor product like this after the local transformations to ensure consistency with Equation (11.39) (notice that if it did not, then there would exist a correlation-revealing measurement contradicting Theorem 2). But the vectors $\mathbf{p}^\gamma_a \otimes \mathbf{p}^\delta_a$ span the space of possible $\check{\mathbf{p}}_a$ vectors, and so Equation (11.49) must be true generally.

If system $d$ is the null system (so the transformation on system 2 is an effect), then the $\tilde{Z}$ matrix has no elements, and we can write

$$\mathbf{p}^{\varepsilon\alpha\beta}_c = (^cZ^\alpha_a \otimes {}^-Z^\beta_b)\check{\mathbf{p}}^\varepsilon_{ab}. \tag{11.50}$$

(Note that if the $\tilde{Z}$ matrix did have elements, then the two sides of this equation would be column vectors of different lengths.) In particular, the reduced state of system 1 is given by

$$\mathbf{p}_a^\varepsilon \equiv \mathbf{p}_a^{\varepsilon 0-} = [{}^a Z_a^0 \otimes (\mathbf{r}_b^-)^T] \check{\mathbf{p}}_{ab}^\varepsilon \tag{11.51}$$

(recall that ${}^a Z_a^0$ is the identity). Hence, the reduced state at either end depends only on the elements in the $\check{\mathbf{p}}_{ab}^\varepsilon$ part of $\mathbf{p}_{ab}^\varepsilon$.

If both systems $c$ and $d$ are null, then Equation (11.50) becomes

$$p^{\varepsilon\alpha\beta} = (\mathbf{r}_a^\alpha \otimes \mathbf{r}_b^\beta) \cdot \check{\mathbf{p}}_{ab}^\varepsilon. \tag{11.52}$$

This also follows directly from Equation (11.39) and the fact that the vectors $\mathbf{p}_a^\gamma \otimes \mathbf{p}_a^\delta$ span the space of possible $\check{\mathbf{p}}_a$ vectors. This equation tells us that all local effects are linear combinations of the fiducial effects corresponding to the $\check{\Omega}_{ab}$ part of $\Omega_{ab}$. Hence, all effects $\mathbf{r}_{ab}^\gamma$ with $\gamma \in \tilde{\Omega}_{ab}$ are nonlocal; the corresponding circuit fragments cannot consist of disjoint parts acting separately on $a$ and $b$. Theories in which the state can be entirely determined by local measurements are called *locally tomographic*. This gives us an important theorem:

**Theorem 5.** Theories having $K_{ab} = K_a K_b$ are locally tomographic and vice versa.

This corresponds to the case in which $\tilde{\Omega}_{ab}$ is the null set.

### 11.7.5 Homogeneity and Uncorrelatability Are Equivalent Notions

We define:

**Definition. An uncorrelatable state** is one having the property that a system in this state cannot be correlated with any other system (so that any joint probabilities factorize).

Let $\mathbf{p}_a^\gamma$ be an uncorrelatable state. Let $\mathbf{p}_{ab}^\gamma$ be a state for a bipartite system having the property that its reduced state is $\mathbf{p}_a^\gamma$. If the bipartite system is prepared in any such state, then the joint probabilities will, by definition, factorize. We will prove the following:

**Theorem 6.** If we allow arbitrary mixtures, then all homogeneous states are uncorrelatable and all uncorrelatable states are homogeneous.

That is, homogeneity and uncorrelatability are equivalent notions. It follows immediately from Theorem 2 that homogeneous states are uncorrelatable. To prove that uncorrelatable states are homogeneous, we assume the converse. Thus, assume that the heterogeneous state $\mathbf{p}_a^\gamma = \mathbf{p}_a^{\gamma_1} + \mathbf{p}_a^{\gamma_2}$ (where the two terms are nonparallel) is uncorrelatable. The state is assumed to be uncorrelatable, so we must be able to write the state of the composite system as

$$\mathbf{p}_{ab}^\gamma = \mu_\gamma [\mathbf{p}_a^\gamma \otimes \mathbf{p}_b^\gamma] \oplus \tilde{\mathbf{p}}_{ab}^\gamma \tag{11.53}$$

because otherwise the probability does not factorize for all $\alpha\beta \in \Omega_a \times \Omega_b$. Hence,

$$\mathbf{p}_{ab}^\gamma = \mu_\gamma [\mathbf{p}_a^{\gamma_1} \otimes \mathbf{p}_b^\gamma] \oplus \tilde{\mathbf{p}}_{ab}^\gamma + \mu_\gamma [\mathbf{p}_a^{\gamma_2} \otimes \mathbf{p}_b^\gamma] \oplus \tilde{\mathbf{p}}_{ab}^\gamma. \tag{11.54}$$

But, using Equation (11.51), we see that the reduced state, $\mathbf{p}_a^\gamma$, for system 1 of Equation (11.54) is parallel to the reduced state, $\mathbf{p}_a^{\gamma'}$, of

$$\mathbf{p}_{ab}^{\gamma'} = \mu_{\gamma_1\delta_1}[\mathbf{p}_a^{\gamma_1} \otimes \mathbf{p}_b^{\delta_1}] \oplus \tilde{\mathbf{p}}_{ab}^{\gamma_1\delta_1} + \mu_{\gamma_2\delta_2}[\mathbf{p}_a^{\gamma_2} \otimes \mathbf{p}_b^{\delta_2}] \oplus \tilde{\mathbf{p}}_{ab}^{\gamma_2\delta_2}, \qquad (11.55)$$

where we choose any two distinct $\mathbf{p}_b^{\delta_1}$ and $\mathbf{p}_b^{\delta_2}$ having normalization such that $\mu_{\gamma_1\delta_1} = \mu_{\gamma_2\delta_2}$. It is possible to choose two distinct states like this if there exist systems that are nontrivial in the sense that they require more than one fiducial effect. (If all systems are trivial, then all states are homogeneous and so, by Theorem 2, all states are uncorrelatable in any case.) We allow arbitrary mixtures and so can take mixtures with the null state to make sure $\mathbf{p}_b^{\delta_1}$ and $\mathbf{p}_b^{\delta_2}$ have normalization so that $\mu_{\gamma_1\delta_1} = \mu_{\gamma_2\delta_2}$. The state Equation (11.55) is preparable by taking a mixture of the preparations $\gamma_1\delta_1$ and $\gamma_2\delta_2$. This state is clearly correlated. By taking a mixture with the null state for the longer of $\mathbf{p}_a^\gamma$ and $\mathbf{p}_a^{\gamma'}$, we obtain two equal states, one of which is uncorrelatable by assumption and one of which is correlatable by the previous proof. Hence, our assumption was false, and it follows that uncorrelatable states are homogeneous.

### 11.7.6 Probabilities for Disjoint Circuit

If we have two disjoint setting-outcome-specified circuits, $\alpha$ and $\beta$, then expect the joint probability to factorize

$$p^{\alpha\beta} = p^\alpha p^\beta. \qquad (11.56)$$

A simple application of Bayes's rule shows that Equation (11.56) is equivalent to demanding that the probability associated with a circuit is independent of the outcomes seen at other disjoint circuits. This is an extremely natural condition because otherwise we would have to take into account all the outcomes seen on all other disjoint circuits in the past that form a part of our memory before writing down a probability for the circuit. Conversely, one can easily envisage a situation in which the probability is not independent of outcomes elsewhere. For example, the eventual outcome of a spinning coin might be correlated with the outcome of an apparently disjoint experiment, which is, incidently, influenced by photons scattered from the coin while it spins. More generally, if there are hidden variables, then there may be correlations between outcomes, even though the marginals are independent of what happens at the other side. It is not clear that disjointness of the circuits is enough to prevent such correlations. In view of this, it is interesting that the following theorem holds.

**Theorem 7.** If there exists at least one type of system that can be prepared in a pure state of norm one, then the probability associated with any circuit is independent of the settings on any other disjoint circuit.

Let the preparation associated with this pure state of norm one be $\gamma$. Consider the circuit $(\gamma-)(\gamma-)(\alpha)(\beta)$ consisting of two instances of the circuit obtained by performing the trace effect on a preparation $\gamma$ and two more disjoint circuits $\alpha$ and $\beta$. Thus, we have four disjoint circuits in total. We can regard this circuit as consisting of the effect $-\alpha$ on one of the $\gamma$ preparations and the effect $-\beta$ on the other $\gamma$ preparation. Then, we have

$$p^{(\gamma-\alpha)(\gamma-\beta)} p^{(\gamma-)(\gamma-)} = p^{\gamma-\alpha} p^{\gamma-\beta}, \qquad (11.57)$$

by Theorem 2 (Equation (11.35) in particular). But $p^{\gamma-} = 1$ because the state is of norm one. Hence, using Bayes's rule, Equation (11.56) follows and the theorem is proved.

We say that a set of operations is *fully closable* if it is closable and if the probability for a circuit is independent of the outcomes seen at other disjoint circuits. It follows from Theorem 7 that closable sets of operations admitting at least one pure state of norm one are fully closable. In the case that we have a fully closable set of operations, it is clear that we can write the $\check{\mathbf{p}}_{ab}$ part of the state associated with disjoint preparations as

$$\check{\mathbf{p}}_{ab}^{\gamma\delta} = \mathbf{p}_a^{\gamma} \otimes \mathbf{p}_b^{\delta}. \tag{11.58}$$

It is interesting to note that Equation (11.48) is an example of this with $v_{\gamma\delta} = 1$, which clearly follows from Equation (11.40) when probabilities for disjoint circuits factorize.

### 11.7.7 Examples of the Relationship among $K_{ab}$, $K_a$, and $K_b$

If $N_a$ is the number of states that can be distinguished in a single-shot measurement, then it is reasonable to suppose $N_{ab} = N_a N_b$. This is true in all of the examples we discuss. In classical probability theory, $K_a = N_a$. In quantum theory, $K_a = N_a^2$. Hence, $K_{ab} = K_a K_b$, and so, by Theorem 5, we have local tomography in these theories. In real Hilbert space quantum theory, where the state is represented by a positive density matrix with real entries, we have $K_a = N_a + N(N-1)/2!$ This has $K_{ab} > K_a K_b$, which is consistent with Corollary 2. However, quaternionic quantum theory has $K_a = N_a + 4N(N-1)/2!$, which has $K_{ab} < K_a K_b$. This is inconsistent with Corollary 2 and, hence, quaternionic quantum theory cannot be formulated in this framework. Because we have made very minimal assumptions (only that we have closable sets of operations in an operational framework), it seems that quaternionic quantum theory is simply an inconsistent theory (at least, for the finite $K_a$ case considered here).

## 11.8 Theories for Which $K_{ab} = K_a K_b$

### 11.8.1 Motivation for Local Tomography

Of the examples we just considered, the two corresponding to real physics are both locally tomographic having $K_{ab} = K_a K_b$. This is a natural property for a theory to have (it is one of the axioms in [23]). It says that from a counting point of view, no new properties come into existence when we put two systems together. It allows a certain very natural type of locality so that it is possible to characterize a system made from many parts by looking at the components. It implies that the full set of states for a composite system requires the same number of parameters for its specification as the separable states (formed by taking mixtures of states prepared by disjoint circuits). Given that this is such a natural constraint, we will study it a little more closely. We will also give axioms for classical probability and quantum theory because they are examples of this sort.

#### *11.8.1.1 Operation Locality*

An extremely useful property of locally tomographic theories is that they are local in the sense that the state is updated by the action of local matrices at each operation. We call this property *operation locality*. We see from Equation (11.47) that if $\tilde{\Omega}_{ab} = \emptyset$, then $\tilde{\mathbf{p}}^\gamma_{ab} = 0$ and $\mathbf{p}^\gamma_{ab} = \check{\mathbf{p}}^\gamma_{ab}$ and, hence, according to Equation (11.49), we see that under local transformations at each end (corresponding to disjoint circuit fragments), the state will update as

$$\mathbf{p}^{\gamma\alpha\beta}_{cd} = (^c Z^\alpha_a \otimes {}^d Z^\beta_b) \mathbf{p}^\gamma_{ab}. \tag{11.59}$$

Hence,

$$^{cd}Z^{\alpha\beta}_{ab} = {}^c Z^\alpha_a \otimes {}^d Z^\beta_b, \tag{11.60}$$

for transformations corresponding to disjoint circuit fragments. In particular, this implies

$$^{cb}Z^{\alpha 0}_{ab} = {}^c Z^\alpha_a \otimes {}^b Z^0_b. \tag{11.61}$$

This is Equation (11.30) that we speculated about in Section 11.7.2. This means an operation has a trivial effect on systems that do not pass through it. If we have a fully closable set of transformations (as long as there exists at least one state of norm one, this follows from Theorem 7), then we can specialize this equation to the case of null input states (where $a = -$ and/or $b = -$) as the state prepared by disjoint preparations is a product state. We assume this in what follows.

The great thing about Equation (11.60) is that it can be used to calculate the probability for any circuit using a $Z$ matrix for each operation. To do this, we choose a complete foliation and then use the tensor product to combine operations at each time step. One way of calculating the probability $p^{\alpha\beta\gamma\delta\epsilon\zeta}$ for the example shown in Figure 11.10 is

$$Z^\zeta_{fg}({}^f Z^0_f \otimes {}^g Z^\epsilon_{ed})({}^f Z^0_f \otimes {}^e Z^\gamma_{bc} \otimes {}^d Z^0_d)({}^f Z^\delta_a \otimes {}^b Z^0_b \otimes {}^c Z^0_c \otimes {}^d Z^0_d)({}^{ab}Z^\alpha \otimes {}^{cd}Z^\beta).$$

$$\tag{11.62}$$

Although it is very satisfying that the calculation can be broken down like this, it is unfortunate that we have to pad out the calculation with lots of identity matrices like ${}^c Z^0_c$. This means that there are more matrices than operations in this calculation. Relatedly, we have to be very careful in what order we take the product of all these matrices (it has to correspond to some complete foliation). In the causaloid approach [24–26], we have neither of these problems. We simply take what is called the causaloid product of a vector associated with each operation without regard for the order and without having to pad out the calculation with identity matrices.

### 11.8.2 Classical Probability Theory

It is easy to characterize classical probability theory in this framework. It is fully characterized by the following two axioms:

**Composition.** $K_{ab} = K_a K_b$.
**Transformations.** Transformation matrices, ${}^c Z^\alpha_a$, have the property that the entries are nonnegative and the sum of the entries in each column is less than or equal to 1.

To see that this is equivalent to usual presentations of classical probability theory, note the following. We can interpret $K_a \equiv N_a$ as the maximum number of distinguishable

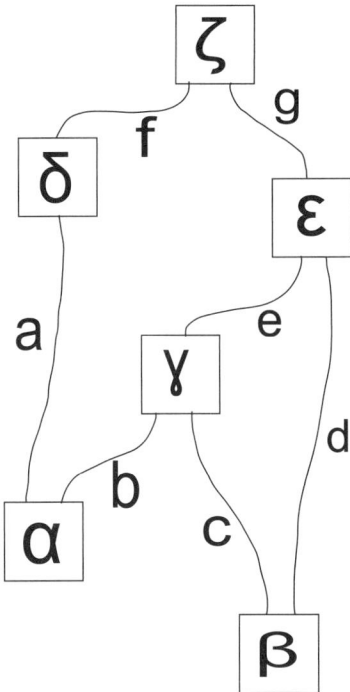

**Figure 11.10.** We show how to calculate the probability associated with this simple example in the text using a $Z$ matrix for each operation.

states for this classical system (for a coin, we have $N_a = 2$; for a die, $N_b = 6$). The state is given by a $\mathbf{p}_a^\alpha = {}^aZ^\alpha$ and is a column vector. The sum of the entries in this vector must be less than or equal to 1. They can be interpreted as the probabilities associated with each of the distinguishable outcomes (for a die, they are the probabilities associated with each face). The trace effect is given by $(\mathbf{r}_a^-) = Z_a^-$. It is a row vector. The value of each entry is 1, and this is consistent with the constraint that the sum of the columns cannot be greater than 1. Norm-preserving transformations are stochastic matrices. Because $K_{ab} = K_a K_b$, we have the operation locality property, and so we can calculate the probability for an arbitrary circuit from matrices for the operations that constitute it.

### 11.8.3 Quantum Theory

To give the rules for quantum theory, we need a few definitions first. Let $\mathcal{H}_{N_a}$ be a complex Hilbert space of dimension $N_a$. Let $\mathcal{V}_{N_a}$ be the space of Hermitian operators that act on this. All positive operators are Hermitian. Furthermore, it is possible to find a set of $N_a^2$ linearly independent positive operators that span $\mathcal{V}_{N_a}$. Let $\widehat{P}_a^\alpha$ for $\alpha \in \Omega_a$ be one such set. Define

$$\widehat{\mathbf{P}}_a = \begin{pmatrix} \vdots \\ \widehat{P}_a^\alpha \\ \vdots \end{pmatrix} \quad \alpha \in \Omega_a. \tag{11.63}$$

A *positive map* $^c\$_a$ from $\mathcal{V}_{N_a}$ to $\mathcal{V}_{N_c}$ is one that acts on a positive operator $\rho_a \in \mathcal{V}_{N_a}$ and returns a positive operator $\rho'_c \in \mathcal{V}_{N_c}$ for any positive operator $\rho_a$. The map $^c\$_a$ is *completely positive* if $^c\$_a \otimes {}^bI_b$ is a positive map from $\mathcal{V}_{N_aN_b}$ to $\mathcal{V}_{N_cN_b}$ for any $b$ where $^bI_b$ is the identity map on $\mathcal{V}_{N_b}$. Further, we want our maps to have the property that they do not lead to probabilities greater than 1. We demand that they must be completely trace nonincreasing when they act on density matrices. This means that $^c\$_a \otimes {}^bI_b$ must be trace nonincreasing for any $b$. Quantum theory is fully characterized by the following two axioms.

**Composition.** $K_{ab} = K_a K_b$

**Transformations.** Transformation matrices are of the form

$$^cZ_a^\alpha = \text{Trace}\left(\widehat{\mathbf{P}}_c \, {}^c\$_a^\alpha(\widehat{\mathbf{P}}_a^T)\right)\left[\text{Trace}(\widehat{\mathbf{P}}_a\widehat{\mathbf{P}}_a^T)\right]^{-1}, \qquad (11.64)$$

where $^c\$_a^\alpha$ is completely positive and completely trace nonincreasing and $T$ denotes transpose.

This is a much more compact statement of the rules of quantum theory than is usually given. We will make a few remarks to decompress this. First, note that $\text{Trace}(\widehat{\mathbf{P}}_c \, {}^c\$_a^\alpha(\widehat{\mathbf{P}}_a^T))$ is a $K_c \times K_a$ matrix having $\beta\gamma$ element $\text{Trace}(\widehat{P}_c^\beta \, {}^c\$_a^\alpha(\widehat{P}_a^\gamma))$ with $\beta \in \Omega_c$ and $\gamma \in \Omega_a$. By defining

$$\mathbf{p}_a^\delta = \text{Trace}(\widehat{\mathbf{P}}_a \rho_a^\delta), \qquad (11.65)$$

where $\widehat{\rho}_a^\delta$ is the usual quantum state, and using $\widehat{\rho}_a^\alpha = \widehat{\mathbf{P}}_a \cdot \mathbf{s}^\alpha$ (because $\widehat{\rho}_a$ must be given by some sum of the linearly independent spanning set), it can be shown after a few lines of algebra that

$$\mathbf{p}_c^{\delta\alpha} = {}^cZ_a^\alpha \mathbf{p}_a^\delta \quad \Leftrightarrow \quad \widehat{\rho}_a^{\delta\alpha} = {}^c\$_a^\alpha(\widehat{\rho}_a^\delta). \qquad (11.66)$$

Hence, we get the correct transformations with Equation (11.64). Note that as in the classical case, we can write a state as $\mathbf{p}_a^\alpha = {}^aZ^\alpha$. This state is associated with a completely positive map $^a\$^\alpha$ where the absence of an input label implies that we have a null input system that corresponds to a one-dimensional Hilbert space. This must have trace less than or equal to one because otherwise $^a\$ \otimes {}^bI_b$ would be trace increasing (this is the reason we impose that the map should be completely trace nonincreasing rather than just trace nonincreasing). Also note that for $^a\$_a^0$ (the identity map), we get the identity for $^aZ_a^0$ in Equation (11.64) as we must. The composition rule $K_{ab} = K_a K_b$ implies that $N_{ab} = N_a N_b$ (we see from inspection of the rank of the matrices that $|\Omega_a| = N_a^2$); hence, the tensor product structure for Hilbert spaces corresponds to the tensor product structure discussed in Section 11.8.1. The fact that we have $K_{ab} = K_a K_b$ means that we have the operation locality property. We can calculate the probability for a general circuit using these $Z$ matrices for each of the operations. Hence, our list of postulates for quantum theory is complete.

### 11.8.4 Reasonable Postulates for Quantum Theory

The objective of this chapter has been to set up a general probabilistic framework. It is worth mentioning that we can give the following reasonable postulates that enable us to reconstruct quantum theory within this framework.

**Information.** Systems having, or constrained to have, a given information carrying capacity have the same properties.
**Composites.** Information-carrying capacity is additive, and local tomography is possible (i.e., $N_{ab} = N_a N_b$ and $K_{ab} = K_a K_b$).
**Continuity.** There exists a continuous reversible transformation between any pair of pure states.
**Simplicity.** Systems are described by the smallest number of probabilities consistent with the other postulates.

We can show from the first two postulates that $K = N^r$ where $r = 1, 2, \ldots$. The continuity postulate rules out the classical probability case where $K = N$. The simplicity postulate then implies that we have $K = N^2$. We construct the Bloch sphere for the $N = 2$ case using, in particular, the continuity postulate. Then, the information postulate and composites postulate are used to obtain quantum theory for general $N$. We refer the reader to [23, 27] for details.

## 11.9 Conclusions

We have exhibited a very natural framework for general probabilistic theories in an operational setting. We represent experiments by circuits and have been particularly careful to give operational interpretations to the elements of these circuits (operations are single uses of an apparatus and wires represent apertures being placed next to one another). By considering closable sets of operations, we are able to introduce probabilities and then set up the full theory wherein $Z$ matrices are associated with circuit fragments. The special case of locally tomographic theories has the operation locality property so that we can combine $Z$ matrices corresponding to circuit fragments that are in parallel using the tensor product. This enables us to break down a calculation into smaller parts. The framework here is still lacking. Most crucially, it is only able to take into account one particular way in which operations can be connected (corresponding to placing apertures next to each other), but there are many other ways. A more general theory is under development to allow more for other types of connections (see [26]). The framework is discrete and hence is not readily adaptable to quantum field theory. It would be interesting either to develop a continuous version or to show how quantum field theory can be fully understood in such a discrete framework.

Algebraic quantum field theory can be understood in operational terms (see, for example, Haag [22]). However, putting the issue of discreteness aside, it is a rather less general operational theory than that presented in this chapter, and so there may be advantages to studying quantum field theory in the framework presented here. It is worth saying that there is a tension between operationalism and use of the continuum in physics. From an operational point of view, the continuum is best understood as a mathematical tool enabling us to talk about a series of ever more precise experiments. It is possible that such a series may, eventually, be better described with other mathematical tools.

There are two types of motivation for considering general probabilistic theories. First, we may be able to formulate and understand better our present theories within these frameworks. It may be possible to write down a set of postulates or axioms that can be used to reconstruct these theories within such a framework. For the case of quantum theory, there has been considerable work of this nature already. It would be interesting to see something similar for general relativity. In particular, there ought to be a simple and elegant formulation of general relativity for the case where there is probabilistic ignorance of the value of quantities that might be measured in general relativity (let us call this *probabilistic general relativity*). Such a theory might be best understood in a general probabilistic framework (although probably more general than the one presented in this chapter) [26]. The second reason to consider general probabilistic theories is to try to go beyond our present theories. The most obvious application would be to work toward a theory of quantum gravity (see [24, 25]). The program of constructing general probabilistic theories and then constraining them using some principles or postulates may free us from the hidden mathematical obstacles to formulating quantum gravity that stand in the way of the more standard approaches such as string theory and loop quantum gravity.

## Acknowledgments

Research at Perimeter Institute for Theoretical Physics is supported in part by the Government of Canada through NSERC and by the Province of Ontario through MRI. I am grateful to Vanessa Hardy for help with the manuscript.

## References

[1] S. Abramsky and B. Coecke. A categorical semantics of quantum protocols. *Proceedings of the 19th Annual IEEE Symposium on Logic in Computer Science (LICS 2004)*, 14–17 July 2004, Turku, Finland. IEEE, 2004, pp. 415–25.

[2] H. Araki. On a characterization of the state space of quantum mechanics. *Communications of Mathematical Physics*, **75** (1980), 1–24.

[3] H. Barnum, J. Barrett, M. Leifer, and A. Wilce. Cloning and broadcasting in generic probabilistic models. arXiv.org:quant-ph/0611295 (2006).

[4] H. Barnum, J. Barrett, M. Leifer, and A. Wilce. A general no-cloning theorem. *Physical Review Letters*, **99** (2007), 240501.

[5] H. Barnum, J. Barrett, M. Leifer, and A. Wilce. Teleportation in general probabilistic theories. arXiv:0805.3553 [quant-ph] (2008).

[6] H. Barnum and A. Wilce. Information processing in convex operational theories. *Electronic Notes in Theoretical Computer Science*. **270** (2011) 1:3–15. arXiv:0908.2352 [quant-ph].

[7] H. Barnum and A. Wilce. Ordered linear spaces and categories as frameworks for information-processing characterizations of quantum and classical theory. arXiv:0908.2354 [quant-ph] (2009).

[8] J. Barrett. Information processing in generalized probabilistic theories. *Physical Review A*, **75** (2007), 032304.

[9] R. F. Blute, I. T. Ivanov, and P. Panangaden. Discrete quantum causal dynamics. *International Journal of Theoretical Physics*, **42** (2003), 2025–41.

[10] G. Chiribella, G. M. D'Ariano, and P. Perinotti. Probabilistic theories with purification. arXiv:0908.1583 [quant-ph] (2009).

[11] B. Coecke. Where quantum meets logic, ... in a world of pictures! Perimeter Institute Recorded Seminar Archive (PIRSA), PIRSA:09040001 (2009). Available at http://pirsa.org/09040001.

[12] G. M. D'Ariano. How to derive the Hilbert-space formulation of quantum mechanics from purely operational axioms. arXiv:quant-ph/0603011 (2006).

[13] E. B. Davies and J. T. Lewis. An operational approach to quantum probability. *Communications of Mathematical Physics*, **17** (1970), 239–60.

[14] A. Einstein. Zur Elektrodynamik bewegter Körper. *Annalen der Physik*, **17** (1905), 891–921.

[15] D. J. Foulis and C. H. Randall. Empirical logic and tensor products. In H. Neumann (ed.). *Interpretations and Foundations of Quantum Theory*. Mannheim: Bibliographisches Institut, Wissenschaftsverlag, 1981.

[16] C. A. Fuchs. Quantum mechanics as quantum information (and only a little more). arXiv:quant-ph/0205039 (2002).

[17] M. Gell-Mann and J. B. Hartle. Classical equations for quantum systems. *Physical Review D*, **47** (1993), 3345–82.

[18] R. B. Griffiths. Consistent histories and the interpretation of quantum mechanics. *Journal of Statistical Physics*, **36** (1984), 219–72.

[19] D. Gillies. *Philosophical Theories of Probability*. London and New York: Routledge, 2000.

[20] D. Gross, M. Mueller, R. Colbeck, and O. C. O. Dahlsten. All reversible dynamics in maximally non-local theories are trivial. arXiv:0910.1840v1 (2009).

[21] S. Gudder, S. Pulmannová, S. Bugajski, and E. Beltrametti. Convex and linear effect algebras. *Reports on Mathematical Physics*, **44** (1999), 359–79.

[22] R. Haag. *Local Quantum Physics: Fields, Particles, Algebras*. Berlin Heidelberg: Springer-Verlag, 1992.

[23] L. Hardy. Quantum theory from five reasonable axioms. arXiv:quant-ph/0101012 (2001).

[24] L. Hardy. Probability theories with dynamic causal structure: A new framework for quantum gravity. arXiv:gr-qc/0509120 (2005).

[25] L. Hardy. Towards quantum gravity: A framework for probabilistic theories with non-fixed causal structure. *Journal of Physics*, **A40** (2007), 3081–99.

[26] L. Hardy. Operational structures as a foundation for probabilistic theories. Perimeter Institute Recorded Seminar Archive (PIRSA), PIRSA:09060015 (2009). Available at http://pirsa.org/09060015.

[27] L. Hardy. Operational structures and natural postulates for quantum theory. Perimeter Institute Recorded Seminar Archive (PIRSA), PIRSA:09080011 (2009). Available at http://pirsa.org/09080011.

[28] J. B. Hartle. Spacetime quantum mechanics and the quantum mechanics of spacetime. *Gravitation and Quantizations: Proceedings of the Les Houches Summer School, 1992*, eds. B. Julia and J. Zinn-Justin, Amsterdam: North Holland, 1995.

[29] M. Leifer. Quantum causal networks. Perimeter Institute Recorded Seminar Archive (PIRSA), PIRSA:06060063 (2006). Available at http://pirsa.org/06060063.

[30] G. Ludwig. *An Axiomatic Basis of Quantum Mechanics*, vols. 1 and 2. Berlin: Springer-Verlag, 1985, 1987.

[31] G. Mackey. *Mathematical Foundations of Quantum Mechanics*. New York: W. A. Benjamin, 1963.

[32] F. Markopoulou. Quantum causal histories. *Classical and Quantum Gravity*, **17** (2000), 2059–77.

[33] R. Omnès. *The Interpretation of Quantum Mechanics*. Princeton, NJ: Princeton University Press, 1994.

[34] P. Panangadan. Discrete quantum causal dynamics. Perimeter Institute Recorded Seminar Archive (PIRSA), PIRSA:09060029 (2009). Available at http://pirsa.org/09060029/.

[35] M. Pawlowski, T. Paterek, D. Kazlikowski, et al. A new physical principle: Information causality. *Nature*, **461** (2009), 1101.

[36] Perimeter Institute Recorded Seminar Archive (PIRSA). Reconstructing Quantum Theory, PIRSA:C09016. Conference organized by P. Goyal and L. Hardy (2009). Available at http://pirsa.org/C09016.

[37] S. Popescu and D. Rohrlich. Nonlocality as an axiom. *Foundations of Physics*, **24** (1994), 379.

[38] R. Sorkin. Spacetime and causal sets. In *Relativity and Gravitation: Classical and Quantum. SILARG VII. Proceedings of the 7th Latin American Symposium on Relativity and Gravitation*, 2–8 December 1990, Cocoyoc, Mexico, eds. J. C. D'Olivo, E. Nahmad-Achar, M. Rosenbaum, et al. Singapore: World Scientific, 1991.

[39] W. K. Wootters. Local accessibility of quantum states. In *Complexity, Entropy and the Physics of Information*, SFI Studies in the Sciences of Complexity, vol. 8, ed. W. H. Zurek. Reading, PA: Addison-Wesley, 1990, pp. 39–46.

[40] W. K. Wootters. Quantum mechanics without probability amplitudes. *Foundations of Physics*, **16** (1986), 391.

# CHAPTER 12
# The Strong Free Will Theorem

John H. Conway and Simon Kochen

## 12.1 Introduction

The two theories that revolutionized physics in the twentieth century, relativity and quantum mechanics, are full of predictions that defy common sense. Recently, we used three such paradoxical ideas to prove "The Free Will Theorem" (strengthened here), which is the culmination of a series of theorems about quantum mechanics that began in the 1960s. It asserts, roughly, that if indeed we humans have free will, then elementary particles already have their own small share of this valuable commodity. More precisely, if the experimenter can freely choose the directions in which to orient his apparatus in a certain measurement, then the particle's response (to be pedantic—the universe's response near the particle) is not determined by the entire previous history of the universe.

Our argument combines the well-known consequence of relativity theory, that the time order of spacelike separated events is not absolute, with the EPR paradox discovered by Einstein, Podolsky, and Rosen in 1935 and the Kochen–Specker Paradox of 1967 (see [5]). We follow Bohm in using a spin version of EPR and Peres in using his set of thirty-three directions, rather than the original configuration used by Kochen and Specker. More contentiously, the argument also involves the notion of free will, but we postpone further discussion of this to the last section of the chapter.

Note that our proof does not mention "probabilities" or the "states" that determine them, which is fortunate because these theoretical notions have led to much confusion. For instance, it is often said that the probabilities of events at one location can be instantaneously changed by happenings at another spacelike separated location, but whether that is true or even meaningful is irrelevant to our proof, which never refers to the notion of probability.

For readers of the original version [3] of our theorem, we note that we have strengthened it by replacing the axiom FIN together with the assumption of the experimenters'

---

This chapter was previously published in *Notices of the American Mathematical Society*, **56** (2009), 226–32. It is reprinted in this volume with permission from the American Mathematical Society.

free choice and temporal causality by a single weaker axiom MIN. The earlier axiom FIN of [3], that there is a finite upper bound to the speed with which information can be transmitted, has been objected to by several authors. Bassi and Ghirardi asked in [1]: What precisely is "information," and do the "hits" and "flashes" of GRW theories (discussed in the Appendix) count as information? Why cannot hits be transmitted instantaneously but not count as signals? These objections miss the point. The only information to which we applied FIN is the choice made by the experimenter and the response of the particle, as signaled by the orientation of the apparatus and the spot on the screen. The speed of transmission of any other information is irrelevant to our argument. The replacement of FIN by MIN has made this fact explicit. The theorem has been further strengthened by allowing the particles' responses to depend on past half-spaces rather than just the past light cones of [3].

## 12.2 The Axioms

We now present and discuss the three axioms on which the theorem rests.

### 12.2.1 The SPIN Axiom and the Kochen–Specker Paradox

Richard Feynman once said that "If someone tells you they understand quantum mechanics, then all you've learned is that you've met a liar." Our first axiom initially seems easy to understand, but beware—Feynman's remark applies! The axiom involves the operation called "measuring the squared spin of a spin 1 particle," which always produces the result 0 or 1.

**SPIN Axiom.** *Measurements of the squared (components of) spin of a spin 1 particle in three orthogonal directions always give the answers* 1, 0, 1 *in some order.*

Quantum mechanics predicts this axiom because for a spin 1 particle, the squared spin operators $s_x^2, s_y^2, s_z^2$ commute and have sum 2.

This "101 property" is paradoxical because it already implies that the quantity that is supposedly being measured cannot in fact exist before its "measurement." For otherwise there would be a function defined on the sphere of possible directions taking each orthogonal triple to 1, 0, 1 in some order. It follows from this that it takes the same value on pairs of opposite directions and never takes two orthogonal directions to 0.

We call a function defined on a set of directions that has all three of these properties a "101 function" for that set. But, unfortunately, we have:

**The Kochen–Specker Paradox.** *There does not exist a* 101 *function for the* 33 *pairs of directions of Figure 12.1 (the Peres configuration).*

> PROOF. We shall call a node even or odd according as the putative 101 function is supposed to take the value 0 or 1 at it, and we progressively assign even or odd numbers to the nodes in Figure 12.1(b) as we establish the contradiction.

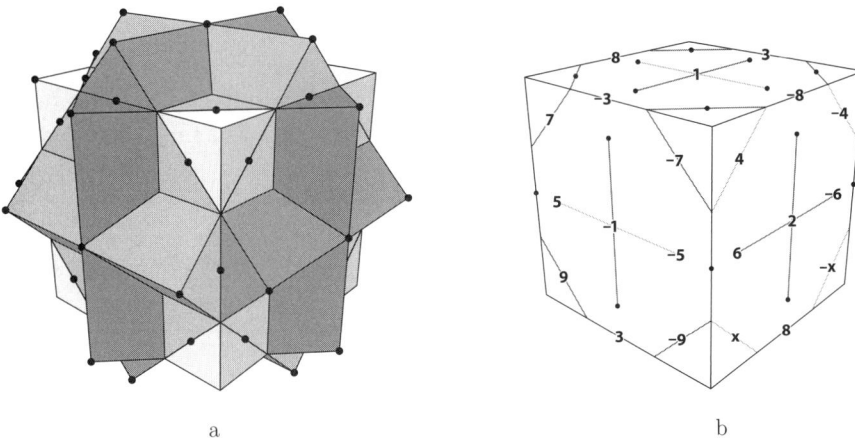

**Figure 12.1.** The three colored cubes in Figure 12.1(a) (represented here by the three darker shades of gray) are obtained by rotating the white cube through 45° about its coordinate axes. The 33 directions are the symmetry axes of the colored cubes and pass through the spots in Figure 12.1(a). Figure 12.1(b) shows where these directions meet the white cube.

We shall use some easily justified orthogonalities—for instance, the coordinate triple rotates to the triple $(2, 3, −3)$ that starts our proof, which in turn rotates (about $−1$) to the triples $(8, −7, 9)$ and $(−8, 7, −9)$ that finish it.

Without loss of generality, nodes 1 and $−1$ are odd and node 2 even, forcing 3 and $−3$ to be odd. Now nodes 4 and $−$x form a triple with 3, so one of them (w.l.o.g. 4) is even. In view of the reflection that interchanges $−4$ and x, while fixing 4 and $−$x, we can w.l.o.g. suppose that $−4$ is also even.

There is a 90° rotation about 1 that moves 7, 5, 9 to 4, 6, x, showing that 5 is orthogonal to 4, while 1, 5, 6 is a triple, and also that 6 is orthogonal to both 7 and 9. Thus, 5 is odd, 6 even, and 7, 9 odd. A similar argument applies to nodes $−5, −6, −7, −9$.

Finally, 8 forms a triple with $−7$ and 9, as does $−8$ with 7 and $−9$. So both of these nodes must be even and, because they are orthogonal, this is a contradiction that completes the proof. ∎

Despite the Kochen–Specker paradox, no physicist would question the truth of our SPIN axiom because it follows from quantum mechanics, which is one of the most strongly substantiated scientific theories of all time. However, it is important to realize that we do not in fact suppose all of quantum mechanics, but only two of its testable consequences—namely, this axiom SPIN and the axiom TWIN of the next section.

It is true that these two axioms deal only with idealized forms of experimentally verifiable predictions because they refer to exact orthogonal and parallel directions in space. However, as we have shown in [3], the theorem is robust in that approximate forms of these axioms still lead to a similar conclusion. At the same time, this shows that any more accurate modifications of special relativity (e.g., general relativity) and of quantum theory will not affect the conclusions of the theorem.

### 12.2.2 The TWIN Axiom and the EPR Paradox

One of the most curious facts about quantum mechanics was pointed out by Einstein, Podolsky, and Rosen in 1935. This says that even though the results of certain remotely separated observations cannot be individually predicted ahead of time, they can be correlated.

In particular, it is possible to produce a pair of "twinned" spin 1 particles (by putting them into the "singleton state" of total spin zero) that will give the same answers to the previous squared spin measurements in parallel directions. Our "TWIN" axiom is part of this assertion.

**The TWIN Axiom.** *For twinned spin 1 particles, suppose experimenter A performs a triple experiment of measuring the squared spin component of particle a in three orthogonal directions $x, y, z$, while experimenter B measures the twinned particle b in one direction, $w$. Then, if $w$ happens to be in the same direction as one of $x, y, z$, experimenter B's measurement will necessarily yield the same answer as the corresponding measurement by A.*

In fact, we will restrict $w$ to be one of the 33 directions in the Peres configuration of the previous section, and $x, y, z$ to be one of 40 particular orthogonal triples—namely, the 16 such triples of that configuration and the 24 further triples obtained by completing its remaining orthogonal pairs.

### 12.2.3 The MIN Axiom, Relativity, and Free Will

One of the paradoxes introduced by relativity was the fact that temporal order depends on the choice of inertial frame. If two events are spacelike separated, then they will appear in one time order with respect to some inertial frames but in the reverse order with respect to others. The two events we use will be the previous twinned spin measurements.

It is usual tacitly to assume the temporal causality principle that the future cannot alter the past. Its relativistic form is that an event cannot be influenced by what happens later in any given inertial frame. Another customarily tacit assumption is that experimenters are free to choose between possible experiments. To be precise, we mean that the choice an experimenter makes is not a function of the past. We explicitly use only some very special cases of these assumptions in justifying our final axiom.

**The MIN Axiom.** *Assume that the experiments performed by A and B are spacelike separated. Then, experimenter B can freely choose any one of the 33 particular directions $w$, and a's response is independent of this choice. Similarly and independently, A can freely choose any one of the 40 triples $x, y, z$, and b's response is independent of that choice.*

It is the experimenters' free will that allows the free and independent choices of $x, y, z$, and $w$. But in one inertial frame—call it the "$A$-first" frame—$B$'s experiment will only happen some time later than $A$'s, and so $a$'s response cannot, by temporal causality, be affected by $B$'s later choice of $w$. In a $B$-first frame, the situation is reversed, justifying the final part of MIN. (We shall discuss the meaning of the term *independent* more fully in the Appendix.)

## 12.3 The (Strong) Free Will Theorem

Our theorem is a strengthened form of the original version of [3]. Before stating it, we make our terms more precise. We use the words *properties*, *events*, and *information* almost interchangeably: whether an event has happened is a property and whether a property obtains can be coded by an information bit. The exact general meaning of these terms, which may vary with some theory that may be considered, is not important because we use them only in the specific context of our three axioms.

To say that $A$'s choice of $x, y, z$ is free means more precisely that it is not determined by (i.e., is not a function of) what has happened at earlier times (in any inertial frame). Our theorem is the surprising consequence that particle $a$'s response must be free in exactly the same sense, that it is not a function of what has happened earlier (with respect to any inertial frame).

**The Free Will Theorem.** *The axioms SPIN, TWIN, and MIN imply that the response of a spin 1 particle to a triple experiment is free—that is to say, is not a function of properties of that part of the universe that is earlier than this response with respect to any given inertial frame.*

**PROOF.** We suppose to the contrary—this is the "functional hypothesis" of [3]—that particle $a$'s response $(i, j, k)$ to the triple experiment with directions $x, y, z$ is given by a function of properties $\alpha, \ldots$ that are earlier than this response with respect to some inertial frame $F$. We write this as

$$\theta_a^F(\alpha) = \text{one of } (0, 1, 1), (1, 0, 1), (1, 1, 0)$$

(in which only a typical one of the properties $\alpha$ is indicated).

Similarly, we suppose that $b$'s response 0 or 1 for the direction $w$ is given by a function

$$\theta_b^G(\beta) = \text{ one of 0 or 1}$$

of properties $\beta, \ldots$ that are earlier with respect to a possibly different inertial frame $G$.

1. If either one of these functions, say $\theta_a^F$, is influenced by some information that is free in the previous sense (i.e., not a function of $A$'s choice of directions and events F-earlier than that choice), then there must be an an earliest ("infimum") F-time $t_0$ after which all such information is available to $a$. Because the nonfree information is also available at $t_0$, all of these information bits, free and nonfree, must have a value 0 or 1 to enter as arguments in the function $\theta_a^F$. So, we regard $a$'s response as having started at $t_0$.

If indeed there is *any* free bit that influences $a$, the universe has by definition taken a free decision near $a$ by time $t_0$, and we remove the pedantry by ascribing this decision to particle $a$. (This is discussed more fully in Section 12.5.)

2. From now on, we can suppose that no such new information bits influence the particles' responses, and therefore that $\alpha$ and $\beta$ are functions of the respective experimenters' choices and of events earlier than those choices.

Now an $\alpha$ can be expected to vary with $x, y, z$ and may or may not vary with $w$. However, whether or not the function varies with them, we can introduce all of $x, y, z, w$ as new arguments and rewrite $\theta_a^F$ as a new function (which, for convenience, we give the same name)

$$\theta_a^F(x, y, z, w; \alpha') \qquad (\star)$$

of $x, y, z, w$ and properties $\alpha'$ independent of $x, y, z, w$.

To see this, replace any $\alpha$ that does depend on $x, y, z, w$ by the constant values $\alpha_1, \ldots, \alpha_{1320}$ it takes for the $40 \times 33 = 1{,}320$ particular quadruples $x, y, z, w$ we shall use. Alternatively, if each $\alpha$ is some function $\alpha(x, y, z, w)$ of $x, y, z, w$, we may substitute these functions in $(\star)$ to obtain information bits independent of $x, y, z, w$.

Similarly, we can rewrite $\theta_b^G$ as a function

$$\theta_b^G(x, y, z, w; \beta')$$

of $x, y, z, w$ and properties $\beta'$ independent of $x, y, z, w$.

Now, there are values $\beta_0$ for $\beta'$, for which

$$\theta_b^G(x, y, z, w; \beta_0)$$

is defined for whatever choice of $w$ that $B$ will make and, therefore, by MIN, for all the 33 possible choices he is free to make at that moment (as $B$ can choose independently of $\beta$).

We now define

$$\theta_0^G(w) = \theta_b^G(x, y, z, w; \beta_0),$$

noting that because by MIN the response of $b$ cannot vary with $x, y, z$, $\theta_0^G$ is a function just of $w$.

Similarly, there is a value $\alpha_0$ of $\alpha'$ for which the function

$$\theta_1^F(x, y, z) = \theta_a^F(x, y, z, w; \alpha_0)$$

is defined for all 40 triples $x, y, z$, and it is also independent of $w$, which argument we have therefore omitted.

But now by TWIN, we have the equation

$$\theta_1^F(x, y, z) = (\theta_0^G(x),\ \theta_0^G(y),\ \theta_0^G(z)).$$

However, because by SPIN the value of the left-hand side is one of $(0, 1, 1)$, $(1, 0, 1)$, $(1, 1, 0)$, this shows that $\theta_0^G$ is a 101 function, which the Kochen–Specker paradox shows does not exist. This completes the proof. ∎

## 12.4 Locating the Response

We now provide a fuller discussion of some delicate points.

1. Because the observed spot on the screen is the result of a cascade of slightly earlier events, it is hard to define just when "the response" really starts. We shall now explain why one can regard $a$'s response (say) as having already started at any time after $A$'s choice when all the free information bits that influence it have become available to $a$.

Let $N(a)$ and $N(b)$ be convex regions of spacetime that are just big enough to be "neighborhoods of the respective experiments," by which we mean that they contain the chosen settings of the apparatus and the appropriate particle's responses. Our proof has shown that if the backward half-space $t < t_F$ determined by a given F-time $t_F$ is disjoint from $N(a)$, then the *available* information it contains is not enough to determine $a$'s response. Conversely, if each of the two such half-spaces contains the respective neighborhood, then of course they already contain the responses. By varying $F$ and $G$, this suffices to locate the free decisions to the two neighborhoods, which justifies our ascribing it to the particles themselves.

2. We remark that not all of the information in the $G$-backward half-space (say) need be available to $b$ because MIN prevents particle $b$'s function $\theta_b^G$ from using experimenter $A$'s choice of directions $x, y, z$. The underlying reason is, of course, that relativity allows us to view the situation from a $B$-first frame, in which $A$'s choice is made only later than $b$'s response, so that $A$ is still free to choose an arbitrary one of the 40 triples. However, this is our only use of relativistic invariance—the argument actually allows any information that does not reveal $A$'s choice to be transmitted superluminally, or even backwards in time.

3. Although we have precluded the possibility that $\theta_b^G$ can vary with $A$'s choice of directions, it is conceivable that it might nevertheless vary with $a$'s (future!) response. However, $\theta_b^G$ cannot be affected by $a$'s response to an unknown triple chosen by $A$ because the same information is conveyed by the responses $(0, 1, 1)$, to $(x, y, z)$, $(1, 0, 1)$ to $(z, x, y)$, and $(1, 1, 0)$ to $(y, z, x)$. For a similar reason, $\theta_a^F$ cannot use $b$'s response because $B$'s experiment might be to investigate some orthogonal triple $u, v, w$ and discard the responses corresponding to $u$ and $v$.

4. It might be objected that free will itself might in some sense be frame dependent. However, the only instance used in our proof is the choice of directions, which because it becomes manifest in the orientation of some macroscopic apparatus, must be the same as seen from arbitrary frames.

5. Finally, we note that the new proof involves four inertial frames: $A$-first, $B$-first, $F$, and $G$. This number cannot be reduced without weakening our theorem because we want it to apply to arbitrary frames $F$ and $G$, including, for example, those in which the two experiments are nearly simultaneous.

## 12.5 Free Will Versus Determinism

We conclude with brief comments on some of the more philosophical consequences of the free will theorem (abbreviated to FWT).

Some readers may object to our use of the term *free will* to describe the indeterminism of particle responses. Our provocative ascription of free will to elementary particles is deliberate because our theorem asserts that if experimenters have a certain freedom, then particles have exactly the same kind of freedom. Indeed, it is natural to suppose that this latter freedom is the ultimate explanation of our own.

The humans who choose $x$, $y$, $z$ and $w$ may, of course, be replaced by a computer program containing a pseudo-random number generator. If we dismiss as ridiculous the idea that the particles might be privy to this program, our proof would remain valid. However, as we remark in [3], free will would still be needed to choose the random number generator because a determined determinist could maintain that this choice was fixed from the dawn of time.

We have supposed that the experimenters' choices of directions from the Peres configuration are totally free and independent. However, the freedom we have deduced for particles is more constrained because it is restricted by the TWIN axiom. We introduced the term *semi-free* in [3] to indicate that it is really the pair of particles that jointly makes a free decision.

Historically, this kind of correlation was a great surprise, which many authors have tried to explain away by saying that one particle influences the other. However, as we argue in detail in [3], the correlation is relativistically invariant, unlike any such explanation. Our attitude is different: following Newton's famous dictum "Hypotheses non fingo," we attempt no explanation but rather accept the correlation as a fact of life.

Some believe that the alternative to determinism is randomness and go on to say that "allowing randomness into the world does not really help in understanding free will." However, this objection does not apply to the free responses of the particles that we have described. It may well be true that classically stochastic processes such as tossing a (true) coin do not help in explaining free will but, as we show in the Appendix and in Section 10.1 of [3], adding randomness also does not explain the quantum mechanical effects described in our theorem. It is precisely the "semi-free" nature of twinned particles and, more generally, of entanglement that shows that something very different from classical stochasticism is at play here.

Although the FWT suggests to us that determinism is not a viable option, it nevertheless enables us to agree with Einstein that "God does not play dice with the Universe." In the present state of knowledge, it is certainly beyond our capabilities to understand the connection between the free decisions of particles and humans, but the free will of neither of these is accounted for by mere randomness.

The tension between human free will and physical determinism has a long history. Long ago, Lucretius made his otherwise deterministic particles "swerve" unpredictably to allow for free will. It was largely the great success of deterministic classical physics that led to the adoption of determinism by so many philosophers and scientists, particularly those in fields remote from current physics. (This remark also applies to "compatibalism," a now unnecessary attempt to allow for human free will in a deterministic world.)

Although, as we show in [3], determinism may formally be shown to be consistent, there is no longer any evidence that supports it, in view of the fact that classical physics has been superseded by quantum mechanics, a nondeterministic theory. The import of the FWT is that it is not only current quantum theory but the world itself that is nondeterministic, so that no future theory can return us to a clockwork universe.

# Appendix. Can There Be a Mechanism for Wave Function Collapse?

Granted our three axioms, the FWT shows that nature itself is nondeterministic. It follows that there can be no correct relativistic deterministic theory of nature. In particular, no relativistic version of a hidden variable theory such as Bohm's well-known theory [2] can exist.

Moreover, the FWT has the stronger implication that there can be no relativistic theory that provides a mechanism for reduction. There are nonlinear extensions of quantum mechanics, which we shall call collectively GRW theories (after Ghirardi, Rimini, and Weber; see [4]) that attempt to give such a mechanism. The original theories were not relativistic, but some newer versions make that claim. We shall focus here on Tumulka's theory rGRWf (see [6]), but our argument below applies, *mutatis mutandis*, to other relativistic GRW theories. We disagree with Tumulka's claim in [7] that the FWT does not apply to rGRWf, for reasons we now examine.

1. As it is presented in [6], rGRWf is not a deterministic theory. It includes stochastic "flashes" that determine the particles' responses. However, in [3], we claim that adding randomness, or a stochastic element, to a deterministic theory does not help:

> To see why, let the stochastic element in a putatively relativistic GRW theory be a sequence of random numbers (not all of which need be used by both particles). Although these might only be generated as needed, it will plainly make no difference to let them be given in advance. But then the behavior of the particles in such a theory would in fact be a function of the information available to them (including this stochastic element).

Tumulka writes in [7] that this "recipe" does not apply to rGRWf:

> Since the random element in rGRWf is the set of flashes, nature should, according to this recipe, make at the initial time the decision where-when flashes will occur, make this decision "available" to every space-time location, and have the flashes just carry out the pre-determined plan. The problem is that the distribution of the flashes depends on the external fields, and thus on the free decision of the experimenters. In particular, the correlation between the flashes in $A$ and those in $B$ depends on both external fields. Thus, to let the randomness "be given in advance" would make a big difference indeed, as it would require nature to know in advance the decision of both experimenters, and would thus require the theory either to give up freedom or to allow influences to the past.

Thus, he denies that both our "functional hypothesis" and, therefore, also the FWT, apply to rGRWf. However, we can easily deal with the dependence of the distribution of flashes on the external fields $F_A$ and $F_B$, which arise from the two experimenters' choices of directions $x, y, z$ and $w$.[1] There are $40 \times 33 = 1,320$ possible fields in question. For each such choice, we have a distribution $X(F_A, F_B)$ of flashes; that is, we have different distributions $X_1, X_2, \ldots, X_{1320}$. Let us be given "in advance" all such random sequences, with their different weightings as determined by the different fields. Note that for this to be given, nature does not have to know in advance the actual

---

[1] This unfortunately makes rGRWf nonpredictive—it can only find the flash distribution that "explains" either particle's behavior when both experimenters' fields are given.

free choices $F_A$ (i.e., $x, y, z$) and $F_B$ (i.e., $w$) of the experimenters. Once the choices are made, nature need only refer to the relevant random sequence $X_k$ in order to emit the flashes in accord with rGRWf.

If we refer to the proof of the FWT, we can see that we are here simply treating the distributions $X(F_A, F_B)$ [$= X(x, y, z, w)$] in exactly the same way we treated any other information-bit $\alpha$ that depended on $x, y, z, w$. There, we substituted all the values $\alpha_1, \ldots, \alpha_{1320}$ for $\alpha$ in the response function $\theta_a(x, y, z, w; \alpha)$. Thus, the functional hypothesis does apply to rGRWf, as modified in this way by the recipe.

Tumulka [7] grants that if that is the case, then rGRWf acquires some nasty properties: In some frame $\Lambda$, "[the flash] $f_y^\Lambda$ will entail influences to the past." Actually, admitting that the functional hypothesis applies to rGRWf has more dire consequences—it leads to a contradiction. For if, as we just showed, the functional hypothesis applies to the flashes, and the first flashes determine the particles' responses, then it also applies to these responses that, by the FWT, leads to a contradiction.

2. Another possible objection is that in our statement of the MIN axiom, the assertion that $a$'s response is independent of $B$'s choice was insufficiently precise. Our view is that the statement must be true whatever precise definition is given to the term *independent* because in no inertial frame can the past appearance of a macroscopic spot on a screen depend on a future free decision.

It is possible to give a more precise form of MIN by replacing the phrase "particle $b$'s response is independent of $A$'s choice" by "if $a$'s response is determined by $B$'s choice, then its value does not vary with that choice." However, we actually need precision only in the presence of the functional hypothesis, when it takes the mathematical form that $a$'s putative response function $\theta_a^F$ cannot, in fact, vary with $B$'s choice. To accept relativity but deny MIN is therefore to suppose that an experimenter can freely make a choice that will alter the past, by changing the location on a screen of a spot that has already been observed.

Tumulka claims in [7] that because in the twinning experiment the question of which one of the first flashes at $A$ and $B$ is earlier is frame dependent, it follows that the determination of which flash influences the other is also frame dependent. However, MIN does not deal with flashes or other occult events, but only with the particles' responses as indicated by macroscopic spots on a screen, and these are surely not frame dependent.

In any case, we may avoid any such questions about the term *independent* by modifying MIN to prove a weaker version of the FWT, which nevertheless still yields a contradiction for relativistic GRW theories, as follows.

**MIN'**. *In an A-first frame, B can freely choose any one of the 33 directions $w$ and $a$'s prior response is independent of B's choice. Similarly, in a B-first frame, A can independently freely choose any one of the 40 triples $x, y, z$, and b's prior response is independent of A's choice.*

To justify MIN' note that $a$'s response, signaled by a spot on the screen, has already happened in an $A$-first frame and cannot be altered by the later free choice of $w$ by $B$; a similar remark applies to $b$'s response. In [7], Tumulka apparently accepts this justification for MIN' in rGRWf: "... the first flash $f_A$ does not depend on the field $F_B$ in a frame in which the points of $B$ are later than those of $A$."

This weakening of MIN allows us to prove a weaker form of the FWT:

**FWT'.** *The axioms SPIN, TWIN, and MIN' imply that there exists an inertial frame such that the response of a spin 1 particle to a triple experiment is not a function of properties of that part of the universe that is earlier than the response with respect to this frame.*

This result follows without change from our present proof of the FWT by taking $F$ to be an $A$-first frame and $G$ a $B$-first frame, and applying MIN' in place of MIN to eliminate $\theta_a^F$'s dependence on $w$ and $\theta_b^G$'s dependence on $x, y, z$.

We can now apply FWT' to show that rGRWf's first flash function ($f_y^\Lambda$ of [7]), which determines $a$'s response, cannot exist, by choosing $\Lambda$ to be the frame named in FWT'.

The FWT thus shows that any such theory, even if it involves a stochastic element, must walk the fine line of predicting that for certain interactions, the wave function collapses to some eigenfunction of the Hamiltonian, without being able to specify which eigenfunction this is. If such a theory exists, the authors have no idea what form it might take.

# References

[1] A. Bassi and G. C. Ghirardi. The Conway-Kochen argument and relativistic GRW models. *Foundations of Physics*, **37** (2007), 169–85.

[2] D. Bohm. A suggested interpretation of the quantum theory in terms of "hidden" variables. I and II *Physical Review*, **85** (1952), 166–79, 180–93.

[3] J. Conway and S. Kochen. The free will theorem. *Foundations of Physics*, **36** (2006), 1441–73.

[4] G. C. Ghirardi, A. Rimini, and T. Weber. Unified dynamics for microscopic and macroscopic systems. *Physical Review D*, **34** (1986), 470–91.

[5] S. Kochen and E. Specker. The problem of hidden variables in quantum mechanics. *Journal of Mathematics and Mechanics*, **17** (1967), 59–88.

[6] R. Tumulka. The point processes of the GRW theory of wave function collapse (2007 October 31). arXiv:0711.0035v1 [math-ph].

[7] R. Tumulka. Comment on "The Free Will Theorem." *Foundations of Physics*, **37** (2007), 186–97.

# Index

Alexandrov topology, 272, 275, 277, 288n1, 301
algebraic quantum field theory (AQFT)
    Bell inequalities and, 324
    $C^*$-algebras and. *See* $C^*$-algebras
    causality and, 335, 350, 355, 359
    entanglement and, 324, 359
    heuristic theory and, 318
    independence in, 143, 354
    locality and, 343–361
    observables and, 134–135, 319, 350
    operationalism and, 439
    relativity theory and, 343, 350, 354
    separability and, 344, 357, 359
    spacetime and, 350, 354
    vacuum state and, 324, 327
    von Neumann algebras and, 318–319
    *See also specific topics*
antonymous functions, 227–235, 260–261
anyons, 62
AQFT. *See* algebraic quantum field theory
Ashtekar, A., 80
associahedra, 31, 33, 96, 105
associativity, 31, 75
Atiyah, M., 25, 67

Baez, J. C., 4, 48
Baez–Dolan theory, 108–114. *See* n-category theory
Bar-Natan model, 117
Barrett, J. W., 91, 171
Batanin model, 111
Bayes's rule, 416, 430, 434, 435
Bell, J., 8
    AQFT and, 324
    entanglement and, 324, 359, 386
    hidden variables and, 393, 394, 405
    noncontextuality constraint, 399
    nonlocality and, 7, 325, 399
    observables and, 325, 386
    Popescu–Rohrlich correlations and, 413
    probabilistic theory and, 366
    realism and, 386
    vacuum and, 7, 325, 326
    von Neumann proof and, 393
bialgebras, 77–79
bicategories, 15, 33–35, 55
    braiding and, 84, 99, 107
    categorification and, 99, 102, 103
    examples of, 34
    Hopf categories and, 102, 103
    linear natural transformations, 84
    MacLane coherence theorem and, 107–108
    monoidal categories and, 50, 82, 98, 99, 102, 107
    morphisms in, 34
    representation theory and, 99, 103
    spin foams and, 81
    sylleptic, 110
    TQFTs and, 97, 99, 102
Birkhoff, G., 5
Birkhoff–von Neumann logic, 130, 132, 208, 211, 220
Bloch representation, 372
Bohm, D., 134
    Broglie–Bohm theory, 385
    configuration variables and, 201
    EPR and, 443. *See* EPR experiment
    hidden variables and, 404–405, 451. *See also* hidden variable theory

Bohrification, 271–313
  $C^*$-algebras and, 286–292, 291
  Gelfand spectra and, 291, 297
  probability and, 301
  quantum logic and, 298
  Rickart algebra and, 297, 305
  state space and, 291
Boolean algebras
  classical theory and, 193–194, 210
  excluded middle and, 193–194, 210
  Hasse diagrams and, 301
  Heyting algebra and, 196, 197, 273, 299
  independence and, 148
  lattice theory and, 193–194, 263, 273, 299
  partial, 298
  probability valuation and, 305
  Sasaki hooks and, 300
  sheafs and, 299
  spectral families and, 267
  Stone spectrum and, 294
Borel functions, 210, 212
Born, M., 20
Bose field, 328
bosons, 41, 73, 150
braiding, 32, 52, 54
  bicategories and, 84, 99, 107
  duals and, 100
  Eckmann–Hilton argument, 109
  free, 52
  hexagon identities and, 32
  interchange law, 97
  Joyal–Street theorem and, 63–64
  knot theory and, 51. See knot theory
  monoidal categories and, 29, 64, 85, 97–99, 109, 112
  n-categories and, 109
  natural isomorphisms and, 53
  string diagrams and, 29–32, 38, 58, 90–93
  symmetry and, 29, 32, 54
  tangles and, 58, 112
  tensor products and, 29
  TQFT and, 67, 85
  weakening and, 53
  Yang–Baxter equation and, 64
Broglie–Bohm theory, 385
Brouwer, L. E. J., 1
Brukner, C., 8
Bruns–Lakser completion, 300, 301
Brussels school, 5
Bub, J., 9
Butterfield–Isham theory, 5–6. See topos theory

$C^*$-algebras, 214, 334
  annihilators, 293, 294
  AQFT and, 134, 135, 140, 141, 273. See also algebraic quantum field theory
  axioms for, 135
  Bohrification and, 286–292. See also Bohrification
  category theory and, 282–283
  commutativity and, 287, 288, 290
  conditional expectation and, 350
  definitions for, 281–283, 293
  diagonal matrices and, 288
  duality and, 283
  f-algebras and, 284, 285
  faithful states and, 302
  Gelfand spectrum and, 214, 271, 283, 291, 306. See also Gelfand spectra
  Hausdorff spaces and, 271
  independence and, 351–353
  infinite, 290
  information theory and, 135
  isotony and, 350
  lattice theory and, 267, 285–286. See also lattice theory
  locales and, 281, 290
  monoidal categories and, 135, 135n11
  nets and, 333, 350
  norm and, 282
  observables and, 135, 140, 308
  probability measure and, 304–305
  quantum logic and, 273
  quasistates and, 302
  rank zero, 294
  Rickart algebras, 273, 292, 294
  simple, 290
  spectral, 293
  state spaces and, 302
  subalgebras of, 287, 348
  topos theory and, 272, 281
  von Neumann algebras and, 254, 292, 294, 319, 319n2, 357. See von Neumann algebras
categorical quantum mechanics (CQM) program, 129
categories, category of, 155n26
category errors, 190
category theory, 3, 13–128, 23. See n-category theory; specific topics
Cauchy sequences, 290
causality, 8, 148, 151–152, 324, 346, 446
  causaloid theory, 436
  classical theory and, 165, 191. See also classical theory
  Common Cause Principle, 359
  determinism and. See determinism

Einstein causality, 321, 335, 337, 343, 350, 355–356
  entanglement and. *See* entanglement
  evolutions and, 150
  graph structure and, 141
  independence and, 143
  indeterminacy and, 8, 156, 411
  information and, 366, 413, 419
  lattices and, 258, 268
  locality and, 321, 350. *See also* locality
  microcausality, 350
  nonisolation and, 165
  probability and, 436–437. *See* probability theory
  process and, 141, 144. *See also* processes
  relativistic, 147, 446
  spacetime and, 446. *See also* spacetime
  vacuum and, 324
  variable structure, 8, 156
  *See also specific topics*
CGMA. *See* Condition of Geometric Modular Action
Chern–Simons theory, 62, 80, 85, 99
choice, axiom of, 279
Cirelson limit, 366
classical theory
  abelian algebras and, 212
  causality and. *See* causality
  complementarity and, 374
  contexts and, 212
  daseinisation and, 209, 226
  determinism and, 169. *See* determinism
  Frobenius algebras and, 165–166, 176
  hidden variables and. *See* hidden variables theory
  logical structure of, 193
  measurement and, 165–178. *See also* measurement
  Newtonian physics and, 192, 365
  observables and, 222, 239–269, 374
  power sets in, 219
  processes and, 130
  propositions in, 217
  quantization and, 130, 193, 214, 251, 267
  quantum theory and, 130, 193, 214, 251, 267. *See also specific topics*
  real numbers on, 235
  realism and, 191–193, 210. *See also* realism
  relativity and, 187. *See also* relativity theory
  representation in, 209–211
  spectral families and, 268
  state spaces and, 210, 236
  topos theory and, 211
  two kinds of, 166

vacuum and, 322. *See also* vacuum state
  valuation and, 226–227
  von Neumann model and, 171–172
cloning operations, 167–168, 175
coalgebras, 77
coarse-graining, 213, 218, 222, 224
cobordisms, 70, 71, 101
  cobordism hypothesis and, 113
  conformal field theory and, 66
  n-dimensional, 66
  Pachner moves and, 91
  spherical categories and, 94
  string diagrams and, 86, 93
  three-dimensional, 93, 94
  TQFTs and, 67, 86, 100
coherence laws, 54, 82
Common Cause Principle, 359
commutativity, 175–176
compactness, 156
compatibalism, 450
complementarity, 176, 386
complex numbers, 35, 44, 83, 88, 190, 201, 284
composition, 130, 137, 139–154
  braiding and, 52
  causality and, 153, 156
  correlatability and, 433–435
  dependent, 142
  foliations and, 426–427
  FUNC and, 200
  functors and, 24, 69
  groups and, 25
  independence constraint, 143–144
  interaction rule for, 144
  measurement and, 431–435
  of morphisms, 23–24, 37–38, 47
  probability theory and, 369, 426–427
  process and, 141
  sufficient isolation, 144
  systems and, 139–154
  tangles and, 62
  tensors and, 70, 175, 176
Condition of Geometric Modular Action (CGMA), 330
configuration variables, 201
conformal field theory, 50, 66, 67, 102
consistent histories approach, 271
constructionism, 279
contextuality, 202, 212, 224, 263–266
continuum. *See* real numbers
contravariant functors, 69, 202
convex set approach, 130
Conway, J., 9. *See also* free will theorem
cosmological constant, 86
cosmology, 86, 188, 200

covariance, 236, 320, 328, 332, 334, 350, 427
CQM. *See* categorical quantum mechanics program
Crane–Frenkel model, 102, 103
Crane–Yetter construction, 96
cryptography, 386

dagger categories, 72, 161
   closed, 134, 160
   dagger functor, 160
   duality and, 69, 162
   monoidal categories and, 73, 140
   observables, 140
Dakic, B, 8
daseinisation
   classical contexts and, 209
   coarse-graining and, 213
   defined, 220
   Hilbert spaces and, 208
   locales and, 306, 308
   map for, 271, 273
   ontological commitment and, 192
   probability and, 301
   projections and, 207, 217–221, 235
   self-adjoint operators and, 207, 208, 222–235
   spectral presheaves and, 219, 235
   topos theory and, 203, 207–238
   two processes of, 208
de Sitter space, 330, 335
decoherence
   broadcasting and, 168
   classicality and, 171
   environment and, 168
   hyperdecoherence, 367n2
   open systems and, 141
Dedekind real line, 280. *See* real numbers
Deligne conjecture, 111
determinism, 168, 169
   causality and. *See* causality
   Einstein and, 450
   free will theorem and, 9, 443–453
   probability and. *See* probability theory
   randomness and, 450. *See also* randomness
   relativity theory and, 451
   uniqueness and, 145
Deutsch–Josza problems, 374
Dijkgraaf, R., 70–72, 77
dimensions, 15, 64, 366–369. *See also* spacetime; *specific topics*
Dirac, P. A. M., 191n4, 345
discreteness, 131, 138–139. *See also specific topics*
distillability, 327
distribution postulate, 5

Dolan, J., 48
Donaldson theory, 67, 103, 117, 118. *See also* Seiberg-Witten theory
Doplicher–Roberts theorem, 72, 74
Drinfel'd, V., 64–65
duals, 68–72
Dyson, F. J., 27

Eckmann–Hilton theory, 97, 98, 107, 108
Eilenberg, S., 18, 23–25
Eilenberg–Moore algebras, 281
Einstein, A.
   AQFT and, 355
   causality and, 321, 335, 350, 355
   determinism and, 450
   *Dialectica* paper, 343–345, 346, 359
   EPR paradox and, 345–346, 443, 446. *See also* entanglement
   field theoretical paradigm, 347
   general theory and, 16, 19. *See* general relativity
   indeterminism and, 7
   nonlocality and, 7, 350
   quantum theory and, 7
   separability and, 6, 7
   special theory and, 8
   von Neumann and, 343–361
Einstein summation convention, 35
electrons, 40, 394
elementary particles, 22, 26, 28, 449–450. *See also specific types, properties*
energy, 22, 403
entanglement, 8, 165, 365–391
   algebraic quantum field theory and, 359
   Bell inequality and, 324, 381, 386
   distillability and, 327
   EPR paradox and, 345–346, 443, 446
   generalized bits and, 378
   hidden variables and, 400. *See* hidden variable theory
   measurement and, 369, 376, 381–382
   mirror states and, 381
   no-signalling property and, 418
   nonlocality and, 460. *See* locality
   paradoxical behavior, 323
   probability theory and, 367, 370, 386
   quantum information theory and, 323, 325
   Reeh–Schlieder Theorem and, 323
   Schrödinger and, 359
   vacuum state and, 324, 326, 327
   von Neumann and, 359
   weak additivity and, 326
environment, system and, 140, 143–145

EPR experiment, 346, 443, 446. *See also* entanglement; locality
Euler characteristic, 115, 116
excluded middle, 197, 279
experiment
  closability and, 418
  free will and, 446
  inertial frames in, 449
  linear compressions and, 424
  MIN axiom, 446
  observables and. *See* observables
  operationalism and. *See* operationalism
  scientific method and, 144
  wave function collapse, 452

Fermat's last theorem, 107
fermions, 41, 73
Feynman diagrams, 4, 13, 15, 26
  Dyson and, 27
  intertwining operators, 26
  loops and, 40
  morphisms and, 27
  Penrose diagrams, 37
  Poincaré group and, 26–27, 28
  spin networks and, 42, 80
  string theory and, 50
  topology of, 50
Feynman, R., 444
Fibonacci series, 401n7
films, spacetime and, 62
filters, 227–228, 230, 240
Fock space, 320n5
foliable structures
  commutativity and, 428–429
  hypersurfaces in, 416
  Lorentz transformations and, 409–410
  measurement and, 424, 431–437
  mixtures and, 424–426
  no-signalling axiom, 418–419
  operationalism and, 409–410, 414
  Popescu–Rohrlich correlations, 413
  probability and, 416–418
  quantum theory and, 437–439
  r-p framework, 412
  spacetime and, 411–412
  states in, 421–422
  system, notion of, 411, 420
  trace measurement, 424
Foulis–Randall school, 132
fractional quantum Hall effect, 62
frames. *See* locales
free will theorem, 9, 443–453
Freed, D. S., 104
Frenkel, I., 102
Freyd–Yetter theory, 58–64
Frobenius algebras, 70, 71, 77, 167
  classicality and, 165–166, 176
  coalgebras and, 71
  commutativity and, 71, 106, 174–176
  complementarity and, 176
  conformal field theories and, 102
  Frobenius law, 169n39
  monoidal category on, 101, 106
  open strings and, 106–107
  string diagrams and, 90, 106–107
Fukuma–Hosono–Kawai model, 91–97, 101
functors, 16–17
  adjoint, 133
  Bohrification and, 272
  categories and, 24–25
  cobordisms and, 94
  covariance and, 69, 202, 236, 272
  defined, 23
  equivalence and, 25
  f-algebras, 284–285, 304
  invariants and, 23
  inverses of, 25
  linear, 93–95
  monoidal, 53–55, 159
  presheaves and, 235–236
  semantics of, 24, 33
  subfunctors, 292–293, 297
  topos theory and, 275–279
  transformations of, 24, 84
  weakening and, 25, 53
  *See also specific types, topics*

Galilean transformations, 410
Galileo, 18
Galileo–Lorentz group, 148
Galois adjoints, 133
gauge theories, 17, 27–28, 188
Gaussian rationals, 281
Gelfand spectra, 203, 215–217, 241, 247, 293
  Bohrification and, 291, 297
  $C^*$-algebras and, 214, 272, 283, 285, 286, 291
  duality and, 271, 283
  filters and, 228–231
  Gelfand-Neimark theorem, 282
  lattice theory and, 271
  locales and, 283
  localic, 107
  presheaves and, 214–217, 264
  representation for, 215
  Rickart algebra and, 297
  self-adjoint operators and, 262

Gelfand spectra (*cont.*)
  spectral transformation, 203, 215, 246, 262–264, 283
  Stone spectrum and, 240, 242, 294
general relativity, 268, 445
  classical physics and, 187
  Einstein and, 16, 19
  gravity and, 16. *See also* quantum gravity
  metric on, 187
  observables and, 268
  probabalistic, 440
  quantum theory and, 187, 203
  spacetime and. *See* spacetime
  vacuum state and, 317
Geneva school, 131, 132, 134, 139, 142
geometric modular action, 330
Ghirardi-Rimini-Weber (GRW) theories, 451
Girard linear logic, 135
Gleason theorem, 136n13, 241–242, 264, 265, 403n8
GNS representation, 320
golden ratio, 401n1
Gordon–Power–Street model, 107–108
Gottesman–Chuang theory, 134
graph theory
  categories and, 139
  causal structure and, 141, 412
  combinatorics and, 111, 139
  complexes and, 93
  composition and, 139
  connectedness and, 139, 143
  four-color theorem, 3–4
  hypersurfaces in, 415
  Kuratowski theorem, 146
  Markopoulou model and, 412
  planar graphs, 146
  undirected graphs, 152
Gray categories, 108
Grothendieck, A., 18, 47–49
Grothendieck group, 115
group theory, 16, 25, 64–65, 162
  Abelian groups, 287
  algebras and, 76, 78, 79
  braiding. *See* braiding
  canonical basis in, 81
  category theory and, 4, 81, 99. *See specific topics*
  class of, 154
  complex forms of, 78
  Feynman diagrams and, 28
  fundamental group. *See* Poincaré group
  homology groups, 25, 104–107, 115–118
  Hopf algebra and, 65. *See* Hopf algebras
  intertwiners and, 4

  isometry groups, 335–336
  Jones polynomial and, 115
  lattices and. *See* lattice theory
  Lie groups, 37, 40, 43, 78–79, 369, 371
  n-groupoids, 18
  orthogonal representation, 371
  perverse sheaves and, 81
  Poincaré group. *See* Poincaré group
  quantum groups, 43, 64, 73–82, 99, 104, 105. *See specific topics*
  SU(2) subgroup, 27, 78, 86
  symmetry groups, 16, 18, 21, 139, 150–151
  tensors and, 85
  TQFT and, 84–85
  triangular decompositions and, 81
  unitary representation and, 21
  universal properties, 52
  Yang–Baxter equation and, 83
  *See also specific types, topics*

hadrons, 187
Hakeda-Tomiyama theory, 147
Hall effect, 62
Hardy, L., 8, 137, 151
Hardy model, 366, 367, 369
Hasse diagrams, 301
Hausdorff spaces, 214, 215, 242, 271, 283, 293
Heegaard-Floer homology, 118
Heidegger, M., 191, 192, 203
Heisenberg, W., 2, 20
Hermitian operators, 394, 395, 403
hexagon identities, 32
Heyting algebras, 195–197, 280
  Boolean algebras and, 196, 197, 273, 299
  distributivity of, 273
  locales and, 274
  morphisms and, 273
  spectral presheaves and, 219, 235
hidden variable theory, 8–9, 344, 385, 434
  Bell and, 394, 403, 405
  Bohm and, 404, 451
  collapse models, 365
  contextuality and, 212, 263–266, 394–395, 404
  entanglement and, 400
  free will theorem and, 451
  Hilbert space operators and, 404
  locality and, 399, 404
  measurement and, 365, 385, 395, 403, 404, 405
  probability theory and, 393–407
  qubits and, 385
  relativity and, 451
  spin and, 402, 405

stochastic states and, 399
von Neumann and, 393–407
Hilbert, D., 2
Hilbert spaces, 20, 21
   AQFT and, 348. *See also* algebraic quantum field theory
   $C^*$-algebras and. *See* $C^*$-algebras
   category theory and, 4, 14, 80, 105, 158, 204
   completeness and, 8
   dagger compactness and, 162
   daseinisation and, 207–208
   dimension two, 264
   Feynman diagrams and, 26. *See also* Feynman diagrams
   first principles of, 191
   formalism of, 2–9, 208
   free will and, 9
   Frobenius algebra and, 176
   graph structure and, 412
   Hermitian operators on, 395, 403, 404, 406
   hidden variables and, 404. *See* hidden variables theory
   infinite-dimensional, 68, 214, 217, 235
   Kraus theorem and, 349
   lattices and, 211. *See also* lattice theory
   linear operators and, 16, 68. *See also* linear operators
   loop quantum gravity and, 80
   Minkowski spacetime and, 419
   n-category theory and. *See* n-category theory
   naturality and, 290
   noncontextuality property, 290
   observables and, 2, 200, 226, 239, 243, 395, 397
   operation locality and, 438
   operators on. *See* operator algebras
   orthonormal basis on, 175–176
   photon and, 26
   physical quantities in, 403
   probability theory and, 435. *See also* probability theory
   pseudo-states and, 203
   quantum theory and, 4, 20, 64, 132, 165, 190, 204, 208, 345, 348, 435, 437. *See specific topics*
   real numbers and, 211
   self-adjoint operators on, 200, 238. *See* self-adjoint operators
   spin and, 8, 140. *See also* spin; *specific topics*
   Stinespring theorem and, 349
   Stone spectra and, 241
   tensor product for, 66, 134, 158, 438
   topos theory and, 200, 207, 208. *See* topos theory
   vacuum state and, 327
   valuations in, 200
   von Neumann algebras and. *See* von Neumann algebras
   weak closedness and, 214
   *See also specific parameters, types, topics*
Hilbert–Poincaré series, 116
Hiley, B. J., 134
HOMFLY-PT polynomial, 58
homotopy theory, 18
   category theory and, 49
   chain homotopies, 105, 115
   framing in, 112
   Grothendieck and, 49
   groupoids and, 105
   homology and, 25, 104–107, 115–118
   homotopy types, 49
   n-categories, 47, 49
Hopf algebras, 177n40
   bialgebras and, 79
   defining equation of, 177n40
   groups and, 65
   Hopf categories, 103
   Planck constant and, 65
   quasitriangular, 74, 75, 79, 100, 104
   representations and, 102
   three-dimensional, 103
   TQFTs and, 102, 103, 104
   trialgebras and, 103

independence, 148, 344
   commutativity and, 351
   spacetime and, 354
   von Neumann algebras and, 352–354
information theory, 419
   bits in, 372–375
   capacity in, 368
   causality and, 366, 413, 419
   computer science and, 4, 13, 73
   constraints for, 366
   dimension and, 366, 368
   distillability and, 327
   matter and, 129–186
   measurement and, 386. *See* measurement
   no-signalling axiom and, 419
   probability and, 366, 368. *See also* probability theory
   properties and, 447
   quantum information theory, 4, 73, 129, 323, 325–327
   qubits and. *See* qubits
   spacetime and, 449
   transmission of, 444

instrumentalism, 187, 201, 366
    laboratory devices, 370–372
    measurement and, 368, 370. *See* measurement
    operationalism and. *See* operationalism
    probability and. *See* probability theory
    quantum correlations and, 366
    quantum gravity and, 208
    relative-frequency interpretation and, 190
intuitionistic logic, 138, 197, 279
Isham, C., 1, 5–6. *See* topos theory

Jones polynomial, 55, 80, 115–116
Joyal, A., 50
Joyal–Street theorem, 63

Kan complexes, 48
Kant, I., 1
Kapranov, M., 82
Kauffman bracket, 56–58, 60, 79
Khovanov homology theory, 115–118
Klyachko model, 397
KMS-states, 329
knot theory, 129
    braiding and, 51
    framing in, 112
    Jones polynomial and, 55, 80
    Kauffman bracket and, 56
    links in, 55, 57
    monoidal categories and, 56
    quantum invariants and, 56, 57, 73
    Reidemeister moves, 51, 57, 82
    skein relations, 56
    stablization hypothesis and, 112
    tangles and, 51, 112
    tricategories and, 107
    Yang–Baxter equation and, 65
*Knots and Physics* (Kauffman), 58
Kochen, S., 9
Kochen–Specker theorem
    contextuality and, 200
    functional relation constraint, 395
    Gleason theorem and, 136n13
    Klyachko on, 394–399
    observables and, 222
    paradox and, 443–446, 448
    physical quantities and, 193, 210, 222, 267
    reformulation of, 290
    spectral presheaf and, 202–203, 226
    topos theory and, 198, 271, 290
    von Neumann algebras and, 210–211
Kontsevich, M., 99–100, 104–107, 111
Kraus operators, 355
Kripke toposes, 280, 283, 284, 286, 295
Kripke–Joyal semantics, 291

Kuratowski theorem, 146

Lambek model, 135n12
Landauer's principle, 167
Langlands program, 107
language, structure of, 135n12
lasers, 386
lattice theory, 132, 193–194
    algebraic restriction and, 259–260
    antitone functions, 296
    Bohrification of, 296
    Boolean algebra and, 263, 299. *See* Boolean algebras
    Bruns–Lakser completion and, 300, 301
    $C^*$-algebras and, 285
    complementarity in, 296, 298
    dual ideal in, 240, 262
    filters and, 228, 240. *See* filters
    finite types and, 241
    Gelfand spectrum and, 271. *See also* Gelfand spectra
    Heyting algebra and. *See* Heyting algebras
    ideals and, 285, 300. *See* ideals
    locales and, 271
    normality and, 286
    observables and, 267
    orthomodular, 240, 241, 298
    paradigm of, 132
    presheaves and, 239
    projections and, 203, 211, 242
    quasipoints and, 242
    regularity condition, 285
    Reisz space and, 284
    semilattices, 275
    spacetime and, 258
    spectral families and, 239–241, 245, 247, 251–252, 256, 259
    Stone spectrum and, 239–241, 240
Lauda, A. D., 4
Laughlin, R. B., 62
Lawvere, F. W., 33
Leibniz–Mach theory, 7
Lie groups, 37, 78–79
light, speed of, 189, 321, 366, 410
Lindenbaum algebra, 280
linear operators
    $C^*$-algebras and, 282, 290. *See also* $C^*$-algebras
    category theory and, 16, 28
    coarse-graining and, 213
    Feynman diagrams and, 13
    Gelfand spectra and. *See* Gelfand spectra
    homotopy and, 105. *See also* homotopy theory

measurement and. *See* measurement
observables and, 2, 200, 226, 239, 243, 395
Pachner moves and, 87, 94, 100
projections. *See* projection operators
topos theory and, 290. *See also* topos theory
TQFTs and, 67–68, 105
*See also specific types, topics*
local tomography assumption, 171
locales, 138, 189, 273
  $C^*$-algebras and, 281
  compact, 282–283
  daseinisation and, 306, 308
  Gelfand spectrum and, 283
  logical aspect, 280
  points and, 271
  sheaf over, 277
  spacetime and, 274, 275, 280
  topos theory and, 276, 277, 279
locality, 215, 318, 335, 386
  algebraic quantum field theory and, 343–361
  Bell inequality and, 7, 325, 399. *See also* Bell, J.
  Einstein causality and, 7, 350
  entanglement. *See* entanglement
  EPR problem, 346, 443, 446
  f-algebras and, 284, 285
  local systems, 355
  logic-gate teleportation, 134
  microcausality and, 350
  nonlocality and, 325
  operationalism and, 343–361
  probabilistic theory and, 368
  quantum information theory and, 325
  quasilocality, 350
  relativity theory and, 343
  separatedness and, 355
  spacetime and, 321
  vacuum state and, 7
logic, 5
  Boolean. *See* Boolean algebra
  classical, 5
  intuitionistic, 6, 138, 197, 279
  linear, 135, 135n12
  quantum. *See* quantum logic
Loomis-Sikorski theorem, 254
loop quantum gravity, 14, 80, 187, 440
Lorentz group, 19, 328
Lorentz, H. A., 18, 406
Lorentz transformations, 18, 409, 410
Lucretius, 450
Ludwig school, 132

Mach's principle, 136
Mach–Zehnder interferometer, 386

MacLane coherence theorem, 107
MacLane, S., 18, 23, 28–32
MacLane strictification theorem, 160
mass, of particles, 22
Mathematical Fondations of Quantum Mechanics (von Neumann), 413
*Mathematical Foundations of Quantum Mechanics* (von Neumann), 393
mathematics, foundations of, 48
Maxwell equations, 409–410
Maxwell, J. C., 17–18
MBQC. *See* measurement-based quantum computational model
measurement
  Bohm theory of, 404–406
  classical theory and, 165–178
  coarse-graining and, 213, 218, 222, 224
  collapse models, 365, 385, 451–452
  complementarity and, 374
  composite systems and, 431–433 (XX) JIM
  contextuality and, 202, 212, 224, 263–266
  continuity axiom and, 380
  determinism and, 395, 399
  Deutsch-Josza problems, 374–375
  dimension and, 369, 384–385
  energy and, 403
  entanglement and, 369, 376, 381–382
  experiment and, 366, 368–372
  fiducial sets and, 372, 374, 431–433
  foundation principles and, 368
  generalized bits and, 378–380
  Hardy model and, 380
  hidden variables and, 365, 385, 395, 403, 404, 405
  independence and, 400, 452
  information theory and, 386
  instrumentalism and, 368, 370. *See* instrumentalism
  laboratory operations, 366, 370–372
  linear operators and. *See* linear operators
  locality and, 376, 399, 409, 433–435. *See* locality
  MBQC. *See* measurement-based quantum computational model
  measurable functions, 253–254
  measurement problem, 136
  mirror theory and, 380
  mixed states and, 372, 373, 376
  no signaling condition, 399
  noncontextuality constraint and, 399
  observables and, 169, 253–254, 319, 348, 349, 405. *See* observables
  operators in, 131, 366, 409, 413–416. *See also* operator algebras

measurement (*cont.*)
  preparations and, 421
  probability and, 187, 371, 372, 373, 377, 399.
    *See* probability theory
  as process, 130, 133–134
  projections and. *See* projection operators
  pure states and, 377
  qubits and, 380, 385
  realism and, 385
  Sorkin theory, 368
  spatiotemporality and, 188, 348, 402
  spectral family and, 253
  spin and, 397, 398, 400, 405
  state and, 371, 384–385
  topos and, 201
  truth values and, 209
  update rule, 384–385
  vacuum and, 322. *See also* vacuum state
  valuation and, 215, 216, 226–227
  von Neumann algebras and, 319, 349. *See also* von Neumann algebras
  von Neumann model, 402
  wave function collapse and, 385, 405
measurement-based quantum computational model (MBQC), 134
Mermin, D., 8
microcosm principle, 287
Mills, R., 27
Milnor conjecture, 117
Minkowski space, 18, 19, 317
  absolute geometry and, 333
  algebraic conditions on, 334–335
  fixed-background and, 7
  metric for, 19
  metric in, 19
  modular covariance in, 334
  no-signalling axiom and, 419
  Poincaré group and, 317. *See* Poincaré group
  quantum field theory and, 330
  spacetime and, 18, 330, 334
mirror mechanics, 106, 381–384
Mitchell-Bénabou theory, 278
modular stability, 328, 332
modular tensor categories, 96
monoidal categories, 4–5, 28, 53, 85, 97
  6*j* symbols, 46–47
  algebraic properties, 77–78
  associator and, 30, 31
  bicategories and, 34, 55
  braiding and, 15, 29, 32, 51–54, 64, 97. *See* braiding
  coherence for, 160
  comonoids and, 172–174
  connectedness and, 136

  dagger compact, 73
  definition of, 30
  duals and, 72, 73
  hexagon identities and, 32
  Joyal–Street theory, 51–52
  knot theory and, 56
  locale morphisms, 280
  monoidal equivalence, 55
  pentagon identity and, 31, 32
  reindexing, 34
  Schur's lemma and, 37
  spherical, 92
  strict, 32, 55, 154–157, 160
  string diagrams and, 28
  symmetric, 15, 29, 32, 50, 55, 59, 129, 136, 154–158, 160
  tangles and, 74
  tensor product and, 30
  topological quantum field theory and, 68
  TQFTs and, 47, 68
  triangle identity, 31
  unit laws, 30, 31
  vector spaces and, 36
  Yang–Baxter equation and, 64
morphisms, 14, 23
  2-morphisms, 35
  associative laws, 23
  bicategories and, 34, 82, 108
  braiding and, 51–52, 73, 112, 117
  cobordisms and, 67–68, 99–101
  coherence laws, 82
  composition of, 14, 23–25, 34, 66
  duals and, 38, 68, 69, 72, 73
  endomorphisms and, 51
  Feynman diagrams and, 27
  frame morphisms, 281
  Frobenius algebras and, 101–102
  functors and, 25. *See also* functors
  geometric, 285
  Heyting algebra and, 273, 274
  homomorphisms, 24
  homotopy theory and, 105
  identity morphisms, 23, 39
  intertwining operators, 37
  invertible, 16
  irreducible representations, 37, 38
  isomorphisms, 30, 31, 34, 43–45, 53, 82, 84
  *j*-morphisms and, 48
  locales and, 280, 303
  monoidal categories and, 30, 35, 136, 154–158, 174
  n-category theory and, 14, 17, 107–109
  networks, 41
  objects and, 14, 23

Poincaré group and, 37, 320
presheaves and, 263
processes and, 14
pullback and, 276
purification of, 163
relationalism and, 136
Shum's category and, 62
simplices, 48
symmetry and, 164
tangles and, 58, 59, 112, 117
tensors and, 28, 35
time and, 68
topos theory and, 278–279
unitary, 23, 59, 69
weak equivalences, 49
zig-zag identities and, 39
Morse theory, 70
Murray, J., 343

n-category theory, 49
braiding and, 52
category theory and, 49
cobordisms and, 113, 114
computer science and, 13
definition of, 49
development of, 13–128
duals and, 63, 112
equivalence and, 49
homotopy theory and, 49
opetopic approach, 47–48
periodic table for, 97, 108, 110
semistrict, 108
stablization hypothesis, 110
string theory and, 50
topology and, 118
TQFTs and, 68, 101, 108, 114
Newtonian physics. *See* classical theory
Newton's constant, 189
no-broadcasting theorem, 365
no-cloning theorem, 365
no-go theorems, 136, 236, 344
no-signalling assumption
closability and, 418
Einstein causality and, 355–356, 366
entanglement and, 418
information causality and, 419
Minkowski spacetime and, 419
Popescu-Rohrlich correlations and, 413
Noether, E., 115
noncontextuality constraint, 399. *See* contextuality
nonlocality, 5, 7, 325. *See* locality

observables

antonymous functions, 227–235, 260–261
AQFT and, 350
Bell's inequality and, 325, 386
Bohmian particles, 404
$C^*$-algebras and. *See* $C^*$-algebras
classical, 168, 222, 239–269. *See also* classical theory
complementarity and, 374, 386
contextuality and, 212, 224, 263–266
dagger symmetry and, 140
defined, 229
determinism and, 395. *See also* determinism
elementary propositions and, 306
energy and, 22
experiments and, 134, 333, 395. *See* experiment
general relativity and, 268
Hilbert spaces and, 2, 200, 226, 239, 243, 404
Kochen–Specker theorem and, 222
lattice theory and, 267
locality and, 201, 344. *See also* locality
measurement and, 169, 201, 253–254, 319, 348, 349, 393n1, 405. *See* measurement
modular involutions and, 329
monoidal categories and, 140
nets and, 329–330, 331, 335, 350
observable functions, 243–251, 255–256, 261
physical quantities and, 131, 207, 222, 394–396. *See* measurement; *specific topics*
real numbers and, 239
representation of, 222–235
self-adjoint operators and, 21, 239, 348. *See* self-adjoint operators
spacetime and, 318. *See* spacetime
spin, 395. *See* spin
state and, 301–309, 318
topological theory and, 255–256
vacuum and, 322
ontological commitment, 146, 192, 385. *See also* classical theory; *specific theories*
operationalism, 131, 385, 410
AQFT and, 439
background time in, 410f, 411
causal structure and, 411
closable sets in, 417, 439
continuum and, 439
experiment and, 413–414
foliable structures. *See* foliable structures
independence and, 344
local tomography and, 435–436
locality and, 343–361
measurement and, 131, 413–416. *See* measurement
operator algebras. *See* operator algebras

operationalism (*cont.*)
  probability and, 409–410, 439. *See* probability theory
  processes and, 142
  quaternionic quantum theory, 435
  realism and, 132
  systems and, 420
operator algebras, 55–56, 200, 318
  $C^*$-algebras and. *See* $C^*$-algebras
  duals in, 348
  foliable structures. *See* foliable structures
  Gelfand spectra and. *See* Gelfand spectra
  Hilbert space and, 6
  independence, 357–358
  intertwining operator, 76–77
  linear operators. *See* linear operators
  locality and, 318
  operads, 33
  projection and. *See* projection operators
  self-adjoint. *See* self-adjoint operators
  separability and, 357
  Tomita–Takesaki theory and, 327
  von Neumann algebras. *See* von Neumann algebras

Pachner moves, 87
  cobordisms and, 91–92
  invariance under, 100
  pentagon identity and, 95f
  string diagrams and, 87
parity, 22
particles, elementary
  Bohm and, 405
  free will and, 449–450
  locales and, 322
  momentum of, 405
  phase multiplication and, 73
  Poincaré groups and, 26
  Standard Model, 28
  *See also specific types, properties*
Pauli operators, 175, 395
PCT theorem, 329
Penrose, R., 35–43, 60, 188
pentagon identity, 31, 32, 95
Peres configuration, 444, 446, 450
perturbation theory, 330
phase groups, 162
phase space, 210
photons, 368, 411
Piron's theorem, 137
Planck scale, 65, 189, 375
Poincaré group, 40, 80, 317
  elementary particles and, 26
  Feynman diagrams and, 26–27, 37, 80

infinite dimensionality and, 40
morphisms and, 37
observables and, 40
Poincaré dual, 87, 89, 93
positive-energy representations, 22, 28, 67
relativity theory and, 21–22
SU(2) and, 81
vacuum state and, 330
Poincaré, H., 18–19
Poincaré invariants, 320
Poincaré spin network, 44
pointfree topology, 271
point, primacy of, 64, 72, 114, 117, 366
Ponzano–Regge model, 44–45, 80, 85
Popper, K., 188
potentiality, 139–140
power sets, 195
probability theory, 20, 131, 145, 416
  Bayes rule, 416, 430, 434, 435
  Bell inequalities and, 366. *See* Bell, J.
  Bohrification and, 301
  Boolean algebra and, 305
  $C^*$-algebras and, 304, 305. *See also* $C^*$-algebras
  causality and, 411
  classical, 436–437
  commutativity and, 319
  continuity and, 190–191, 370, 386
  correlation and, 433
  daseinisation and, 301
  degrees of freedom and, 366, 367
  determinism and, 394, 398, 400
  dimension and, 368
  dispersion-free ensembles, 402–403
  entanglement and, 367f, 370, 386
  expectation and, 350, 394, 403
  fiducial measurements and, 371, 374, 431
  foliable structures and, 416–418. *See* foliable structures
  generalized truth values and, 191
  hidden probabilities, 399
  hidden variables and, 393–407
  information and, 366, 368
  interpretations of, 416
  Klyachko model and, 396
  latency, 190
  local tomography, 412, 413
  locales and, 443
  locality and, 368, 437
  matrix theory and, 423
  measurement and, 187, 371, 372, 373, 376, 377. *See also* measurement
  mixed states and, 372, 426

operationalism and, 409–411, 437, 439
potentiality and, 190
probability vector, 372
problems with, 443
projection and. *See* projection operators
purification of, 163–165
quantum gravity and, 142, 411, 440
r-p framework and, 412
randomness and. *See* randomness
real numbers and, 190–191, 370, 386
relative frequency and, 190, 416
reversibility and, 20, 368
Schrödinger on, 345
state space and, 367, 371
topos theory and, 273
trace and, 403, 424
transition amplitude, 20
transpose and, 423
uncorrelatability and, 433
update rules, 384–385
von Neumann algebras and. *See* von Neumann algebras
weights and, 142, 144–145
processes
classicality and, 130, 166–169, 168
composition of, 141–143
environment and, 143
examples of, 154
feed-into-the-environment, 144
graphical representation of, 145–146
identical, 149–152
interactions, logic of, 135–136
inverses to, 149
matter and, 141
open systems and, 144
operations and, 142
probes and, 160
relations and, 130, 136–137
reversibility and, 20
stochastic, 169. *See* probability theory
structure and, 129–186
symmetry relations, 20, 141
systems from, 134–135
virtual, 148
projection operators, 203, 292
abelian algebras and, 213–214
$C^*$-algebras and. *See* $C^*$-algebras
clopeness in, 216
coarse-graining and, 218
complex numbers and, 201, 201n14
daseinisation and, 207, 208, 217–221, 219
Klyachko model and, 396
lattice theory and, 211, 220
maximal filters, 227

measurement and, 142, 144. *See also* measurement
operators and, 292. *See also* operator algebras
orthogonal bases for, 213–214
physical quantities and, 218. *See also* observables
projection postulate, 349
projective geometry, 133n33
propositions and, 218
self-adjoint, 224. *See* self-adjoint operators
spectral theorem and, 211, 242
von Neumann algebra and, 241–243, 292. *See* von Neumann algebras
proof theory, 135–136
properties, physical, 131–133, 222. *See* observables; *specific topics*
*Pursuing Stacks* (Grothendieck), 49
Putman, H., 5

QFT. *See* quantum field theory
quantization, 111, 130, 239–269
quantum cosmology, 188, 200
quantum field theory (QFT), 6, 86, 140, 328
AQFT. *See* algebraic quantum field theory
Atiyah and, 25
Chern–Simons theory and, 62
Donaldson theory and, 118
functorial semantics, 33
gravity and. *See* quantum gravity
infinities and, 40
locality and, 7. *See also* locality
Milnor conjecture and, 117
Minkowski space and, 330. *See also* Minkowski space
quantum groups and, 99
renormalization in, 137
Seiberg–Witten theory, 118
spacetime in. *See* spacetime
spin-statistics theorem, 62
string theory and, 50, 118
system in, 131
Tomita–Takesaki theory and, 7, 318, 327–329
TQFTs and, 25, 47, 53, 67. *See* topological quantum field theory
two-dimensional, 64
vacuum state and, 330
quantum gravity, 15
conventional formalism and, 203
gravity and, 44, 188
instrumentalism and, 208
loop quantum gravity, 14, 80, 187, 440
matter and, 46
Ponzano–Regge model and, 44, 45–46
probabalistic theory and, 440

quantum gravity (cont.)
  problems of, 187, 189
  quantum theory and, 189
  real numbers and, 190–191
  spacetime and, 44, 45. See also spacetime
  topos methods and, 187–207
quantum information theory
  distillability and, 327
  entanglement and, 323, 325–326
  locality and, 325
  logic and. See quantum logic
  quantum computing, 4, 73, 129
  qubits and. See qubits
quantum logic, 130, 344, 345
  Bohrification and, 298
  $C^*$-algebras and, 273
  classical logic and, 5
  conceptual problems of, 220
  implication relation, 218n12
  information theory. See quantum information theory
  operator algebras and, 292
  proof theory and, 136
  qubits and. See qubits
  topos theory and, 207
quantum teleportation, 156–157
quasicategories, 114
quasiparticles, 62
quasipoints, 240, 242, 253, 258, 267
quasistates, 302
quaternions, 374, 409, 435
qubits, 374
  hidden variables and, 385
  information and, 385
  logic of interactions, 418
  measurement and, 380, 385
  Spekkens toy theory, 130, 162, 177

r-p framework, 411–413
Radiohead, 137
randomness, 386
  determinism and, 450. See also determinism
  free will and, 450
  probability and. See probability theory
  relativity theory and, 451
  stochastic elements, 451
Raussendorf–Briegel model, 134
real numbers, 211
  continuum and, 21, 138, 436
  observables and, 239
  probability and, 190–191
  quantum gravity and, 190–191
  spacetime and, 190
  spectral theory and, 203

topos theory and, 199, 203, 209n3, 225
realism, 187, 191–194, 210
  Bell's inequality and, 386
  classical physics and, 191–193, 210
  measurement and, 385
  neorealism and, 196–197
  operationalism and, 132
  physical properties and, 132
  quantum theory and, 193, 194
Reeh–Schlieder theorem, 318, 327, 356
  entanglement and, 322, 323
  vacuum state and, 351
Reichenbach, H., 359
Reidemeister moves, 51, 57, 82. See also knot theory
relationalism, 7
  causality and, 165
  classical theory and, 165–169
  composition and, 139–145
  continuum and, 138–139
  measurement and, 169–177
  physical theory and, 154–162
  processes and, 136–138, 145–146
  symmetry and, 148–152
  systems and, 140–141
  vacuous relations, 152–154
  von Neumann model and, 162–165
relativity theory, 136
  determinism and, 451
  Einstein and, 19. See Einstein, A.
  general theory. See general relativity
  geometry and, 5
  locality and, 343. See locality
  Maxwell and, 18
  quantum field theory and, 343, 346
  randomness and, 451
  spacetime and. See spacetime
  special theory, 8, 445
representations, theory of, 74, 76, 81, 129
Reshetikhin, N., 73, 84–85
reversibility, 20, 368, 386
ribbon structure, 63
Rickart $C^*$-algebras, 273, 292–297, 305
Riesz space, 284, 302
Roberts, J., 46, 72
Robertson–Walker theory, 335, 336
Rovelli, C., 80
Russell's paradox, 154

Sasaki hooks, 133n4, 300
scattering, 329, 335
Schlieder property, 325
Schmidt decomposition, 377
Schrödinger equation, 345, 359, 365

Schur–Auerbach lemma, 371, 377
scientific method, 144
Segal spaces, 114
Seiberg-Witten theory, 103, 118
self-adjoint operators, 21, 239
    algebras and, 202
    antonymous functions and, 260
    approximations and, 224
    Borel functions and, 212
    context and, 223
    daseinisation and, 207, 208, 225–227, 235
    eigenstate-eigenvalue link and, 232–235
    evolving, 335
    FUNC principle and, 215
    Galois adjoints and, 133
    Gelfand transform and, 262
    global sections and, 264
    Hilbert space and. *See* Hilbert spaces
    linear order, 223
    observables and, 348. *See* observables;
      *specific types, topics*
    order on, 223
    projection and, 224. *See* projection operators
    Riesz theorem. *See* Riesz theorem
    spatial restriction and, 257
    spectral families and, 223, 251, 258, 259
    Stone's theorem. *See* Stone's theorem
    von Neumann algebras and. *See* von Neumann algebras
semisimple algebras, 86, 88–89, 92, 102
separability, 6, 7, 344, 357–358
sequential composition, 142
sheaf theory
    étalé space and, 240
    Gelfand spectra and, 214–217
    Kochen–Specker theorem and, 202–203, 226
    locales and, 277
    presheaves and, 202–203, 214–217, 263, 266–267
    stalks and, 240
    Stone spectrum and, 268
    von Neumann algebras and, 263
Shum's theorem, 62–63
simplicity axiom, 367
Smolin, L., 1, 80
Sorkin measure theory, 368
spacetime
    absolute geometry and, 333
    algebraic conditions on, 334–335
    AQFT and. *See* algebraic quantum field theory
    background, 7, 187, 200
    causal principle and, 446
    CGMA and, 330, 336
    circuit model and, 415
    combinatorial topology for, 100
    cosmological constant and, 86
    covariance and, 350–351
    curvature of, 16, 187
    differential geometry and, 191
    dimensions in, 15, 100
    discrete, 138
    experiments in, 134
    films and, 62
    foliable structures and, 411, 414
    gauge theories and, 17, 27–28, 188
    hypersurfaces in, 415–416
    independence and, 347, 354
    information and, 449
    lattice theory and, 258. *See also* lattice theory
    local operation, 347
    local operations and, 347, 354–357
    locales and, 274
    locality and, 321. *See* locality
    matter in, 136
    Minkowski space and, 19, 330, 334
    modular stability and, 336
    Newtonian, 191
    nonlocality and, 7. *See* locality
    nonpathological models, 354
    nonstandard models, 189
    observables and, 318, 332–335. *See also* observables
    points in, 187
    quantum gravity and, 44. *See* quantum gravity
    real numbers and, 21, 138, 190–191, 436
    relativity and. *See* relativity theory
    Sorkin model, 412
    spins and, 44, 62
    states and, 332–335
    string theory and. *See* string theory
    systems and, 140
    temporal order, 446
    three-dimensional, 44, 62
    topos theory and. *See* topos theory
    TQFTs and. *See* topological quantum field theory
    vacuum state and, 7, 318
spectral families, 244, 247–248, 251, 254
    admissible domain of, 251–252
    algebraic restriction of, 258–262
    antonymous functions and, 260
    Boolean algebra and, 267
    continuous functions and, 255–256
    Gelfand spectra, 203, 214–217. *See* Gelfand spectra

spectral families (*cont.*)
   Kochen–Specker theorem. *See*
      Kochen–Specker theorem
   measurable functions and, 253
   noncommutative, 203
   observables and, 245, 248, 250, 268
   presheaves and, 263
   projections and, 242. *See* projection operators
   real numbers and, 203. *See* real numbers
   self-adjoint operators and, 251, 258, 259. *See* self-adjoint operators
   sheaf theory and, 257, 266, 267
   spectral theorem and, 200, 211, 242
   spectrum condition, 320, 351
   Stone spectrum and, 267. *See* Stone spectrum
   vacuum and, 351
Spekkens toy theory, 130–131, 162, 177
spherical categories, 92
spin, 40, 42
   $6j$ symbols, 44–45
   Bell inequalities. *See* Bell, J.
   binor identity, 43
   closed, 42
   entanglement and. *See* entanglement
   EPR paradox and, 446. *See* EPR experiment
   Feynman diagrams and, 80
   free will theorem, 447–448
   graph theory and. *See* graph theory
   gravity and, 44
   hidden variables and, 402, 405
   Kauffman bracket and, 56, 57–58
   Klyachko model and, 397
   measurement and, 397, 398, 400, 405
   observables and, 395
   Penrose and, 60
   Peres configuration and, 444
   Poincaré dual, 44–45
   projections and, 374
   spin axiom, 444
   spin foams, 14, 15, 22, 80, 94
   spin-statistics, 329, 335
   topos theory and, 220–221, 233
   twinned spin, 446
   zig-zag identities and, 42
Standard Model, 22, 27, 28, 449–450. *See specific topics*
Stasheff associahedra, 31, 33, 96, 105
Stinespring theorem, 349
stochastic processes, 169. *See also* probability theory; *specific topics*
Stone spectrum, 240, 246, 294
   clopen subsets and, 293
   Gelfand spectrum and, 240, 242, 294
   lattice theory and, 239–241
   representation theorem, 210, 293
   Rickart $C^*$-algebra and, 294
   sheaves and, 268
   spectral families and, 267
   von Neumann algebras and, 241, 243, 266
Stone–von Neumann theorem, 21, 65–67
Störmer, H., 62
Street, R., 50
string theory, 15, 50, 75, 106
   blackboard framing, 61
   categorical structures and, 104–106. *See specific topics*
   closed strings, 50
   cobordisms and, 86, 93
   conformal structure and, 50
   duals and, 38
   Feynman diagrams and, 50. *See also* Feynman diagrams
   Frobenius algebras and, 90. *See also* Frobenius algebras
   higher-dimensional, 96
   loop quantum gravity and, 14, 80, 187, 440
   monoidal categories and, 28. *See also* monoidal categories
   n-categories and, 50. *See* n-category theory
   open stings, 50
   Pachner moves and, 87. *See* Pachner moves
   Poincaré duality and, 87, 89
   projections and, 90. *See* projection operators
   quantization and, 111
   string diagrams and, 29–32, 38, 58, 90–93
   TQFTs and, 90. *See* topological quantum field theory
   zig-zag identities and, 39
superposition, 188, 266, 365
superselection, 6
supersymmetry, 41
syllepsis, 110–112
symmetry relations, 16, 141
   compactness and, 156
   examples of, 154
   groups and, 16, 18, 150–151
   monoids and, 15, 29, 32, 55, 59, 129, 136, 154–158, 160
   natural isomorphisms, 155–156
   processes and, 148–152
   reversibility and, 20
   reversible process and, 20

tangles, 58, 79
   Baez–Dolan tangle hypothesis, 112
   braiding and, 58, 74, 112. *See* braiding
   dimension of, 112
   framing, 61, 112

invariants of, 79, 116
isotopic, 59
knot theory and, 51. *See also* knot theory
monoids and, 74. *See also* monoidal categories
oriented, 58, 59
particles and, 61
Shum's category, 62
tangle hypothesis, 112
Yang–Baxter equation and, 83
Tannaka–Krein reconstruction theorem, 74, 100, 102
thermodynamic behavior, 329, 335
three-slit experiment, 368
time reversal, 22
Tomita–Takesaki theory, 318, 327–329
topological quantum field theory (TQFT), 97, 100, 102, 103, 129
  bicategories, 103
  braiding and, 85. *See also* braiding
  $C^*$-algebras and. *See* $C^*$-algebras
  cobordisms and, 67, 86, 100, 114
  diffeomorphism and, 103
  dimensions in, 67, 71, 91–92, 96, 101–103
  duals and, 68
  extended, 100–102, 114
  Frobenius algebra and, 71
  homology theory, 25, 104–107, 115–118
  homotopy and. *See* homotopy theory
  Hopf algebras and, 102, 103, 104
  Lagrangian formulation of, 104
  modular tensor categories, 96, 99
  monoidal categories and, 68
  n-categories and, 101, 108, 114
  Pachner moves and, 87, 89
  pentagon identity, 94
  piecewise-linearity, 100
  quantum field theory and, 67
  quantum groups and, 84–85
  semisimple algebras and, 102
  spherical categories, 102
  string theory and, 90. *See also* string theory
  SU(2) and, 96
  topos theory and. *See* topos theory
  Turaev–Viro model and, 85
  Witten–Reshetikhin–Turaev theory and, 85
topos theory, 5, 132, 138, 194–200, 264, 273, 278
  base category in, 236
  basic structures in, 211
  Butterfield–Isham idea and, 6
  $C^*$-algebras and, 271–272, 281
  category theory and, 202
  characteristic arrows in, 195
  classical theory and, 211, 218
  constructive theory and, 271
  contextuality and, 204
  contravariance in, 236
  daseinisation and, 203, 207–238
  definition of, 194, 199, 207, 222
  Döring–Isham theory, 6
  exponentiation and, 195
  global elements and, 194–195, 199, 202, 204
  Heunen–Landsman–Spitters approach, 6
  Hilbert spaces and, 207, 208
  internal logic of, 197, 279
  Kochen–Specker theorem and, 290
  local valuations in, 201
  locales and, 276, 277
  mathematical foundations and, 197
  measurement and, 201
  natural transformations in, 222
  order-reversing, 225
  physical quantities in, 236
  propositions in, 217–221
  pseudo-states and, 197–200, 203, 204, 226–227, 235
  pure state and, 226–227
  quantum gravity and, 187–207, 211, 216
  real numbers and, 199, 203, 209n3, 225
  representation in, 217–221
  sheaves and, 202–203, 216, 222, 226, 235–236
  spin and, 208, 220–221, 233
  state spaces and, 290
  truth values in, 196, 197, 198
  ultraweak topology, 295
toy qubit theory, 162, 177
TQFT. *See* topological quantum field theory
trace, 145, 266, 403, 424, 434, 438
triangle identity, 31
triangular decomposition, 81
tricategories, 55, 107–108
Tsui, D., 62
Tumulka theory, 451
Turaev–Viro model, 73, 84–86, 91
Turaev–Viro–Barrett–Westbury model, 94, 96, 102
twinning experiment, 452

Unruh effect, 329

vacuum state, 7, 317
  AQFT and, 327
  Bell's inequalities and, 7, 325, 326
  CGMA and, 333
  concept of, 317
  correlations in, 323–327

vacuum state (*cont.*)
  cyclicity of, 322
  distillability and, 327
  energy and, 323
  entanglement and, 318, 324, 326, 327
  GNS representation, 351
  intrinsic characterization of, 328–332
  mathematical framework for, 318
  measurement and, 322
  Minkowski space and, 330
  modular covariance, 329, 332–334
  nonlocality and, 7
  observables and, 322
  perturbation theory and, 330
  Poincaré group, 320, 330
  quantum field theory and, 330
  Reeh–Schlieder Theorem and, 351
  as reference state, 330
  relativistic, 317
  spacetime and, 7. *See* spacetime
  spectrum condition and, 320, 330, 332, 351
  superluminal correlations, 324
  von Neumann algebras and, 328
  weak additivity and, 321, 323
  wedge algebras and, 329
vector spaces, 28, 92
Veneziano model, 187
Viro, O. Y., 85
Voevodsky, V., 82
von Neumann algebras, 104, 201n14, 208, 210, 211, 299, 345
  abelian, 256
  Bell correlation of, 324
  Bohrified state space, 309
  $C^*$-algebras and. *See* $C^*$-algebras
  continuity properties, 352
  duality and. *See* duals
  filters and, 227–230
  Gelfand spectrum and, 247. *See also* Gelfand spectra
  global sections and, 265
  Hakeda–Tomiyama and, 147
  hyperfinite, 350–351, 354
  independence and, 352, 353, 354
  lattices and, 241–243, 258
  measurement and, 319, 349. *See also* measurement
  normal states, 352
  observables and, 246. *See* observables
  operator algebras and, 319
  pairing formula in, 309
  probability theory and, 319n3
  projectors in, 241–243, 258, 292. *See* projection operators
  Rickart $C^*$-algebras and, 273, 292
  Schlieder property and, 325
  self-adjoint operators and, 212, 246, 251. *See also* self-adjoint operators
  sheaves and, 217, 257, 263, 265. *See also* sheaf theory
  split property, 352–353
  Stone spectra and, 241, 243, 266–267, 294
  subalgebras of, 230, 352
  Tomita–Takesaki theory, 327
  trace problem, 267
  vacuum and, 328
von Neumann, J., 1, 2, 5, 19–20
  classical theory and, 171–172
  Einstein and, 343–361
  electron and, 394
  entanglement and, 359
  hidden variables and, 8, 393–407. *See also* hidden variables theory
  on Hilbert spaces, 132
  von Neumann algebras. *See* von Neumann algebras

$W^*$-algebras. *See* von Neumann algebras
wave function, collapse of, 385, 451–452
wedge algebras, 328, 329
Wess–Zumino–Witten model, 80
Westbury, B. W., 91
Weyl, H., 79
Wheeler, J., 191n4
Whitehead, A. N., 133
Wightman field, 321, 328, 329
Wigner, E., 21–22
Wigner's theorem, 133n3
Wilczek, F., 62
Wiles, A., 107
Witten, E., 80
Witten–Reshetikhin–Turaev theory, 85, 96, 102

Yang–Baxter equation, 64, 82
  braiding and, 64
  knot theory and, 65
  monoidal categories and, 64
  quantum groups and, 83
  Reidemeister moves and, 82
  solutions of, 65
  tangles and, 83
Yang–Mills theory, 27–28
Yorke, T., 137
Young tableaux, 288

Zamolodchikov equation, 82, 83, 84
Zeilinger principle, 368
zig-zag identities, 39, 42, 59